D0145909

DISCRETE MATHEMATICS WITH ALGORITHMS

DISCRETE MATHEMATICS WITH ALGORITHMS

Michael O. Albertson

Joan P. Hutchinson

Smith College

WILEY

JOHN WILEY & SONS
New York Chichester Brisbane Toronto Singapore

Library of Congress Cataloging in Publication Data:

Albertson, Michael O.
Discrete mathematics with algorithms.

Includes indexes.
1. Mathematics—1961– . 2. Electronic data
processing—Mathematics. I. Hutchinson, Joan P.
II. Title.
QA39.2.A43 1988 510 88-235
ISBN 0-471-84902-2

Printed in the United States of America

10 9 8 7 6 5 4 3 2 1

Dedicated to

Margot, Matt, Nick, and Beth,
and to
Neal H. McCoy

PREFACE

Discrete mathematics is playing an increasingly central role in the development of computer science and in both pure and applied mathematics. Consequently, there is pressure to teach courses in discrete mathematics earlier in the college curriculum. This makes sense. The material is accessible to students at this level, probably more accessible than calculus. Furthermore, a background in discrete mathematics is necessary for study in mathematics and computer science and is helpful throughout a broad spectrum of the sciences.

Historically, Finite Mathematics as a freshman course has had two fatal flaws. First, it did not lead anywhere in the curriculum. Second, the material was not a coherent entity. Within the mathematics curriculum, Discrete Mathematics can be a prerequisite for Linear Algebra and Number Theory and should be a prerequisite for Combinatorics and Graph Theory, Linear Programming, and Probability. Within the computer science curriculum it would be desirable to study Discrete Mathematics either before, or concurrently with, Data Structures. Both the Foundations of Computer Science and the Design and Analysis of Algorithms should have Discrete Mathematics as a prerequisite. Thus the influence of computer science has firmly placed Discrete Mathematics in the mainstream of both mathematics and computer science.

To cure the second flaw, we adopt algorithmic reasoning as our unifying theme. We are problem solvers by nature and want efficient algorithmic solutions in preference to existential results. Our paradigm begins with a specific mathematical problem that is transparent and easy to solve in small instances. The naive algorithm that works in small cases may require unimaginably large amounts of computation when the problem size is increased (sometimes only modestly). Then the search is on for an efficient algorithm. For each major problem considered, we construct an algorithm whose performance is markedly better than the naive algorithm. Discrete mathematics provides the tools to understand these algorithms, prove their correctness, and analyze their efficiency.

This book has no calculus prerequisite and no calculus is used. However, four years of high school mathematics are required for technique and sophistication. Some exposure to computers is advisable, but no programming experience is needed. The algorithms are presented in English in a format compatible with the Pascal programming language. Most could easily be turned into Pascal programs. If further technical details are needed to implement the algorithm as a program, these are included in the exercises. We have found it instructive to have the students observe computer implementations of many of the algorithms, especially to see the differences in efficiencies of different algorithms.

The text is meant to be *read* **by the student**. We mean the whole text including examples, questions, and proofs. Mathematics is not a spectator sport! Acquiring an understanding of the definitions, theorems, examples, and algorithms requires participation. Designated questions are placed throughout the text. These involve checking examples, exploring newly introduced theory, and working through algorithms by hand. The questions should be required work. Specifically, every question should be attempted before the material following it is read. We believe that it is crucial for the student to work out the solutions to these questions in order to become involved with the material as it is presented. To facilitate this end, solutions to all questions are supplied at the end of the book. Detailed solutions are given for the more difficult questions. With these, students can check their work and improve their understanding. After completing a section, the student should attempt a substantial number of the exercises at the end of the section. (In later references an exercise labeled $x.y$ is the yth exercise in Section x. If a chapter reference is necessary, it is indicated.) The exercises are not necessarily listed in the order of increasing difficulty; however, at the end of each chapter the Supplementary Exercises contain challenging problems on material from the entire chapter.

Discrete Mathematics with Algorithms is intended as a textbook for a one-semester course at the freshman-sophomore level. All the material cannot be covered in one semester so that choices must be made. In our 13-week semester we typically teach most of Chapters 1–5, several sections of Chapter 8, and an introduction to Chapter 6. A one-quarter course might cover Chapters 1–3 and one additional chapter. A two-quarter course could cover the entire text. Chapters 1, 2, and the beginning of 3 are necessary for the rest of the book.

Chapter 1, which introduces set theory, the definition of an algorithm, and the basic properties of functions and Boolean functions, can be covered in two weeks (less if the students' backgrounds warrant). Some material on functions can be deferred until needed later; the Satisfiability Problem can be omitted or deferred. There are three substantive topics in Chapter 2. First, the students meet induction proofs. We recommend ample written work. Submitting and resubmitting induction proofs until they are correct is an effective technique. Second, the big oh concept is not easy to grasp. It is the discrete mathematics analogue of the definition of limit in the calculus. Computer science students tend to see a rather casual development of big oh in their intermediate and advanced courses. It is our experience that the big oh concept becomes easier to understand after the students see it applied. Finally, proofs by contradiction are introduced with a concurrent discussion of logical reasoning. A total of 3–4 weeks should be spent on Chapter 2. The material in Chapter 3 on subsets and binomial coefficients goes more rapidly. Two weeks suffice. It is possible to omit the algorithm JSET, the game of Mastermind, and harder applications of the binomial theorem if time is tight.

With the semester approximately half over it is time to embark on the optional material. Each of the remaining five chapters can be thoroughly covered in two or three weeks. Each contains material that can be omitted. We enjoy the material on Fibonacci numbers and the Euclidean algorithm in Chapter 4. It is very different in content from most of what gets taught in discrete mathematics courses. Students find the application to public key encryption memorable. This application necessitates an introduction to relations, modular arithmetic, and basic results of number theory. Complete induction is first explained in this chapter. Although the Fibonacci numbers reappear for motivation in Chapter 7, no subsequent chapter depends on Chapter 4.

Chapter 5 forms a brief introduction to graph theory. The goal is a minimum-weight spanning tree algorithm, and only those graph theory definitions and results necessary for this end are introduced here. Students frequently implement such an algorithm in a Data Structures course. Here they understand how, why, and how efficiently it works. If time is short, the final section on greedy algorithms can be safely omitted. The material on trees reappears in the middle of Chapter 6. We believe that at least one additional graph theory application from Chapter 8 should be covered.

Chapter 6 covers sorting and searching algorithms. These algorithms are implemented in Data Structures, but the analysis is usually omitted. Proving, for instance, that the binary insertion sort is essentially best possible is important for the computer science student. This material is covered in five sections and forms a complete unit. If time permits, the final sections provide an introduction to recursion and recursive algorithms. Many earlier algorithms are reformulated using this technique and merge sorting is presented as an efficient recursive sorting algorithm.

Many algorithms in Chapter 6 lead naturally into the material of Chapter 7 on recurrence relations. The topics of sequences and recurrence relations in Chapter 7 are classic combinatorics, essential for the analysis of recursive algorithms. The recurrence relations are motivated by topics introduced earlier and are solved first by common-sense approaches, then by the general theory of linear homogeneous recurrence relations with constant coefficients, and finally by divide-and-conquer methods.

Chapter 8 contains an algorithmic development of some of the most important problems in graph theory. Each of the five sections introduces a real-world problem, modeled by graphs. The underlying abstract graph theory is developed, and then algorithms, derived from this theory, are applied to solve the problem at hand. Each section is independent of the others (except that Section 8.5 uses material from part of Section 8.3) although each uses and reinforces material from Chapter 5. In some cases a best-possible solution to the graph theory and real-world problem is produced. In one section an efficient approximation algorithm is used to search for an optimal solution. In the instances of three well-known NP-complete problems, the unsettled state of affairs is discussed, since there are no known polynomial-time solutions and yet no one has proved that there cannot be such a polynomial solution.

There is no consensus among either mathematicians or computer scientists concerning the curriculum of discrete mathematics. We believe the essentials of such a course must include an introduction to proofs, especially by induction, and an introduction to algorithmic problem solving. Otherwise, students are best served by working on a few realistic problems, developed in depth. This book includes material that we have been teaching for four years in a freshman-level Discrete Mathematics course. We have found the material and its presentation effective and relevant for further work in mathematics and computer science. Our colleagues in computer science appreciate the improved understanding their students in Data Structures exhibit and the increased amount of material the foundations course can cover. We have found Discrete Mathematics to be an excellent beginning college mathematics course for well-prepared freshmen and an appropriate transition for all students from the computational approach of low-level mathematics courses to more problem-oriented and abstract intermediate courses.

We especially thank our colleagues David Berman, Jim Henle, Marjorie Senechal, and Patricia Sipe for teaching from preliminary versions of this book and for their helpful suggestions. We also thank the Mathematics 153 students of Smith College for their patience and perseverance as this book was being prepared. We are grateful to Ellen Gethner for her help in preparing solutions to the book's questions. In addition, we have benefited from insightful comments from Larry Carter, Jeanne Ferrante, Andrew Pasquale, Margot Thomas, and Stan Wagon. Finally, we thank Gene Davenport, our editor, Deborah Herbert, Elaine Wetterau, copy editors, and Pam Pelton, production supervisor, at John Wiley & Sons for their help in preparing the manuscript.

Michael O. Albertson
Joan P. Hutchinson

CONTENTS

Chapter 1	***SETS AND ALGORITHMS: AN INTRODUCTION***	***1***
1:1	Introduction	1
1:2	Binary Arithmetic and the Magic Trick Revisited	5
1:3	Algorithms	8
1:4	Between Decimal and Binary	15
1:5	Set Theory and the Magic Trick	20
1:6	Pictures of Sets	25
1:7	Subsets	29
1:8	Set Cardinality and Counting	35
1:9	Functions	40
1:10	Boolean Functions and Boolean Algebra	51
1:11	A Look Back	59
Chapter 2	***ARITHMETIC***	***65***
2:1	Introduction	65
2:2	Exponentiation, A First Look	68
2:3	Induction	71
2:4	Three Inductive Proofs	80
2:5	Exponentiation Revisited	88
2:6	How Good Is Fast Exponentiation?	91

2:7 How Logarithms Grow 97
2:8 The "Big Oh" Notation 102
2:9 $2^n \neq O(p(n))$: Proof by Contradiction 110
2:10 Good and Bad Algorithms 118
2:11 Another Look Back 122

Chapter 3 ***ARITHMETIC OF SETS*** ***127***

3:1 Introduction 127
3:2 Binomial Coefficients 131
3:3 Subsets of Sets 141
3:4 Permutations 153
3:5 An Application of Permutations: The Game of Mastermind 161
3:6 The Binomial Theorem 168
3:7 Important Subsets 176

Chapter 4 ***NUMBER THEORY*** ***181***

4:1 Greatest Common Divisors 181
4:2 Another Look at Complexities 186
4:3 The Euclidean Algorithm 190
4:4 Fibonacci Numbers 197
4:5 The Complexity of the Euclidean Algorithm 206
4:6 Congruences and Equivalence Relations 211
4:7 An Application: Public Key Encryption Schemes 222
4:8 The Dividends 234

Chapter 5 ***GRAPH THEORY*** ***239***

5:1 Building the LAN 239
5:2 Graphs 241
5:3 Trees and the LAN 251
5:4 A Good Minimum-Weight Spanning Tree Algorithm 263
5:5 An Ode to Greed 272
5:6 Graphical Highlights 279

Chapter 6 ***SEARCHING AND SORTING*** ***283***

6:1 Introduction: Record Keeping 283
6:2 Searching a Sorted File 290
6:3 Sorting a File 295
6:4 Search Trees 302
6:5 Lower Bounds on Sorting 310
6:6 Recursion 317
6:7 MERGESORT 325
6:8 Sorting It All Out 331

Chapter 7 ***RECURRENCE RELATIONS*** ***339***

7:1 Beginnings of Sequences 339
7:2 Iteration and Induction 346
7:3 Linear Homogeneous Recurrence Relations with Constant Coefficients 353
7:4 LHRRWCCs with Multiple Roots: More About Rabbits 364
7:5 Divide-and-Conquer Recurrence Relations 372
7:6 Recurring Thoughts 381

Chapter 8 ***MORE GRAPH THEORY*** ***389***

8:1 Minimum-Distance Trees 389
8:2 Eulerian Cycles 399
8:3 Hamiltonian Cycles 410
8:4 Minimum-Weight Hamiltonian Cycles 425
8:5 Graph Coloring and an Application to Storage Allocation 431

SOLUTIONS TO QUESTIONS ***451***

INDEX ***538***

ALGORITHMS AND PROCEDURES ***543***

NOTATIONS ***545***

1

SETS AND ALGORITHMS: AN INTRODUCTION

1:1 INTRODUCTION

The four cards labeled *A*, *B*, *C*, and *D* in Figure 1.1 are part of a magic trick played by Player 1 upon Player 2. The trick is played as follows:

A		*B*		*C*		*D*	
8	9	4	5	2	3	1	3
10	11	6	7	6	7	5	7
12	13	12	13	10	11	9	11
14	15	14	15	14	15	13	15

Figure 1.1

Player 1	*Player 2*
Pick a whole number between 0 and 15.	
Got it?	Yes.
Is it on card *A*?	Yes.
Is it on card *B*?	No.
Is it on card *C*?	No.
Is it on card *D*?	Yes.
The number you picked is 9.	That's amazing! How did you do that? And so fast!

Now let's play again only this time you'll be player 1. I have a whole number between 0 and 15. It appears on cards A, C, and D and does not appear on card B. What number am I thinking of?

Question 1.1. (Figure it out before you read any further.) If you are at a loss for what to do, ask yourself the following questions. Can it be 0? Can it be 1? . . . Can it be 15?

Now it can't be 0 because 0 doesn't appear on any of the cards and the number I'm thinking of appears on three cards. It can't be 1 because even though 1 does appear on card D, it does not appear on cards A or C, and the number I'm thinking of appears on both cards A and C. If this magic trick is well designed, meaning that it is always possible for player 1 to guess player 2's number correctly, then there must be a unique number that corresponds with any possible sequence of answers provided by player 2. In this case the number I am thinking of is 11. It is easy to check that 11 appears on the cards A, C, and D but does not appear on the card B. It seems less obvious that 11 is the only such number.

Question 1.2. What would you need to do to check that this trick will always work?

Question 1.3. Design a pair of cards that will serve to distinguish the numbers 0, 1, 2, and 3. Is there more than one way to do this? Why can't two cards distinguish the numbers 0, 1, 2, 3, and 4?

Understanding why two cards can distinguish four numbers and why four cards can distinguish 16 numbers is fundamental to seeing how to design this game as well as how to play it well. Each card that player 1 shows to player 2 elicits one of two responses, either a "yes" or a "no." A game with two cards has four possible responses from player 2. These are "no, no," "no, yes," "yes, no," and "yes, yes." How many responses has a game with four cards? Justice seems to suggest that you respond 16. That is correct. Now let's see why.

Multiplication Principle. Suppose that a counting procedure can be divided into two successive stages. If there are r outcomes for the first stage, and if for each of these outcomes for the first stage, there are s outcomes for the second stage (where r and s are positive integers), then the total number of possible outcomes equals the product of r and s, rs.

Example 1.1. At tea one afternoon you are offered your choice of a bagel, a corn muffin, or a croissant with either cream cheese or lightly salted butter. How many different choices do you have? (Reread the multiplication principle.) At the first stage you can choose whether to have a bagel, muffin, or croissant. There are three different outcomes ($r = 3$). At the second stage you can choose cheese or butter. There are two different outcomes ($s = 2$). By the multiplication principle as well as by a direct count you have 6 ($=rs$) choices.

Example 1.2. In the magic trick, how many different responses are there to the four cards? (Reread the multiplication principle.) First consider cards A and B. As we've already seen, there are four distinct responses to these two cards ($r = 4$). Next look at cards C and D. It doesn't matter what the responses to the A and B cards were. There are four distinct responses to these two cards ($s = 4$). Thus there are 16 ($=rs$) distinct responses in all to the four cards. Note that these 16 responses could have been counted in four stages with two responses at each of these stages. The multiplication principle works analogously for any number of stages. (See Exercises 7 and 8.)

Question 1.4. How many different seven-digit telephone numbers are there beginning with the digits 584?

Now returning to the magic trick, you see that player 1 could perform the trick by memorizing the 16 different responses that player 2 might give in order to successfully "guess" player 2's number. The possible responses are listed in Table 1.1.

Table 1.1

Player 2's Number	*Responses*			
	Card A	*Card B*	*Card C*	*Card D*
0	no	no	no	no
1	no	no	no	yes
2	no	no	yes	no
3	no	no	yes	yes
4	no	yes	no	no
5				
6				
7				
8				
9				
10				
11				
12				
13				
14				
15				

Question 1.5. Complete Table 1.1.

EXERCISES FOR SECTION 1

1. Design a set of three cards that will distinguish the numbers $0, 1, 2, \ldots, 7$. Suppose that we only wished to distinguish the numbers $0, 1, 2, \ldots, 5$. Could your three cards be modified to play this game? Could your three cards be modified to play the game with the numbers $0, 1, 2, \ldots, 8$?

2. Suppose that the local ice cream store offers 12 different flavors of ice cream and 5 different types of topping (chocolate, butterscotch, strawberry, blueberry, and raspberry). How many different dishes of ice cream plus topping are possible? Suppose that you can turn these dishes of ice cream plus topping into special sundaes by adding one kind of nuts (walnuts, almonds, or hazelnuts) and whipped cream if you like. How many different types of special sundaes can you order at this ice cream store?

3. A certain fast food chain offers a one-price meal consisting of a burger, an order of potatoes, a salad, a dessert, and a beverage. There are seven different kinds of burgers, three different kinds of potatoes, five different kinds of salads, and four different kinds of desserts. The restaurant advertises that you can eat one meal here every day for four years without ever having the same meal twice. What can you say about the number of beverage choices that the restaurant offers?

4. Often, when you sign onto a time-sharing computer, you are asked to specify the room you are in and the kind of terminal that you are using. Suppose that there are 13 different room categories and 16 different kinds of terminals. How many different pairs of answers is it possible to give as you sign on?

5. In the context of the preceding problem it is typically the case that not all answers are possible, since there are not 16 different kinds of terminals in every room. If every room contains four different kinds of terminals, how many different answers are possible?

6. Even the idea in the last problem might not be correct, since the kind and number of terminal types may vary from place to place. Suppose that we consider only five rooms and that they contain the following kinds of terminals: Every room contains a Digital VT terminal; Tektronix machines are located in the Social Science Room and in the Science Lab; IBM PCs are found in the Graphics Lab and in the Library Terminal Room; and Apple Macintoshes are available in the Library Terminal room and in the Humanities Computer Room. How many pairs of responses are now possible to send to the computer when you sign on?

7. Here is an extension of the multiplication principle: Suppose that a counting procedure can be divided into four successive stages. If there are p outcomes for the first stage, if for each of these outcomes for the first stage, there are r outcomes for the second stage, if for each pair of these first two outcomes,

there are s outcomes for the third stage, and finally if for each of these first three outcomes, there are t outcomes for the fourth stage (where p, r, s, and t are positive integers), then the total number of possible outcomes equals the product $prst$. Explain why this is valid, using the original form (two-stage) of the multiplication principle.

8. State and explain a multiplication principle that is valid for three stages, and then do the same for five stages.

9. Suppose that we have a rather primitive computer that can receive only strings of zeros and ones as input. Furthermore, these strings must contain exactly eight digits. How many different input strings are there?

10. Suppose that the machine in the preceding problem can receive strings with one to eight digits, and suppose that the machine disregards initial zeros. Thus, for instance, 1001 is the same input as the string of eight digits, 00001001. Now how many different input strings are possible?

11. How many different seven-digit phone numbers are there that begin 584 and contain no zero? How many phone numbers are there that begin 584 and contain at least one zero?

12. How many different seven-digit phone numbers are there that begin 58_-____ and contain seven different digits? How many of these contain no zero? How many do contain a zero? How many different phone numbers are there that begin with 58_, but contain no two identical consecutive digits?

13. Recently, a new telephone area code was introduced for the area of New York that contains Brooklyn and Queens because all seven-digit phone numbers had been used up. Assuming that none of the first three digits in a phone number can be either a 0 or a 1, what can you say about the number of phone lines in this area?

14. In the lottery game called Megabucks a player selects six different numbers between 0 and 35. How many different such selections are there? Before answering, specify when two selections are the same and when they are different.

1:2 BINARY ARITHMETIC AND THE MAGIC TRICK REVISITED

The magic trick of Section 1 was based on each of four questions receiving either a "yes" or a "no" answer. Thus a seemingly complex task, in this case deciding which number player 2 had chosen, could be broken down into a sequence of smaller tasks associated with each of the cards. This fundamental yes-no, true-false, or on-off dichotomy pervades most of the mathematics associated with computers. It even is fundamental to how computers "think" about numbers. We now model how a computer stores an integer using binary numbers.

A number in **binary notation** is just a finite list (sequence or string) of zeros and ones. For example, 1, 101, 111001, and 1011 are all binary numbers. For the moment don't be concerned about which numbers these sequences are. We'll get to that shortly. Rather, think in the familiar decimal system. The number 37, for instance, can be thought of as 3 tens together with 7 ones,

$$37 = 3 \cdot 10 + 7.$$

The number 234 is 2 hundreds plus 3 tens plus 4 ones,

$$234 = 2 \cdot 100 + 3 \cdot 10 + 4.$$

The decimal number system is so named because successive columns when reading from the right represent consecutive powers of ten. We'll use the same device for the binary system. Specifically, let the rightmost column of a number written in binary represent the ones. We call this the 0th column or the one's column. The next column to the left represents the twos. This is called the 1st column or the two's column. The third column from the right represents the fours, the fourth column from the right represents the eights, and so on. So the pth column from the right (starting with $p = 0$) in the binary representation of a number represents the pth power of 2.

Example 2.1. The binary number 1 is the same number as the decimal number 1. The binary number 101 consists of 1 one, no twos, and 1 four. It is equivalent to the (decimal) number 5. Similarly, the binary number 111001 is equivalent to the decimal number 57, since $57 = 1$ one $+ 0$ twos $+ 0$ fours $+ 1$ eight $+ 1$ sixteen $+ 1$ thirty-two (see Figure 1.2).

BINARY NUMBER		$\longleftrightarrow$		DECIMAL NUMBER
101	$=$	$1 \cdot 4 + 0 \cdot 2 + 1$	$=$	5
111001	$=$	$1 \cdot 32 + 1 \cdot 16 + 1 \cdot 8 + 1 \cdot 1$	$=$	57

Figure 1.2

Question 2.1. Find the decimal equivalents of the following binary numbers: (a) 10101, (b) 100101, and (c) 11010. Given a binary number, how would you decide whether it is an even number or an odd number?

Question 2.2. Construct a table with all the numbers from 0 to 15 written in binary. Compare your results with the table that you completed for Question 1.5.

Question 2.3. List all the numbers from 0 to 15 that have a 1 in the four's column of their binary representation.

After doing the previous question, you should note that the numbers you obtain appear familiar. In fact they are just the numbers that appear on card B in the magic trick. You should check that the numbers that appear on card A are just those numbers between 0 and 15 whose binary representation has a 1 in the eight's column, those on card C are just those with a 1 in the two's column, and those on card D are just those with a 1 in the one's column. This suggests a quick way for player 1 to perform the magic trick. Player 1 should remember the number in the upper left-hand corner of every card to which player 2 says "yes" and add these numbers up. Thus in our original play the yes to card A produces an 8, and the yes to card D produces a 1 for a total of 9.

At the moment we have used binary representations of numbers to produce a simple procedure for player 1 to perform the magic trick. It is easy to proceed from the binary representation of a number to its decimal equivalent. What about the other direction, that is, given a number in decimal form how should we arrive at its binary equivalent?

Question 2.4. Write the following (decimal) numbers in binary: (a) 6, (b) 19, (c) 52, (d) 84, and (e) 232.

You have already done part (a) in previous questions. Part (b) you probably see. To write 52 and 84 in binary, you might require pencil and paper. By the time you get to 232, you will be glad the question stops. Surely you realize that given any positive integer of moderate size, you could find its binary representation with enough time, motivation, and paper. Still if the method that you have used is basically trial and error, you might wish for an alternative. What you need in the jargon of discrete mathematics is a "good algorithm." We shall introduce you to this language in the next section.

EXERCISES FOR SECTION 2

1. List all numbers from 0 to 15 that have a 1 in the eight's column. Then list all of these numbers that have a 1 in the two's column. Why is it precisely the odd numbers that have a 1 in the zero's column?
2. Without listing all the numbers, give a characterization of the numbers from 0 to 31 that have a 1 in the sixteen's column. Then characterize those numbers from 0 to 31 that have a 1 in the eight's column. Finally, describe all numbers from 0 to 31 whose binary expansion ends with the digits 01.
3. What decimal numbers do the following binary numbers represent? (***a***) 11011, (***b***) 101011, (***c***) 10001, and (***d***) 11000.

4. Find the binary representation of the following decimal numbers: **(*a*)** 28, **(*b*)** 43, **(*c*)** 100, and **(*d*)** 81.

5. Suppose that you are given a decimal number m that has the property that when m is divided by 8 there is a remainder of 2. What can you say about the binary representation of m?

6. Suppose that you are given a decimal number m that has the property that when m is divided by 4 there is a remainder of 3. What can you say about the binary representation of m?

7. Suppose that you are given a decimal number m that has the property that when m is divided by 16 there is a remainder of 6. What can you say about the binary representation of m?

8. What is the maximum number of integers that a six-card magic trick could distinguish? If the six cards were designed as in the original trick from Section 1 and the responses were yes, no, yes, yes, no, no; what number was selected?

9. Given two binary numbers, how could you tell (without converting them into decimal) which is bigger?

10. A **binary fraction** is a finite sequence of zeros and ones that follows what is called the binary point. For example 0.101 is a binary fraction. The column immediately to the right of the binary point represents the halves, the next column the quarters, the third column the eighths, and so on. Thus $0.101 = \frac{1}{2} + \frac{1}{8} = \frac{5}{8}$. Make a table of all four-bit binary fractions.

11. Express each of the following in binary: **(*a*)** $\frac{27}{32}$, **(*b*)** $\frac{36}{64}$, **(*c*)** $\frac{31}{8}$, **(*d*)** $\frac{53}{4}$, and **(*e*)** $\frac{127}{16}$.

1:3 ALGORITHMS

Definition. An **algorithm** is a finite sequence of well-described instructions with the following properties.

1. There is no ambiguity in any instruction.
2. After performing a particular instruction, there is no ambiguity about which instruction is to be performed next.
3. The instruction to stop is always reached after the execution of a finite number of instructions.

Example 3.1. Here are the instructions inscribed on a metal plate attached to the front of a video game:

STEP 1. Insert quarter into slot on side of machine.

STEP 2. Press green button on top of machine when ready to begin.

This is a finite sequence of well-described instructions. There is no ambiguity in any instruction. It is always clear what to do next. There is no explicit instruction to stop, but we overlook this, since it is clear that each instruction is executed just once for each play. Thus we may call this an algorithm: It is an algorithm to begin playing a video game.

Example 3.2. Suppose that we add an instruction to the previous example.

STEP 3. When each game is over, type your initials and press red button on top of machine to record score for posterity.

Step 3 fulfills the role of a stop instruction provided games don't go on forever and provided you cannot play an additional game on your initial quarter.

Example 3.3. Suppose we insert the following instruction.

STEP 4. For each 10,000 points you accumulate you will win a free game. When current game is over, if you have won a free game, go to step 2.

This has changed the nature of the instructions. The first two examples are known as sequential algorithms. By that we mean that each step (or instruction) is executed exactly once and that the next step on the list is the next instruction to be executed. Step 4 adds the possibility of executing an instruction many times. Indeed since there is no reason to suppose that the winning of free games couldn't go on forever the list of instructions in Example 3.3 is not an algorithm. You might think of this by reinterpreting instruction 4 as saying,

"If score $\geq$ 10,000, then go to instruction 2."

This logical structure is known as a loop. Although this particular loop has made the sequence of instructions fail to be an algorithm, a loop need not force the execution of an infinite number of steps.

Example 3.4. Suppose that we modify instruction 4 so that our instructions now say:

STEP 1. Insert quarter into slot on side of machine.

STEP 2. Press green button on top of machine when ready to begin.

STEP 3. When each game is over, type your initials and press red button on top of machine to record score for posterity.

STEP 4. For each 10,000 points you accumulate, you will win a free game up to a maximum of 10 free games for each paid game. When current play is over, if you have won a free game, go to step 2.

The modification in step 4 can be described as adding a counter to the loop in order to insure that the loop is executed a finite number of times.

The logical structure of Example 3.3 illustrates one of the most common mistakes made by beginning programmers, that of an infinite loop. Example 3.4 shows a typical quick fix.

Reread the definition of algorithm! Given a sequence of instructions, how could it fail to be an algorithm? First, it might be the case that at least one of the instructions is ambiguous. In other words, some instruction might be poorly specified so that you, the reader, do not know how to carry out the instruction or so that a computer programmer does not know how to translate the instruction into a suitable computer language. An instruction that is clear to one person may be full of ambiguities for another. For example, an instruction like, "Start the airplane engine and take off on runway 4," might be unambiguous to trained pilots.

A sequence of instructions might fail to be an algorithm because after executing a particular instruction, it might not be clear which instruction is to be executed next. Some instructions will clearly indicate the next instruction, such as ". . . and then go to step—." If no such direction is given, we always move to the next step as given in the sequence of instructions. Finally, we would not have an algorithm if the execution of the sequence of instructions did not terminate in all instances.

If a sequence of instructions does satisfy the definition of an algorithm, it still might not be a correct algorithm to perform the desired task. To illustrate this possibility, we introduce an example from the kitchen. (It seems that most expositions about algorithms revert to cooking recipes at some time.)

Example 3.5. Consider the following sequence of instructions.

STEP 1. Place one cup of water in the top of a double boiler.

STEP 2. Place one cup of quick oatmeal in the bottom of a double boiler.

STEP 3. Turn on a stove burner to medium.

STEP 4. Place double boiler on burner and heat for 10 minutes.

STEP 5. Remove pot.

STEP 6. Turn off burner.

You may verify that the instructions satisfy the definition of algorithm. However, you would be unlikely to enjoy eating the results of this recipe. This example illustrates that a particular algorithm is designed in response to a particular problem. In this case the problem (which while not explicitly stated) can reasonably be inferred to be to make oatmeal. This sequence of instructions is not a correct algorithm for making oatmeal.

Question 3.1. Rewrite the steps in Example 3.5 so that the resulting algorithm correctly instructs us to make oatmeal.

Notice that writing a correct algorithm may be significantly more difficult than checking whether a given sequence of instructions is a valid algorithm. Creating an algorithm requires expertise with the subject matter. To give a correct algorithm for making quick oatmeal, you need to know (or read on the box) the proper proportions of water and oatmeal, the cooking time, and so on. To play a video game requires knowledge of the rules and object of the game. For the more mathematical algorithms of this book we need to develop the language and techniques of the subject before we can use, let alone create, new algorithms.

Sometimes it may be difficult to decide if a given sequence of instructions will necessarily terminate in a finite number of steps. The obvious cases might always stop, but how can we know if we have tested all possibilities?

Example 3.6. Consider the following algorithm.

Algorithm(?) COLLATZ

STEP 1. Input z a positive integer

STEP 2. If z is even, replace z by $z/2$

STEP 3. If $z = 1$, then output z and stop.

STEP 4. If z is odd, replace z by $3z + 1$

STEP 5. Go to step 2

This will be an algorithm if it stops. This will happen if z eventually equals 1. Whether or not this will happen for every positive integer z is a famous unsolved problem known as the Collatz Problem.

Question 3.2. Run COLLATZ for the following initial values of z: (a) $z = 1$, (b) $z = 20$, and (c) $z = 7$.

An algorithm will typically need input to begin and will produce output at the end. Input is the data or material needed to start the algorithm, like a quarter, water and oatmeal, or a positive integer z. The output is the result of the algorithm, derived from the particular input, like a score stored on a video game, burned oatmeal, or the number $z = 1$. General-purpose algorithms, which are the most useful, will draw input values from a set of possibilities, like all positive integers or all quick-cooking hot cereals. For each input there will be exactly one set of output. (The word *set* will be explained more fully in Section 5.)

Given a finite sequence of instructions that is in fact an algorithm, we now address the question of what qualities make this algorithm "good." Algorithms are created to solve problems. The term *correct* is used to label algorithms that produce correct solutions to a particular problem. Thus the sine qua non of a good algorithm is that the output must be correct for all possible input data. A great deal of effort within computer science is expended proving that programs (implementations of algorithms) are correct.

What properties might distinguish two correct algorithmic solutions to a particular problem? One might be easier to understand. One might provide internal consistency checks to assure that the algorithm was being carried out correctly. One might be easier to implement in your favorite programming language. Most commonly, one algorithm is said to be better than another if it requires fewer resources to implement. These resources may consist either of time (the number of steps required until the algorithm terminates) or space (the amount of memory required to implement the algorithm). In the next section we illustrate these notions with several different responses to the problem of taking a number written in decimal and translating it into binary.

EXERCISES FOR SECTION 3

1. Here are two approaches to making whipped cream. Discuss whether these are or are not algorithms.

Approach 1

STEP 1. Buy a pint of whipping cream.

STEP 2. Chill cream and beaters until cold.

STEP 3. Add a small amount of sugar and vanilla extract to cream.

STEP 4. Whip cream until stiff but not dry.

STEP 5. Wash dishes and stop.

Approach 2

STEP 1. Buy a can of Readi-Whip.

STEP 2. Shake 30 times.

STEP 3. Invert can.

STEP 4. Press nozzle.

STEP 5. Stop.

2. Here are two algorithms that calculate the sum of the integers from 1 to 100. Comment on the relative efficiency of the two responses.

Response 1

STEP 1. Sum the integers from 1 to 100 and store the result in the variable Answer.

STEP 2. Print out the value of Answer.

STEP 3. Stop.

Response 2

STEP 1. Store each of the integers from 1 to 200 in different memory locations.

STEP 2. Sum the integers from 1 to 200 and store the result in the variable Answer.

STEP 3. Sum the integers from 101 to 200, subtract the sum from the number in Answer, and retain the difference in Answer.

STEP 4. Print out the values of all 200 integers and also the value of Answer.

STEP 5. Stop.

3. Each of the following fails to be an algorithm; what rule or rules do they violate?

Attempt 1

STEP 1. Set Sum equal to 1

STEP 2. Set X equal to 2

STEP 3. Give Sum the value Sum + X

STEP 4. If Sum is even, then go to Step 5;
if Sum is odd, go to Step 3

STEP 5. Print out the value of Sum and stop.

Attempt 2

STEP 1. Set Sum equal to 1

STEP 2. Set X equal to 2

STEP 3. Give Sum the value Sum + X

STEP 4. If Sum is even, then stop.

4. Give a finite sequence of well-described instructions that never stops; that is, give an example of a pseudo-algorithm that fails to meet the third property needed to be an algorithm.

5. In each of the following decide whether or not the sequence of instructions is an algorithm. If not, explain why not. If yes, figure out what the algorithm produces as an answer.

Algorithm A

STEP 1. Set $j = 0$

STEP 2. Set Answer = 0

STEP 3. Give Answer the value Answer + 2^j

STEP 4. Add 1 to j

STEP 5. If j is less than 5, go to step 3

STEP 6. Output Answer

STEP 7. Stop.

Algorithm B

STEP 1. Set $j = 0$

STEP 2. Set Answer = 0

STEP 3. Give Answer the value Answer + 2^j

STEP 4. Add 1 to j

STEP 5. If j is not zero, go to step 3

STEP 6. Output Answer

STEP 7. Stop.

Algorithm C

STEP 1. Set $j = 0$

STEP 2. Set Answer = 0

STEP 3. Give Answer the value (Answer) (2^j)

STEP 4. Add 1 to j

STEP 5. If j is less than 5, go to step 3

STEP 6. Output Answer

STEP 7. Stop.

6. Find a more efficient algorithm that produces the same output as the one given here.

Algorithm FUN

STEP 1. Input z, an integer between 100 and 300
STEP 2. Let $u = 3z$
STEP 3. Let w be u written twice {so 345 becomes 345,345}
STEP 4. Let y be w with an extra zero on the end
STEP 5. Let a be y divided by 2
STEP 6. Let b be a divided by 3
STEP 7. Let c be b divided by 5
STEP 8. Let d be c divided by 7
STEP 9. Let e be d divided by 11
STEP 10. Let f be e divided by 13
STEP 11. Let g be f plus 1
STEP 12. Output g and stop.

1:4 BETWEEN DECIMAL AND BINARY

In Section 2 we discussed the problem of changing numbers from decimal to binary, and vice versa. Example 2.1 outlines a procedure for changing a binary number into its decimal equivalent. We shall soon make this procedure precise. First we digress to introduce some convenient notation that we shall use to present various algorithms.

The symbol ":=" is used for assignment. Specifically, the statement $a := b$ means that the current value of the variable b is assigned to the variable a. We can use this to form instructions that do not represent equalities in the normal arithmetic sense. For instance, the statement

$$a := a + 1$$

does not mean that $a = a + 1$, an assertion that is never true. Rather it means that the value assigned to the variable a should have 1 added to it, and the resulting sum should be reassigned to the variable a.

The symbol "$*$" is used for multiplication. Specifically, the expression $a * b$ means the product of the numbers a and b. The symbol "/" is used for division. Specifically, a/b means that a is divided by b. Thus the statement

$$a := a/b + 2^3$$

will divide the value of a by the value of b, add 8 to this quotient, and store the result in the variable a. For this instruction to make sense, the variables a and b must have been previously assigned values.

Problem. Given a binary number s as a string of zeros and ones, convert this into its decimal equivalent.

(To add clarity to our algorithmic instructions, we shall often insert comments inside braces as in {COMMENT: ...}).

Algorithm BtoD

STEP 1. Set $j := 0$
{j will stand for the binary column with which we are currently working. We label the columns from right to left beginning with 0.}

STEP 2. Set $m := 0$
{m will contain the final decimal number.}

STEP 3. If there is no jth entry of s, then stop.

STEP 4. If the jth entry of s is a 1, add 2^j to m
{We write this as $m := m + 2^j$.}

STEP 5. Increase the value of j by 1, that is, set $j := j + 1$

STEP 6. Go to step 3

Table 1.2

	Value Assigned to the Given Variable		
Step	*s*	*j*	*m*
1	1011	0	
2	1011	0	0
3	1011	0	0
4	1011	0	1
5	1011	1	1
6	1011	1	1
4	1011	1	3
5	1011	2	3
4	1011	2	3
5	1011	3	3
4	1011	3	11
5	1011	4	11
3	STOP		

Example 4.1. Table 1.2 is a detailed look at what happens when this algorithm is applied to the binary number 1011. Such a tabulation is called a **trace** of the algorithm. We record the values of the variables at the end of each step.

Question 4.1. Apply BtoD to the binary numbers 10101, 11010, and 100101. Do you get the same answers and are you carrying out the same procedure as in Question 2.1?

Next we return to the harder Question 2.4: Given a number written in decimal, how should we find its binary equivalent? Several sequences of instructions listed in order of increasing quality follow. First the problem is formalized.

Problem. Find the binary representation of a positive integer m.

Response 1

STEP 1. Write down a finite sequence of zeros and ones.

STEP 2. Take the binary number you wrote down in step 1 and translate it into decimal using Algorithm BtoD, given above.

STEP 3. If the number you obtain in step 2 equals m, then stop. Otherwise go to step 1.

Question 4.2. Why is Response 1 not an algorithm?

Response 2

STEP 1. Set $k := 1$
{k will denote the number of binary digits in the number under consideration.}

STEP 2. List all possible sequences of zeros and ones with k digits in increasing order
{For example, if $k = 1$, $0 < 1$, and if $k = 2$, $00 < 01 < 10 < 11$.}

STEP 3. For each sequence from step 2, find its decimal equivalent using Algorithm BtoD

STEP 4. If one of the numbers that you obtain in step 3 equals m, then stop.

STEP 5. $k := k + 1$

STEP 6. Go to step 2

The binary digits, zeros and ones, in a binary number are called **bits**.

Question 4.3. Why is the sequence of steps listed in Response 2 a correct algorithm for solving the problem? Use this algorithm to find the binary representation of 19. What makes this algorithm low quality?

Response 3

STEP 1. Find the largest power of 2 that is less than or equal to m. If this is the rth power of 2, place a 1 in the rth column (reading from the right and beginning with 0)

STEP 2. Subtract the power of 2 obtained in step 1 from m and set the result equal to m, or in symbols, set $m := m - 2^r$. If m equals zero, fill in the remaining columns with zeros and stop.

STEP 3. Go to step 1

Question 4.4. Why does the sequence of steps listed in Response 3 necessarily stop? Use this algorithm to find the binary representation of 182.

Response 4 (ALGORITHM DtoB)

STEP 1. Set $j := 0$
{j will indicate the column in the binary representation of m on which we are working.}

STEP 2. Divide m by 2 to obtain the quotient q and the remainder r {necessarily either 0 or 1}; place r in the jth column of the answer {reading from the right}

STEP 3. If $q = 0$, then stop.

STEP 4. Set $m := q$

STEP 5. Set $j := j + 1$

STEP 6. Go to step 2

Example 4.2. Table 1.3 is a trace of the algorithm DtoB, run on the decimal number 21.

Question 4.5. Why does the sequence of steps listed in Response 4 necessarily stop? Use this algorithm to find the binary representation of 395. Compare the algorithms in Responses 3 and 4.

The algorithm given in Response 4 is one that we shall use again and so we have named it Algorithm DtoB. In the exercises you are asked to work with the algorithms and ideas of this section. Here and throughout the book you will be asked to write algorithms. How should you create an algorithm from scratch? Here are some ideas, but there is no all-purpose algorithm to create an algorithm! First

Table 1.3 Values Assigned to the Variable After the Execution of the Given Step.

Step	*j*	*m*	*q*	*r*	*Answer*
1	0	21			
2	0	21	10	1	1
4	0	10	10	1	
5	1	10	10	1	
2	1	10	5	0	01
4	1	5	5	0	
5	2	5	5	0	
2	2	5	2	1	101
4	2	2	2	1	
5	3	2	2	1	
2	3	2	1	0	0101
4	3	1	1	0	
5	4	1	1	0	
2	4	1	0	1	10101
3	STOP				

figure out how to solve the problem at hand. Then ask yourself what your steps were and try to write them down so that another person or a computer could understand and follow them. Then analyze these steps, as we have in this section, to see whether your steps are a correct algorithm. This task is always challenging. Sometimes you will have seen algorithms in the text and exercises that you can modify and build upon; other times you need to jump in and follow your own logical path to a solution.

EXERCISES FOR SECTION 4

1. Apply BtoD to the following binary numbers: **(*a*)** 11, **(*b*)** 101, **(*c*)** 1101, **(*d*)** 1011, **(*e*)** 1111, and **(*f*)** 10101010.
2. Suppose that the decimal number D is expressed in binary as a sequence S of zeros and ones. If a zero is placed at the right end of S, how does the decimal value of the resulting number compare with D? If a one were placed at the right end of S, how would the decimal value change?
3. Suppose that S is a string of zeros and ones that corresponds with the even decimal number D. If the last entry of S on the right is erased, express the value of the new decimal number in terms of D. Repeat if D is odd.
4. Apply DtoB to the decimal numbers 17 and 59. Then apply BtoD to the resulting binary numbers. Next apply BtoD to the binary numbers 10001 and 110110. Then apply DtoB to the resulting decimal numbers.

5. A 16-bit computer allocates 16 spaces or bits to store an integer. The first bit designates whether the number is positive or negative and the remaining bits contain either a zero or a one, expressing the integer in binary. How many different integers can you store in 16 bits? What is the decimal value of the largest and of the smallest integer that can be stored using 16 bits?

6. Numbers written in base 3 can use only the digits 0, 1, and 2. Thus the decimal numbers 0, 1, and 2 are expressed in base 3 in the same way, but to write 3 in base 3 we must write $10 = 1 \cdot 3^1 + 0 \cdot 3^0$. Write the (decimal) numbers from 4 to 12 in base 3. The base 3 representation of a number is also called its **ternary representation.**

7. For each of the following numbers written in base 3, determine its decimal equivalent: **(*a*)** 22, **(*b*)** 20102, **(*c*)** 12121, **(*d*)** 20010, and **(*e*)** 1121.

8. Translate each of the following decimal numbers into a base 3 representation: **(*a*)** 13, **(*b*)** 15, **(*c*)** 21, **(*d*)** 27, **(*e*)** 30, and **(*f*)** 80.

9. Find the base 3 representation of the number 20 and then translate that base 3 number back into decimal notation. Similarly, begin with the base 3 number 111, find the decimal number that it represents, and change that decimal number back into base 3.

10. Given a number s expressed in base 3 as a string of 0s, 1s and 2s, write down an algorithm that will convert s into its decimal equivalent. (*Hint:* Look at Algorithm BtoD.)

11. How can you tell if a given number written in base 3 is even?

12. Given a decimal number n, write down an algorithm that will express n in base 3 as a string of 0s, 1s, and 2s.

13. What can you say about the base 3 representation of a decimal number that when divided by 9 leaves a remainder of 7?

14. Create an algorithm that will input a binary fraction (see Exercise 2.10) and output the fraction in standard form. Run your algorithm with input **(*a*)** 0.1101, **(*b*)** 0.00101, and **(*c*)** 0.101010.

15. Create an algorithm that will input a positive integer n and a rational number z with $0 < z < 1$ and output n bits of z's binary representation. Run your algorithm for $n = 6$ and **(*a*)** $z = \frac{5}{16}$, **(*b*)** $z = \frac{7}{32}$, **(*c*)** $z = \frac{1}{3}$, and **(*d*)** $z = \frac{1}{10}$.

16. Why can't there be an algorithm to input a fraction z with $0 < z < 1$ and output z's binary representation?

1:5 SET THEORY AND THE MAGIC TRICK

When we think about the magic trick presented in Section 1, the numbers $0, 1, 2, \ldots, 15$ are all of the objects with which we are concerned. The totality of these objects we call our **universe** or underlying set. The eight numbers that appear on card A are called a set of objects. More generally, given any universe of objects we use the word **set** to denote any well-specified collection of objects from the universe. We have included the descriptive "well-specified" in order to make it clear that there can be no ambiguity as to what is in and what is not in the set.

Given two sets, say A and B, of objects from the same universe, A is said to be a **subset** of B, denoted by $A \subseteq B$, provided every object that is contained in A is also contained in B. Every set is necessarily a subset of the universal set, and the **empty set** or null set, the set with no objects, is a subset of every set. The empty set is often denoted by $\emptyset$. The objects in a set are also called **elements** of the set. Two sets A and B are said to be **equal**, written $A = B$, if they contain precisely the same objects (or elements). Note that $A = B$ if and only if $A \subseteq B$ and $B \subseteq A$.

Example 5.1. Let Z denote the integers. We can list the objects in Z by writing $0, 1, -1, 2, -2, \ldots$. For this example Z will be the underlying set or universe.

Let N denote the natural numbers, that is, N can be listed as $0, 1, 2, 3, \ldots$. The natural numbers are objects in Z and there is no ambiguity about which objects are in N, so the natural numbers are a set in the universe of the integers.

Let P denote the set of prime numbers. An integer greater than one is called **prime** if it cannot be factored into the product of two smaller positive integers. Thus 5 is prime while $6 = 2 \cdot 3$ is not prime. P can be listed as $2, 3, 5, 7, 11, \ldots$. P is a subset of N. Note that for a large integer it might be computationally difficult to decide whether the integer is prime or not. Nevertheless, the elements in P are well specified.

Example 5.2. Let the universe consist of the students enrolled in mathematics courses this semester. If S denotes the set of students who are enrolled in Discrete Mathematics and A denotes the set of students who will earn As in Discrete Mathematics, then A is (we hope) a nonempty subset of S.

Typically, we shall use capital letters to denote sets and small letters to denote the objects, when this is possible. If A is a set and s an element of A, we write $s \in A$; read "s is an element of A." If s is not an element of A, we write $s \notin A$. When specifying a set, we shall occasionally list all objects within a pair of curly braces; however, the order in which the objects are so listed is immaterial. The set $\{1, 2, 3\}$ is the same set as $\{3, 1, 2\}$ and $\{1, 3, 2\}$, since each contains precisely the same objects. More frequently, we shall place within the curly braces the property or properties that specify the set.

Example 5.3. The numbers that appear on card A in the magic trick form a set that we could denote by $A = \{8, 9, 10, 11, 12, 13, 14, 15\}$ or by

$$A = \{x\colon 0 \le x \le 15 \text{ and the binary form of } x \text{ contains a 1 in the third column, reading from the right and beginning with } 0\}.$$

Read the above line as "A equals the set of x such that zero is less than or equal to x, which is less than or equal to 15, and the binary form of x contains a 1 in the third column . . . ." It is often the case that some contextual information is left out of the specification of the set. Here, we did not note that the objects in the universe are all integers. Such information may be omitted provided that it does not lead to any confusion on your part.

Question 5.1. Let the universe consist of the positive integers less than 30. Below we list several sets that are well specified by the properties that their elements must satisfy. For each such set list the elements in the set.
(a) $A = \{x: x \text{ is not prime}\}$.
(b) $B = \{x: x \text{ is a square, that is, for some integer } y,\ x = y^2\}$.
(c) $C = \{x: x \text{ is divisible by a square greater than one}\}$.

Given a set A consisting of some objects from the universe U, the **complement** of A, denoted by A^c, is the set of objects from the universe that are not elements of A. In the curly brace notation

$$\begin{aligned} A^c &= \{x \text{ is in } U: x \text{ is not in } A\} \\ &= \{x \in U: x \notin A\}. \end{aligned}$$

Question 5.2. Find the complements of each of the sets from the preceding question.

Question 5.3. For the six sets you found in the preceding two questions determine which are subsets of each other.

From two sets of elements in the same universe, say A and B, we can derive two new sets, the union of A and B and the intersection of A and B. The **union** of A and B, denoted by $A \cup B$, consists of all the elements of U that are either in A or in B (or in both). Note that in English "or" often means exclusive or: For lunch I shall eat a pizza or a grinder (but not both). Here we use "or" in the inclusive sense: mathematics majors usually study statistics or computer science (or both). The **intersection** of A and B, denoted by $A \cap B$, consists of all the elements of U that are in both A and B. In curly brace notation

$$A \cup B = \{x: x \text{ is in } A \text{ or } x \text{ is in } B \text{ or } x \text{ is in both } A \text{ and } B\}$$

and

$$\begin{aligned} A \cap B &= \{x: x \text{ is in } A \text{ and } x \text{ is in } B\} \\ &= \{x: x \in A \text{ and } x \in B\}. \end{aligned}$$

Question 5.4. Find the pairwise unions and intersections of the sets you found in Question 5.1.

Example 5.4. Let A, B, C, and D denote the sets of numbers on the cards of the magic trick. For instance, suppose that player 2 is thinking of the number 6. This number appears on cards B and C and does not appear on cards A and D. When player 2 says yes to card B and yes to card C, player 1 knows that player 2's number is in the set labeled B and in the set labeled C. In the language just introduced player 2's number is in $B \cap C$. Now

$$B \cap C = \{4, 5, 6, 7, 12, 13, 14, 15\} \cap \{2, 3, 6, 7, 10, 11, 14, 15\} = \{6, 7, 14, 15\}.$$

Let us do the analogous set theory for the no responses. When player 2 says no to card A, player 1 knows that the number is not on card A. If a number is not in A but is in the universe, then it must be in the complement of A. Similarly, player 2's number must be in the complement of D. In the language of set theory, $A^c = \{0, 1, 2, 3, 4, 5, 6, 7\}$, $D^c = \{0, 2, 4, 6, 8, 10, 12, 14\}$, so

$$A^c \cap D^c = \{0, 2, 4, 6\}.$$

Player 2's responses mean that the number chosen must be in A^c, B, C, and D^c. What number is it?

$$A^c \cap B \cap C \cap D^c = (B \cap C) \cap (A^c \cap D^c) = \{6, 7, 14, 15\} \cap \{0, 2, 4, 6\} = \{6\}.$$

In general, one way to explain why the magic trick always works is to notice that if the set S is either A or A^c, the set T either B or B^c, V either C or C^c, and W either D or D^c, then $S \cap T \cap V \cap W$ contains exactly one number.

EXERCISES FOR SECTION 5

1. Let the universe consist of all two-letter "words," that is, all sequences of two alphabetic characters (which don't have to form a real English word). Let A consist of all of these words that begin with an "a," and B those that end with a "b." Let C be those that contain no "c," D those that contain no vowel, and E those that contain only vowels.
   (a) Describe the complementary set in each case.
   (b) List or describe $A \cap B$, $A \cup B$, $A \cap C$, $A \cup C$, $A \cap D$, $A \cap E$, $B \cap D$, $B \cup D$, $C \cap D$, $C \cup E$, and $D \cup E$.
2. Let the universe consist of all two-digit numbers:

$$\{x\colon 10 \le x \le 99\}.$$

Let A be the set of two-digit numbers that begin with a 1, B those that end with a 9, C those that are multiples of 3, D those in which both digits are even, and E those that are even.

(*a*) Among these sets, find two such that one is contained in the other.

(*b*) Are there three sets whose intersection is empty, but such that the intersection of any two is not empty?

(*c*) Construct a set with half as many elements as A using $A, \ldots, E$, their complements, unions, intersections, and so on.

(*d*) List the following sets: $A \cap (B \cup C)$, $A \cup (D \cap E)$, $(A \cap B) \cup (D \cap E)$, $C \cap D \cap E$, $A \cup C \cup D$, $A \cap B \cap C \cap D \cap E$, $A \cap C \cap E$, and $C \cap (D \cup E)$.

3. Using the universe and sets $A, B, \ldots, E$ of the previous problem, identify the following sets and then show that the indicated identity is valid:

(*a*) $(D \cup E)$, $(C \cap D)$, $(C \cap E)$: $C \cap (D \cup E) = (C \cap D) \cup (C \cap E)$.

(*b*) $(A \cup C)$, $(A \cap E)$, $(C \cap E)$: $(A \cup C) \cap E = (A \cap E) \cup (C \cap E)$.

(*c*) $(B \cup D)$, B^c, D^c: $(B \cup D)^c = B^c \cap D^c$.

(*d*) $(A \cap E)$, A^c, E^c: $(A \cap E)^c = A^c \cup E^c$.

(*e*) $(B \cap E)$, $(A \cup B)$, $(A \cup E)$: $A \cup (B \cap E) = (A \cup B) \cap (A \cup E)$.

4. Suppose that you are designing a version of the magic trick with the numbers $0, 1, \ldots, 7$. Furthermore, you have already constructed two cards labeled E and F where in the notation of set theory

$$E = \{1, 2, 3, 4\} \qquad \text{and} \qquad F = \{3, 4, 5, 6\}.$$

Construct a single card G that will enable you to successfully perform the trick. Is G uniquely determined, that is, is there choice about what numbers can be put on G?

5. Suppose that you are again designing a version of the magic trick with the numbers $0, 1, \ldots, 7$. What set theoretic properties must two cards E and F satisfy so that it is possible to construct a third card G with which the magic trick can be played?

6. From a universe of seven objects, find seven sets each containing three objects such that each object is contained in three sets and the intersection of any pair of sets consists of one object.

7. Let the universe consist of all five-bit binary fractions. Suppose that $A = \{x\colon x \geq \frac{1}{2}\}$, $B = \{x\colon x$ has a 1 in its $\frac{1}{4}$ column$\}$, $C = \{x\colon x$ has an odd number of 1s in its representation$\}$, $D = \{x\colon$ the last two bits of x are $0\}$, and $E = \{x\colon \frac{1}{4} < x < \frac{3}{4}\}$.

(*a*) Describe the complementary set in each case.

(*b*) List or describe $A \cap B$, $A \cup B$, $A \cap C$, $A \cup C$, $A \cap D$, $A \cap E$, $B \cap D$, $B \cup D$, $C \cap D$, $C \cup E$, and $D \cup E$.

8. Using the universe and sets $A, B, \ldots, E$ of the previous problem, identify the following sets and then show that the indicated identity is valid.

(*a*) $(D \cup E), (C \cap D), (C \cap E)$: $C \cap (D \cup E) = (C \cap D) \cup (C \cap E)$.
(*b*) $(A \cup C), (A \cap E), (C \cap E)$: $(A \cup C) \cap E = (A \cap E) \cup (C \cap E)$.
(*c*) $(B \cup D), B^c, D^c$: $(B \cup D)^c = B^c \cap D^c$.
(*d*) $(A \cap E), A^c, E^c$: $(A \cap E)^c = A^c \cup E^c$.
(*e*) $(B \cap E), (A \cup E), (A \cup E)$: $A \cup (B \cap E) = (A \cup B) \cap (A \cup E)$.

1:6 PICTURES OF SETS

Given two sets, A and B, in the same universe we can form their union and intersection as indicated in Section 5. We can also form their **difference**, which is defined as follows:

$$A - B = \{x: x \in A \text{ and } x \notin B\}.$$

Thus $A - B$ is a subset of A and is sometimes called the **relative complement** of B with respect to A. This is because if you narrow your viewpoint to think of A as the whole universe, then it is natural to restrict B to $A \cap B$. Now, in the A-universe, $B^c = A - B$. Notice also that this provides us with an alternative way to represent A^c as $U - A$, where U is the universe under consideration.

Example 6.1. Let A and B be defined as in the original card trick.

$$A - B = \{x: x \in A \text{ and } x \notin B\} = \{8, 9, 10, 11\}.$$

Note that

$$B - A = \{x: x \in B \text{ and } x \notin A\} = \{4, 5, 6, 7\}.$$

We shall study relations between sets and shall want to establish the validity of certain assertions. For example, we claim that it is always true that $A - B$ and $B - A$ are **disjoint**, that is, they have no element in common; in symbols

$$(A - B) \cap (B - A) = \varnothing.$$

How can we be sure that this statement is always true? It was true in the case of Example 6.1, but we need a general proof like the following. Suppose that x is an element of $(A - B)$. Then x is in A, but not in B. Consequently, x is not in B and cannot be an element of $(B - A)$, which is a subset of B. Thus the set $(A - B)$ has no element in common with the set $(B - A)$, and so the intersection is empty.

Writing a correct proof is more complicated than working out a specific example. The reason for this complexity is that it is necessary to chase down all the definitions and all possible cases. English is a somewhat clumsy vehicle with which to do this. These sorts of logical arguments are made easier both to construct and to read when accompanied by a picture.

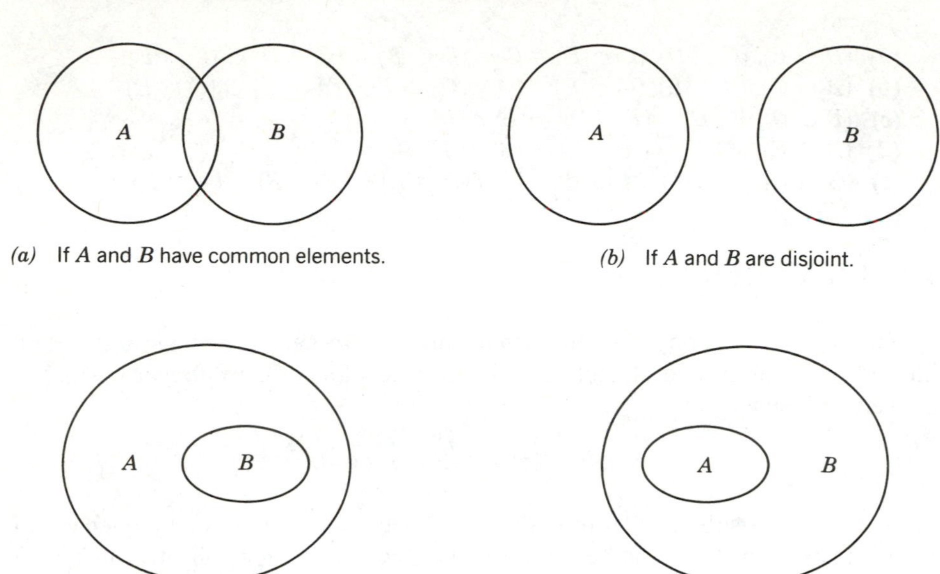

(a) If A and B have common elements.

(b) If A and B are disjoint.

(c) If B is a subset of A.

(d) If A is a subset of B.

Figure 1.3

To obtain a picture of a particular statement concerning sets, we represent the sets in question as regions of the plane. For example, the set A can be conveniently thought of as all points that are inside of or on the boundary of a circular region. Thinking of B in the same way, we can picture these two sets at the same time with one of the diagrams in Figure 1.3.

We concentrate on the first picture, since in some sense it represents the most general situation for two sets. We label the various regions of the plane with the sets they represent (Figure 1.4). These pictures are called **Venn diagrams**. Frequently, they are made more useful by appropriate shading of the basic regions.

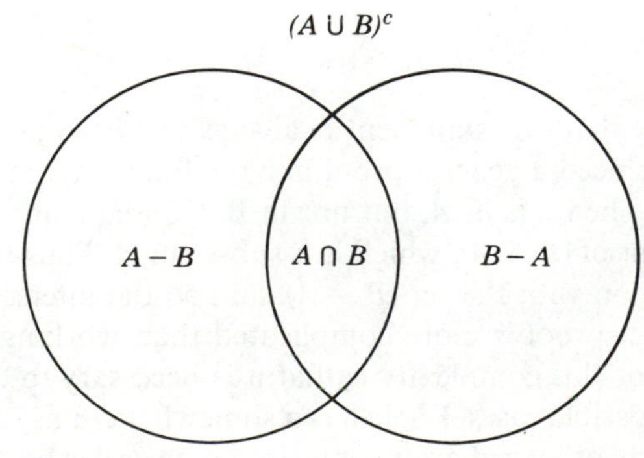

Figure 1.4

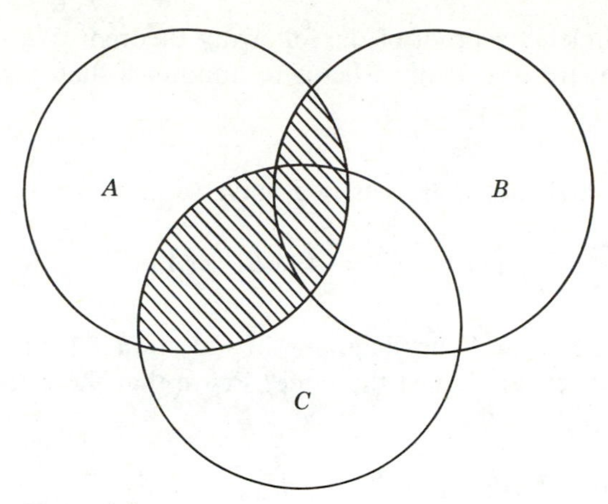

Figure 1.5

Example 6.2. Given sets A, B, and C, Figure 1.5 shows a shaded Venn diagram that highlights the set $A \cap (B \cup C)$.

Question 6.1. Draw shaded Venn diagrams for the following: (a) $(A \cup B)^c$, (b) $A^c \cup B^c$, (c) $(A \cap B)^c$, (d) $A^c \cap B^c$, (e) $A \cup ((B \cap C)^c)$, and (f) $(A \cup B)^c \cap (A \cup C)^c$.

Again we choose to draw the circles as mutually and partially overlapping. If we knew more about specific properties of the sets, for example, that $A \subseteq B$ or that $B \cap C$ is empty, then we could incorporate these properties in the picture, but when the sets are unspecified the drawing in Figure 1.5 is most useful.

From the Venn diagram in Figure 1.5 you might notice, for example, that

$$A \cap (B \cup C) = (A \cap B) \cup (A \cap C).$$

The Venn diagram convinces us of this equality but does not prove the result. For example, does the equality still hold when $A \cap B$ is empty or $B \subseteq C$? To be certain of this statement, we construct an abstract and fully general proof.

There is a straightforward strategy to prove that two apparently different sets are equal. If the two sets are called V and W, first take an element in V and show that it must be in W; thus $V \subseteq W$. Then take an element in W and show that it must be in V. Then $W \subseteq V$, and we conclude that $V = W$.

We follow this strategy here. Let x be in $A \cap (B \cup C)$. Then x is in A and in addition x is in either B or C (or both). Thus either x is in A and B or x is in A and C; that is, x is in $(A \cap B) \cup (A \cap C)$. Conversely, if x is in $(A \cap B) \cup (A \cap C)$, then x is in $A \cap B$ or x is in $A \cap C$. Since both B and C are subsets of $B \cup C$, we have x is in $A \cap B \subseteq A \cap (B \cup C)$ or x is in $A \cap C \subseteq A \cap (B \cup C)$ as desired. □

We have completed a proof of the following theorem. We mark the end of a proof with a square box. It often helps to announce that what was supposed to be proved has been proved.

Theorem 6.1. If A, B, and C are sets in a universe U, then

$$A \cap (B \cup C) = (A \cap B) \cup (A \cap C).$$

Question 6.2. From your Venn diagrams of Question 6.1 what other pairs of seemingly different sets are in fact the same? Prove that these sets are equal.

EXERCISES FOR SECTION 6

1. Draw a Venn diagram of your three-card magic trick from Exercise 5.4. Label points with the integers $0, 1, \ldots, 7$ and show in which region each of these points lies.
2. Copy the Venn diagram in Figure 1.4 and explain why

$$(A \cup B) = (A - B) \cup (B - A) \cup (A \cap B).$$

Explain why each pair of the sets $(A - B)$, $(B - A)$, and $(A \cap B)$ is disjoint. We say that $A \cup B$ is the **disjoint union** of $(A - B)$, $(B - A)$, and $(A \cap B)$, and that these three sets **partition** $A \cup B$.

3. Draw a Venn diagram that portrays arbitrary sets A, B, and C, and shade the region that represents the set $A \cup (B \cap C)$. Then on a separate Venn diagram shade in the sets $A \cup B$ and $A \cup C$, and show that

$$A \cup (B \cap C) = (A \cup B) \cap (A \cup C).$$

Finally, give a proof that the last equality holds for all sets A, B, and C.

4. Draw shaded Venn diagrams for each of the following:
   (*a*) $(A \cap B \cap C)^c$. (*b*) $(A \cup B \cup C)^c$.
   (*c*) $A^c \cup B^c \cup C^c$. (*d*) $A^c \cup (B^c \cap C^c)$.
   (*e*) $A^c \cap (B^c \cup C^c)$. (*f*) $A^c \cap B^c \cap C^c$.
   From these diagrams find pairs of sets that are equal.
5. Draw shaded Venn diagrams for each of the following:
   (*a*) $(A - B) \cap C$. (*b*) $A \cup (B - C)$.
   (*c*) $A \cap (B - C)$. (*d*) $(A - B) \cup C$.
   (*e*) $(A \cap C) - (B \cap C)$. (*f*) $(A \cup C) - (B \cup C)$.
   (*g*) $(A \cup B) - (A \cup C)$. (*h*) $(A \cap B) - (A \cap C)$.

6. Suppose that A, B, and C are three sets such that $A \subseteq B$ and $B \cap C$ is the empty set. Draw a Venn diagram that illustrates this situation. Next draw a Venn diagram that illustrates the case when $A \subseteq B$ and $A \cap C$ is the empty set.
7. Suppose that A, B, C, and D are four sets in the same universe with the property that A and D do not intersect while both B and C intersect both A and D as well as each other. Draw a Venn diagram that pictorially represents this situation; label it and shade in the region $(A \cup D) \cap (B \cap C)$.
8. Given sets A, B, and C in the same universe, draw a Venn diagram to indicate the set D, where

$$D = (B \cap (A \cup C)) \cup (C \cap A).$$

Give a proof that $A \cap B \cap C \subseteq D$.

1:7 SUBSETS

In the previous section we saw how two or three sets, say A, B, and C, could be combined to form new sets like $A - B$, $A \cap (B \cup C)$, and $(A \cap B)^c$. Beginning with only one set A, we can also derive a variety of different sets, for example, A^c and the subsets of A.

If $A = Z$, the integers, there are infinitely many subsets of Z. For instance, consider all sets of the form $\{i, i+1\}$ for i in Z. But if the set A contains a finite number of objects, then there can only be a finite number of different subsets of A. If A contains n objects, A is called an ***n*-set**. We investigate now how many subsets an n-set has and how we can go about finding all of them.

Example 7.1. Suppose that F is the set of fruit in my refrigerator at this moment; F consists of one apple, one banana, and one cantaloupe. Expressed more briefly,

$$F = \{a, b, c\}.$$

Then the subsets of F consist of the choices I have in selecting fruit for dessert, and these choices vary from the extremes of the empty set (i.e., no fruit for dessert) to the whole set F (i.e., eating all three pieces of fruit.)

Question 7.1. Show that there are eight possible fruit desserts in Example 7.1 by listing all subsets of F.

A set A with only one element has exactly two subsets: itself and the empty set. But if $A = \{a, b\}$, then we find four subsets: A, $\{a\}$, $\{b\}$, and the empty set $\varnothing$.

From your work in Question 7.1 it should seem plausible that if A is any set containing three objects, then there are eight subsets of A. In general if A is an n-set, then A has 2^n subsets, a result we shall demonstrate later in this section and prove rigorously in Chapter 2. We want to study such a general n-set A, where n is a positive integer. Typically, we describe the set by

$$A = \{a_1, a_2, \ldots, a_n\},$$

where $a_1, a_2, \ldots, a_n$ stand for the n elements in A. These may be the numbers $\{1, 2, \ldots, n\}$, or they may stand for a different set of numbers like $\{1, 5, 17, 25, \ldots, 94\}$, or they may stand for names of pieces of fruit as in {apple, banana, . . . , quince}. Here are two methods to list all subsets of a general set like A.

Problem. Given an n-set $A = \{a_1, a_2, \ldots, a_n\}$, list all 2^n subsets of A.

Response 1

STEP 1. List the empty set

STEP 2. Set $j := 1$

STEP 3. List all subsets of A that contain j elements

STEP 4. If $j < n$, set $j := j + 1$ and go to step 3; otherwise, go to step 5

STEP 5. Stop.

Is it clear how to carry out step 3? For example, if A contains 17 elements and $j = 9$ we must list all subsets that contain exactly 9 elements. But how? (A subset containing j elements is known as a **j-subset**. In Chapter 3 we'll study the problem of counting and constructing j-subsets of an n-set, for j an arbitrary integer between 0 and n.) For now Response 1 is too imprecise for us to call it an algorithm or to use it effectively.

The idea of the next response is to begin by listing all subsets of $\{a_1\}$; we've seen above how to do this. Then we add to the list those subsets of $\{a_1, a_2\}$ that have not been listed so far, namely those subsets that contain a_2. Then we list additional subsets of $\{a_1, a_2, a_3\}$ that contain a_3 and therefore have not been listed so far. More generally, once we've listed all subsets of $\{a_1, a_2, \ldots, a_j\}$, then we just need to add in the subsets of $\{a_1, \ldots, a_j, a_{j+1}\}$ that contain the last element a_{j+1}.

Response 2 (Algorithm SUBSET)

STEP 1. List the empty set

STEP 2. Set $j := 1$

STEP 3. For each subset B listed so far, create and list the subset $B \cup \{a_j\}$
$\{$At this point we've listed all subsets of $\{a_1, \ldots, a_j\}$.$\}$

STEP 4. If $j < n$, set $j := j + 1$ and go to step 3; otherwise go to step 5

STEP 5. Stop.

Example 7.2. We apply the algorithm SUBSET to the set $A = \{a_1, a_2\}$ with $n = 2$. In Table 1.4 we trace for each step the value of the variable j and show the subsets as they are produced.

Table 1.4

Step	*Value Assigned to j*	*List of Subsets*
1		$\varnothing$
2	1	
3		$\varnothing \cup \{a_1\} = \{a_1\}$
4	2	
3		$\varnothing \cup \{a_2\} = \{a_2\}$
		$\{a_1\} \cup \{a_2\} = \{a_1, a_2\}$
4	2	
5	STOP	

Question 7.2. With $A = \{a_1, a_2, a_3\}$ use SUBSET to list all of A's subsets.

How do we know SUBSET is a correct solution? We don't. We only have evidence from the examples we've done, $n = 2$ and 3. Perhaps we have confidence that this approach will continue to work for larger n. To be certain of the validity of SUBSET, we need a proof of its correctness; however, we must defer the proof until Chapter 2 where we'll develop more proof techniques.

One additional important set that is constructed from a single set is known as the Cartesian product. We have been stressing that order in sets and subsets does not matter, but there are many instances when order does matter. For example, in the magic trick an answer "yes, yes, no, yes" gives us information about a number on cards A, B, C, and D (in that order), and it is important to the game that the responses are given in the correct order. Similarly, in the binary number 1101 and the decimal number 29 the order in which the digits are presented is crucial to our representation of the numbers. The Cartesian product is exactly the idea that we need.

If A is any set, we define the **Cartesian product** $A \times A$ to be the set of all ordered pairs (s, t) such that s and t are elements of A. Symbolically,

$$A \times A = \{(s, t): s \text{ and } t \in A\}.$$

Notice that we have called these **ordered pairs**; that is, we consider (s, t) to be different from the pair (t, s) because the order of the elements matters in this setting. Also notice that we may have a pair of the form (s, s) as well as (s, t) with $s \neq t$.

Similarly, we define

$$A \times A \times A = \{(s, t, u): s, t, \text{ and } u \in A\},$$

and, in general, for any positive integer n we define the n-fold Cartesian product

$$\begin{aligned} A \times A \times \cdots \times A &= A^n \\ &= \{(a_1, a_2, \ldots, a_n): a_i \in A \text{ for } i = 1, \ldots, n\}. \end{aligned}$$

An element of $A \times A \times A$ is often called an **ordered triple** and an element of A^n is known as an (ordered) **n-tuple**.

Example 7.3. Let $A = \{0, 1\}$. Then $A \times A$ consists of the four ordered pairs $(0, 0)$, $(0, 1)$, $(1, 0)$, and $(1, 1)$. We may associate with each of these a binary number with exactly two digits. A^n consists of all n-tuples of zeros and ones and thus corresponds with all binary numbers of length n.

Without listing elements, we could have seen that $A \times A$ contains four pairs by the multiplication principle: In each pair (s, t) there are two choices for s and, independent of these choices, there are two choices for t. The multiplication principle also tells us $A \times A \times A$ contains $2 \cdot 2 \cdot 2 = 8$ triples and that $A \times A \times A \times A$ contains 16 4-tuples. Generalizing this process to A^n, A^n contains 2^n n-tuples. From this we conclude also that there are 2^n binary numbers of length n.

Question 7.3. If $A = \{0, 1, 2, 3, 4, 5, 6, 7, 8, 9\}$, describe $A \times A$, $A \times A \times A$, and A^n for n an arbitrary positive integer.

Question 7.4. If $A = \{a, b, c\}$, list all elements of $A \times A$. If A contains r objects, how many objects does $A \times A$ contain? How many does $A \times A \times A$ contain? A^n?

Example 7.4. Let $A = Z$. Then $Z \times Z$ is the set of ordered pairs (i, j), where i and j are integers. We often think of these as points in the coordinate plane with both coordinates being integers. If $A = R$, the real numbers, then $R \times R$ gives us the real, 2-dimensional coordinate plane and $R \times R \times R$ or R^3 is (real) 3-dimensional coordinate space.

When $A = \{0, 1\}$, the Cartesian products $A \times A$, $A \times A \times A, \ldots$, and A^n are, perhaps surprisingly, related to subsets of a 2-set, subsets of a 3-set, ..., and subsets of an n-set. Suppose that S is an n-set, and that its elements are listed in a fixed order, say $x, y, z, \ldots$. For each subset T of S we may construct an n-tuple of 0s and 1s (an element of A^n) whose first entry is a 1 if and only if x, the first element of S, is contained in T. Next, the second entry in the n-tuple is a 1 if and only if y, the second element of S, is in the subset T, and so on. Thus each subset is described by the n-tuple or the string of binary digits as derived above. (The n-tuple is also called the **bit vector** or the **characteristic function** of the subset.) A subset containing exactly i elements corresponds to an n-tuple containing i ones and $(n - i)$ zeros. Conversely, every n-tuple of A^n corresponds to a unique subset of S: For $j = 1, 2, \ldots, n$ the subset contains the jth element of S if and only if the jth entry of the n-tuple equals 1.

Question 7.5. Given the universal set $S = \{u, v, w, x, y, z\}$, explain why the bit vector of $T = \{u, v, y\}$ is 110010. Then find the bit vectors of the subsets $\{u, w, z\}$ and $\{x, y, z\}$. Explain why the subset corresponding to the string 000001 is $\{z\}$. Then find the subsets corresponding to the strings 000010 and 111100.

Now we see why there are 2^n subsets of an n-set. We know from Example 7.3 that there are 2^n elements in A^n. Each n-tuple of A^n corresponds to one subset of an n-set, and every subset of an n-set corresponds to exactly one n-tuple. Thus there are as many subsets of an n-set as there are n-tuples in $A^n = \{0, 1\}^n$.

Cartesian products also generalize to different sets. Specifically, we define

$$A \times B = \{(s, t): s \in A \text{ and } t \in B\}$$

and

$$A \times B \times C = \{(s, t, u): s \in A, t \in B, \text{ and } u \in C\}.$$

An ordered list of elements taken from one or several sets is called by a variety of names. We have previously called a binary number a string and a sequence of zeros and ones. We shall also call an n-tuple a **vector** and an **array**, depending upon the convention in the particular example.

The Cartesian product is familiar from coordinate geometry of real 2- and 3-dimensional space. This product is important in computer science and arises, for example, in presentations of abstract conceptions of computers, called Turing machines.

EXERCISES FOR SECTION 7

1. Let $A = \{a, b, c, d, e\}$. List all subsets that contain one element, then all subsets with two elements, next all subsets containing three elements, and finally all subsets that contain four elements. How many subsets have you listed?

2. Given a set with n elements, $A = \{a_1, a_2, \ldots, a_n\}$, explain why A has exactly n subsets that contain one element. Then write an algorithm that will list all subsets of A containing just one element.

3. A subset of an n-set A that contains $n - 1$ elements can be formed by omitting just one element of A. Explain why A contains n subsets with $n - 1$ elements and then explain how to list them.

4. List and count all subsets of A that contain two elements when $A = \{a_1, a_2, a_3\}$, when $A = \{a_1, a_2, a_3, a_4\}$, and when $A = \{a_1, a_2, a_3, a_4, a_5\}$.

5. Explain why, for a positive integer n, the set $A = \{a_1, a_2, \ldots, a_n\}$ contains $n(n - 1)/2$ subsets with two elements.

6. Given the set A as in Exercise 5, design an algorithm that will list all subsets of A that contain exactly two elements.

7. If $A = \{a_1, a_2, \ldots, a_n\}$, determine a formula for the number of subsets of A that contain $n - 2$ elements.

8. If A is an n-set, determine a formula for the number of subsets of A that contain exactly three elements.

9. In Exercise 5.1 we considered the universe of all two-letter "words." Find a set S such that this universe can be described also as $S \times S$. Then find a set T such that the subset D of that problem can be described as $T \times T$. Is there a set V such that the subset A can be described as $V \times V$?

10. List all four-digit binary numbers. Then associate each with a subset of $A = \{a, b, c, d\}$ and check that all subsets of A have been listed once and only once.

11. Suppose that $A = \{d, e, f\}$ and $B = \{g, h\}$. List all elements of $A \times B$. If A contains r elements and B contains s elements, how many elements are contained in $A \times B$? Justify your answer. If, in addition, C contains t elements, determine the number of elements in $A \times B \times C$.

12. Canadian zip codes are always six symbols long, alternating between letters and digits, beginning with a letter; for example, H4V-2M9 is a valid zip code. Find sets A and B and express the universe of all possible Canadian zip codes as the Cartesian product of As and Bs. How many zip codes are possible?

13. Massachusetts license plates are six symbols long and each symbol may be either a letter or a digit. Express the universe of all possible license plates as a Cartesian product of an appropriate set or sets. How big is this universe?

14. Suppose that my refrigerator contains four pieces of fruit: two apples, one banana, and one cantaloupe. List all possible fruit desserts. By a fruit dessert I mean a collection of fruit (possibly empty). Next suppose that I have five pieces of fruit: two apples, one banana, one cantaloupe, and one damson plum.

List and count all possible fruit desserts. In general, given the so-called **multiset**

$$A = \{a_1, a_1, a_2, a_3, \ldots, a_n\}$$

with the first element repeated, find a formula for the number of different subsets of A. The two copies of a_1 are considered to be indistinguishable.

1:8 SET CARDINALITY AND COUNTING

A set A is said to be **finite** if it consists of a finite number of objects. An m-set A is a finite set containing m objects for m, a positive integer. We also say that A has **cardinality** m and write $|A| = m$. In our original magic trick the universe has cardinality 16 while each of the sets A, B, C, and D has cardinality 8.

Example 8.1. Let the universe consist of the positive integers less than 30. If $A = \{x: x \text{ is even}\}$, $B = \{x: x \text{ is divisible by } 3\}$, and $C = \{x: x \text{ is divisible by } 5\}$, then $|A| = 14$, $|B| = 9$, and $|C| = 5$.

Example 8.2. Suppose that we wanted to know the cardinality of $A \cup B$ from the previous example. $A \cup B = \{x: x \text{ is divisible by 2 or 3}\}$. We list the elements of $A \cup B$.

$$A \cup B = \{2, 3, 4, 6, 8, 9, 10, 12, 14, 15, 16, 18, 20, 21, 22, 24, 26, 27, 28\}.$$

Note that the cardinality of $A \cup B$ equals 19. This, at first, may seem somewhat strange. We took the union of a 14-element set and a 9-element set and came up with a 19-element set.

Question 8.1. In the context of Example 8.1, find $|A \cup C|$ and $|B \cup C|$.

Question 8.2. In the original magic trick what is the cardinality of each union of two cards?

We shall frequently have occasion to count the number of elements in particular sets, and often these sets can be written as the union (or intersection) of simpler sets. The counting question becomes: Given the cardinalities of two sets A and B, what is the cardinality of their union (or intersection)?

Question 8.3. If possible, find a 5-set A and a 3-set B whose union is a set of cardinality (a) 4, (b) 5, (c) 6, (d) 7, (e) 8, and (f) 9.

In trying to answer the above question, you probably came up with some conclusions about the possible cardinalities of the union of two sets. If you look back at Example 8.2 and Questions 8.1 and 8.2, the following result seems plausible.

Theorem 8.1. Given sets A and B in the same universe,

$$|A \cup B| = |A| + |B| - |A \cap B|.$$

Example 8.2 (continued). We know that $|A| = 14$ and $|B| = 9$.

$$\begin{aligned} A \cap B &= \{x\colon x \text{ is divisible by both 2 and 3}\} \\ &= \{x\colon x \text{ is divisible by 6}\} = \{6, 12, 18, 24\}. \end{aligned}$$

Thus $|A \cap B| = 4$. In this particular instance we verify Theorem 8.1 by noting that $14 + 9 - 4 = 19$.

Proof of Theorem 8.1. We must show that each element of $A \cup B$ contributes exactly one to $|A| + |B| - |A \cap B|$ (see Figure 1.6).

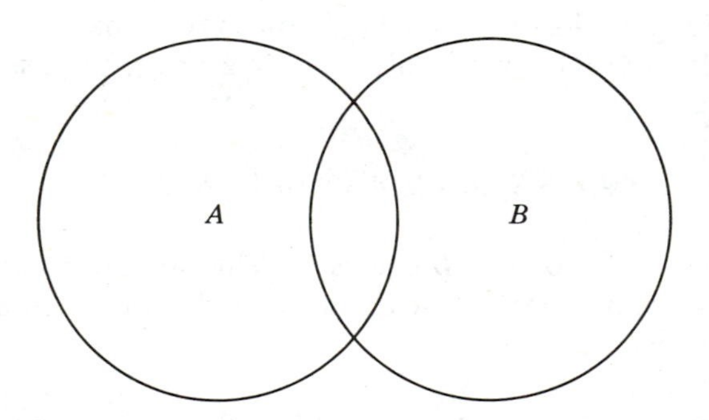

Figure 1.6

Case i. Suppose that x is a member of only one set, say A. Then x is counted in $|A|$, but not in $|B|$ nor in $|A \cap B|$, since $A \cap B \subseteq B$. Thus x contributes a count of one to $|A| + |B| - |A \cap B|$. If x is in B, but not in A, a similar argument suffices.

Case ii. Suppose that x is an element of both A and B. Then x is counted once in $|A|$, once in $|B|$, and once in $|A \cap B|$. Thus it contributes $1 + 1 - 1 = 1$ to $|A| + |B| - |A \cap B|$, just as we wanted. □

Example 8.3. How many seven-digit telephone numbers are there that begin 584- and contain at least one 0 and at least one 1? Suppose that we let A be the set of all such numbers that contain a 0 and B be the set of all numbers that contain a 1. The telephone numbers we're looking for are in both A and B. Thus we want $|A \cap B|$. By Theorem 8.1, $|A \cap B| = |A| + |B| - |A \cup B|$. First let's find $|A|$. (See Exercise 1.11.) There are 10^4 possible telephone numbers, since there are 10 choices for each of 4 numbers. If we exclude 0, then there are 9 choices for each of 4 numbers. Thus there are 9^4 numbers that contain no 0. Thus $|A| = 10^4 - 9^4$. Similarly, $|B| = 10^4 - 9^4$. What about $|A \cup B|$? There are 8^4 numbers that contain neither a 0 nor a 1 and so $|A \cup B| = 10^4 - 8^4$. Thus

$$|A \cap B| = (10^4 - 9^4) + (10^4 - 9^4) - (10^4 - 8^4) = 974.$$

Question 8.4. A joint meeting of Discrete Mathematics and Introductory Computer Science had 232 students. If 146 students are enrolled in the mathematics course and 205 students are enrolled in the computer science course, how many students are enrolled in both courses at once?

We present a formula that generalizes the pattern of Theorem 8.1 from two sets to three sets; the proof is similar. See Exercise 12.

Theorem 8.2. Given sets A, B, and C in the same universe, then

$$|A \cup B \cup C| = |A| + |B| + |C| - (|A \cap B| + |A \cap C| + |B \cap C|) + |A \cap B \cap C|.$$

(See Figure 1.7.)

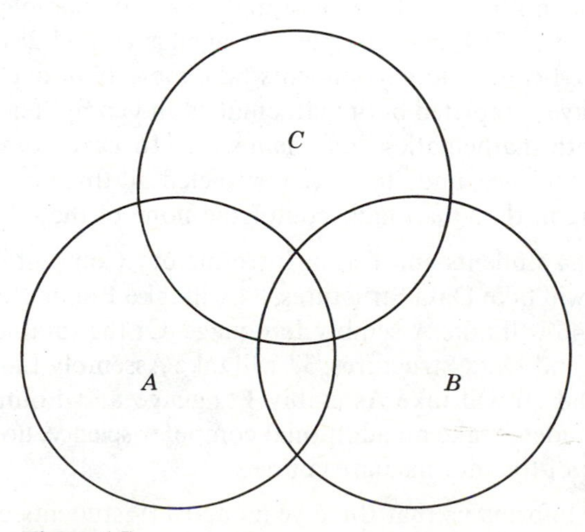

Figure 1.7

Example 8.4. Let A, B, and C be the sets from the original magic trick. Then $A \cup B \cup C = \{2, 3, 4, 5, 6, 7, 8, 9, 10, 11, 12, 13, 14, 15\}$ contains 14 elements. As we have seen $|A| = |B| = |C| = 8$. We have also noted that the intersection of any two of these sets has cardinality equal to 4. You can check that $A \cap B \cap C = \{14, 15\}$. Then

$$|A \cup B \cup C| = 8 + 8 + 8 - (4 + 4 + 4) + 2 = 14.$$

The Principle of Inclusion and Exclusion. Theorems 8.1 and 8.2 are both instances of a general counting result known as the Principle of Inclusion and Exclusion or P.I.E. for short. The same type of counting formula applies for 4 or 5 or k, an arbitrary number, sets. See Exercise 13.

EXERCISES FOR SECTION 8

1. Refer to the sets in Exercise 5.1. Determine the indicated set cardinalities and then check the validity of Theorem 8.1:
   (*a*) $|A \cup B| = |A| + |B| - |A \cap B|$.
   (*b*) $|A \cup C| = |A| + |C| - |A \cap C|$.
   (*c*) $|B \cup D| = |B| + |D| - |B \cap D|$.
   (*d*) $|D \cup E| = |D| + |E| - |D \cap E|$.
2. Refer to the sets in Exercise 5.2. Determine the indicated set cardinalities and then check the validity of Theorem 8.2:
   (*a*) $|A \cup C \cup D| = |A| + |C| + |D| - |A \cap C| - |A \cap D| - |C \cap D| + |A \cap C \cap D|$.
   (*b*) $|C \cup D \cup E| = |C| + |D| + |E| - |C \cap D| - |C \cap E| - |D \cap E| + |C \cap D \cap E|$.
3. Of the 876 students living in the Quadrangle, 530 have completed Introductory Computer Science, 364 have completed Calculus II, and 287 have completed Chemistry I. Of course, lots of students take more than one of these courses. In fact, 213 have completed both mathematics and computer science, 164 have completed both mathematics and chemistry, 116 have completed chemistry and computer science, and 103 have completed all three courses. How many students living in the Quad have completed none of these three courses?
4. There were 184 students enrolled in Introductory Computer Science last fall. Of these 112 will take Data Structures, 84 will take Foundations of Computer Science, and 46 will take Assembly Language. Of the total, 66 will take both Foundations and Data Structures, 37 will take Assembly Language and Data Structures, and 30 will take Assembly Language and Foundations. If 45 of the original students take no additional computer science, how many students take all three of the intermediate courses?
5. The registrar informs us that three years ago 119 students enrolled in Introductory Computer Science. In the following semester, of these 119 students, 96

took Data Structures, 53 took Foundations, and 39 took Assembly Language. Also 38 took both Data Structures and Foundations, 31 took both Foundations and Assembly Language, 32 took both Data Structures and Assembly Language, and 22 brave souls took all three courses. We claim that the registrar must have made an error. Why?

6. How many seven-digit telephone numbers are there that begin with 584- and that contain a 0, a 1, and a 2?

7. Suppose that A and B are sets in the universe U. Find a way to express the cardinality of the set $A^c \cup B^c$ in terms of the cardinality of U, A, B, and combinations of these sets.

8. Consider the universe of all possible strings of six letters made from the letters a, b, c, d, e, f with no repetitions of letters. How many such strings are there in total? How many of these are such that the first letter is neither "a" nor "b" and the last letter is neither "e" nor "f"? (*Hint:* Let A be the set of all these words that do begin with an a or b and B the set of those that end with an e or f. Then we are looking for the cardinality of $A^c \cap B^c$.)

9. A group of 100 students was surveyed to determine the students' interest in winter sports. It was found that 69 liked downhill skiing, 38 liked cross-country skiing, 75 liked skating, and 15 liked none of these sports. On further questioning it was determined that 35 liked both kinds of skiing, 30 liked cross-country skiing and skating, and 42 liked downhill skiing and skating. How many like to ski, either downhill or cross-country? How many like all three sports?

10. In a sample of 100 students, 43 like avocados, 71 like radishes, and 36 like olives in their salad. Each student liked at least one vegetable. If 26 students like both avocados and radishes, 16 students like avocados and olives while 22 like radishes and olives, how many students like all the ingredients in an avocado, radish, and olive salad?

11. Here is an alternative proof of Theorem 8.1. Give reasons for each of the following steps:

STEP 1. If $A \cap B = \varnothing$, then $|A \cup B| = |A| + |B|$
$= |A| + |B| - |A \cap B|$.

STEP 2. $B = (A \cap B) \cup (B - A)$.

STEP 3. $|B| = |A \cap B| + |B - A|$.

STEP 4. $|B - A| = |B| - |A \cap B|$.

STEP 5. $A \cup B = A \cup (B - A)$.

STEP 6. $|A \cup B| = |A| + |B - A|$.

STEP 7. $|A \cup B| = |A| + |B| - |A \cap B|$.

12. Prove Theorem 8.2. (*Hint:* Let x be an element of $A \cup B \cup C$. Show that x's contribution is exactly one. Divide the proof into three cases depending on how many of A, B, and C contain x.)

13. Find an expression for the cardinality of the union of four sets in terms of the cardinalities of the sets and various intersections.

1:9 FUNCTIONS

The concept of a function is fundamental to both mathematics and computer science and will be used throughout this book.

Definition. A **function** f is a mapping from a set D to a set T with the property that for every element d in D, f maps d to a unique element, denoted $f(d)$, of T. Here D is called the **domain** of f, and T is called the **target** of f. We write $f:D \to T$. We also say that $f(d)$ is the **image** of d under f, and we call the set of all images the **range** R of f. In set notation

$$R = \{f(d): d \in D\}.$$

Note that $R \subseteq T$.

A mapping might fail to be a function if it is not defined at every element of the domain or if it maps an element of the domain to two or more elements in the range. Figure 1.8 illustrates these ideas.

To define a function f, we must specify its domain D and a rule for how it operates. If the domain is changed, we consider that a new function is formed.

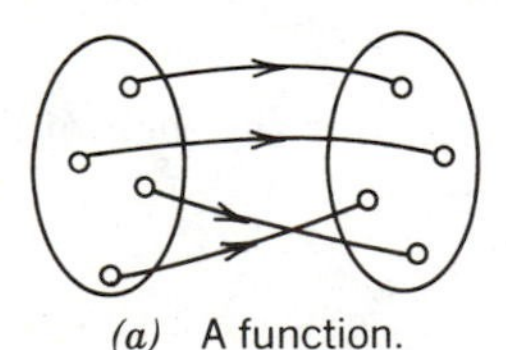

(a) A function.

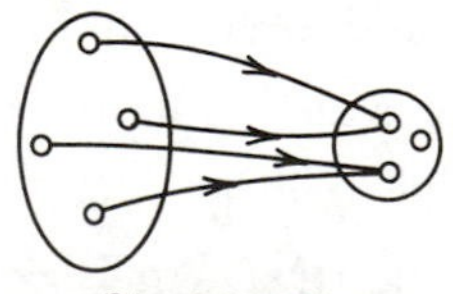

(b) A function.

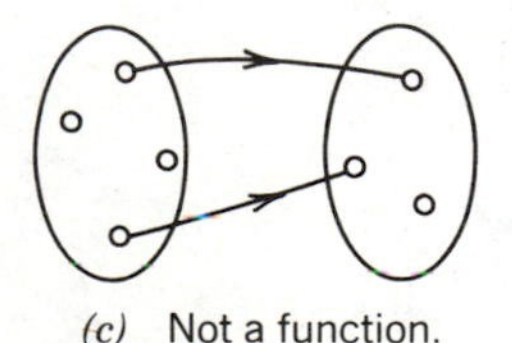

(c) Not a function.

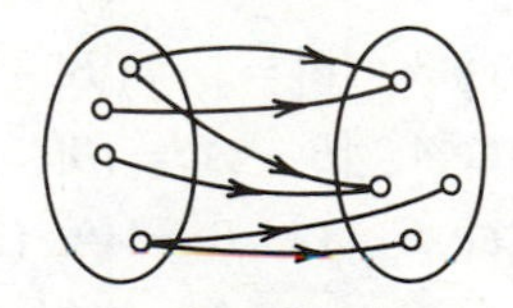

(d) Not a function.

Figure 1.8

For example, $f(x) = x^2$ with D equal all real numbers and $g(x) = x^2$ with D equal all reals greater than one are different functions. They have different graphs, for instance. Given a function f with domain D, the range R is determined. However, in a specific instance, it might take some effort to decide what the range is. In the preceding example the range of f is all nonnegative real numbers and the range of g is all reals greater than one. In contrast the target of a function is not uniquely determined. It is often useful to designate a target T that is a large and familiar set containing the range. Thus we stated that the functions f and g above have the real numbers as target. We could also have picked the nonnegative reals as the target.

In this section we shall consider repeatedly the functions that map a binary number to its decimal equivalent and a natural number to its binary equivalent. We don't have formulas for these functions, like $f(x) = x^2$, but we can think about these mappings, and we have algorithms to compute these functions whenever necessary.

Example 9.1. Let B be the set of all binary numbers, or equivalently all finite strings of zeros and ones, and let N be the set of all natural numbers expressed in decimal notation. Then f, g, h, and j given below are functions from B to N. For s in B,

$f(s)$ equals the decimal equivalent of s,

$g(s)$ equals the number of bits in s,

$h(s)$ equals the number of ones in s,

and

$j(s)$ equals the ones bit of s.

For instance, if $s = 110010$, then $f(s) = 50$, $g(s) = 6$, $h(s) = 3$, and $j(s) = 0$. The range of f, g, and h is in each case all of N. To see this, let m be any decimal number in N. If s is m's binary equivalent, then $f(s) = m$. If r is the binary number consisting of m ones, then $g(r) = h(r) = m$. However, the range of j is $\{0, 1\}$.

Here are two mappings from B to N that are not functions. For s in B,

$k(s)$ equals the fifth bit in s, counting from the left,

and

$$l(s) = \begin{cases} 1 \text{ if } s \text{ ends with } 1 \\ 2 \text{ if } s \text{ ends with } 0 \\ 4 \text{ if } s \text{ ends with } 00. \end{cases}$$

The mapping k is not defined on all of B, only on those with five or more digits, and l specifies two different images for strings ending with two zeros.

Question 9.1. Suppose that B and N are as in the preceding example. Define $b: N \to B$ by $b(r) = s$ if s is the binary equivalent of the decimal number r. Explain why the range of b is all of B.

Question 9.2. Which of the following is a function from N to B, where N and B are as defined in Example 9.1? For r in N,

$$f_1(r) \text{ equals the number of digits of } r,$$

$$f_2(r) = \begin{cases} 0 & \text{if } r \text{ is even} \\ 1 & \text{if } r \text{ is odd,} \end{cases}$$

$$f_3(r) \text{ equals the string of } r \text{ ones,}$$

and

$$f_4(r) = \begin{cases} 0 & \text{if 2 divides } r \\ 1 & \text{if 3 divides } r \\ 1 & \text{if neither 2 nor 3 divides } r. \end{cases}$$

For each that is a function, specify its range.

A function is said to be **onto** or an onto function if its range equals its target, $R = T$. Thus functions f, g, and h of Example 9.1 are onto, but j is not. In Figure 1.8 the first function shown is onto whereas the second is not onto. We also say that two functions f and g are **equal** if they have the same domain D and $f(d) = g(d)$ for every d in D.

Sometimes the domain of a function is a set of sets.

Example 9.2. If $U = \{a_1, a_2, \ldots, a_n\}$, let $P(U)$ be the set of all subsets of U. We know that $P(U)$ contains 2^n subsets. We define the complementation function $c: P(U) \to P(U)$ by $c(A) = A^c$ for every A in $P(U)$. Then c is an onto function, because for every set B in $P(U)$ we have

$$c(B^c) = (B^c)^c = B.$$

If we define $n: P(U) \to P(U)$ by $n(A) = \varnothing$ for every A in $P(U)$, then n is not onto. It is evident that the functions c and n have the same domain, but they are not equal.

Question 9.3. Which of the functions in Questions 9.1 and 9.2 are onto?

A function is said to be **one-to-one** (or 1-1) if it maps distinct elements of the domain to distinct elements of the range. In other words, if $d \neq d'$, then $f(d) \neq f(d')$. A diagram of different function properties is shown in Figure 1.9.

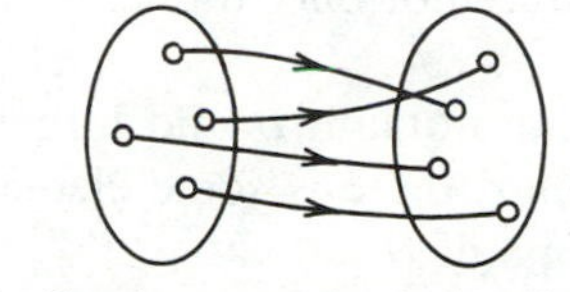

(a) One-to-one and onto function.

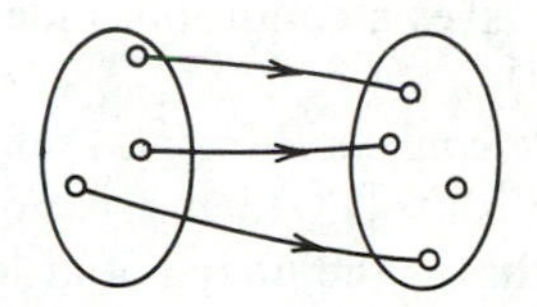

(b) One-to-one but not onto function.

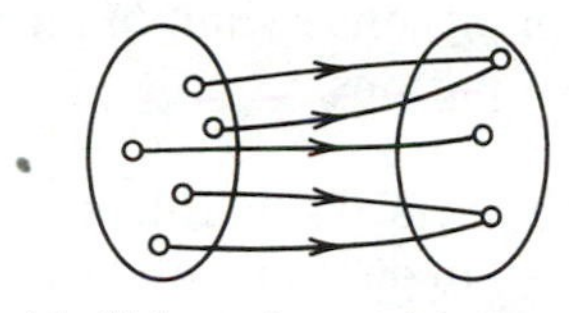

(c) Not one-to-one function.

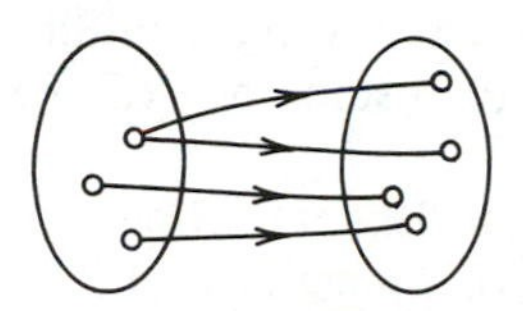

(d) Not a function.

Figure 1.9

Example 9.1 (continued). Here again are the functions defined in Example 9.1. For s in B,

$f(s)$ equals the decimal equivalent of s,

$g(s)$ equals the number of bits in s,

$h(s)$ equals the number of ones in s,

and

$j(s)$ equals the ones bit of s.

The function f is one-to-one because if $s \neq s'$, then their decimal equivalents will be different. However, g is not one-to-one because, for example, $g(101) = 3 = g(111)$. Neither h nor j is one-to-one, since $h(101) = h(110) = 2$, and $j(101) = j(11) = 1$.

Question 9.4. Let the functions b, f_2, and f_3 be as defined in Questions 9.1 and 9.2. Which of these functions is one-to-one and why?

Question 9.5. Let $U = \{a_1, a_2, \ldots, a_n\}$. Is $c: P(U) \to P(U)$ defined by $c(A) = A^c$ a one-to-one function?

Suppose that a function $f: D \to T$ is not one-to-one. Then we know that there is an element t and two elements $d \neq d'$ in D such that $f(d) = f(d') = t$. The next result gives a condition under which a function is surely not one-to-one.

The Pigeonhole Principle. If f is a function with finite domain D and target T, where $|D| > |T|$, then f is not one-to-one. In particular, there is some element t in T that is the image of at least two different elements of D.

Why is this so? The function f is either one-to-one or it isn't. Since $R \subseteq T$ for any function f, it follows that $|R| \leq |T|$. If $|D| > |T|$ and f were one-to-one, then $|R| = |D|$ and so $|R| > |T|$, a contradiction. Thus f cannot be one-to-one.

This principle has far-reaching applications in combinatorics. Its name derives from the following flightful application: If more than n pigeons fly into n pigeonholes, then some pigeonhole must contain at least two pigeons.

Example 9.3. Let S be a set of 11 or more binary numbers. Then at least 2 elements of S must have the same last digit when expressed in decimal notation. The pigeonhole principle shows why: Define f^* to be the function with domain S and target $T = \{0, 1, \ldots, 9\}$, where for s in S, $f^*(s)$ equals the last digit of s when s is expressed in decimal notation. Since $|S| > 10 = |T|$, there must be two numbers in S that map to the same element of T.

For further examples see Exercises 20–23 and Supplementary Exercises 11 and 12.

Often a function maps one set into the same set, that is, $T = D$. This was the case in Example 9.2 and Question 9.5. When $T = D$ and the function is one-to-one and onto, it is called a **permutation**. The examples in Figure 1.10 illustrate per-

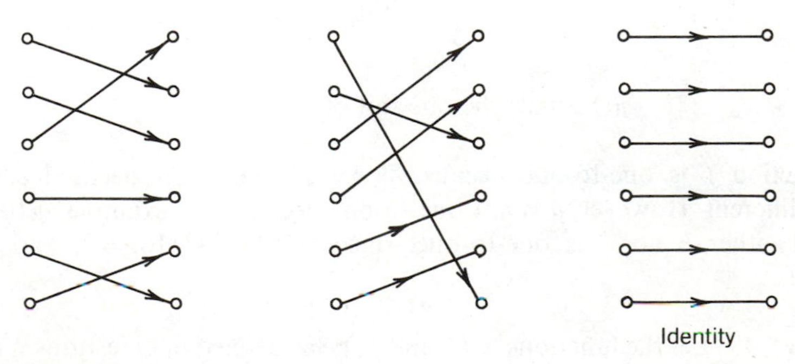

Figure 1.10 Permutations

mutations. One important, but easy, permutation is called the **identity map** or **identity permutation**. We define it by $i:D \to D$, where $i(d) = d$ for every d in D. This map doesn't do much, but it is one-to-one and onto; soon we'll see that it plays an important role.

When $T = D$, the properties of being one-to-one and onto are closely related.

Theorem 9.1. Let D be a finite set and let $f:D \to D$ be a function. Then f is one-to-one if and only if f is onto.

Proof. If D has n elements and f is one-to-one, then R, the range of f, has n elements also. Since $R \subseteq D$ and they have the same (finite) cardinalities, $R = D$ and so f is onto. On the other hand, if f is onto but not one-to-one, then some pair of elements in D gets mapped by f to the same element. Consequently, the remaining $n - 2$ elements of the domain must be mapped to $n - 1$ elements of the range. Then some element must have two images, contradicting the definition of a function. Thus f must be one-to-one. (See Exercise 24 for another look at this idea.) □

Just as we can combine sets to form new sets, so can we combine functions to form new functions. Suppose that $f:D \to T$ and $g:T \to W$ are such that the range of f is contained in the domain of g. Then we define the **composite** of g with f, denoted $g \circ f$, to be the function

$$g \circ f:D \to W, \qquad \text{where } g \circ f(d) = g(f(d))$$

for all d in D. In words this means that for d in D, we first map d to T using f, and then we map the result, $f(d)$, to W using g. This process of combining two functions is known as **composition**. Notice that in the composition $g \circ f$, it is f that gets performed first even though it is g that is on the left and thus read first. This idea is illustrated in Figure 1.11.

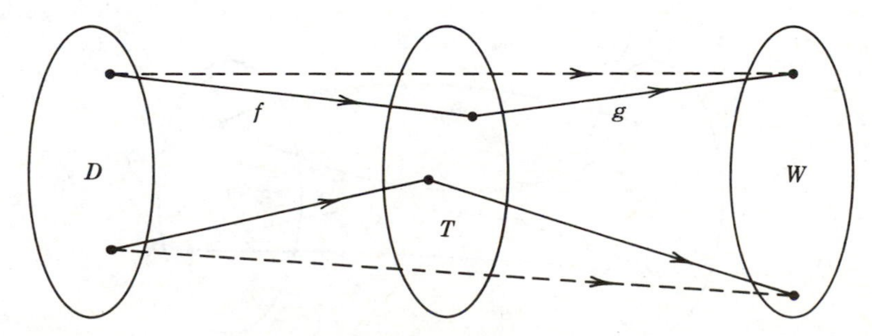

Figure 1.11 $g \circ f$ is shown in dashed lines.

Example 9.4. If $f:B \to N$ is defined as before with $f(s) = m$, where m is the decimal equivalent of s, and if $g:N \to N$ is defined by

$$g(t) = \begin{cases} 0 & \text{if } t \text{ is even} \\ 1 & \text{if } t \text{ is odd,} \end{cases}$$

then we can define $g \circ f:B \to N$ by

$$\begin{aligned} g \circ f(s) &= g(f(s)) \\ &= \begin{cases} 0 & \text{if } f(s) \text{ is even} \\ 1 & \text{if } f(s) \text{ is odd.} \end{cases} \end{aligned}$$

Thus $g \circ f = j$, where j is defined in Example 9.1.

Now suppose that $f:D \to T$ and $g:T \to D$; that is, the domain of g is the target of f, and vice versa. Then

$$g \circ f:D \to D,$$

and the composite function takes us back where we started from.

Sometimes $g \circ f$ does even more than that. If $g \circ f = i$, the identity map $i(d) = d$, then g is called the **inverse** of f. Specifically, if

$$g \circ f(d) = g(f(d)) = d$$

for every d in D, then g undoes the work of f, and g is the inverse of f. Similarly, if we compose the other way around, $f \circ g:T \to T$ and get

$$f \circ g(t) = f(g(t)) = t$$

for all t in T, then f is called the inverse of g. (See Figure 1.12, where f is shown with a solid line, its inverse by a dashed line.)

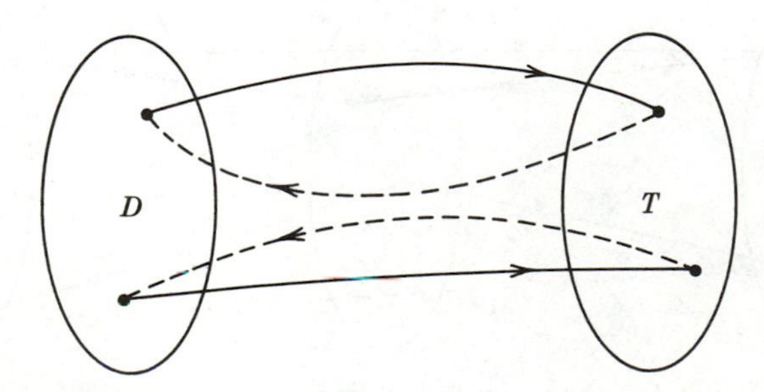

Figure 1.12

Example 9.1 (continued again). Let $f(s) = m$ and $b(m) = s$ be defined as in Example 9.1 and Question 9.1. Then $b \circ f: B \to B$ maps a binary number to its decimal equivalent and back to its binary equivalent. Thus $b \circ f = i$, and b is the inverse of f. Also $f \circ b = i$, since translating decimals to binary and then back to decimals returns the original decimal number. Thus f is also the inverse of b. Note that the two identity functions of this example are different, since they have different domains.

Example 9.4 (continued). Let f be as already given, and $g: N \to B$ be defined by $g(t) = 0$ or 1 according as t is even or odd. Then $g \circ f: B \to B$ and $f \circ g: N \to N$, but neither composite mapping is the identity. For instance, $g \circ f(101) = g(5) = 1$, and $f \circ g(2) = f(0) = 0$.

It is not by chance that in the latest continuation of Example 9.1 f and g were inverses of each other.

Theorem 9.2. Suppose that both $f: D \to T$ and $g: T \to D$ are onto functions. Then f is the inverse of g if and only if g is the inverse of f.

Proof. Suppose that f is the inverse of g, that is, $f \circ g = i$. We must show that g is the inverse of f, that is, we must show that $g \circ f(d) = d$ for all d in D. For any element d in D, since $g: T \to D$ is onto, there is an element t in T such that $g(t) = d$. Then

$$
\begin{aligned}
g \circ f(d) = g \circ f(g(t)) &= g(f(g(t))) && \text{by definition of } g \circ f \\
&= g(f \circ g(t)) && \text{by definition of } f \circ g \\
&= g(i(t)) && \text{since } f \circ g = i \\
&= g(t) = d.
\end{aligned}
$$

The remaining proof, that if g is the inverse of f, then f is the inverse of g, goes the same way. □

Question 9.6. Let U be a finite set, $P(U)$ the set of all subsets of U, and $c: P(U) \to P(U)$ be defined by $c(A) = A^c$. Then explain why c is its own inverse, that is, why $c \circ c = i$.

In summary, a function, like an algorithm, has input and output, elements of the domain and range, respectively. So what is the difference between a function and an algorithm? An algorithm may have as input values for several variables, say x, q, and n. If x can be from the set A, q from B, and n from C, then the input to the algorithm can be thought of as one element in the Cartesian product

$A \times B \times C$. Similarly, the output of the algorithm can be thought of as an element from a Cartesian product. In short, every algorithm is a function. Conversely, the functions in this book (though not all functions) are mappings that can be computed by algorithms, sometimes by several different algorithms.

There is, however, a striking difference in the contexts in which these concepts get used. When we think of an algorithm, we are vitally concerned with the mechanism by which a domain element d gets mapped to its corresponding range element $f(d)$. In contrast, when we think of a function it is the correspondence itself that matters, not how $f(d)$ is computed. One of the main goals of discrete mathematics is to supply the tools which enable a rational choice among various algorithms that evaluate a function.

EXERCISES FOR SECTION 9

1. Let N be the set of all nonnegative integers. Which of the following are functions with domain and target N?

(*a*) $f(n) = n + 1$. (*b*) $f(n) = 2n + 1$.

(*c*) $f(n) = \dfrac{n}{2} + 1$. (*d*) $f(n) = n - 1$.

(*e*) $f(n) = n^2 + 1$. (*f*) $f(n) = \sqrt{n} + 1$.

(*g*) $f(n) = \dfrac{1}{n} + 1$. (*h*) $f(n) = n^3$.

(*i*) $f(n) = 2^n$. (*j*) $f(n)$ equals the remainder when n is divided by 3.

Of those that are functions with domain and target N, find their range and determine whether or not they are onto.

2. Let R be the set of all real numbers. Which of the following are functions with domain and target R?

(*a*) $f(x) = 2x + 1$. (*b*) $f(x) = \dfrac{x}{2} + 1$.

(*c*) $f(x) = x - 1$. (*d*) $f(x) = \sqrt{x} + 1$.

(*e*) $f(x) = \dfrac{1}{x} + 1$. (*f*) $f(x) = x^2 - 3x + 2$.

(*g*) $f(x) = |x|$, where $|x|$ stands for the absolute value of x.

Of those that are functions, find those that are one-to-one.

3. Give an example of a function, with domain and target the positive integers, that is onto and is not the identity map.

4. If two functions are not equal, they are called different. Suppose that $A = \{a_1, a_2, \ldots, a_n\}$. How many different functions are there with domain A and target $\{0, 1\}$? How many of these are onto and how many of these are one-to-one?

5. For each of the following conditions give an example of a function $f:Z \to Z$, where Z is the set of all integers, that satisfies the condition:
   (i) f is onto and one-to-one.
   (ii) f is onto but not one-to-one.
   (iii) f is one-to-one but not onto.
   (iv) f is neither one-to-one nor onto.
   (v) Every integer is the image of exactly two integers.
   (vi) f has an inverse g.
   (vii) f does not have an inverse g.

6. Let $U = \{a_1, a_2, a_3\}$ and let $S = \{a_1\}$. For every set A in U we define the map $f:P(U) \to P(U)$ by $f(A) = A \cap S$ and the map $g:P(U) \to P(U)$ by $g(A) = A \cup S$. Write down the image $f(A)$ and $g(A)$ for every subset A. Is either f or g onto?

7. Let $U = \{a_1, \ldots, a_n\}$, and for S a fixed subset of U define $f(A) = A \cap S$ and $g(A) = A \cup S$ for every subset A of U. For what sets S is f an onto function and for what sets S is g onto? Is either f or g one-to-one?

8. Fix a finite universal set U. The size function $s:P(U) \to N$, where N is the set of all nonnegative integers, is given by $s(A) = |A|$ for all subsets A of U. What is the range of s? Is it one-to-one?

9. The characteristic function of a set S, a fixed subset of the universe U, is given by $\chi_S(x): U \to N$, where

$$\chi_S(x) = \begin{cases} 0 & \text{if } x \text{ is not in } S \\ 1 & \text{if } x \text{ is in } S. \end{cases}$$

   For a fixed subset S of U, let $h:U \to N$ be given by $h(x) = |\{x\} \cap S|$. Explain why χ_S and h are equal functions.

10. Suppose that A and B are finite sets and $A \subseteq B$. Explain why $A = B$ if and only if $|A| = |B|$.

11. Suppose that $f:D \to D$ is a function with the same domain and target. Then we can define $f^2 = f \circ f$ as the composition of f with itself. For each of the following, write down a simple expression for f^2:
    (i) $f:R \to R, f(x) = x$.
    (ii) $f:R \to R, f(x) = x^2$.
    (iii) $f:N \to N, f(i) = i + 1$.
    (iv) $f:N \to N, f(j) = 2j + 1$.

(v) $f: P(U) \to P(U), f(A) = A^c$.
(vi) $f: P(U) \to P(U), f(A) = \varnothing$.

12. If $f: D \to D$, then we can also define $f^3 = f \circ (f^2) = f \circ (f \circ f)$ and $f^4 = f \circ (f^3) = f \circ [f \circ (f \circ f)]$. Determine f^3 and f^4 for each of the functions in Exercise 11.

13. If $f: D \to D$, we define f^n, where n is a positive integer, by $f^1 = f, f^2 = f \circ f$, and, in general, $f^n = f \circ (f^{n-1})$. For each of the functions in Exercise 11 find an expression for f^n in terms of n.

14. Let $A = \{a_1, a_2, \ldots, a_n\}$ and form the Cartesian product $A \times A$. Define two projection functions, P_1 and P_2: $A \times A \to A$ by $P_1(a_i, a_j) = a_i$ and $P_2(a_i, a_j) = a_j$. Is either of these functions one-to-one or onto?

15. Suppose that we define a function $b: A \to A \times A$ by $b(a_i) = (a_i, a_1)$. Then is b the inverse of either of the projection functions P_1 or P_2? Is either P_1 or P_2 the inverse of b?

16. Show by example that Theorem 9.1 is false if D is not finite.

17. Suppose that f and g are functions such that $f: D \to T$, $g: T \to D$, and $f \circ g = i$. Prove that f is onto.

18. Suppose that $f: D \to T$ is a one-to-one function and its domain D is finite. Then prove that there is a function g that is f's inverse.

19. For each of the following, find the inverse of f. Let Z stand for the integers, N the nonnegative integers, R the real numbers, and $U = \{a_1, a_2, \ldots, a_n\}$.

(i) $f: Z \to Z, f(x) = x + 1$.

(ii) $f: R - \{0\} \to R, f(x) = \dfrac{1}{x}$.

(iii) $f: N \to R, f(x) = \sqrt{x}$.

(iv) $f: U \to N, f(a_i) = i$.

(v) $f: U \to U, f(a_i) = a_{n-i}$.

20. Suppose that $f: D \to T$, where $|D| > |T|$. Can you conclude any of the following? Explain.

(*a*) There are at least two elements t_1 and t_2 of T that are each the image of two or more domain elements.
(*b*) Every t in T is the image of at least two domain elements.
(*c*) There is an element t in T and three distinct elements d_1, d_2, and d_3 in D such that $f(d_1) = f(d_2) = f(d_3) = t$.

21. Explain why a set of 16 numbers selected from $\{2, \ldots, 50\}$ must contain two with a common divisor greater than one.

22. Explain why a subset of 51 numbers taken from $\{1, 2, \ldots, 100\}$ must contain two numbers, where one is a divisor of the other.

23. Let f be a function with domain D and target T. If $|D| = d$ and $|T| = n$, then explain why there is an element t of T that is the image of at least d/n elements of D.

24. Here is a stronger version of Theorem 9.1. If $f:D \to T$ is a function, where $|D| = |T|$, then f is one-to-one if and only if f is onto. Prove that this is so and explain why this is more general than Theorem 9.1.

25. Suppose that f and g are functions such that $f:D \to T$ and $g:T \to D$, but one of them is not onto. Then is the conclusion of Theorem 9.2 still true; that is, is it still the case that f is the inverse of g if and only if g is the inverse of f?

1:10 BOOLEAN FUNCTIONS AND BOOLEAN ALGEBRA

We return to the dichotomies mentioned in Section 2 and concentrate on functions with two-element targets: $T = \{0, 1\}$ or $\{\text{True}, \text{False}\}$ or $\{\text{yes}, \text{no}\}$, and so on; without loss of generality we assume that $T = \{0, 1\}$. Such functions are central to computer science and related mathematics. In this section we develop algebraic properties of these functions, characterize their fundamental forms, and consider an easily stated, but unresolved, research problem, known as the Satisfiability Problem.

Definition. Let $Z_2 = \{0, 1\}$. A function $f:(Z_2)^n \to Z_2$, with domain a Cartesian product of Z_2 and target Z_2, is called a **Boolean function**.

Example 10.1. Here are three fundamental Boolean functions.

1. **NOT:** $Z_2 \to Z_2$ defined by $\qquad \text{NOT}(x) = \begin{cases} 0 & \text{if } x = 1 \\ 1 & \text{if } x = 0. \end{cases}$

   This is usually written $\text{NOT}(x) = \sim x$.

2. **AND:** $(Z_2)^2 \to Z_2$ defined by $\qquad \text{AND}(x, y) = \begin{cases} 1 & \text{if } x = y = 1 \\ 0 & \text{otherwise.} \end{cases}$

   This is usually written $\text{AND}(x, y) = x \wedge y$.

3. **OR:** $(Z_2)^2 \to Z_2$ defined by $\qquad \text{OR}(x, y) = \begin{cases} 0 & \text{if } x = y = 0 \\ 1 & \text{otherwise.} \end{cases}$

   This is usually written $\text{OR}(x, y) = x \vee y$.

These three functions represent so-called "logical operations." If we associate "False" with 0 and "True" with 1, then NOT(x) is "True" (or 1) if and only if x is "False" (or 0). Furthermore, AND(x, y) is "True" precisely when both x and y are "True," and OR(x, y) is "True" precisely when either x or y is "True" or both are "True."

Example 10.2. The functions of Example 10.1 can be combined to make more complex functions. For example,

$$f(x, y) = \sim(x \wedge y) = \begin{cases} 0 & \text{if } x = y = 1 \\ 1 & \text{otherwise,} \end{cases}$$

and

$$g(x, y, z) = (x \wedge y) \wedge z = \begin{cases} 1 & \text{if } x = y = z = 1 \\ 0 & \text{otherwise.} \end{cases}$$

How can we check that functions like those of Example 10.2 take on the values claimed? Or how can we determine the values of a new function? One foolproof method is to construct a table of all domain values and the resulting function values.

Example 10.3. The values of $f^*(x, y) = (\sim x) \vee (\sim y)$ are listed in Table 1.5.

Table 1.5

x	y	$\sim x$	$\sim y$	$f^*(x, y) = (\sim x) \vee (\sim y)$
0	0	1	1	1
0	1	1	0	1
1	0	0	1	1
1	1	0	0	0

Notice that the function f^* agrees with the function f of Example 10.2 and so the two functions are equal. In other words, we have shown that

$$\sim(x \wedge y) = (\sim x) \vee (\sim y).$$

Question 10.1. Show that (a) $\sim(x \vee y) = (\sim x) \wedge (\sim y)$ and (b) $(x \wedge y) \wedge z = x \wedge (y \wedge z)$.

The equalities of Example 10.3 and Question 10.1(a) are known as **de Morgan's laws.**

Example 10.4. Another useful function is known as "**exclusive or,**" abbreviated XOR. It is a function that is "True" if x or y is "True", but not both. In Boolean notation

$$\mathbf{XOR}:(Z_2)^2 \to Z_2 \text{ is defined by} \qquad \mathrm{XOR}(x, y) = \begin{cases} 0 & \text{if } x = y \\ 1 & \text{otherwise.} \end{cases}$$

The function XOR is also written as $\mathrm{XOR}(x, y) = x \oplus y$.

Question 10.2. By checking all domain elements, verify that XOR can also be expressed by

$$\begin{aligned} \mathrm{XOR}(x, y) &= (x \vee y) \wedge \sim(x \wedge y), \text{ or equivalently} \\ &= [x \wedge (\sim y)] \vee [(\sim x) \wedge y]. \end{aligned}$$

Boolean functions model digital networks and electronic circuits well. In these models, the voltage is either high (1) or low (0), and switches are either ON (1) or OFF (0). For example, consider the problem of adding two binary numbers, say $0 + 1$. In a computer this is carried out by a circuit in which high voltage (1) together with low voltage (0) is combined to produce high voltage (1). Such circuit combinations can be imitated by Boolean functions as follows. Define

$$\mathrm{ADD}(x, y) = \begin{cases} 0 & \text{if } x = y = 0 \text{ or } x = y = 1 \\ 1 & \text{otherwise,} \end{cases}$$

and

$$\mathrm{MULT}(x, y) = \begin{cases} 1 & \text{if } x = y = 1 \\ 0 & \text{otherwise.} \end{cases}$$

Notice that MULT is simply the AND function, $x \wedge y$, and ADD is the XOR function, $x \oplus y$. Then for one-digit binary numbers x and y, $\mathrm{MULT}(x, y) = x \cdot y$ and $\mathrm{ADD}(x, y) = x + y$ except that $1 + 1 = 10$ in binary. A "carry" bit is needed to complete the latter addition.

Example 10.5. Here are Boolean functions that produce the sum of 2 two-digit binary numbers. Let x_1, x_0, y_1 and y_0 be in Z_2 so that x_1x_0 and y_1y_0 both represent two-digit binary numbers. Then their sum is $z_2z_1z_0$, where

$$z_0 = \mathrm{ADD}(x_0, y_0) = x_0 \oplus y_0.$$

$$z_1 = \begin{cases} (x_1 \oplus y_1) \oplus 1 & \text{if } x_0 = y_0 = 1 \\ (x_1 \oplus y_1) \oplus 0 & \text{otherwise.} \end{cases}$$

(In other words, the two's digit in $z_2z_1z_0$ is the sum of x_1 and y_1 plus 1 if the 1 is "carried over" from the addition of x_0 and y_0.) There is a carry if and only if $x_0 = y_0 = 1$, and so

$$z_1 = [(x_1 \oplus y_1) \oplus (x_0 \wedge y_0)].$$

Finally, z_2, the four's digit of the sum, is 0 unless there is a carry from the second addition. There will be a carry from the second addition if either $x_1 = y_1 = 1$ or if there is a carry from the first addition and either x_1 or y_1 is 1. In symbols

$$z_2 = [x_1 \wedge y_1] \vee [(x_0 \wedge y_0) \wedge (x_1 \vee y_1)].$$

There are a few more rules of arithmetic concerning Boolean functions that are summarized in the next result. Two other useful rules of arithmetic are de Morgan's laws (see after Question 10.1).

Theorem 10.1. The functions AND and OR satisfy the following properties:

1. $x \wedge y = y \wedge x$ Commutative law
2. $x \vee y = y \vee x$ Commutative law
3. $(x \wedge y) \wedge z = x \wedge (y \wedge z)$ Associative law
4. $(x \vee y) \vee z = x \vee (y \vee z)$ Associative law
5. $x \wedge (y \vee z) = (x \wedge y) \vee (x \wedge z)$ Distributive law
6. $x \vee (y \wedge z) = (x \vee y) \wedge (x \vee z)$ Distributive law

Notice that part 3 has been verified in Question 10.1. A consequence of parts 3 and 4 is that we may write $x \wedge y \wedge z$ and $x \vee y \vee z$ (i.e., without parentheses) without ambiguity.

Question 10.3. Verify parts 1 and 4.

Proof of part 5. The straightforward way to check this is to substitute all eight possible choices of 0s and 1s for x, y, and z. For example, $0 \wedge (0 \vee 1) = 0 \wedge 1 = 0 = 0 \vee 0 = (0 \wedge 0) \vee (0 \wedge 1)$.

Here is another type of proof. Think of the $\{\text{True}, \text{False}\} \leftrightarrow \{1,0\}$ correspondence. Then the statement $x \wedge (y \vee z)$ is true if and only if x is true and either y or z is true, that is, if and only if either x and y are true or x and z are true, that is, if and only if $(x \wedge y) \vee (x \wedge z)$ is true. □

Optional Material

You might notice that every Boolean function considered in this section has been expressed in terms of NOT, AND, OR, and/or XOR (and XOR can be expressed in terms of the first three by Question 10.2). This is the case for all Boolean functions; in addition, the form of the expressions can be specified.

Definition. If $x_1, x_2, \ldots, x_n$ are variables, then for $i = 1, 2, \ldots, n$ both x_i and $\sim x_i$ are called **literals**. An expression of the form $L_1 \vee L_2 \vee \cdots \vee L_j$, where each L_i is a literal for $i = 1, 2, \ldots, j$ is called a **clause**. A function f: $(Z_2)^n \to Z_2$ is said to be in **conjunctive normal form** (or CNF) if it is of the form

$$f(x_1, x_2, \ldots, x_n) = C_1 \wedge C_2 \wedge \cdots \wedge C_k,$$

where C_i is a clause for $i = 1, 2, \ldots, k$.

Example 10.2 (reexamined.) $f(x, y) = \sim(x \wedge y)$ is not in CNF, but we know that $f(x, y) = f^*(x, y) = (\sim x) \vee (\sim y)$ by Question 10.1, which is in CNF with $f^*(x, y) = C_1$, where $C_1 = L_1 \vee L_2$, $L_1 = \sim x$ and $L_2 = \sim y$. Then $g(x, y) = (x \wedge y) \wedge z$ is in CNF with $C_1 = x$, $C_2 = y$, and $C_3 = z$.

Question 10.4. Which of the functions that determine z_2, z_1, and z_0 in Example 10.5 are in CNF?

Although the function f of Example 10.2 was not in CNF, we found that it was equal to a function, f^*, that was in CNF.

Theorem 10.2. Every Boolean function is equal to a function in CNF.

Proof. Let $f:(Z_2)^n \to Z_2$. Imagine writing out the table of all 2^n elements of $(Z_2)^n$ and the corresponding values of f. Suppose that $b = (b_1, b_2, \ldots, b_n)$ in $(Z_2)^n$ is such that $f(b) = 0$. Create a clause C of the form $C = L_1 \vee L_2 \vee \cdots \vee L_n$, where

$$L_i = \begin{cases} x_i & \text{if } b_i = 0 \\ \sim x_i & \text{if } b_i = 1. \end{cases}$$

Notice that when, for $i = 1, 2, \ldots, n$, b_i is substituted for x_i in the clause C, the value is 0. Furthermore, b is the only element of $(Z_2)^n$ which when substituted yields 0; since $C = L_1 \vee L_2 \vee \cdots \vee L_n$, C assumes the value of one when some L_i is one and some L_i is one unless all are zero, precisely the value at b.

Next we form clauses $C_1, C_2, \ldots, C_k$, one for each b in $(Z_2)^n$ for which $f(b) = 0$. Then we let g be defined by

$$g(x_1, x_2, \ldots, x_n) = C_1 \wedge C_2 \wedge \cdots \wedge C_k.$$

We claim that $f = g$. Suppose that b in $(Z_2)^n$ is such that $f(b) = 0$. Then there is a clause C_i in g corresponding to b; when b is substituted in C_i, a value of 0 results.

From the formula for g we see that $g(b) = 0$. Suppose that $b = (b_1, b_2, \ldots, b_n)$ is such that $f(b) = 1$. Consider what happens when, for $i = 1, 2, \ldots, n$; b_i is substituted for x_i in any clause C. As we have already seen, a 1 results, since C corresponds to an element $d \neq b$ for which $f(d) = 0$. Since every clause in g equals 1 upon input of b, by definition of g, $g(b) = 1$. □

Example 10.6. We use the technique of the proof of Theorem 10.2 to find a function equal to f, but in CNF, where

$$f(x, y, z) = \sim[(x \wedge y) \vee (x \wedge z)].$$

The elements of $(Z_2)^3$ on which f is 0 are precisely $(1,1,0)$, $(1,0,1)$ and $(1,1,1)$. We form the clauses $C_1 = (\sim x) \vee (\sim y) \vee z$, $C_2 = (\sim x) \vee y \vee (\sim z)$, and $C_3 = (\sim x) \vee (\sim y) \vee (\sim z)$, and let $g = C_1 \wedge C_2 \wedge C_3$.

Two types of Boolean functions merit special names. A Boolean function is called a **tautology** if its range is $\{1\}$; in other words, the function assumes the value "True" upon all input. A Boolean function is called a **contradiction** if its range is $\{0\}$; that is, it is always "False." For example, the function

$$f(x) = x \vee (\sim x)$$

is a tautology whereas

$$g(x) = x \wedge (\sim x)$$

is a contradiction.

Question 10.5. Decide if the following are tautologies, contradictions, or neither.

(a) $(x \vee y) \wedge [(\sim x) \wedge y] \wedge [x \wedge (\sim y)] \wedge [(\sim x) \vee (\sim y)]$.
(b) $(x \wedge y) \vee [(\sim x) \wedge y] \vee [x \wedge (\sim y)] \vee [(\sim x) \wedge (\sim y)]$.
(c) $(x \vee y \vee z) \vee (\sim y)$.

A Boolean function is not a contradiction if upon some input b the value of the function is 1. Then we say that f is **satisfied** by b and f is **satisfiable**. Suppose that we are given a function, expressed as a huge combination of ANDs, ORs, and NOTs, like

$$f(v, w, x, y, z) = ((((\sim x) \wedge y) \vee z) \wedge ((\sim x) \wedge v) \wedge ((\sim y) \vee v) \wedge w).$$

It certainly could be a lot of work to check whether this is a tautology or a contradiction, but suppose that all we want to know is whether there is some value b for which $f(b) = 1$. This is an instance of the following important problem.

The Satisfiability Problem. Given a Boolean function f, is there a b in the domain for which $f(b) = 1$?

How might we proceed to solve the Satisfiability Problem? For a specific function f we could check all domain elements in f. If the domain of f is $(Z_2)^n$, this checking might require all 2^n binary strings before we find, say, that the last string satisfies f or that no string satisfies f. Using the ideas of Chapter 2, we shall make this process precise, as an algorithm, and we shall see that this particular process is a very slow and time-consuming one. But surely there must be some "tricks of the trade," some techniques with which to attack this problem without trying all possibilities. In fact, it is an open research problem whether there is a "fast" and "efficient" process by which to determine whether an arbitrary Boolean function is satisfiable; these terms will be defined precisely in Chapter 2. For the time being we repeat that this simply stated problem is at the heart of some very hard, unresolved research questions that are actively being studied by the research community.

EXERCISES FOR SECTION 10

1. Which of the following tables represent a Boolean function?

Function	
Domain	Value
(0, 0)	1
(0, 1)	1
(1, 0)	1

Function	
Domain	Value
(0, 1)	0
(0, 2)	1
(1, 0)	0
(2, 0)	1

Function	
Domain	Value
(0, 0)	0
(0, 1)	0
(1, 0)	0
(1, 1)	0

2. Give an example of a Boolean function from $(Z_2)^3$ onto Z_2. How many Boolean functions from $(Z_2)^n$ to Z_2 are onto? How many are one-to-one?

3. Write out a table of values for the following functions. Then identify all pairs of functions that are equal.

(a) $f_1(x, y) = (\sim x) \vee y$. **(b)** $f_2(x, y) = ((\sim x) \vee y) \wedge x$.
(c) $f_3(x, y) = y \wedge x \wedge (\sim y)$. **(d)** $f_4(x, y) = \sim(x \vee (\sim x))$.
(e) $f_5(x, y) = x \wedge (\sim y)$. **(f)** $f_6(x, y) = \sim(x \wedge (\sim y))$.
(g) $f_7(x, y) = y \vee x$. **(h)** $f_8(x, y) = x \wedge y$.

4. Show that the function AND $(x, y) = x \wedge y$ is equal to a function using only ORs and NOTs. Show that the function OR $(x, y) = x \vee y$ is equal to a function using only ANDs and NOTs. Is the function $f(x, y) = \sim(x \wedge y)$ equal to a function that uses only ANDs and ORs, but no NOTs?

5. If x_1x_0 and y_1y_0 are two-digit binary numbers, find the functions that give their product in binary.

$$z_2z_1z_0 = (x_1x_0) \cdot (y_1y_0).$$

6. Verify parts 2 and 6 of Theorem 10.1.

7. For each of the following, find an equal function expressed in CNF.
   (*a*) $f_1(x, y, z) = (\sim x) \vee (y \wedge z)$.
   (*b*) $f_2(x, y, z) = x \vee ((\sim y) \wedge z)$.
   (*c*) $f_3(x, y, z) = \sim(x \vee y \vee z)$.
   (*d*) $f_4(x, y, z) = \sim(x \wedge y \wedge z)$.
   (*e*) $f_5(x, y, z) = (x \wedge y) \vee (x \wedge z)$.
   (*f*) $f_6(x, y, z, w) = (x \wedge y) \vee (z \wedge w)$.

8. A function is said to be in **disjunctive normal form** (or DNF) if it is in the form

$$f(x_1, x_2, \ldots, x_n) = D_1 \vee D_2 \vee \cdots \vee D_k,$$

where, for $i = 1, 2, \ldots, k$, D_i is of the form $L_1 \wedge L_2 \wedge \cdots \wedge L_j$ with each L_i a literal for $i = 1, 2, \ldots, j$. Find all functions in Exercises 3 and 7 that are in DNF.

9. Find an example of a function that is in neither CNF nor DNF.

10. For each of the following, find an equal function expressed in DNF.
   (*a*) $f_1(x, y, z) = (\sim x) \wedge (y \vee z)$.
   (*b*) $f_2(x, y, z) = x \wedge [(\sim y) \vee z]$.
   (*c*) $f_3(x, y, z) = \sim(x \wedge y \wedge z)$.
   (*d*) $f_4(x, y, z) = \sim(x \vee y \vee z)$.
   (*e*) $f_5(x, y, z) = (x \vee y) \wedge (x \vee z)$.
   (*f*) $f_6(x, y, z, w) = (x \vee y) \wedge (z \vee w)$.

11. Is each of the following satisfiable?
   (*a*) $f(x, y, z) = y \vee \sim[z \vee (\sim x)]$.
   (*b*) $f(x, y, z, w) = [x \vee \sim(y \vee \sim z)] \vee \sim\{x \vee \sim[y \vee \sim(z \wedge w)]\}$.
   (*c*) $f(x, y) = (x \wedge y) \vee ((\sim x) \wedge y)$.
   (*d*) $f(x, y) = x \vee \sim\{x \vee \sim[y \vee (\sim x)]\}$.
   (*e*) $f(x, y) = (x \wedge y) \wedge (x \wedge (\sim y))$.

12. For $n = 2$, 3, and 4 find Boolean functions $f:(Z_2)^n \to Z_2$ that are tautologies.

13. For $n = 2$, 3, and 4 find Boolean functions $f:(Z_2)^n \to Z_2$ that are contradictions.

14. Identify which of the following are tautologies.
   (*a*) $g(x, y) = y \vee \sim\{[y \vee (\sim x)] \wedge (y \vee x)\}$
   (*b*) $g(x, y) = [(\sim x) \vee y] \vee (y \vee x)$.
   (*c*) $g(x, y) = x \vee \sim\{y \vee [(\sim x) \wedge y]\}$.
   (*d*) $g(x, y) = x \vee [\sim(x \vee y)]$.
   (*e*) $g(x, y) = (\sim x) \vee \{y \wedge [(\sim y) \vee (\sim x)]\}$.
   (*f*) $g(x, y) = [x \wedge (\sim y)] \vee [y \vee (\sim x)]$.
   (*g*) $g(x, y, z) = [z \vee (\sim y)] \vee \sim[z \vee \sim(x \wedge y)]$.
   (*h*) $g(x, y, z) = z \vee (\sim x) \vee [(\sim z) \wedge (x \wedge y)]$.

15. A mathematical statement of the form "If A, then B" is said to be true if whenever A is true, then so is B. When A is false, then it doesn't matter whether B is true or false. Thus "If A, then B" is logically equivalent to the statement that either A is false or B is true. We define a new function

$$\textbf{IMPLIES } (x, y):(Z_2)^2 \rightarrow Z_2 \text{ defined by IMPLIES } (x, y) = (\sim x) \vee y.$$

Determine the value of $(Z_2)^2$ on which IMPLIES is 1. Interpreting 1 as "True," explain why this is analogous to "If A, then B."

16. A mathematical statement of the form "A **if and only if** B" is true if whenever A is true then so is B and whenever A is false then so is B. Find a Boolean function EQUIVALENT (x, y) that is the appropriate analogue of "A if and only if B."

1:11 A LOOK BACK

We began this chapter with a particular problem (admittedly one of only small import) and introduced some substantial mathematics in order to understand the problem and its solutions. There are a number of ideas that you should be comfortable with before you proceed to the next chapter. Foremost among the important concepts discussed is the notion of algorithm. This course will be oriented toward the solution of problems. Typically, a problem will be to find a mathematical object with a particular property. Often the problem is fairly easy in small instances. However, larger cases may be quite difficult. Such problems frequently have algorithmic solutions and it will be on these that we concentrate. The algorithmic solutions provide us with systematic ways to solve problems, so systematic that we could easily turn the algorithm over to a computer, or more realistically to a computer programmer. Although computers can handle many large numbers quickly, there are limits to computer size and speed. One of the principal themes of the rest of this course will be the analysis of the correctness and efficiency of algorithms and the search for such effective algorithms.

This chapter also contains an introduction to set theory, an important tool because our mathematical objects will usually be described in the language of sets. We need to distinguish between contexts where the order of objects is not important, as with subsets, versus those where order is important, as in Cartesian products. We have also begun to develop counting techniques, the multiplication principle and P.I.E. This material is significant because of its applications to the analysis of algorithms. In particular, these mathematical techniques will be necessary tools in the evaluation of the quality of particular algorithms.

Functions are also important mathematical tools because they describe transformations and relations between sets. Often it will be important to know whether a particular function is one-to-one or onto or whether it has an inverse. For example, in Chapter 3 we shall study the one-to-one functions called permutations in depth. We have noted that algorithms and functions are similar. They take input and domain elements, respectively, and transform these in a well-defined and unique manner into output and range elements. We shall use algorithms to study functions, and vice versa. In the next chapters two central problems will be to find algorithms that compute functions efficiently and to find and study functions that measure the efficiency of algorithms. The special case of Boolean functions is sufficiently important and applicable to warrant extra study. These functions model well how computers work. They may be thought of as mapping from complicated sets to the values of "True" and "False," and so they also model how logical thinking and proofs work. The Satisfiability Problem is introduced not only because it is an unsolved research problem, but also because it is computationally equivalent to other famous problems, some of which will arise in Chapters 5 and 8.

We have begun to see proofs. Understanding these proofs is crucial to the development of this course. We have left some important facts unproved in this chapter, for example, that the algorithms BtoD, DtoB, and SUBSET work correctly in all cases. One of the main goals of this book is to develop proof techniques and skill in their application to mathematical and algorithmic problems.

The concept of algorithm is pervasive throughout computer science and is increasingly important in abstract mathematics. A beginning computer science student is tempted to attack problems by writing computer code directly, trying out the resulting program on reasonable examples of data and correcting obvious problems as they arise. Instead, especially with complicated problems and programs, it will become essential to plan ahead carefully, to outline the entire attack on the problem. An algorithm is just such a precise outline of the approach to be made on the problem. In mathematical work students are always searching for concrete guidelines for ways to solve specified problems. In the past mathematicians would often prove that a problem could be solved, but then fail to address explicitly how to solve the problem. Now many mathematicians are revising their philosophical ideas about mathematics and are turning to algorithmic approaches to problems. Indeed, some would say that a problem hasn't been solved even in theory unless an algorithmic solution has been found. Thus although the problems that we consider in this book are beginning ones in the study of discrete mathematics and algorithms, the philosophy is central to current approaches and research in these areas.

SUPPLEMENTARY EXERCISES FOR CHAPTER 1

1. Suppose that a computer program contains two Boolean variables, Done and Correct. That is, each of these variables can assume the value either True or False. How many different pairs of values can the two variables (Done, Correct)

assume? Suppose that JEQUALSONE is another Boolean variable. How many different values can the triple (Done, Correct, JEQUALSONE) assume? Suppose that we had n Boolean variables called $X_1, X_2, X_3, \ldots, X_n$. Determine a formula for the number of different values the n-tuple $(X_1, X_2, \ldots, X_n)$ can assume.

2. *Problem.* Given a positive integer n, calculate and print the sum of the integers from 1 to n. Is the following an algorithm that correctly solves this problem?

Response

STEP 1. Read in n, a positive integer

STEP 2. Calculate $n(n + 1)/2$ and print this out

STEP 3. Stop.

3. Describe an algorithm to change a number written in base four to an equivalent number expressed in decimal.

4. Describe an algorithm to change a number from decimal to base four.

5. Describe an algorithm to convert between binary and base four that does not use decimal representations.

6. A numerical representation system often used in calculators is known as BCD or **binary coded decimal**. In this system each digit in a decimal number is converted to binary and stored in four consecutive bits, lined up in the same order as the original decimal digits. Thus the number 139 is stored as 000100111001. Convert the following numbers into BCD: **(*a*)** 12, **(*b*)** 19, **(*c*)** 25, **(*d*)** 28, and **(*e*)** 77.

7. How many bits does an r-digit decimal number require to be stored in BCD? Given 16 bits with the first bit reserved for sign designation, what is the largest decimal number that can be stored in BCD in the remaining 15 bits?

8. Write an algorithm to convert a decimal integer into BCD, and vice versa.

9. Write an algorithm that inputs $A = \{a_1, a_2, \ldots, a_n\}$ and outputs all elements of the Cartesian product $A \times A$.

10. Suppose that $A = \{a_1, \ldots, a_n\}$ and $T = \{t_1, t_2, \ldots, t_m\}$. How many functions are there with domain A and target T? How many of these functions are one-to-one? (*Hint*: Consider separately the cases when $n < m$, $n = m$, and $n > m$.)

11. How many numbers must you pick from the set $\{1, 2, \ldots, n\}$ so that there must be two with a common divisor greater than one?

12. Explain why a subset of $(n + 1)$ numbers taken from $\{1, 2, \ldots, 2n\}$ must contain two numbers where one divides the other. Is the same true if the subset contains only n numbers?

13. A hiker is lost in the mountains but stumbles into an area where it is known that all inhabitants are either True-tellers or Liars, meaning that an individual either always tells the truth or always lies. She meets a man at a fork in the road and wants to learn the way to the nearest village by asking only one question. Explain why she will not learn the way to the village by asking any of the following questions:
(*a*) Are you a truth-teller?
(*b*) Are you a truth-teller and does the left fork lead to the village?
(*c*) Are you a truth-teller or does the left fork lead to the village?
(*d*) If you are a truth-teller, does the left fork lead to the village?
(*e*) Are you a truth-teller if and only if you are a liar?
(*f*) If the left fork leads to the village, are you a liar?

14. Devise one question with which the hiker, in the predicament of the preceding exercise, can determine the way to the village.

15. Deborah receives $1 million from an anonymous friend. She suspects that either Alice, Bob, or Catherine gave it to her. When she asks each of them, they respond as follows:
(*a*) Alice says, "I didn't do it. Bob is an acquaintance of yours, and Catherine is an especially good friend of yours."
(*b*) Bob says, "I didn't do it. I've never met you before, and I've been out of town for the past month."
(*c*) Catherine says, "I didn't do it. I saw Alice and Bob in the bank on the day you received the check so it must be one of them."
Assuming that the two who didn't give Deborah the money are telling the truth and that the donor is lying, who gave Deborah the money?

16. There is a very simple programming language known as TRIVIAL. In this language only the following six types of instructions are allowed:
(*a*) Input X, a natural number
(*b*) Go to step # —
(*c*) Set $X := X + 1$
(*d*) If $X = 0$, then go to step # —
(*e*) If $X > 0$, then set $X := X - 1$
(*f*) Stop.

Suppose that a program must begin with the instruction Input X and end with the instruction Stop. How many different programs are there, written in the language TRIVIAL if the program is two steps long? Three steps long? 12 steps long? n steps long?

17. Suppose that a TRIVIAL program can begin with one or more statements Input X, where X may be any letter of the alphabet; for example, we may begin with Input A and then follow with Input B. Then we allow statements of the form $A := A + 1$, and if $B > 0$, then set $B := B - 1$, and so on, provided that A and B are input variables. If a TRIVIAL program begins by reading in two variables A and B, then how many different TRIVIAL programs are there using three steps? Four steps? n steps?

18. Write a program in the language TRIVIAL that upon input of X and Y, positive integers, calculates the sum $X + Y$ and leaves the result stored in X. Note that $X := X + Y$ is not a valid statement in TRIVIAL.

19. Suppose that we allow two output statements: "X is even" and "X is odd." Write a program in TRIVIAL that upon the input of X, a positive integer, determines whether X is even or odd, outputs the correct message, and stops.

20. Open Mathematical Problem: The Busy Beaver N-game. In this version we eliminate the Input X statement and assume instead that every variable begins with the value 0. Then we can do the equivalent of reading in a positive integer i for X by writing i consecutive "$X := X + 1$" statements. The score of a TRIVIAL program is defined to be the sum of the values of all variables when the program stops or 0 if the program never stops. Then the Busy Beaver n-game is the problem of determining a TRIVIAL program with n instructions with the highest possible score among all TRIVIAL programs with n instructions. We call that maximum score $BB(n)$. Thus $BB(1) = 0$, since the only TRIVIAL program with one line is

STEP 1. Stop.

A two-step program could be

STEP 1. Set $X := X + 1$

STEP 2. Stop.

Thus $BB(2) = 1$, since X begins at 0 and is increased only to 1. Explain why $BB(n) \geq (n - 1)$. Then show that $BB(3) = 2$ and $BB(4) = 3$. Find a value of n such that $BB(n) > (n - 1)$; you will need to use at least two variables. Very little is known about the Busy Beaver n-game for even small values of n. It is known that there is no simple function that expresses $BB(n)$ as a function of n. It has been shown that $BB(20) \geq 4^{(4^{(4^4)})}$, but the actual value of $BB(20)$ is unknown.

2

ARITHMETIC

2:1 INTRODUCTION

In Chapters 2 and 4 we take a new look at arithmetic. Some of the problems we shall consider will be more advanced than that which the word arithmetic usually conjures up. However, our principal goal will be algorithmic thinking. We shall be especially concerned with the quality of the methods discussed and the mathematics that is necessary to understand what makes one method better than another. We concentrate on the question of exponentiation, that is, of calculating x^n. Although this problem is more important than the magic trick that formed the theme of the previous chapter, it is the methods rather than the solutions to any particular problem that are worth learning. We introduce the important proof techniques of induction and contradiction to analyze the correctness and efficiency of algorithms.

Here is an appetizer. Suppose that we have two variables, say x and y, that have been assigned the values 5 and 2, respectively. It is a common task to switch the values of the variables, that is, to arrange it so that $x = 2$ and $y = 5$. We want a procedure that will work no matter what the values assigned to x and y are.

Example 1.1. Using the ":=" notation, we can suggest a way to trade the values of the variables assigned to x and y.

STEP 1. $x := y$

STEP 2. $y := x$

STEP 3. Stop.

At first glance this seems reasonable, since step 1 assigns to x whatever is assigned to y and step 2 assigns to y whatever is assigned to x. If we trace what happens to the values assigned to the variables x and y, we get Table 2.1.

Table 2.1

	Value Assigned to x	*Value Assigned to y*
Before step 1	5	2
After step 1	2	2
After step 2	2	2

What happened to the 5? After step 1 it has been forgotten. We wanted to assign the old value of x to the variable y; however, the old value of x has been written over.

Question 1.1. Here is a sequence of instructions that will perform the desired switch of values.

STEP 1. $xold := x$

STEP 2. $yold := y$

STEP 3. $y := xold$

STEP 4. $x := yold$

STEP 5. Stop.

As in the preceding example trace what happens when x is initially assigned the value 5 and y is initially assigned the value 2.

It turns out that only four of the five steps just listed are necessary. Exercise 1 asks you to figure out which step can be safely omitted. Thus there is a four-step algorithm that will switch values between two variables; this algorithm requires the use of a supplementary storage location.

Example 1.2. Here is an algorithm that will switch the values of x and y without the use of an extra storage location. Note that there are more steps and that we are required to do some arithmetic.

STEP 1. $x := x + y$

STEP 2. $y := y - x$

STEP 3. $x := x + y$

STEP 4. $y := -y$

STEP 5. Stop.

As in our previous example we trace what is assigned to each of the variables (Table 2.2).

Table 2.2

	Value of x	*Value of y*
Before step 1.	5	2
After step 1.	7	2
After step 2.	7	−5
After step 3.	2	−5
After step 4.	2	5

We have just seen two algorithmic solutions to the problem of switching the values assigned to two variables. The first requires four steps and a supplementary memory location. The second requires five steps and no extra storage. Furthermore, the steps are arithmetic operations rather than just assignment statements. This is typical. Frequently, a problem will have various algorithmic solutions, and one will be faster while a second might require less space. In general, we shall favor algorithms that have fewer steps over those that require less storage. In Sections 2 and 5 we shall consider exponentiation algorithms that are similarly related.

EXERCISES FOR SECTION 1

1. Which of the following four-step algorithms succeed in trading the values of x and y? For each, trace what happens.

(*a*) STEP 1. $yold := y$
STEP 2. $y := x$
STEP 3. $x := yold$
STEP 4. Stop.

(*b*) STEP 1. $xold := x$
STEP 2. $y := xold$
STEP 3. $x := y$
STEP 4. Stop.

(*c*) STEP 1. $xold := x$
STEP 2. $yold := y$
STEP 3. $y := xold$
STEP 4. Stop.

2. Suppose that $x \geq y \geq 0$. Find an algorithm that interchanges the values of x and y and has the properties that no supplementary storage is required and at no stage is a negative number stored in either location.

3. Find a four-step algorithm that interchanges the values of x and y and does not require a supplementary storage location.

4. Suppose that the three variables x, y, and z are each assigned values. Find an algorithm that will cyclically switch the assigned values so that the old value of x will be assigned to y, the old value of y will be assigned to z, and the old value of z will be assigned to x. Can this be accomplished with no arithmetic and just one extra storage location?

5. Suppose that you are given n variables labeled $x_1, x_2, \ldots, x_n$ each of which has been assigned a value. Construct an algorithm that will take the value assigned to x_n and store it in x_1, take the old value of x_1 and store it in x_2, take the old value of x_2 and store it in $x_3, \ldots$, and take the old value of x_{n-1} and store it in x_n.

6. Here is an algorithm whose goal is, upon input of numbers x and y, to calculate their sum and store it in x, and calculate their difference and store it in y. Is this algorithm correct?

   STEP 1. Input x and y

   STEP 2. $x := x + y$

   STEP 3. $y := x - y$

   STEP 4. Stop.

   If it is incorrect, rewrite the algorithm correctly.

7. Design a four-step algorithm that upon input of numbers x and y calculates xy and x/y and stores these in the variables x and y, respectively. If possible, use no supplementary storage location.

2:2 EXPONENTIATION, A FIRST LOOK

We begin with the problem of computing the nth power of x, given a real number x and a natural number n. How do computers calculate powers of numbers? The basic arithmetic operations of computers are addition, subtraction, multiplication, and division. Using these operations, we want to develop correct and efficient ways to calculate x^n. There is a straightforward solution to this problem, namely, multiply x by itself $(n - 1)$ times. If n were fixed, then there would be no difficulty in writing down a correct algorithm to solve this problem. For example, if n were always 5 we could calculate

$$\text{answer} := x \cdot x \cdot x \cdot x \cdot x.$$

The situation when n is allowed to vary is slightly more complicated. Now we list an algorithm that finds x^n.

Algorithm EXPONENT

STEP 1. Input x, n $\{n$ a natural number$\}$
STEP 2. $i := 0$, ans $:= 1$

STEP 3. While $i < n$ do
Begin
STEP 4. ans := ans $*$ x
STEP 5. $i := i + 1$
End {step 3}
STEP 6. Output ans and stop.

COMMENTS. In Section 1.4 we used "Go to" statements to create loops within algorithms. Here we use the more modern "**While . . . do**" construction, wherein the steps within the loop are executed in order as long as the while condition is satisfied. Note that the loop is indented for ease of reading. When a loop has more than one step, we will signal the beginning of the loop by "Begin" and the end of the loop by "End." The statements within the loop will not be executed if $n = 0$. However, the output of the algorithm will still be correct since $x^0 = 1$ for all x.

This algorithm is sufficiently complicated that you might wonder whether it does what it claims (in all cases).

A Fundamental Problem. How can you be sure that the algorithm listed above (or, more generally, any algorithmic solution to a problem) is correct?

A first response to this problem is to implement the algorithm. If the algorithm is written in a programming language, then the algorithm could be tested by running the program. If, as above, the algorithm is written in a pseudo-code (i.e., in English with a structure similar to a programming language), then the algorithm can be tested by a person tracing the algorithm's commands. In this setting the person acts as a computer. We shall do this for the algorithm listed above, but before we do, it is appropriate to consider what will be learned from such a test.

Presumably, we shall run the algorithm on input for which we already know the answer. If the algorithm's answer is wrong, then we can discard the algorithm. In one sense this is satisfactory, since we know without a doubt that the algorithm is incorrect. If the algorithm's answer agrees with the already known answer, what does that prove? It does show that the algorithm performs correctly on at least one set of input data. If we run the algorithm for a variety of inputs and each run gives a correct answer, then we can increase our confidence in the algorithm. Unless the algorithm only runs on a finite set of inputs, such a strategy cannot demonstrate that the algorithm will always work.

Example 2.1. If $x = 5$ and $n = 3$, then $x^n = 5^3 = 125$. In Table 2.3 we trace EXPONENT with this input.

Table 2.3

Current Step Completed	i	Answer
1	?	?
2	0	1
3	0	1
4	0	5
5	1	5
4	1	25
5	2	25
4	2	125
5	3	125
6 STOP	3	125

Question 2.1. Trace the executive of EXPONENT for $x = 3$ and $n = 4$.

After completing the above question, you probably believe that EXPONENT correctly produces the value of x^n when n is a positive integer. Your belief is based upon the fact that you have witnessed one experiment and performed one additional experiment. This is similar to what you might do in a chemistry class. In the next section we discuss the principle of mathematical induction that will enable us to prove that the algorithm EXPONENT works for all input.

EXERCISES FOR SECTION 2

1. Trace Algorithm EXPONENT for
   (*a*) $x = 17$ and $n = 1$. (*b*) $x = 2$ and $n = 5$.
   (*c*) $x = -2$ and $n = 3$. (*d*) $x = 5$ and $n = 0$.
2. Construct an algorithm that will input a real number x and an integer n, which may be positive, negative, or zero, and output x^n. Trace your algorithm for
   (*a*) $x = 3$ and $n = 3$. (*b*) $x = 3$ and $n = -3$.
   (*c*) $x = 0$ and $n = -5$. (*d*) $x = -1$ and $n = 0$.
3. If Algorithm EXPONENT is run with $x = 5$ and $n = 7$, how many multiplications are performed? If $n = 132$, how many multiplications are done?
4. Suppose that we have a computer that can perform addition but not multiplication. Devise an algorithm that, upon input of a real number x and an integer $n \geq 0$, calculates and outputs the product nx. How many additions does your algorithm use?
5. Construct an algorithm NEWEXP to compute x^n for n, a positive integer. The first step, after x and n are input, should be to compute and store $z = x^2$. At subsequent steps the variable ans (which will contain the answer) is multiplied by z unless that would make ans too large, in which case ans is multiplied by x.

Trace NEWEXP for
(*a*) $x = 3$ and $n = 7$. (*b*) $x = 3$ and $n = 16$.
(*c*) $x = 3$ and $n = 10$.

6. Determine how many multiplications NEWEXP requires to find (*a*) x^6, (*b*) x^{12}, (*c*) x^{13}, and finally (*d*) x^n, where $n = 2t$ for t, any positive integer.
7. Construct an algorithm NEWEREXP whose first two steps, after x and n are input, is to compute and store $z = x^2$ and $w = z^2$. At subsequent steps the variable ans (which will contain the answer) is multiplied by w unless that would make ans too large, in which case ans is multiplied by x until ans is x^n. Trace NEWEREXP for
(*a*) $x = 3$ and $n = 12$. (*b*) $x = 3$ and $n = 16$.
(*c*) $x = 3$ and $n = 19$.
8. Determine how many multiplications NEWEREXP requires to find (*a*) x^{10}, (*b*) x^{16}, (*c*) x^{25}, and finally (*d*) x^n, where $n = 4t$ for t, any positive integer.
9. Write an algorithm that calculates nx (as requested in Exercise 4) using fewer than $n - 1$ additions.

2:3 INDUCTION

Mathematics has distinguished itself from other human endeavors because the truths of mathematics are known with greater certainty than the truths of other subjects. This is because assertions in mathematics must be proved before they are regarded as valid. Although this does not eliminate the possibility of error (mathematicians are, after all, human beings who can and do make mistakes), the necessity of providing proofs greatly diminishes the potential for error.

Example 3.1. The most famous open problem in all of mathematics concerns the positive integers and is very simply stated. It goes as follows: Do there exist positive integers x, y, z, and n with $n > 2$ such that

$$x^n + y^n = z^n?$$

The restriction that $n > 2$ is there because, for instance,

$$\text{if } n = 1, \text{ then } 2^1 + 3^1 = 5^1$$

and

$$\text{if } n = 2, \text{ then } 3^2 + 4^2 = 5^2$$

provide easy affirmative answers to the question. The assertion that for n bigger than 2, there do not exist integers with the required relationship is known as

"**Fermat's Last Theorem.**" The name is misleading, since the assertion is still unproved; however, Fermat scribbled in the margin of a book that he had found a proof but had no room to write it down. This note was found after his death! It has been proved that no such integers exist if $n = 3$, that is, there do not exist integers x, y, and z with $x^3 + y^3 = z^3$. Indeed, Fermat's Last Theorem has been proved to be true for all exponents of reasonable size, in fact for $n \leq 125{,}080$. The philosophical question

"Is Fermat's Last Theorem true?"

suggests itself. Certainly, mathematicians would be surprised if it turned out to be false. However, no mathematician would consider the assertion of Fermat's Last Theorem to be a mathematical truth until a valid proof is found, no matter how much numerical evidence is marshaled in support of it.

The complex programs and structures of computer science have certain features analogous to the state of our knowledge concerning Fermat's Last Theorem. It is a common occurrence for programs (or algorithms) to work correctly on some inputs without working correctly on all inputs. Even if a program always has worked correctly, that is no guarantee that it always will work correctly. Because of this, computer scientists frequently require proofs that their assertions and programs are correct. How does one construct a proof that an algorithm (or a program) is correct? The most common technique is mathematical induction.

Principle of Mathematical Induction. Suppose that we wish to prove that a certain assertion or proposition is true. Further suppose that the statement of the proposition explicitly depends on a positive integer, say n. We denote the proposition emphasizing the dependence on the integer n by P_n. Our goal will be to prove that P_n is true for all values of n.

Frequently, it will be easy to prove that the proposition P_n is true for certain (usually small) values of n. The first step in applying an induction proof is to verify the proposition directly for one value of n. This is called checking the **base case**. This one value will usually be $n = 1$ or $n = 0$. The next step in applying an induction proof is analogous to climbing a ladder. We must demonstrate that whenever the proposition P_k is true, then so is the proposition P_{k+1}.

Here is a formal statement of the Principle of Mathematical Induction:

Let P_n be a proposition that depends upon the integer n. Then P_n is true for all positive n provided that

(i) P_1 is true,

and

(ii) if P_k is true, then so is P_{k+1}.

Soon you will see examples of proofs using induction. Before looking at these, let's see why the principle is valid. Note that when we say the principle is valid, we mean that if assertions (i) and (ii) are both verified, then the proposition P_n is proved true for all n.

To begin with, P_1 is true by (i). Since P_1 is true, setting $k = 1$ in (ii) shows that P_2 is true. Then we can repeat (ii) with $k = 2$. Since we've just demonstrated that P_2 is true, we get the result that P_3 is true, and so on. Is P_{17} true? We won't do all the details, but we could work our way up to 17 using (ii) repeatedly, building upon the known results P_j for smaller values of j. In general, we can work our way up to the truth of P_n for any integer n.

The way that we go about establishing an assertion by induction is quite algorithmic. In fact, here are the key steps.

Algorithm INDUCTION

STEP 1. (The base case). Verify that P_1 is true.

STEP 2. (The **inductive hypothesis**). Assume that P_k is true for an arbitrary value of k.

STEP 3. (The **inductive step**). Verify that P_{k+1} is true, using the assumption that P_k is true.

We illustrate proofs by induction with three typical examples.

Example 3.2. The following is an important identity that will reappear several times in this book.

$$1 + 2 + 3 + \cdots + n = \frac{n(n+1)}{2}. \tag{A}$$

Suppose that we wish to prove this using induction. The statement already has an explicit dependence on the positive integer n: P_n is the statement that equation (A) is true.

First we check the base case when $n = 1$. For $n = 1$ the left-hand side of equation (A) is 1, while the right-hand side equals $1(1 + 1)/2 = 1$. (Although we've done the base case, we'll get a better feel for the problem if we check at least one more case. If the equation were not true, it could be a real time-sink to try to prove it!) For $n = 2$ the left-hand side of (A) equals $1 + 2 = 3$, while the right-hand side equals $2(2 + 1)/2 = 3$. For $n = 3$ we check that the two sides of (A) both equal 6.

Now we come to the inductive hypothesis (step 2). We assume that

$$1 + 2 + \cdots + k = \frac{k(k+1)}{2}. \tag{B}$$

We want to prove that

$$1 + 2 + \cdots + k + (k + 1) = \frac{(k + 1)(k + 2)}{2}. \tag{B$'$}$$

Note that equation (B′) agrees with (A) after substituting $k + 1$ for n. How can we proceed? We want the sum of $k + 1$ integers and what we have to build upon is the sum of k integers. So we use associativity to obtain

$$\begin{aligned}(1 + 2 + \cdots + k) &+ (k + 1)\\ &= \frac{k(k + 1)}{2} + (k + 1) && \text{by inductive hypothesis}\\ &= (k + 1)\left(\frac{k}{2} + 1\right) && \text{by factoring}\\ &= \frac{(k + 1)(k + 2)}{2} && \text{by algebra.}\end{aligned}$$

□

Question 3.1. Prove by induction that $2 + 4 + \cdots + 2n = n(n + 1)$. (*Hint:* Mimic the preceding example.)

Example 3.3. The following formula gives the sum of a (finite) **geometric series**:

$$1 + x + x^2 + \cdots + x^n = \begin{cases} \dfrac{1 - x^{n+1}}{1 - x} & \text{if } x \neq 1\\ (n + 1) & \text{if } x = 1.\end{cases}$$

It is a polynomial identity that holds true for every real number x and for every integer $n \geq 0$. This can be proved by induction on n. Instead we prove the slightly more complex formula for the sum of an **alternating geometric series** in the next example. You will be asked to imitate the latter proof in Question 3.3 to verify Example 3.3.

Example 3.4. Here we use induction to verify that for every integer $n > 0$ the following polynomial identity is valid.

$$1 - x + x^2 - x^3 + \cdots + (-x)^n = \begin{cases} \dfrac{1 - (-x)^{n+1}}{1 + x} & \text{if } x \neq -1\\ n + 1 & \text{if } x = -1.\end{cases} \tag{C}$$

By a **polynomial identity** we mean an equation that is valid for every substitution of a real number for the variable x.

If $x = -1$, the left-hand side of (C) consists of $n + 1$ terms, each of which is 1. Thus the sum is $n + 1$. If $x \neq -1$, then the identity has an explicit dependence on the positive integer n: P_n is the statement that equation (C) is true for all positive integers n when x is any real number other than -1. First we check the base case when $n = 1$. For $n = 1$, the left-hand side of (C) equals $1 - x$. With $n = 1$, the right-hand side of (C) is

$$\frac{1 - (-x)^2}{1 + x} = \frac{1 - x^2}{1 + x} = 1 - x.$$

Question 3.2. Check that (C) holds for both $n = 2$ and $n = 3$.

Now we come to the inductive hypothesis (step 2). We assume that

$$1 - x + \cdots + (-x)^k = \frac{1 - x^{k+1}}{1 + x}. \tag{D}$$

We want to prove that

$$1 - x + \cdots + (-x)^{k+1} = \frac{1 - (-x)^{k+2}}{1 + x}. \tag{D$'$}$$

Note that equation (D′) agrees with (C) by substituting $k + 1$ for n. We use associativity to obtain

$$\begin{aligned}
&1 - x + \cdots + (-x)^k + (-x)^{k+1} \\
&\quad = \frac{1 - (-x)^{k+1}}{1 + x} + (-x)^{k+1} && \text{by inductive hypothesis} \\
&\quad = \frac{1 - (-x)^{k+1} + (-x)^{k+1} + x(-x)^{k+1}}{1 + x} && \text{by making common denominators} \\
&\quad = \frac{1 + x(-x)^{k+1}}{1 + x} && \text{by algebra} \\
&\quad = \frac{1 - (-x)(-x)^{k+1}}{1 + x} && \text{by algebra} \\
&\quad = \frac{1 - (-x)^{k+2}}{1 + x} && \text{by algebra.} \qquad \square
\end{aligned}$$

Question 3.3. Prove by induction on n that the sum of a geometric series is as given in Example 3.3.

Example 3.3 (another look). Here is another way to verify this identity. We have established that the identity of line (C) is valid for all x. Thus we may substitute the expression $-x$ in every instance of the variable x and the identity remains valid. Upon substitution we get

$$1-(-x)+(-x)^2-(-x)^3+\cdots+[-(-x)]^n$$
$$=\begin{cases}\dfrac{1-[-(-x)]^{n+1}}{1+(-x)} & \text{if } -x\neq -1\\ n+1 & \text{if } -x=-1\end{cases}$$

This simplifies to

$$1+x+x^2+x^3+\cdots+x^n$$
$$=\begin{cases}\dfrac{1-x^{n+1}}{1-x} & \text{if } x\neq 1\\ n+1 & \text{if } x=1.\end{cases}$$ □

Example 3.5. Choose $n + 2$ distinct points from the circumference of a circle. If consecutive points along the circle are joined by line segments creating a polygon with $n + 2$ sides, then the sum of the interior angles of the resulting polygon equals $180n$ degrees (see Figure 2.1). Even though the assertion to be proved depends on n in an obvious way, we still need to be careful with the statement P_n. Specifically, we insist that the assertion holds no matter which $n + 2$ points are selected.

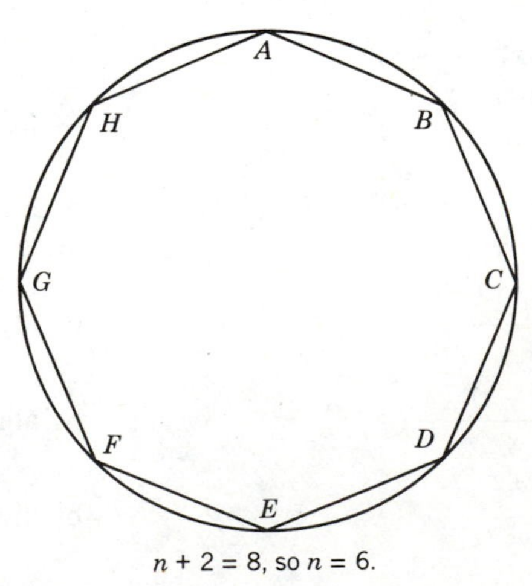

Figure 2.1 The sum of angles $A + B + \cdots + H = 180 \cdot 6° = 1080°$.

The base case, when $n = 1$, asserts that no matter how three points are selected from the circumference of a circle, then the resulting triangle contains three interior angles that total $180°$, a result known from plane geometry.

The inductive hypothesis asserts that no matter how $k + 2$ points are selected from the circumference of a circle, then the resulting polygon contains $k + 2$ interior angles that total $180k$ degrees. Suppose that we are given $k + 3$ [note that $k + 3 = (k + 1) + 2$] points chosen from the circumference of a circle. The inscribed $k + 3$ sided polygon P is shown in Figure 2.2(*a*). To use the inductive hypothesis, we need to ignore one of the selected points and consider the resulting $k + 2$ sided polygon P' shown in Figure 2.2(*b*).

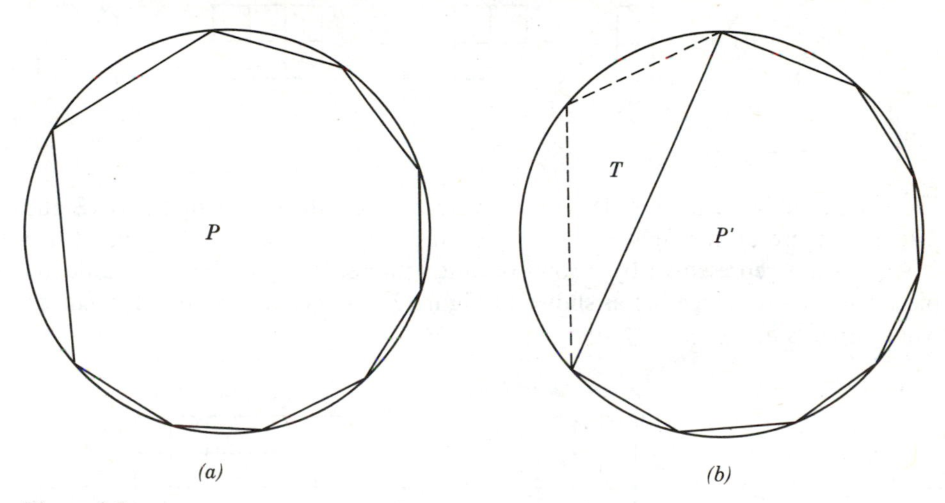

Figure 2.2

P can be obtained from P' by attaching the triangle T as shown. The sum of the interior angles of P equals the sum of the interior angles of P' together with the interior angles of T. Thus

$$\begin{aligned}\text{angles of } P &= \text{angles of } P' + \text{angles of } T \\ &= \quad 180k \quad + \quad 180 \\ &= 180(k + 1).\end{aligned}$$

□

Example 3.2 (revisited). There are numerous proofs of equation (A) that do not require the technique of mathematical induction. We present here two especially beautiful ones for your pleasure.

First a geometric proof: We shall represent the left-hand side of (A) as the area of a plane figure. For example, if $n = 2$ we think of the sum $1 + 2$ as the area of two rows of unit squares one with one square and one with two squares as shown in Figure 2.3.

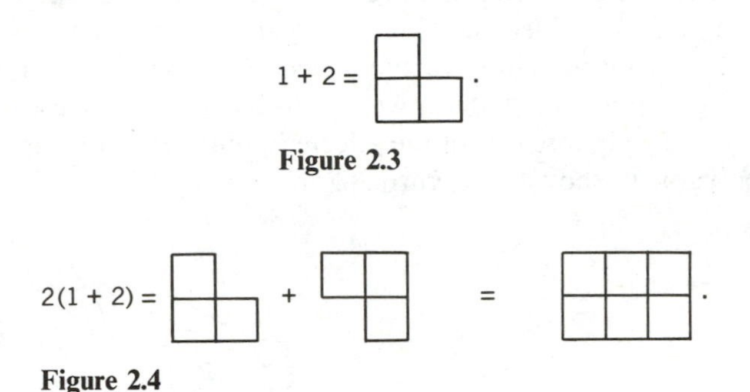

Figure 2.3

Figure 2.4

If we double the area under consideration, we get the following figure (Figure 2.4). The figure on the right is a 2×3 rectangle whose area is 6. In general, $1 + 2 + \cdots + n$ is represented by n rows of unit squares. If we double this value and piece the two areas together as shown in Figure 2.5, we get an $n \times (n + 1)$ rectangle whose area is $n(n + 1)$. □

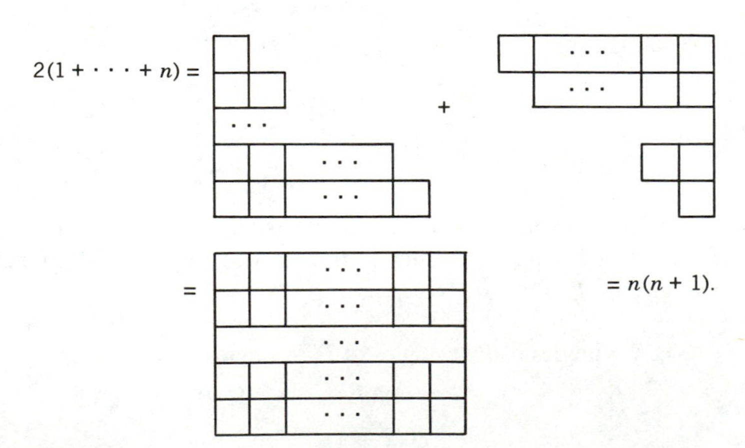

Figure 2.5

The third proof is even simpler although it is divided into two cases. The first case is when n is even, that is, $n = 2r$ for some integer r. We begin with the left-hand side of equation (A), where $2r$ has been substituted for n. Next we add the largest and smallest terms, then the second largest and second smallest, and so on, as shown in the following equation.

$$\begin{aligned} 1 + 2 + 3 + \cdots + (2r - 2) + (2r - 1) + 2r \\ &= (1 + 2r) + (2 + 2r - 1) + (3 + 2r - 2) + \cdots && \text{by regrouping} \\ &= (2r + 1) + (2r + 1) + (2r + 1) + \cdots && \text{by arithmetic} \\ &= r(2r + 1) = \frac{(2r)(2r + 1)}{2} && \text{by algebra} \\ &= \frac{n(n + 1)}{2} && \text{by substitution.} \quad \square \end{aligned}$$

Exercise 4 asks you to complete this proof for the case when n is odd.

EXERCISES FOR SECTION 3

1. Give a noninductive proof that $2 + 4 + \cdots + 2r = r(r + 1)$.

2. Show that $1 + 3 + 5 + \cdots + (2r - 1) = r^2$ in two different ways, one of which must be induction.

3. Prove that $1 + 4 + 7 + \cdots + (3n - 2) = (3n^2 - n)/2$.

4. Complete the third proof of equation (A) for the case $n = 2r + 1$. You can do this by temporarily ignoring either the last term or the middle term.

5. Use induction to show that

$$1 + 5 + 9 + \cdots + (4n + 1) = (n + 1)(2n + 1).$$

6. Here is another argument that proves the result in Exercise 5.

$$\begin{aligned} 1 + 5 + 9 + \cdots + (4n + 1) &= (4 \cdot 0 + 1) + (4 \cdot 1 + 1) + (4 \cdot 2 + 1) + \cdots \\ &\quad + (4 \cdot n + 1) \\ &= (4 \cdot 0 + 4 \cdot 1 + 4 \cdot 2 + \cdots + 4 \cdot n) \\ &\quad + (1 + 1 + \cdots + 1) \\ &= 4 \cdot (0 + 1 + 2 + \cdots + n) + (n + 1) \cdot 1. \end{aligned}$$

Use known results to simplify the right-hand side and to deduce that the sum is the same as is given in Exercise 5.

7. Prove, by any method you like, that
(*a*) $2 + 6 + 10 + \cdots + (4n + 2) = 2n^2 + 4n + 2$.
(*b*) $2 + 5 + 8 + \cdots + (3r - 1) = (3r^2 + r)/2$.

8. Express the following sums as simply as possible:
(*a*) $1 + 3 + 9 + \cdots + 3^n$.

(*b*) $\frac{1}{2} + \frac{1}{4} + \frac{1}{8} + \cdots + \frac{1}{2^n}$.

(*c*) $1 - 4 + 16 - 64 + \cdots + (-1)^n \cdot 4^n$.

9. Find a formula for the following sums:
(*a*) $1 + x^2 + x^4 + \cdots + x^{2n}$.

(*b*) $1 + \frac{1}{x} + \frac{1}{x^2} + \cdots + \frac{1}{x^n}$.

10. Use induction to show that $n^2 - n$ is always even.

11. Use induction to show that $n^4 + n^3 - 2n^2$ is always even.

12. Use induction to show that $n^4 + n^3 - 2n^2$ is always divisible by 4.

13. Find a formula for the sum $m + (m + 1) + (m + 2) + \cdots + n$; your answer will be in terms of the variables m and n. Use induction to show that your formula is correct.

14. Suppose that n distinct points are selected from the circumference of a circle. Let C_n denote the maximum number of line segments joining two distinct points that can be drawn so that no two segments intersect. Find a formula for C_n and verify your guess by induction.

2:4 THREE INDUCTIVE PROOFS

It would be difficult to overemphasize the importance of mathematical induction to the mathematician and computer scientist. Believing that the best way to learn is by studying and doing, we offer some more examples and problems.

Example 4.1. Look back at the assertions and algorithms about the number of subsets of a set given in Chapter 1. We claimed that if A is a set with n elements, then there exist 2^n distinct subsets of the set A. We can prove this assertion now by induction.

Precisely, we formulate P_n to be the statement that a set with n elements has exactly 2^n subsets. First we check the base case with $n = 1$, P_1. But we have already checked that P_1, P_2, and P_3 are true in Chapter 1. We saw there that a 1-set has two subsets, a 2-set has four subsets, and a 3-set has eight subsets. Thus we have accomplished step 1.

Now we get to steps 2 and 3, which are, of course, the heart of the matter. We assume the truth of the statement P_k: specifically that any set with k elements has 2^k subsets. Then we move to step 3 and examine P_{k+1}.

Suppose that A is a set with $k + 1$ elements. We want to show that there exist exactly 2^{k+1} subsets of A. What we are allowed to use is the assumption that P_k is true. The trick here is to overlook one of the elements of the set A in order to obtain a set with k elements. Let x be an element of A and define B to be $A - \{x\}$. We illustrate in Table 2.4 where, with $A = \{x, y, z, w\}$, we list the subsets of B and the subsets of A that contain x.

Table 2.4

$A = \{x, y, z, w\}$ Subsets of A That Contain x	$B = A - \{x\} = \{y, z, w\}$ Subsets of B
$\{x\}$	$\varnothing$
$\{x, y\}$	$\{y\}$
$\{x, z\}$	$\{z\}$
$\{x, w\}$	$\{w\}$
$\{x, y, z\}$	$\{y, z\}$
$\{x, y, w\}$	$\{y, w\}$
$\{x, z, w\}$	$\{z, w\}$
$\{x, y, z, w\}$	$\{y, z, w\}$

In the specific example above $k = 3$. B has 2^3 subsets, each of which is a subset of A. A has 2^3 subsets containing the specific element x. Each of these subsets could be obtained from a subset of B by adding the element x. Each of the 2^4 subsets of A can be constructed in this manner.

The general argument is the same. If A has $k + 1$ elements, B has k elements. Thus we know by step 2, the inductive hypothesis, that there exist 2^k subsets of B. Each of these subsets of B is also a subset of A. Thus we know that there exist 2^k subsets of A, none of which contains the element x. From each of these subsets that don't contain x we can create a new subset of A that does contain x. Specifically, if $S \subseteq B$, define S' to be $S \cup \{x\}$. Now S' is a subset of A. Furthermore, since S' contains x, it is not a subset of B.

Finally, every subset of A either contains x or it doesn't. If a subset, say T, of A does not contain x, then T is a subset of B. If, on the other hand, T does contain x, then $T - \{x\}$ is a subset of B. Either way T has been accounted for. Then we see that

$$\begin{aligned}
\#(\text{subsets of } A) &= \#(\text{subsets containing } x) + \#(\text{subsets not containing } x) \\
&= 2^k + 2^k \qquad \text{by inductive hypothesis} \\
&= (1 + 1) \cdot 2^k \qquad \text{by factoring} \\
&= 2 \cdot 2k \qquad \text{by algebra} \\
&= 2^{k+1} \qquad \text{by laws of exponents.} \qquad \square
\end{aligned}$$

We have proved the following theorem using the Principle of Induction.

Theorem 4.1. If A is a set containing n objects, then A has 2^n subsets.

The proof of Theorem 4.1 suggests an algorithm for listing all subsets of a given set. Suppose that we want to list all subsets of the set $A = \{a_1, a_2, \ldots, a_{10}\}$. Then the idea is, picking $x = a_{10}$, to list all subsets of $B = \{a_1, a_2, \ldots, a_9\}$ and then to repeat each subset with element a_{10} added in. How do we get all subsets of B? We could list all of the subsets of $\{a_1, \ldots, a_8\}$ and .... Wait a minute. Let's go forward rather than backward. We know that the set $\{a_1\}$ has two subsets, the empty subset and $\{a_1\}$. From these subsets we can get all subsets of $\{a_1, a_2\}$ by repeating those just listed and adding a_2 to get the additional sets $\{a_2\}$ and $\{a_1, a_2\}$. This procedure should sound familiar. Reread algorithm SUBSET in Chapter 1 and see that the algorithm uses exactly this idea of adding in elements a_j to previously formed subsets.

QUESTION 4.1. A set is said to be **even** if it has an even number of elements. Note that the empty set has zero elements and is thus an even set. If A is a set with n elements, guess a formula for the number of even subsets of A. Prove your formula by induction. (*Hint:* How many odd subsets does A have?)

Example 4.2. We now return to the algorithm presented at the beginning of Section 2 and use induction to prove that the algorithm does compute x^n. Here is the algorithm listed again with a comment between step 5 and step 6. (To avoid two different uses of the integer n the algorithm now calculates x^r.)

Algorithm EXPONENT

STEP 1. Input x, r $\{r$ a natural number$\}$
STEP 2. $i := 0$, ans $:= 1$
STEP 3. While $i < r$ do
Begin
STEP 4. ans $:=$ ans $* x$
STEP 5. $i := i + 1$
$\{$*Comment:* Right now ans has the value x^i.$\}$
End $\{$step 3$\}$
STEP 6. Output ans and stop.

This example is more complicated than previous ones because the proposition we need to verify, the P_n, is not explicitly presented. What we will do is use induction to show that the comment inserted between step 5 and step 6 is true.

Before we do this, note that if the comment is always true, then it will be true the last time it is encountered. The variable i is assigned the value 0 at step 2, and this value is incremented by 1 each time through the loop. This continues until $i = r$, when the algorithm, upon returning to step 3, discovers that the condition "While $i < r$" is no longer true. Hence the algorithm proceeds to step 6 in which it outputs the value of ans. If the comment is true at $i = r$, the value of ans equals x^r, which is what the algorithm was supposed to produce.

The P_n then is the statement that the nth time the comment is encountered it is true. First we verify the base case. If $n = 1$, then we have just finished step 5 for the first time. At step 2 the value of 0 was assigned to the variable i. This remains unchanged until step 5 when the value assigned to i was increased by 1. Thus the first time the comment is encountered the value of 1 is assigned to the variable i. Similarly, at step 2 the value of 1 was assigned to the variable ans. This remains unchanged until step 4 when the value of ans is multiplied by x. Thus the first time the comment is encountered the value of $x = x^1$ is assigned to the variable ans and thus P_1 is true.

Now for the inductive hypothesis. We assume that the comment is true the kth time it is encountered. To accomplish the inductive step, we must use this assumption and show that the $(k + 1)$st time the comment is encountered it is still true. The value assigned to i at the $(k + 1)$st encounter with the comment is the value assigned at the kth encounter plus one. This value is $k + 1$. The value assigned to ans at the $(k + 1)$st encounter equals the value assigned at the kth encounter times x. The value assigned to ans at the kth encounter equals x^k by the inductive hypothesis. Thus

$$\begin{aligned} \text{ans } \{\text{after } k + 1 \text{ encounters}\} &= x \cdot \text{ans } \{\text{after } k \text{ encounters}\} \\ &= x \cdot x^k \\ &= x^{k+1}. \end{aligned}$$

□

Question 4.2. Consider the following algorithm:

Algorithm SUM

STEP 1. Input r, set ans := 0
STEP 2. For $j = 1$ to r do
STEP 3. Set ans := ans + j {*Comment:* Right now ans has the value $j(j + 1)/2$.}
STEP 4. Output ans and stop.

COMMENT. Step 2 "**For** $j = 1$ to r **do** Step 3" is similar to the "While · · · do" loop of Example 4.2, only more compactly written. It means that first we set $j = 1$ and execute step 3; then we set $j = 2$ and carry out step 3, . . . ; until finally we set $j = r$ and execute step 3.

Use induction to show that this algorithm outputs $r(r + 1)/2$.

Example 4.3. We now verify that the algorithm DtoB, which finds the binary representation of a number, is correct. For convenience we list the algorithm once again.

Algorithm DtoB

STEP 1. Set $j := 0$
STEP 2. Divide m by 2 to obtain the quotient q and the remainder r; place r into the jth column of the answer (reading from the right and starting at zero)
STEP 3. If $q = 0$, then stop.
STEP 4. Set $m := q$
STEP 5. Set $j := j + 1$
STEP 6. Go to step 2

The proof will be by induction. Thus we need a statement P_n with which to work. Let n denote the number of bits in the binary representation of the integer m. The statement P_n will be that the algorithm correctly finds the binary representation of all integers whose representation has exactly n bits.

First we check the base case P_1. There is just one positive integer whose binary representation has exactly one bit, namely $m = 1$. In this case at step 2, q will equal 0, r will equal 1, and the base case holds. Back in Chapter 1 you undoubtedly checked many other cases, so the result seems reasonable.

Now for the inductive hypothesis. We assume that whenever m is an integer whose binary representation has k bits, then DtoB correctly finds these k bits. The inductive step says that we must prove the same for an integer with $(k + 1)$ bits: Suppose that m is such an integer. Take m and step through the algorithm until we get to step 6 for the first time. If m is even, r will be 0 and q will be $m/2$. If m is odd, r will be 1 and q will be $(m - 1)/2$. Consider the number q. We assert that it has one fewer binary digits than m.

Question 4.3. How many bits are there in the binary representation of (a) 14, (b) 7, (c) 13, and (d) 6? How can you get the binary representation of 14 from that of 7? How can you get the binary representation of 13 from that of 6?

We may assume (by the inductive hypothesis) that the algorithm will correctly produce the binary representation for q, since it has k bits. If we begin working the algorithm on q, we shall get exactly the same sequence of remainders beginning with j equal to 0 that we would have obtained from m beginning at the second encounter of step 2 with $j = 1$.

If m is even, $m = 2q$. When a number in binary is multiplied by 2, its digits are just shifted one space to the left and a zero, the first remainder, is attached at the end. Consequently, the algorithm will produce the correct binary representation for m. If m is odd, then $m = 2q + 1$. Here the binary representation for m can be obtained from the binary representation for q by shifting each digit one space to the left and atttaching a 1 as the last digit. If m is odd, the first remainder is one, so the algorithm will once again produce the correct binary representation. □

EXERCISES FOR SECTION 4

1. Compare the idea for an algorithm contained in Example 4.1 with that of Algorithm SUBSET in Section 1.7. In what ways do they agree and in what ways do they differ?
2. Prove by induction that the number of 2-subsets of an n-set A equals $n(n-1)/2$. [*Hint:* Let x be any object of A and $B = A - \{x\}$. Then a 2-subset of A is either a 2-subset of B or a 1-subset of B. Count the number of subsets in each case.]
3. Prove by induction that the number of 3-subsets of an n-set equals $n(n-1)(n-2)/6$. (*Hint:* Do Exercise 2 first.)
4. Consider the following algorithm:

   Algorithm ODDSUM

   STEP 1. Input n, set ans := 0
   STEP 2. For $j = 1$ to n do
   Begin
   STEP 3. Set $t := 2 * j - 1$
   STEP 4. Set ans := ans + t
   {*Comment:* Right now ans has the value j^2.}
   End
   STEP 5. Output ans and stop.

   Use induction to show that this algorithm outputs n^2.
5. Consider the following algorithm:

   Algorithm FOURSUM

   STEP 1. Input n, set ans := 0
   STEP 2. For $j = 1$ to n do
   Begin
   STEP 3. Set $t := 4 * j - 3$
   STEP 4. Set ans := ans + t
   End
   STEP 5. Output ans and stop.

   Use induction to show that this algorithm outputs $n(2n - 1)$.

6. How many binary digits do each of the following pairs of numbers have?

(i) 6 and 3. **(ii)** 10 and 5.
(iii) 12 and 6. **(iv)** 5 and 2.
(v) 7 and 3. **(vi)** 11 and 5.

7. What is the relationship between the pairs of decimal numbers whose binary representations follow?

(i) 10 and 100. **(ii)** 11 and 110.
(iii) 101 and 1010. **(iv)** 10 and 101.
(v) 11 and 111. **(vi)** 101 and 1011.

8. Suppose that m is a decimal number with binary representation s. Describe the binary representation of the numbers $n = 4m$, $p = 4m + 1$, $q = 4m + 2$, and $r = 4m + 3$ in terms of s.

9. Reread Algorithm SUBSET in Section 1.7. Let P_k be the statement that the kth time the comment after step 3 is encountered, it is correct. Prove that P_k is true for all positive integers k.

10. Design an algorithm to list all even subsets of an n-set and prove by induction that the algorithm is correct.

11. Consider the following algorithm.

Algorithm SQUARESUM

STEP 1. Input n, set ans $:= 0$
STEP 2. For $j = 1$ to n do
Begin
STEP 3. Set $k := j * j$
STEP 4. Set ans $:=$ ans $+ k$
{*Comment:* Right now ans has the value $j(j + 1)(2j + 1)/6$.}
End {step 2}
STEP 5. Output ans and stop.

Use induction to show that this algorithm outputs $n(n + 1)(2n + 1)/6$.

12. Consider the following algorithm.

Algorithm MAX

STEP 1. Input n, a positive integer, and $x_1, \ldots, x_j, \ldots, x_n$, real numbers.
STEP 2. Set max $:= x_1$
STEP 3. For $j = 2$ to n do
STEP 4. If $x_j >$ max, then max $:= x_j$
STEP 5. Output max and stop.

Use induction to show that this algorithm outputs the maximum of the numbers $x_1, x_2, \ldots, x_n$.

13. Consider the following algorithm.

Algorithm BUBBLES

STEP 1. Input n, a positive integer, and $x_1, \ldots, x_j, \ldots, x_n$, real numbers
STEP 2. For $j = 1$ to $n - 1$ do
  STEP 3. If $x_j > x_{j+1}$, then do
    Begin {Trade the values of x_j and x_{j+1}.}
    STEP 4. Set temp $:= x_j$
    STEP 5. Set $x_j := x_{j+1}$
    STEP 6. Set $x_{j+1} :=$ temp
    End {step 3}
STEP 7. Output x_n and stop.

Figure out what this algorithm does. Prove your guess by induction.

14. Find a formula for the sum $1 - 2 + 3 - 4 + \cdots + (-1)^n n$. Then prove that your formula is correct.

15. Here is a general statement of the Multiplication Principle: Suppose a counting procedure can be divided into n independent and successive stages. If there are c_1 outcomes for the first stage, and for each of these there are c_2 outcomes for the second stage, and for each of these initial outcomes there are c_3 outcomes for the third stage, and ..., and finally for each of these there are c_n outcomes for the final stage, then the total number of possible outcomes equals $c_1 \cdot c_2 \cdot \cdots \cdot c_n$. Prove this principle by induction on n.

16. Here is a short algorithm.

STEP 1. Input n (n a positive integer)
STEP 2. Set $i := n$, set ans $:= 0$
STEP 3. While $i > 0$ do
  Begin
  STEP 4. ans $:=$ ans $+ i$
  STEP 5. $i := i - 1$
  End
STEP 6. Output ans and stop.

(*a*) If the input is 4, what answer does this algorithm produce?
(*b*) Explain why this algorithm will always stop regardless of what positive integer n is used as the input.
(*c*) In general, what is the answer (in terms of n) produced by this algorithm? Express this answer as simply as possible.

17. Use induction to prove that 3 divides $n^3 + 3n^2 + 2n$ for every nonnegative integer n.

2:5 EXPONENTIATION REVISITED

The algorithm EXPONENT, verified by induction in the preceding section, correctly computes x^n. We discuss briefly the resources needed for implementation. There are four memory locations required, for x, i, ans, and n. It is possible to construct an exponentiation algorithm similar to EXPONENT that does not require separate memory locations for both n and i; however, the resulting procedure is not as clear as the one we have presented.

Time is the other resource to assess, but what does that mean? If we run an algorithm on a big, fast computer, then we require less time than on a small, slow machine. If we run an algorithm on a time-sharing minicomputer, the time required depends upon the number of users and what they are computing. Maybe if we program the algorithm in FORTRAN, it will run faster than in Pascal. In this course we do not want to be concerned with specific machines (hardware) or specific languages (software). Instead we concentrate on a theoretical (but useful) measure of time.

The fundamental operations that occur in our algorithms are assignment statements (:=), comparisons (If $j < n$, then $\cdots$), additions and subtractions ($+$ and $-$), and multiplications and divisions ($*$ and $/$). In a more advanced course you may study the time required for each of these operations. We'll assume that assignments happen instantaneously, that comparisons, additions, and subtractions each require a modest amount of time, and that multiplications and divisions are the most time-consuming operations. These assumptions certainly hold true when human beings make these calculations! We also assume that one multiplication and one division require the same amount of time. In fact, that is essentially the case for computations on computers.

Now look at Algorithm EXPONENT. We focus on step 4, since it is the only step that involves (time-consuming) multiplications. That is, we estimate the time needed to run this algorithm by counting the number of multiplications. Step 4 is executed once for each integer between 0 and $(n - 1)$ inclusive. Thus exactly n multiplications are executed. At first blush you might think that any algorithm that computes x^n must require this many multiplications, but that is not the case.

Example 5.1. To form x^4 using EXPONENT requires four multiplications. Alternatively, we could form x^2 with one multiplication and then multiply x^2 by itself to obtain x^4 with just two multiplications. We can denote this by

$$x^4 = 1 \cdot x \cdot x \cdot x \cdot x = (x \cdot x) \cdot (x \cdot x) = x^2 \cdot x^2.$$

There are four multiplications in the original algorithm. If you look at the second expression, you see three multiplications; however, the quantities within the two pairs of parentheses are identical. Consequently, we can compute x^4 with two multiplications. Similarly, we can find x^5 with just three multiplications as indicated by

$$x^5 = [(x \cdot x) \cdot (x \cdot x)] \cdot x = x^4 \cdot x.$$

For x^6 we require only three multiplications as follows.

$$x^6 = [(x \cdot x) \cdot (x \cdot x)] \cdot (x \cdot x) = x^4 \cdot x^2.$$

In English, it takes one multiplication to form x^2, one additional multiplication to form x^4, and one final multiplication to combine these two products.

Question 5.1. Using the ideas suggested by the preceding example, how many multiplications do you need to form x^n for $n = 7$, 11, 12, and 16?

Answering the previous question suggests that efficient evaluation of x^n is related to the binary expansion of the integer n. Thus, for example, since $n = 25 = 16 + 8 + 1$ we could write

$$x^{25} = x^{16} \cdot x^8 \cdot x,$$

use three multiplications to obtain x^8, one additional multiplication to obtain x^{16} and two more multiplications to combine the three factors. Thus it seems natural to suspect that there is an efficient algorithm to produce x^n that contains an algorithm to find the binary representation of the integer n. We use the Algorithm DtoB whose validity we have verified at the end of the previous section. For convenience we repeat the algorithm here and trace another instance.

Algorithm DtoB

STEP 1. Set $j := 0$
STEP 2. Divide m by 2 to obtain the quotient q and the remainder r; place r in the jth column of the answer (reading from the right)
STEP 3. If $q = 0$, then stop.
STEP 4. Set $m := q$
STEP 5. Set $j := j + 1$
STEP 6. Go to step 2

Example 5.2. If we use the algorithm DtoB to compute the binary representation of $m = 25$, we get the following table of the values (Table 2.5).

To jazz this algorithm up so it will produce the value of x^m, we need to add three additional steps. First we need

STEP 0. Input x, m, set ans $:= 1$

Table 2.5

Variables	j	m	q	r
Values after step 2	0	25	12	1
	1	12	6	0
	2	6	3	0
	3	3	1	1
	4	1	0	1

to initialize the algorithm. Next, where necessary, we need to multiply the intermediate result into the answer. We insert

STEP 2.5. If $r = 1$, set ans := ans $*\ x$

Finally, we need to insert a step that doubles the exponent on x.

STEP 5.5. Set $x := x * x$

To see how this works, we trace the new algorithm leaving x unspecified.

Example 5.3. The trace of the algorithm to find x^{25} is shown in Table 2.6.

Table 2.6

Variables	j	m	q	r	x	ans
Values after 2.5	0	25	12	1	x	x
	1	12	6	0	x^2	x
	2	6	3	0	x^4	x
	3	3	1	1	x^8	x^9
	4	1	0	1	x^{16}	x^{25}

Notice that four multiplications are required to compute the powers of x corresponding with each power of 2. Three multiplications are required to multiply the various factors together, and there is an additional cost of five divisions to form the binary representation.

Question 5.2. Trace the result of applying the above algorithm to find x^{37}. How many multiplications and divisions does the algorithm make? Do the same for x^{52}.

The straightforward algorithm to produce x^m required m multiplications. This new algorithm based on binary representation seems to do much better. In the next section we shall study just how good this algorithm is and introduce the mathematics and jargon needed to discuss this question.

EXERCISES FOR SECTION 5

1. Modify the algorithm EXPONENT so that upon input x and an integer $n \geq 0$, x^n is computed using only $n - 1$ multiplications. (Be sure to cover the case when $n = 0$.)
2. Modify the algorithm EXPONENT so that it uses only three memory locations for x, n, and ans.
3. Among the integers n with $16 < n < 33$, which integer requires the most multiplications to form x^n using the new algorithm from this section?
4. For each of the following integers n, find a factorization of x^n that will allow its computation by few multiplications **(*a*)** $n = 28$, **(*b*)** $n = 48$, **(*c*)** $n = 53$, and **(*d*)** $n = 56$.
5. Trace the result of applying the algorithm presented in this section to the problem of finding 2^{10}.
6. Notice that $x^{18} = x^{16} \cdot x^2 = (x^6)^3$. Which factorization provides the more efficient evaluation of x^{18}?
7. Notice that $x^6 = (x \cdot x \cdot x)^2$ and that $x^9 = (x \cdot x \cdot x)^3$. Do these factorizations lead to more efficient computations than using the methods of this section?
8. Look back at your algorithm in Exercise 2.4 designed to evaluate nx using only addition. Find a way, using the binary expansion of n, to calculate nx using fewer additions.

2:6 HOW GOOD IS FAST EXPONENTIATION?

In Example 5.3 we saw that computing x^{25} based on binary representation required 12 multiplications and divisions while the traditional method would require 25 multiplications. In this section we investigate just how fast this binary exponentiation method is. We list the algorithm once again, named in anticipation of the analysis of its running time:

Algorithm FASTEXP

STEP 0. Input x, m, set ans $:= 1$
STEP 2. Divide m by 2 to obtain quotient q and remainder r
STEP 2.5. If $r = 1$, set ans $:=$ ans $* x$
STEP 3. If $q = 0$, then stop.
STEP 4. Set $m := q$
STEP 5.5. Set $x := x * x$
STEP 6. Go to step 2

Question 6.1. Why have we omitted steps 1 and 5?

To compare the time of FASTEXP with that of EXPONENT, we count the number of multiplications and divisions. Thus given an integer m, how many multiplications and divisions will FASTEXP require to form x^m? Notice that every time step 2 is executed, there will be exactly one division. Similarly, every time step 5.5 is executed, there will be exactly one multiplication. Now look at step 2.5. Sometimes this step results in a multiplication and sometimes it doesn't. We have two options. We can either think hard and try to figure out exactly how many times step 2.5 requires a multiplication. Or we can be blasé and say that in the most time-consuming case step 2.5 will require a multiplication every time it is executed. The most time-consuming case is also called the **worst case**. Thus for the worst-case analysis we count one multiplication every time that step 2.5 is encountered.

For this particular algorithm we can carry out both the precise and the blasé or worst-case analyses. For more complex procedures the worst-case analysis is commonly used and gives an upper bound on the time of the exact counting analysis.

The Worst-Case Analysis. The number of multiplications and divisions required to implement FASTEXP is no more than the number of times steps 2, 2.5, and 5.5 are encountered. Since we never execute steps 2.5 and 5.5 without having first executed step 2, we know that the number of multiplications and divisions is no more than three times the number of times step 2 is encountered. So the analysis depends on the number of times 2 can be divided into m. This is essentially the logarithm of m.

Logarithms. Given integers p and q with $2^p = q$, then p is said to be the **logarithm** to the base 2 of the number q. We shorten this to $p = \log(q)$. Here is the defining relationship in symbols:

$$p = \log(q) \text{ if and only if } 2^p = q.$$

Note that while base 10 is common in high school algebra and base e is typically used in calculus, in discrete mathematics and computer science logs are always assumed to be base 2 unless otherwise specified.

Example 6.1. If $p = 3$, then $2^3 = 8$. Consequently, $\log(8) = 3$. Similarly, $\log(32) = 5$ and $\log(1) = 0$.

Question 6.2. Calculate the following:

$\log(2^2)$ $\quad\log(2^3)$ $\quad\log(2^5)$ $\quad\log(2^{10})$

$2^{\log(2)}$ $\quad 2^{\log(4)}$ $\quad 2^{\log(6)}$ $\quad 2^{\log(8)}$

Question 6.3. Explain why $\log(2^p) = p$ and $2^{\log(q)} = q$.

Remember that if n is a positive integer, then $2^{-n} = 1/(2^n)$. Thus if $q = 1/(2^n)$, then $\log(q) = -n$.

We can mimic the definition of logarithm given above for some numbers that are not integers. Suppose that y is a rational number; that is, $y = a/b$, where a and b are integers. Then

$$\log(x) = y,$$

where

$$x = 2^y = 2^{a/b} = (2^a)^{1/b}.$$

In English, 2 to the y is the bth root of 2 to the a. For example, $2^{2/3}$ is the cube root of 4 while $2^{3/2}$ is the square root of 8. Notice that $x = 2^y$ is always a positive number for all rational numbers y. Thus the domain of the function $f(x) = \log(x)$ contains only positive numbers.

Here are two rules that help us work with exponents:

$$s^n \cdot s^m = s^{n+m}, \tag{A}$$

and

$$(s^p)^q = s^{pq}. \tag{B}$$

What do these rules tell us about logarithms? They are equivalent to the following properties, which are convenient for manipulating logarithms.

$$\log(ab) = \log(a) + \log(b), \tag{A$'$}$$

and

$$\log(a^b) = b\log(a). \tag{B$'$}$$

Here's why (A′) is true in general. Suppose that $n = \log(a)$ and $m = \log(b)$. Then $a = 2^n$ and $b = 2^m$. Thus

$$\begin{aligned} ab &= 2^n \cdot 2^m \\ &= 2^{n+m}. \qquad \text{by (A).} \end{aligned}$$

This means that

$$\begin{aligned} \log(ab) &= n + m \\ &= \log(a) + \log(b). \end{aligned} \qquad \square$$

Property (B′) follows similarly from (B). (See Exercise 2.)

It would take us too far afield to attempt to define 2^y rigorously for arbitrary real numbers y or $\log(x)$ for arbitrary positive real numbers x. Luckily, it is not necessary. What we are really interested in is the integers that are near to $\log(x)$.

Floor Function. Given a real number x, the **floor function** of x, denoted by $\lfloor x \rfloor$, is defined to be the largest integer that is less than or equal to x. For example, $\lfloor 3 \rfloor = 3$, $\lfloor 3.11 \rfloor = 3$, $\lfloor 15.773 \rfloor = 15$, and $\lfloor -4.15 \rfloor = -5$.

Ceiling Function. Given a real number x, the **ceiling function** of x, denoted by $\lceil x \rceil$, is defined to be the smallest integer that is greater than or equal to x. For example, $\lceil 3 \rceil = 3$, $\lceil 3.11 \rceil = 4$, $\lceil 15.773 \rceil = 16$, and $\lceil -4.15 \rceil = -4$.

Note that for any number x, $\lfloor x \rfloor \le x \le \lceil x \rceil$.

Question 6.4. Find $\lfloor \frac{17}{3} \rfloor$, $\lceil \frac{25}{7} \rceil$, $\lfloor \log(8) \rfloor$, $\lceil \log(13) \rceil$, $\lfloor -\frac{14}{9} \rfloor$, $\lceil \log(25) \rceil$, and $\lfloor \log(13.73) \rfloor$.

We now have the vocabulary to answer the question of how many times we encounter step 2 in the execution of FASTEXP. Essentially, we want to know how many binary digits there are in the representation of the number m. For example, the two binary numbers with exactly two digits are 10 and 11; these represent the decimal numbers 2 and 3, respectively. Similarly, the decimal numbers 4, 5, 6, and 7 are all represented by three-digit binary numbers. In general, the decimal number m will be represented by a binary number with exactly r digits if and only if

$$2^{r-1} \le m < 2^r.$$

Since $0 < c < d$ if and only if $\log(c) < \log(d)$, we can take logs of the previous inequality to get

$$\log(2^{r-1}) \le \log(m) < \log(2^r)$$

which simplifies to

$$r - 1 \le \log(m) < r.$$

The floor function allows the convenient representation

$$r - 1 = \lfloor \log(m) \rfloor \qquad \text{or} \qquad r = 1 + \lfloor \log(m) \rfloor.$$

What we have learned is summarized in Figure 2.6.

Thus in using FASTEXP to find x^m, step 2 will be executed exactly $1 + \lfloor \log(m) \rfloor$ times. So in the worst case the algorithm will require $3(1 + \lfloor \log(m) \rfloor) \le 3 + 3\log(m)$ multiplications and divisions.

Decimal number ⟷	*Number of binary digits*
m	$1 + \lfloor \log(m) \rfloor$
$2^{r-1} \leq m < 2^r$	$r.$

Figure 2.6

There are two differences between the worst-case analysis and the exact count. First FASTEXP terminates at step 3 when $q = 0$. Thus step 5.5 is executed once less than step 2. The second difference occurs because the multiplication in step 2.5 is not executed as often as step 2 is encountered. A little thought will convince you that the number of times this multiplication is made equals the number of 1s in the binary representation of the number m. Thus the exact number of multiplications and divisions required by FASTEXP equals

$$2(1 + \lfloor \log(m) \rfloor) - 1 + \#(1 \text{ bits in } m). \tag{C}$$

At times this exact count may be useful, but on the whole the upper bound given by the worst-case analysis is easier to work with.

Now let us contrast these two algorithms. EXPONENT requires n multiplications to compute x^n. Such an algorithm is called **linear**. The name is appropriate because if one plots the number of multiplications as a function of the exponent, the result is a straight line. Any such linear algorithm has the property that if one doubles the problem size, then the number of steps required (and thus the amount of time required) approximately doubles.

Example 6.2. Suppose that a particular algorithm reads in an integer n and requires $3n + 7$ steps. If $n = 100$, the number of steps is 307. If n is doubled to 200, then the number of steps equals 607. Notice that this is almost double the original 307 steps.

That doubling the input size to a linear algorithm does not exactly double the number of steps is a consequence of the fact that not all straight lines go through the origin. Thus how much the doubling rule is off depends on the y-intercept. In general, if the size of n is large compared to the y-intercept, then the doubling rule will be fairly accurate.

FASTEXP requires no more than $3\log(n) + 3$ multiplications and divisions to compute x^n. [Notice that $3\log(n) + 3$ is more convenient to work with than the more precise result in (C).] Such an algorithm is called **logarithmic**. If we plot the number of steps as a function of the input size, we get a logarithmic curve. A logarithmic algorithm has the significant property that one must square the size of the input before the number of steps (and thus the time) approximately doubles.

Example 6.3. Suppose that a particular algorithm has an integer n as input and requires no more than $9\lfloor\log(n)\rfloor$ steps. As in the preceding example if we input $n = 100$, the number of steps is no more than $9\lfloor\log(100)\rfloor$. Since $2^6 = 64$ and $2^7 = 128$, $\lfloor\log(100)\rfloor = 6$. Thus the number of steps is no more than $9 \cdot 6 = 54$. If we double the input size to $n = 200$, the number of steps is no more than $9 \cdot 7$, since $\lfloor\log(200)\rfloor = 7$. If we square the input size to $n = 10{,}000$, the number of steps is now no more than $9\lfloor\log(10{,}000)\rfloor = 9 \cdot 13 = 117$.

The contrast of the two preceding examples illustrates why logarithmic algorithms are much preferred to linear algorithms. In the next section we shall examine the logarithm function and others in more detail.

EXERCISES FOR SECTION 6

1. Calculate the following. Then match up the answers from the left-hand column that agree with one from the right-hand column.

$\log(2 \cdot 2)$	$\log(2) + \log(4)$
$\log(2 \cdot 4)$	$\log(4) + \log(8)$
$\log(2 \cdot 8)$	$4\log(16)$
$\log(4 \cdot 8)$	$2\log(2)$
$\log(2^2)$	$2\log(8)$
$\log(8^2)$	$\log(2) + \log(8)$
$\log(16^4)$	

2. Explain why $\log(a^b) = b\log(a)$.
3. Find $\lfloor\log(n)\rfloor$ if $n =$ ***(a)*** 10, ***(b)*** 100, ***(c)*** 1000, and ***(d)*** 10,000.
4. Decide whether each of the following is true or false.
   (a) $\log(2^{56}) = 56$. ***(b)*** $2^{\log(3)} = \log(3)$.
   (c) $\log(2 \cdot 3) = \log(3)$. ***(d)*** $\log(\frac{1}{2}) = -1$.
   (e) $\lfloor\log(17)\rfloor = 4$. ***(f)*** $\log(2^3) = 4$.
   (g) $\log(-2) = -1$. ***(h)*** $\log(\frac{3}{8}) = \log(3)/3$.
   (i) $\log(\frac{3}{4}) = -2\log(3)$.
   Correct those that are false.
5. Show that $\log(10^t) < 4t$ for all positive integers t.
6. Find the smallest integer k such that $\log(100t) < kt$ for all positive integers t.
7. Find ***(a)*** $\lfloor\frac{19}{11}\rfloor$, ***(b)*** $\lceil\frac{19}{11}\rceil$, ***(c)*** $\lfloor\frac{23}{12}\rfloor$, ***(d)*** $\lceil\log(73)\rceil$, ***(e)*** $\lfloor\log(73)\rfloor$, ***(f)*** $\lfloor\frac{114}{19}\rfloor$, ***(g)*** $\lceil\log(4^9)\rceil$, and ***(h)*** $\lfloor\log((2^3)^4)\rfloor$.
8. Suppose that $f(n) = 7n + 11$. Find the quotient $f(2n)/f(n)$ if ***(a)*** $n = 100$, ***(b)*** $n = 200$, and ***(c)*** $n = 1000$.

9. Suppose that $f(n) = 3n^2 + 4n + 5$. Find the quotient $f(2n)/f(n)$ if **(*a*)** $n = 100$, **(*b*)** $n = 200$, and **(*c*)** $n = 1000$.

10. Suppose that $f(n) = 4\lfloor \log(n) \rfloor + 13$. Find the quotient $f(n^2)/f(n)$ if **(*a*)** $n = 100$, **(*b*)** $n = 200$, **(*c*)** $n = 10{,}000$, and **(*d*)** $= 20{,}000$.

11. **(*a*)** Show that if $f(n)$ is a linear function of the form $f(n) = a \cdot n$, where a is a constant, then $f(2n) = 2f(n)$. Find an equation expressing $f(n^2)$ in terms of $f(n)$.
(*b*) Suppose that $g(n)$ is a logarithmic function of the form $g(n) = b \log(n)$. Then express both $g(2n)$ and $g(n^2)$ in terms of $g(n)$.

12. Use induction to show that FASTEXP is correct. (*Hint:* Reread the proof that DtoB is correct.)

13. How many multiplications and divisions are performed if FASTEXP is used to compute x^{89}?

14. For each of the following values of m and r, verify that m has r binary digits, where $r = 1 + \lfloor \log(m) \rfloor$:
(*a*) $m = 2, r = 2$. **(*b*)** $m = 3, r = 2$. **(*c*)** $m = 4, r = 3$.
(*d*) $m = 7, r = 3$. **(*e*)** $m = 8, r = 4$. **(*f*)** $m = 15, r = 4$.
(*g*) $m = 37, r = 6$. **(*h*)** $m = 100, r = 7$.

15. Is the following true or false? The number of binary digits in the number m is $r = \lceil \log(m) \rceil$. Explain.

16. Calculate $2^{(3^2)}$ and $(2^3)^2$. Explain, in general, why $a^{(b^c)}$ does not equal $(a^b)^c$.

17. Does $\log(a^b)$ equal $(\log(a))^b$? Explain.

18. The Post Office now charges 25 cents for a letter weighing up to 1 ounce and then 20 cents for each additional partial or whole ounce. If x is the weight of a letter in ounces and $x \geq 1$, then find a formula for the cost of mailing that letter.

19. Suppose that $R(x)$ is the function that takes a real number x and rounds it to the nearest integer. If $x = j + .5$, where j is an integer, then $R(x) = j + 1$; that is, R rounds up. Find a formula for $R(x)$ using the floor and/or ceiling functions.

2:7 HOW LOGARITHMS GROW

In the previous section we distinguished between EXPONENT and FASTEXP by looking at the functions that count the maximum number of divisions and multiplications that each would perform for a given input. This kind of analysis is crucial to any comparison of algorithms.

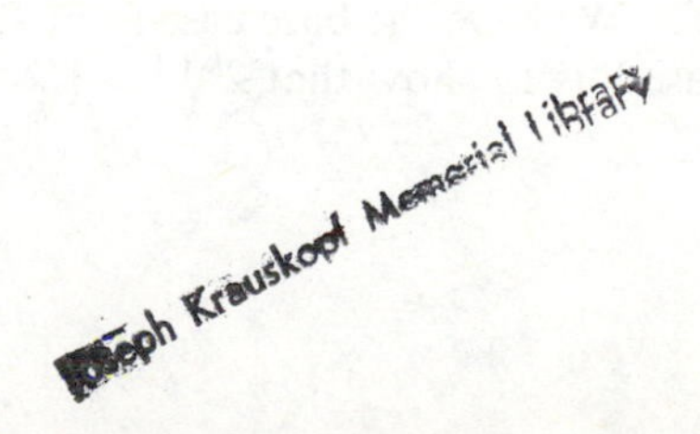

To appreciate fully the advantages of a logarithmic algorithm over a linear algorithm, we must have a thorough understanding of the logarithm function. Our first observation about the logarithm function is that it gets arbitrarily large. By that we mean that given any positive integer M, no matter how large, once n is sufficiently large $f(n) = \log(n)$ will be larger than M. To see this, note that $\log(2^M) = M$ by definition. Thus

$$\text{if } n > 2^M, \text{ then } \log(n) > M.$$

Many functions get arbitrarily large, for example, $f(n) = n$, $g(n) = n^2$ and the square root function $h(n) = \sqrt{n}$. Note that $g(n) = n^2$ grows more rapidly than $f(n) = n$, which grows more rapidly than $h(n) = \sqrt{n}$. How fast does the logarithm function grow in comparison with these functions? Our principal result about the logarithm function is that its growth is very slow, in fact, even slower than the square root function. The remainder of this section is devoted to this property. We begin with the following fact about exponents.

Lemma 7.1. If r is an integer greater than 5, then $2^r > (r + 1)^2$.

Mathematicians call a particular mathematical statement a *lemma* if it doesn't appear very interesting but is useful in proving something else. Notice that the conclusion of the above lemma is false for $r = 1, 2, 3, 4$, and 5 as shown in Table 2.7.

Table 2.7

r	2^r	$(r+1)^2$
1	2	4
2	4	9
3	8	16
4	16	25
5	32	36
6	64	49
7	128	64
8	256	81

Lemma 7.1 will be proved by induction. The proposition P_r will be that $2^r > (r + 1)^2$. The base case will be $r = 6$, and we shall show that the truth of P_k implies the truth of P_{k+1}.

Proof. We have the base case from Table 2.7. Next we assume that $2^k > (k + 1)^2$ and use this to show that $2^{k+1} > [(k + 1) + 1]^2$. Now

$$
\begin{aligned}
2^{k+1} &= 2 \cdot 2^k && \text{by algebra} \\
&> 2(k+1)^2 && \text{by inductive hypothesis} \\
&= 2k^2 + 4k + 2 && \text{by algebra} \\
&= k^2 + 4k + 4 + (k^2 - 2) && \text{by algebra} \\
&= (k+2)^2 + (k^2 - 2) && \text{by algebra} \\
&> (k+2)^2 && \text{by ignoring } (k^2 - 2)\text{, which} \\
& && \text{is positive for } k > 1.
\end{aligned}
$$

□

We use Lemma 7.1 to show that the logarithm function is eventually smaller than the square root function.

Theorem 7.2. If $n \geq 64$, then $\sqrt{n} > \log(n)$.

Proof. Suppose that k is the largest integer with $2^k \leq n$, that is, $k = \lfloor \log(n) \rfloor$. Then by definition

$$2^{k+1} > n \geq 2^k.$$

These two inequalities will produce the proof. First, if

$$2^{k+1} > n,$$

then, by taking logs of both sides, we get

$$k + 1 > \log(n). \tag{A}$$

Second, if $n \geq 64$, then $k \geq 6$ and Lemma 7.1 applies. Thus

$$n \geq 2^k > (k+1)^2.$$

Taking square roots of the two ends, we get

$$\sqrt{n} > k + 1. \tag{B}$$

Combining inequalities (A) and (B) gives $\sqrt{n} > \log(n)$. □

Question 7.1. Using a calculator, find the smallest integer N such that $\sqrt{N} > \log(N)$. Does this contradict Theorem 7.2? (*Comment*: if your calculator does not work in base 2, see Supplementary Exercises 9 and 10.)

Corollary 7.3. If $n \geq 64$, then $n/\log(n) > \sqrt{n}$.

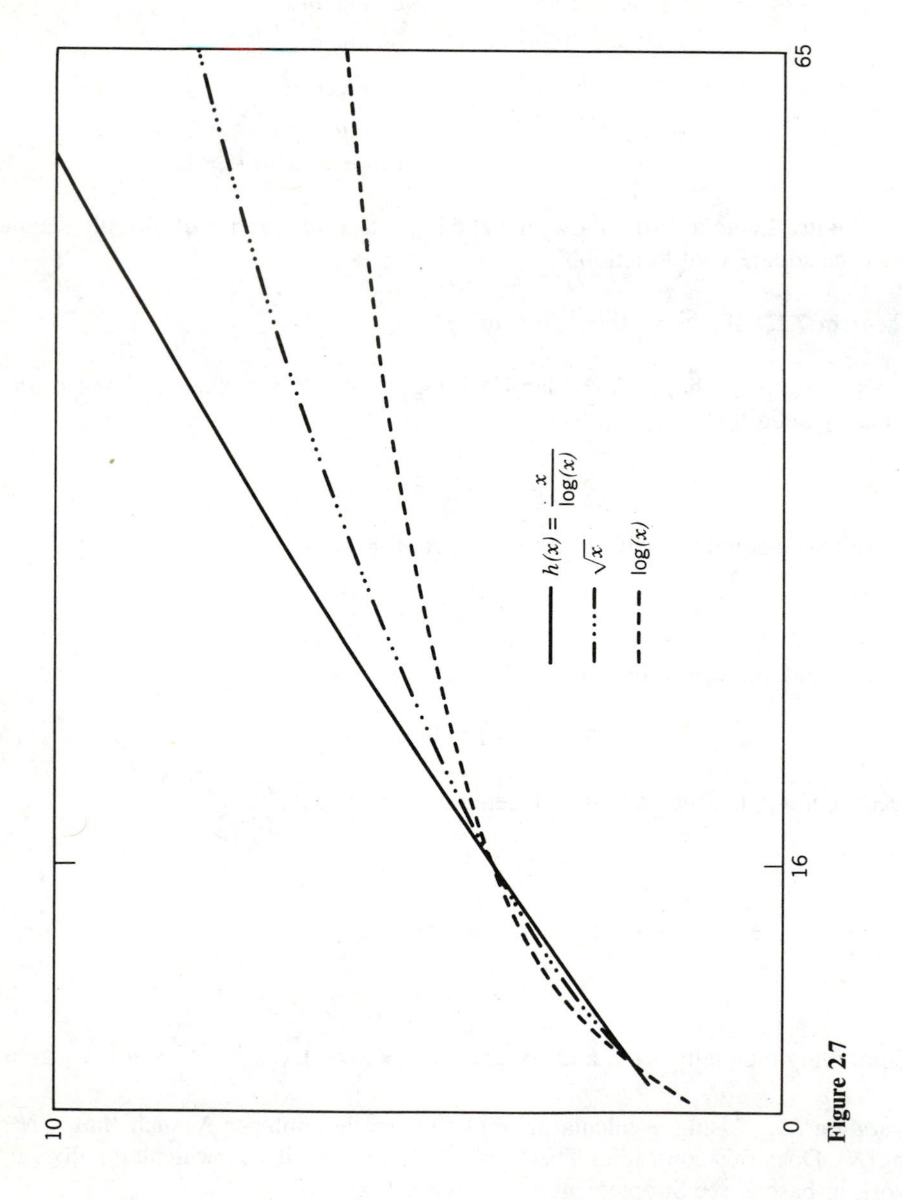

Figure 2.7

Proof. Since $n \geq 64$ implies $\sqrt{n} > \log(n)$, then if $n \geq 64$,

$$\frac{n}{\log(n)} > \frac{n}{\sqrt{n}} = \sqrt{n}. \qquad \square$$

Corollary 7.4. The function $h(n) = n/\log(n)$ gets arbitrarily large.

Proof. As above we mean that given any (large) integer M, if n is big enough, then $h(n)$ is bigger than M. Suppose that $M \geq 8$. Then if $n > M^2$, by the previous corollary,

$$h(n) > \sqrt{n} > \sqrt{M^2} = M. \qquad \square$$

Mathematicians use the word *corollary* for a statement whose proof almost immediately follows from a previous result. Corollary 7.4 looks inconsequential, but we shall have an important need for it later in this chapter.

We illustrate the growth of these functions in Figure 2.7; however, we shift our point of view on the domain of these functions. Until now we have been interested in counting problems, involving integers and functions evaluated at integers. Thus we write functions like $f(n) = \log(n)$, $g(m) = \sqrt{m}$, and $h(r) = r^2$, where n, m, and r represent integers. Furthermore, we think of our functions as having integer domains, usually the nonnegative or positive integers. On the other hand, most of the functions we are using can be defined for domains of real numbers. For example, $f(x) = \log(x)$ has domain all positive reals, $g(x) = \sqrt{x}$ has domain all nonnegative reals, and $h(x) = x^2$ has domain all real numbers. When we specify a function in terms of the variable x, we mean that x may take on any real value in its domain; in general, the context will make the appropriate domain evident. For graphical illustration we choose to consider the functions $x/\log(x)$, $\sqrt{x}$ and $\log(x)$ as functions of real variables and to graph them as continuous curves as in Figure 2.7, rather than just plot the functions at integer values.

In the next section we develop the standard notation for comparing functions and their growth rates.

EXERCISES FOR SECTION 7

1. Find an integer N such that if $n \geq N$, then $n/\log(n) > 100$.
2. Find the smallest integer m such $2^m > m^2$. Then show that if r is any integer at least as large as m, then $2^r > r^2$.
3. Find the smallest integer q such that $2^q > (q + 1)^3$. Use induction to show that if r is any integer with $r \geq q$, then $2^r > (r + 1)^3$.

4. Use the preceding exercise to show that the cube root function is eventually larger than the log function.

5. In Question 7.1 you found the least integer N such that $\sqrt{N} > \log(N)$. Prove that if $n \geq N$, then $\sqrt{n} > \log(n)$.

6. Show that $\sqrt{n}/\log(n)$ gets arbitrarily large.

7. What is the largest value that $2^{\log(n)/\sqrt{n}}$ can achieve for any positive integer n?

8. Suppose that $f(n) = \log(n)$ and $g(n) = \log(f(n))$. The following table lists some values for n, $f(n)$, and $g(n)$. Determine the values that correctly replace the question marks.

n	$f(n)$	$g(n)$
1	0	—
2	1	0
4	2	1
?	4	?
?	?	4
?	?	8

9. With $g(n)$ defined as in Exercise 8 comment on the remark, "For practical purposes, $\log(g(n)) < 8$."

10. The function $g(n)$ of Exercise 8 is sometimes written as $g(n) = \mathbf{\log\log}(n)$. Find a simple expression for $\log\log(2^{2^n})$. Explain why $\log\log(n)$ is defined only for $n > 1$. Then explain why $\log\log(n)$ gets arbitrarily large for large values of n.

11. Which function is larger: $\log(n)$ or $\log\log(n)$? Explain.

12. For what function $m(n)$ is it true that $\log\log(m(n)) = 2\log\log(n)$?

2:8 THE "BIG OH" NOTATION

In Section 2.7 we began to explore a hierarchy of functions that might appear in the analysis of algorithms. We saw that $\log(n) < \sqrt{n} < n/\log(n) < n$, provided that n is large enough. Consequently, we would prefer an algorithm that used $\sqrt{n}$ steps to one that used $n/\log(n)$ steps on a problem with input size n. Since most algorithms are analyzed on a worst-case basis, we don't want our distinctions to be too finely drawn. Computer scientists and mathematicians have come to use what is called the "big oh" notation when discussing the efficiency of algorithms.

Definition 1. Let $f(n)$ and $g(n)$ be functions whose domains are the positive integers and whose ranges are the positive reals. Then we write

$$\boldsymbol{f(n) = O(g(n))} \quad \text{or} \quad \boldsymbol{f = O(g)}$$

(read "**f of n is big oh of g of n**" or "**f is big oh of g**") if there is a positive constant C such that

$$f(n) \leq C \cdot g(n)$$

for all positive integers n.

Example 8.1. Suppose that $f(n) = 5n$ and $g(n) = n$. To show that $f = O(g)$, we have to show the existence of the constant C in Definition 1. The best way to show the existence of such a C is to actually produce it. Here the example has been cooked up so that the constant C is staring us in the face. For if we choose $C = 5$, then $f(n)$ actually equals $C \cdot g(n)$. Notice that we could have selected a larger number for C. For instance, if we choose $C = 6$, then $f(n) = 5n < 6n = C \cdot g(n)$. So we write $f(n) = O(n)$.

Example 8.2. Suppose that $f(n) = 5n + 8$. To show that $f(n) = O(n)$, we must produce a constant C such that $f(n) \leq C \cdot n$ for all n. If we try $C = 5$, this doesn't work, since $f(n) = 5n + 8$ is not less than $5n$ [e.g., $f(1) = 13 > 5 = 5 \cdot 1$]. We need C to be at least 13. To see that $C = 13$ will work, note that $8 \leq 8n$. Thus

$$5n + 8 \leq 5n + 8n = 13n.$$

Let's look at a function that is not linear.

Example 8.3. Suppose that $f(n) = n^2$. We show that $f(n)$ is not big oh of n, denoted $f(n) \neq O(n)$. To accomplish this, we must show that there cannot exist a constant C that satisfies the big oh definition. Suppose that there were such a constant C. We would need

$$n^2 \leq C \cdot n$$

for all n. However, if $n > C$,

$$n^2 > C \cdot n.$$

Combining these two inequalities shows that for $n > C$,

$$n^2 > C \cdot n \geq n^2.$$

Thus no C can work.

Example 8.4. Suppose that $f(n) = n^2 + 3n - 1$. Then

$$\begin{aligned} f(n) &= n^2 + 3n - 1. \\ &< n^2 + 3n, && \text{since subtraction makes things smaller} \\ &\leq n^2 + 3n^2, && \text{since } n \leq n^2 \text{ for integer } n \\ &= 4n^2. \end{aligned}$$

Thus, letting $C = 4$, we have $f(n) = O(n^2)$.

Question 8.1. Show that $f(n) = 12n^2 - 11$ and $h(n) = 3n^2 + 4n + 11$ are both $O(n^2)$.

These examples illustrate the big oh notation for linear and quadratic functions. Each time we find a simple function that is an upper bound on the original function. We emphasize that it is the growth of the functions that is of interest to us. Any linear function f has the property that $f(2n)$ is almost double $f(n)$. Furthermore, the larger n is, the more exact this rule. In contrast a quadratic function h has the property that $h(2n)$ is almost quadruple $h(n)$. Similarly, the larger n is, the more exact this rule.

We call an algorithm **linear** (or **quadratic**) if there is a function $f(n)$ that counts the number of the most time-consuming steps the algorithm performs, given a problem of size n, and $f(n) = O(n)$ (or $O(n^2)$). Let us see how this idea might help us decide which algorithm to select in a given situation.

Example 8.5. Suppose that we have two algorithms, say L and Q that each correctly solves a particular problem. L is linear and takes 20 minutes on a problem of size 10; Q is quadratic and takes 5 minutes on the same problem of size 10. Suppose that we have a problem of size 100. Which should we use? L ought to take about 200 minutes to solve the problem, and Q ought to take about 500 minutes to solve the problem. Why the difference? Because a tenfold increase in problem size ought to produce a hundredfold increase in running time for a quadratic algorithm while only a tenfold increase in running time for a linear algorithm.

Question 8.2. Suppose that L is a linear algorithm that solves a problem of size 100 in 8 minutes while C is a cubic algorithm that solves a problem of size 100 in 2 minutes. For a problem of size 200 which algorithm would you use? How about a problem of size 1000? {First you will have to decide what we mean by a cubic algorithm.}

It might be that, say, an algorithm we have called linear is faster than linear, that is, $f(n) = O(n)$ and $f(n) = O(g(n))$ for some function $g(n)$ smaller than n. It might be that it is our analysis that is weak and only demonstrates $f(n) = O(n)$. We shall discuss this in more detail in the next section.

Since there are linear, quadratic, and cubic algorithms, you should not be surprised to learn that this hierarchy generalizes to arbitrary exponents.

Example 8.6. If $f(n) = n^{17} + 3n^{15} - 7n^{10} + 20n^6 - 10n$, then we show that $f(n) = O(n^{17})$.

$$f(n) < n^{17} + 3n^{15} + 7n^{10} + 20n^6 + 10n,$$

since for $n > 0$, changing negative coefficients to positive makes the function larger;

$$\leq n^{17} + 3n^{17} + 7n^{17} + 20n^{17} + 10n^{17},$$

since making exponents larger makes the function larger;

$$= (1 + 3 + 7 + 20 + 10)n^{17}$$
$$= 41n^{17}.$$

Question 8.3. Show that $f(n) = 2n^7 - 6n^5 + 10n^2 - 5 = O(n^7)$.

The first theorem of this section says that any polynomial is big oh of its term of highest degree. The proof mimics the previous example. We let a_j denote the coefficient of the term of degree j in the following theorem.

Theorem 8.1. Let $f(n) = a_d \cdot n^d + a_{d-1} \cdot n^{d-1} + \cdots + a_1 \cdot n + a_0$, where d is a positive integer and $a_d, a_{d-1}, \ldots, a_1$ and a_0 are constants. Then $f(n) = O(n^d)$.

Proof. First we change all the coefficients of f to positive numbers. This can only increase the value of $f(n)$ for positive integers n. Next we note that $n^j \leq n^d$ if $j \leq d$. Thus

$$\begin{aligned} f(n) &= a_d \cdot n^d + a_{d-1} \cdot n^{d-1} + \cdots + a_j \cdot n^j + \cdots + a_0 \\ &\leq |a^d| \cdot n^d + \cdots + |a_j| \cdot n^j + \cdots + |a_0| \\ &\leq |a_d| \cdot n^d + \cdots + |a_j| \cdot n^d + \cdots + |a_0| \cdot n^d \\ &= (|a_d| + \cdots + |a_j| + \cdots + |a_0|) \cdot n^d \\ &= C \cdot n^d \end{aligned}$$

provided that C is equal to the sum of the absolute values of the coefficients in the original polynomial. □

The big oh notation has certain peculiar features. For example, when we write $f = O(g)$ we are not writing down an equation of the usual sort. It does not express a symmetric relation; that is, we do not write $O(g) = f$ nor does it follow that $g = O(f)$, although it might in some instances. Here are some basic properties about big oh that are helpful in manipulations.

Theorem 8.2. If $f = O(g)$, then for any constant a,

$$\text{(i)}\ a \cdot f = O(g).$$

If, in addition, $h = O(k)$, then

$$\text{(ii)}\ f + h = O(g + k),$$

and

$$\text{(iii)}\ f \cdot h = O(g \cdot k).$$

Finally, if $f = O(g)$ and $g = O(h)$, then

$$\text{(iv)}\ f = O(h).$$

Example 8.7. Suppose that $f(n) = 5n^3 + 2n$. Then $f(n) = O(n^3)$ by Theorem 8.1. Next consider $w(n) = 15n^3 + 6n = 3f(n)$. By (i) of Theorem 8.2 (as well as by Theorem 8.1.), $w(n) = O(n^3)$.

Example 8.8. Suppose that $f(n) = 3n + 7$ and $h(n) = 2n^2 - n + 8$. We know that $f = O(n)$ and that $h = O(n^2)$ by Theorem 8.1. By (ii) of Theorem 8.2 $f + h = O(n + n^2)$. Since $n + n^2 = O(n^2)$, we can use part (iv) of the Theorem 8.2 to obtain $f + h = O(n^2)$. This result also follows from Theorem 8.1, since $f(n) + h(n) = 2n^2 + 2n + 15$.

Example 8.9. Suppose that $f(n) = n^2 + 3n + 7$ and $h(n) = n^3 + 17$. Since $f = O(n^2)$ and $h = O(n^3)$, by part (iii) $f \cdot h = O(n^5)$. Notice that we didn't multiply out $f \cdot h$, but if we had, the result would be a polynomial of degree 5.

Question 8.4. If $f(n) = 3n^5 + 13n^3 - 10$ and $g(n) = 2n^4 + 3n^2$, find $h(n)$ and $k(n)$ such that $f + g = O(h)$ and $f \cdot g = O(k)$. Justify.

Proof of Theorem 8.2. We prove the second assertion leaving the remaining three proofs for Exercise 6. Suppose that C and C' are constants such that $f(n) \leq C \cdot g(n)$ and $h(n) \leq C' \cdot k(n)$. Let D equal the larger of C and C'. Then

$$\begin{aligned} f(n) + h(n) &\leq C \cdot g(n) + C' \cdot k(n) \\ &\leq D \cdot g(n) + D \cdot k(n) \\ &= D \cdot (g(n) + k(n)) \\ &= O(g(n) + k(n)). \end{aligned}$$ □

Here is a summary of what we know about comparisons between functions. First, for all positive n

$$1 \leq \sqrt{n} \leq n \leq n^2 \leq n^3 \leq \cdots \leq n^i \leq \cdots.$$

Thus we also have these big oh results:

$$\begin{aligned} 1 &= O(\sqrt{n}), \\ \sqrt{n} &= O(n), \\ n &= O(n^2), \\ n^2 &= O(n^3), \end{aligned}$$

and in general,

$$n^i = O(n^j) \qquad \text{if } i \leq j.$$

You may have noticed that we have been avoiding the log function. Partly, this is because $\log(1) = 0$ and so the log function doesn't have range the positive reals and thus does not fit into the big oh definition. However, we know that if n is large enough ($n \geq 16$), then $\log(n) \leq \sqrt{n}$. Furthermore, as shown in Table 2.8 and Exercise 8, $\log(n)$ is always less than twice $\sqrt{n}$. Thus we want to say that $\log(n) = O(\sqrt{n})$.

Table 2.8

n	$\log(n)$	$\sqrt{n}$	$\log(n)/\sqrt{n}$
1	0	1	0
2	1	1.4···	0.7···
4	2	2	1
5	2.32···	2.23···	1.03···
6	2.58···	2.44···	1.05···
7	2.80···	2.6···	1.06···
8	3	2.8···	1.06···
9	3.16···	3	1.05···
10	3.32···	3.16···	1.05···
16	4	4	1
32	5	5.6···	0.8···
64	6	8	0.75

We would also like to say that

$$1 = O(\log(n));$$

however, this also doesn't satisfy our present definition of big oh. Since $\log(1) = 0$, there is no constant C such that $1 \leq C\log(n)$ for all n. For $n \geq 2$, it is the case that $1 \leq \log(n)$, and so the constant $C = 1$ will work for n sufficiently large (larger than 1). What this suggests is that we need a stronger definition of big oh.

Definition 2. Let $f(n)$ and $g(n)$ be functions with domain the positive integers. Then we say

$$\boldsymbol{f(n) = O(g(n))} \qquad \text{or} \qquad \boldsymbol{f = O(g)}$$

if there are positive constants C and N such that

$$f(n) \leq C \cdot g(n)$$

for all integers $n \geq N$.

With this more general definition we can say that $1 = O(\log(n))$, since $C = 1$ and $N = 2$ work. Also $\log(n) = O(\sqrt{n})$ using $C = 2$. We shall need this stronger definition only when we want to show that an algorithm performs $f(n) = O(\log(n))$ steps on a problem of size n. Such an algorithm is called **logarithmic**.

Definition 2 is the standard definition of the big oh machine. We have emphasized the simpler form because it is almost as widely applicable and it is easier to use. Furthermore, any pair of functions f and g with $f = O(g)$ in the first definition also has $f = O(g)$ in the second definition.

Our hierarchy of functions now looks like

$$1 \leq \log(n) \leq \sqrt{n} \leq n \leq n^2 \leq n^3 \leq \cdots \leq n^i \leq \cdots.$$

This hierarchy of functions is worth remembering. In a typical analysis you will have a (complicated) function $f(n)$ and will need to find $g(n)$, as small and simple as possible, so that $f = O(g)$. The best candidates for g are the functions listed above.

Example 8.10. Suppose that $f(n) = \log(n) \cdot \sqrt{n^3} \cdot \sin(n)$. We want to find g from the list of basic functions above so that $f = O(g)$. First, recall that the value of the sine function never exceeds 1. Next we note that $\sqrt{n^3} = n \cdot \sqrt{n}$. Thus

$$\begin{aligned} f(n) &\leq \log(n) \cdot \sqrt{n^3} \\ &= \log(n) \cdot n \cdot \sqrt{n} \\ &= O(\log(n) \cdot \sqrt{n} \cdot n) \\ &= O(\sqrt{n} \cdot \sqrt{n} \cdot n) = O(n^2). \end{aligned}$$

In many of the examples of this section when $f = O(g)$, it has also been the case that $g = O(f)$. This is a common occurrence in problems about polynomials but will not generally be true. We know from Example 8.3 that $n = O(n^2)$, but $n^2 \neq O(n)$. Also, in the preceding example, although $f(n) = O(n^2)$, it is not true that $n^2 = O(f(n))$, although we shall not go into the tricky details.

EXERCISES FOR SECTION 8

1. Find a constant C that demonstrates that $f(n) = O(g(n))$ for each of the following.
(a) $f(n) = 17n + 31$, $g(n) = n$.
(b) $f(n) = 21n - 13$, $g(n) = n$.
(c) $f(n) = 12n^2 + 3n + 15$, $g(n) = n^2$.
(d) $f(n) = 3n^2 - 4n + 5$, $g(n) = n^2$.
(e) $f(n) = 2n^2 - n - 1$, $g(n) = n^2$.
(f) $f(n) = 0.2n + 100{,}000$, $g(n) = n$.
(g) $f(n) = n^3 + 3n^2 + 5n + 11$, $g(n) = n^3$.
(h) $f(n) = n^3 + 3n^2 + 5n + 11$, $g(n) = n^4$.

2. Which of the following is True and which False?
(a) $n = O(n)$, **(b)** $n = O(n^2)$, **(c)** $n = O(n^3)$, **(d)** $n^2 = O(n)$, **(e)** $\log(n) = O(n)$, **(f)** $n = O(\log(n))$, **(g)** $n = O(n \cdot \log(n))$, **(h)** $n \cdot \log(n) = O(n^2)$, and **(i)** $1/(n^2 + 1) = O(1/n^2)$.
For each true statement, find a C that demonstrates the big oh definition.

3. Here is the functional hierarchy: for n sufficiently large.

$$1 \leq \log(n) \leq \sqrt{n} \leq n \leq n^2 \leq \cdots \leq n^i \leq \cdots.$$

Add all the following to this hierarchy. **(a)** $1/n$, **(b)** $1/(n^2)$, **(c)** $n^{3/2}$, **(d)** $n/\log(n)$, **(e)** $n^2/\log(n)$, **(f)** $n/(\log(n))^2$, and **(g)** the cube root of n, $n^{1/3}$.

4. Answer the following with explanations for your answers:
(a) Is $f(n) = 0.5n \cdot \log(n) + 3n + 15 = O(n^2)$? $O(n)$?
(b) Is $f(n) = (3n^2 + 5n - 13)^2 = O(n^5)$?
(c) Is $f(n) = (3\log(n) + n)^2 = O(n^2)$?
(d) Is $f(n) = 1/(n^2 + 1) = O(1)$?
(e) Is $4^{\log(n^2)} = O(\log(n) + 1)$?
(f) Is $(4^{\log(n^2)})^2 = O(n^3)$?

5. Some functions $f(n)$ are listed in column A of the following table. Some functions $g(n)$ are listed in column B. For each f in column A find the smallest g in column B with the property that $f = O(g)$.

A	B
$7 + 32n + 14n^2$	1
$17 + \log(n^4)$	$\log(n)$
$\log(n^n)$	$\sqrt{n}$
$(\log(n^n))^2$	n
$\sin(n^2)$	n^2
$2^n/(n+1)^3$	n^3
$(n^2 + n \cdot \log(n))^2$	n^5
$2^n/n^n$	2^n
$(\log(n))^{0.5}$	n^n
$\sqrt{n \cdot \log(n)}$	$n \cdot (\log(n))$
$\sqrt{n^3}$	$1/n$
$\sqrt{n^3 + 3n^2}$	
$4^{\log(n)}$	
$\log(3^{n^2})$	
$n^{\log(n)}$	
$(n + \sqrt{n})^2$	
$(n+1)/(n^2+1)$	
$(\log(n))^4/n$	
$5 + (\log(n))^2 - n/2^n + 0.01 + n^3$	
$1 + 4 + 9 + \cdots + n^2$	
$1 + \frac{1}{2} + \frac{1}{4} + \cdots + \frac{1}{2^n}$	
$(\log(n)/n)^n$	

6. Prove parts (i), (iii), and (iv) of Theorem 8.2.

7. Suppose that g is a function such that $g(n) > 0$ for all n. Express $f(n) - g(n)$ as big oh of some combination of $f(n)$ and/or $g(n)$. Then prove your answer. Can you deduce a similar result for $f(n)/g(n)$?

8. Fill in the values of Table 2.8 for $n = 11, 12, \ldots, 15$. See Supplementary Exercises 9 and 10 about computing $\log(x)$.

9. Show that n^3 is not $O(n^2)$.

10. Show that $2n^2 + 5n$ is not $O(n)$.

11. Is it true that for every pair of functions f and g either $f = O(g)$ or $g = O(f)$?

2:9 $2^n \neq O(p(n))$: PROOF BY CONTRADICTION

We have seen that the polynomials form a natural hierarchy of functions against which we can compare the counting functions that really interest us. Is it the case that all our counting functions are $O(p(n))$ for some polynomial $p(n)$? No! The first natural example is the function 2^n that counts the number of subsets of a set with n elements. Here we have a negative (and significant) result.

Theorem 9.1. If r is a constant bigger than 1, then the function r^n is not $O(p(n))$ for any polynomial $p(n)$.

What does it mean to say that a function $f(n)$ is not big oh of the polynomial $p(n)$? Well, if f were $O(p)$, then there would exist constants C and N such that $f(n) \leq C \cdot p(n)$ for all $n \geq N$. If f is not $O(p)$, then there are no such constants. Specifically, there is no constant C such that $f(n) \leq C \cdot p(n)$ for all sufficiently large n. Such a fact can be harder to demonstrate than the fact that one function is big oh of another.

The proof of Theorem 9.1. will employ the technique of proof by contradiction. This technique is frequently used without notice if the statement to be proved is simple enough. We did it, for instance, in Chapter 1 in Theorem 9.1 and in (this chapter's) Example 8.3. The technique is almost as important as Mathematical Induction, but isn't as completely specified. We digress to explain.

Proof by Contradiction. We pause first to think a bit about what a mathematical proof is. Although a proof can be described precisely in logical terms with axioms, truth tables, and rules of inference, we choose to be more informal. The aim of a proof is to establish the validity of some assertion A. It may be a simple statement like "14 is an even integer," or it may be in the form of an implication, like "If $n \geq 2$, then $\log(n) \geq 1$," or it may be in another form, like "There is an integer that is divisible by both 3 and 7." In a (usual) direct proof, we begin working with known mathematical truths and proceed logically until we deduce the truth of assertion A. For example, look back at the proof that $\log(a \cdot b) = \log(a) + \log(b)$ in Section 6. This was a straightforward, direct proof.

A proof by contradiction follows a different pattern. Suppose that we want to prove an assertion A. In a **proof by contradiction** we begin by assuming that assertion A is false. Then we argue using that assumption until we come to a contradiction. The contradiction will be the denial of some mathematical fact. For example, we might deduce that $0 = 1$ or that 5 is an even number. But now what? The absolute world of mathematics assumes that every statement is either true or false. Thus our original assertion A is either true or false. We assumed that it was false and deduced a contradiction. Thus it must be that assertion A is true.

To begin a proof of assertion A by contradiction, we must formulate the **negation** of A, that is, we must know what it means for A to be false. If assertion A is simple enough, its negation is easily formed. We emphasize that exactly one of assertion A and its negation is true, but for the moment we don't care which. Instead we concentrate on formulating the negation of a given statement A.

Example 9.1. Recall that an integer greater than 1 is called a prime if it cannot be written as the product of two smaller positive integers. The assertion that "7,891,011 is a prime number" is negated by the assertion that "7,891,011 is not a prime number" or "There are integers c and d with $1 < c \leq d$ and $c \cdot d = 7{,}891{,}011$."

Question 9.1. Write down the negation of each of the following assertions.
(a) 353 is not a prime number.
(b) 238 is an even integer.

In contrast, if assertion A is complicated, care might be required to correctly obtain the negation of A. We illustrate with several examples.

Example 9.2. Consider the assertion that "$n^3 + 3n^2 + 2n$ is always divisible by 3." (This can be proved by induction on n: see Exercise 4.17.) What this statement says precisely is that for every integer n, the quantity $n^3 + 3n^2 + 2n$ is a multiple of 3. The negation of this assertion is that there is at least one integer, say m, with the property that $m^3 + 3m^2 + 2m$ is not a multiple of 3.

In general, an assertion of the form "For every instance, something happens" is negated by "There is at least one instance where that something doesn't happen."

Question 9.2. Negate each of the following assertions.
(a) Every integer greater than one has a prime divisor.
(b) Every integer of the form $4n + 1$ is a prime.
(c) Every prime greater than 2 is odd.

Example 9.3. The assertion that "For some integer n, $n^2 + n + 1$ is divisible by 3" can be negated by the assertion that "For every integer n, $n^2 + n + 1$ is not a multiple of 3."

In general, an assertion of the form "There exists at least one instance when something happens" is negated by an assertion of the form "for every instance, that something doesn't happen."

Question 9.3. Negate each of the following assertions.
(a) For some integer n, $3n + 1$ is a prime number.
(b) For some integer n, $\log(n) > n$.
(c) For some integer n, $n^2 > 2^n$.

Many mathematical assertions are of the form "If statement H is true, then statement C is true." We frequently use the shorter form IF H, THEN C. Statement H is called the **hypothesis** and statement C is called the **conclusion**. Some assertions that do not appear to be in this form of hypothesis and conclusion can be rephrased as such. For example, the assertion "For every integer n, $n(n + 1)/2$ is an integer" can be rephrased, "If n is an integer, then $n(n + 1)/2$ is an integer." Here H represents the assertion that n is an integer while C represents the assertion that $n(n + 1)/2$ is an integer.

Example 9.4. The assertion "If n is prime, then $n^2 + 1$ is even" has as its hypothesis that n is prime and as its conclusion that $n^2 + 1$ is even. The negation of this assertion is that there exists a prime n with the property that $n^2 + 1$ is not even.

What is the negation of the assertion

"IF H, THEN C"?

The negation of this is the assertion that statement H is true and statement C is false. We shorten this to

"H AND (NOT C)."

Question 9.4. Identify the hypothesis and conclusion of each of the following assertions. Then negate each.
(a) If n is even, then $n^2 + n + 1$ is prime.
(b) If $n^2 + n + 1$ is prime, then n is even.
(c) The integer n^2 is divisible by 4 whenever n is divisible by 6.

If we want to construct a proof by contradiction for the assertion "IF H, THEN C," we begin by assuming its negation, namely that both statement H is true and statement C is false, "H AND (NOT C)." We then work logically until we reach the sought-after contradiction.

To illustrate the technique of proof by contradiction, we present an easy version of Theorem 9.1: If $r > 2$, then $r^n \neq O(1)$.

Proof. We begin by assuming that "H AND (NOT C)," specifically that $r > 2$ and $r^n = O(1)$. From the (first) definition of big oh there is a constant C such that for all n

$$r^n \leq C \cdot 1.$$

Since $r > 2$,

$$2^n < r^n.$$

Now we combine these two inequalities and choose n to be an integer greater than C. This yields

$$2^n < r^n \leq C < n.$$

This provides a contradiction, since $2^n > n$ for every integer n. □

We now construct a proof by contradiction for Theorem 9.1. To make the arithmetic go a little easier, we shall prove the statement "2^n is not $O(p(n))$ for any polynomial $p(n)$." This is the same statement as in Theorem 9.1, except that $r = 2$ has been substituted. A similar proof would work for any constant $r > 1$.

Proof of Theorem 9.1. The negation of the assertion to be proved is "There exists a polynomial p with $2^n = O(p)$." Thus we begin by assuming this fact. By Theorem 8.1 there is an integer a such that $p = O(n^a)$, and so by part (iv) of Theorem 8.2 we have that $2^n = O(n^a)$. By definition, there is a constant C with

$$2^n \leq C \cdot n^a$$

for all positive integers n. Taking logarithms of both sides of this inequality, we get

$$\log(2^n) \leq \log(C \cdot n^a)$$

or

$$n \leq \log(C) + \log(n^a) \qquad \text{by properties of log,}$$

or

$$n \leq \log(C) + a \cdot \log(n) \qquad \text{by properties of log.}$$

We divide by $\log(n)$ to get

$$\frac{n}{\log(n)} \leq \frac{\log(C)}{\log(n)} + a.$$

Since for $n \geq 2$, $\log(n) \geq 1$,

$$\frac{n}{\log(n)} \leq \log(C) + a.$$

The right-hand side of this last inequality is a constant number, and by Corollary 7.4 the left-hand side is a function that grows arbitrarily large. These two statements form a contradiction, as we had hoped. We conclude that 2^n is not $O(p)$ for any polynomial p. □

In practice, a proof of an "IF H, THEN C" assertion often begins with the logically equivalent "IF (NOT C), THEN (NOT H)." We digress briefly to explain and begin with an analogy.

Example 9.5. Imagine that you are sitting in an ice cream parlour. There is a sign in front of you that says, "Try our scrumptious butterscotch raspberry syrup.

Only on top of creamy French vanilla ice cream." Before you lose too much concentration, let's rephrase this sign in a more mathematically structured way. If you want b.r. syrup, then you have to have F.v. ice cream. More simply, IF B R SYRUP, THEN F V ICE CREAM. Here the syrup is analogous to the mathematical statement H and the truth of H corresponds to your having the special syrup. The ice cream is analogous to the mathematical statement C and the truth of C corresponds to your having the vanilla. What about the statement H AND (NOT C)? This translates into b.r. syrup without F.v. ice cream and negates the original sign.

Example 9.5 (continued). Suppose that as you get up to leave the ice cream parlour you notice a sign on the back wall that says. "If you don't have our French vanilla ice cream, then you can't have our butterscotch raspberry syrup." Recall that the sign in the front says IF B R SYRUP, THEN F V ICE CREAM, while this one say IF NO F V ICE CREAM, THEN NO B R SYRUP. A little reflection will convince you that these signs mean the same thing.

As suggested by the previous example, given the assertion IF H, THEN C we can form what is called its **contrapositive** by negating both statements and reversing their order. Thus the contrapositive of the original

$$\text{IF H, THEN C} \tag{1}$$

is the assertion

$$\text{IF NOT C, THEN NOT H.} \tag{2}$$

As in the ice cream example a statement and its contrapositive are **logically equivalent**, that is, either both are true or both are false.

What other possibilities could there be? Well, it might be that the statement in (1) is true and the statement in (2) is false, or vice versa. Let's see why neither of these cases can happen. First suppose that the statement of line (2) is true and the statement of line (1) is false, that is,

$$\text{H AND (NOT C)} \tag{3}$$

is a true statement. Combining (3) and (2), we get

$$\text{H AND (NOT H),}$$

which is a contradiction. Next suppose that statement (1) is true and statement (2) is false, that is,

$$\text{(NOT C) AND NOT(NOT H)}$$

or

$$\text{(NOT C) AND H} \tag{4}$$

is a true statement. Combining statements (1) and (4) yields

$$\text{(NOT C) AND C,}$$

another contradiction. We conclude that statements (1) and (2) are either both true or both false. In other words, they are logically equivalent.

Example 9.6. Recall that a function is one-to-one if whenever $d \neq d'$, then $f(d) \neq f(d')$. The condition that specifies this one-to-one property is of the form IF H, THEN C, where H is the statement "$d \neq d'$" and C is the statement "$f(d) \neq f(d')$." Thus this property is the same as the property IF NOT C, THEN NOT H. In other words, a function is one-to-one if whenever $f(d) = f(d')$, then $d = d'$. This second, but equivalent, definition is often easier to check, since working with equalities can be easier than with inequalities.

Given an assertion IF A, THEN B, we can also form its **converse**, namely the assertion IF B, THEN A. In the following example we illustrate that the truth of a particular assertion does not determine the truth of the converse.

Example 9.7. The ice cream parlour does allow a customer to have plain French vanilla ice cream, that is, B is true but A is false.

Question 9.5. Form the converse and the contrapositive of each of the assertions from the preceding question as well as Lemma 7.1 and Theorem 7.2.

Sometimes it is the case that both an assertion and its converse are true, that is, both "IF A, THEN B" and "IF B, THEN A" are true. In that event we say that A is true **if and only if** B is true, which is abbreviated A IFF B. We shall see examples of this sort of assertion in Chapter 4.

EXERCISES FOR SECTION 9

1. Write out a detailed proof that 4^n is not $O(p)$ for any polynomial p.
2. Regardless of the truth or falsehood of the following, write the negation of the following assertions.
   (*a*) 14 is even.
   (*b*) 6 is prime.
   (*c*) $1 + 2 + \cdots + n = n(n + 1)/2$.

(*d*) There is an integer divisible by 3 and by 7.
(*e*) For every integer n, $n(n+1)/2$ is also an integer.
(*f*) Every set has more 2-subsets than 1-subsets.
(*g*) Every even number has an even number of 1s in its binary representation.
(*h*) If $f(n) = O(g(n))$, then $f(n) \leq g(n)$.
(*i*) There is a set that is larger than its complement.
(*j*) If x^2 is odd, then x^6 is odd.
(*k*) If x^6 is divisible by 8, then x is divisible by 2.
(*l*) 1728 is a sum of four cubes.
(*m*) Every even composite number is the sum of two primes.

3. Determine for which of the following functions $g(n)$ it is true that $2^n = O(g(n))$.
(*a*) $g(n) = n^2$, (*b*) $g(n) = 2^n$, (*c*) $g(n) = n^{10}$, (*d*) $g(n) = 10^n$, (*e*) $g(n) = 2^{\sqrt{n}}$, (*f*) $g(n) = 2^{\log(n)}$, and (*g*) $g(n) = 2^{(n^2)}$.

4. For each of the following statements identify the hypothesis H and the conclusion C, rewriting the statements if necessary:
(*a*) If $n \geq 2$, then $n \leq n^2 - 2$.
(*b*) A set with n elements has 2^{n-1} even subsets.
(*c*) x^n can be calculated with exactly $\log(n) + 1$ multiplications when n is a power of 2.
(*d*) $1 + 3 + \cdots + (2r - 1) = r^2$.
(*e*) All quadratic polynomials are $O(n^2)$.
(*f*) Every even composite number is the sum of two primes.
(*g*) Every even number has an even number of 1s in its binary representation.

5. Write the negation, the converse, and the contrapositive of each statement in the preceding problem.

6. (Another Analogy) Many serious hikers have a cardinal principle that they won't hike if they aren't wearing two pairs of socks. Break this principle up into statements H and C and rephrase it in the IF H, THEN C paradigm. Form the negation.

7. Form the contrapositive and the converse of the hikers and socks assertion that you created in the preceding exercise.

8. Each of the following allegations can be put into the form IF H, THEN C. After identifying the statements that are H and C, form the converse and the contrapositive of each.
(*a*) If n is even, then $n/2$ is an integer.
(*b*) If $n = p \cdot q$ (where $1 < p, q < n$), then n is not a prime.
(*c*) If r is odd and s is odd, then $r + s$ is odd.
(*d*) x^2 is divisible by 2 provided that x is divisible by 2.
(*e*) x an odd integer implies $(x + 1)/2$ an odd integer.
For each statement above, decide whether it is true or false. Do the same with the converses and the contrapositives.

9. Identify the IFF statements in the preceding problem, that is, those allegations for which both the statement and the converse are true.

10. Beginning with the assertion, "If $r > 2$, then $r^n = O(1)$," use definition 2 of big oh to reach a contradiction.

11. Prove Theorem 9.1 using definition 2 of big oh.

2:10 GOOD AND BAD ALGORITHMS

We have introduced a variety of algorithms, some correct, some incorrect, some efficient, some inefficient, some transparent, some quite complex. We have also begun to talk about what is known as the space and time complexity of an algorithm, that is, the number of variable memories needed and the time needed to run the algorithm (or at least an upper bound on the number of key operations like multiplication). In this section we shall formulate what is meant by a good algorithm.

Our description of "good" will depend upon the efficiency of the algorithm, but we must always remember that an algorithm must be correct to be "good" in any sense of the word. We can easily write efficient algorithms, like the following.

Algorithm SPEEDY

STEP 1. Stop.

This is the world's shortest algorithm and in that sense the most efficient; however, this algorithm has no relevance to any problem in the real world!

Space resources are important. We need enough space to read in the problem addressed by the algorithm, and we shall try to be conservative with additional memory needed for new variables. For example, in Section 2.1 we saw how we could interchange the values of two variables without using an additional memory location. However, our first priority will be to minimize running time and then secondly to minimize storage. The time requirement is a parameter commonly studied and one that leads to a rich and important theory in computer science.

We have already stated that we do not want to analyze time in terms of specific computers or programming languages. We want a procedure for comparing two algorithms that (correctly) solve the same problem. But we want a fair comparison. For example, if we run algorithm EXPONENT with $x = 2$ and $n = 3$, it certainly would be quicker than algorithm FASTEXP with $x = 10{,}000{,}000$ and $n = 123{,}456{,}789$, and yet we have claimed that the latter algorithm is faster or more efficient. Or think of just one algorithm. Of course, it runs faster when we enter few and small pieces of data rather than larger ones. For example, we can much more quickly list all subsets of a 2-set than of a 20-set.

Thus we must compare execution times on the same data set or at least on data sets of the same cardinality. We do this by introducing a parameter that measures the size of the data set. How we measure this size depends upon the problem at hand. For instance, we could describe the size of the data for the algorithms DtoB and BtoD by stating the number of digits in the given decimal or binary number. In SUBSET we could describe the size of the problem by giving the size of the set all of whose subsets we want to list. In EXPONENT and FASTEXP we could specify n, the power to which we are raising x, or the number of bits needed to store n.

Typically, we shall begin by supposing that the data or the input to the algorithm is of size n. This may mean that we have n bits of information or that the integer n is the crucial variable. Then we shall estimate the time of running the algorithm in terms of the variable n. Often we shall denote the time as some function $f(n)$ that counts the number of multiplications or the number of comparisons or the number of some time-consuming operation. This function will be known as the (time) **complexity** of the algorithm. Sometimes we shall be able to determine $f(n)$ explicitly; other times we shall make a worst-case analysis and get an upper bound on $f(n)$. Most frequently, we shall be happy to determine that $f = O(g)$ for some nice function g.

Our goal is simplistic: We want to divide correct algorithms into one pile called "good" and another pile called "bad." Just as philosophers who study language arrive at the meaning of the word *good* through comparisons, so shall we. Suppose that we have two correct algorithms to solve a particular problem, say A and B. Let $a(n)$ denote the complexity of A and $b(n)$ denote the complexity of B. Assume that we know $a = O(f)$ and $b = O(g)$. [Of course, if we know $a(n)$ and $b(n)$ explicitly we can compare them directly.]

Definition. If an algorithm A has complexity $a(n) = O(f(n))$ and an algorithm B has complexity $b(n) = O(g(n))$, we say that A **"appears to be" as efficient as** B if $f = O(g)$. If, in addition, $g = O(f)$ we say that A **"appears to be" equivalent** to B. Otherwise, if $f = O(g)$ and g is not $O(f)$ we say that A **"appears to be" more efficient** than B.

In practice, we shall replace "appears to be" with "is" in the above definitions. You might think that our definition is overly wishy-washy; however, restraint is forced on us by the big oh comparisons. The problem is that it might be difficult to bound the complexity of one or both of the algorithms in question. For example, we might be able to prove that $a(n) = O(n^3)$ and that $b(n) = O(n^4)$. In that instance we would say that A appears to be more efficient than B, since $n^3 = O(n^4)$ and n^4 is not $O(n^3)$. In reality, it might be the case that $a(n) = n^{11/4}$ while $b(n) = n^{5/2}$ in which case B is actually more efficient than A. Thus the quality of our comparison depends on how good a big oh estimate we have.

Example 10.1. Suppose that the time complexity function of algorithm A is given by $a(n) = n^2 - 3n + 6$ and that of algorithm B by $b(n) = n + 2$. Then A is a $O(n^2)$ algorithm while B is a $O(n)$ algorithm. In this case we would naturally say that B is more efficient than A.

Thus we measure the efficiency of algorithms using the hierarchy of functions developed in the preceding sections. An algorithm with complexity n^i will be more efficient than an algorithm with complexity n^j if and only if $i < j$. An $n \log(n)$ algorithm is more efficient than a quadratic algorithm but less efficient than a linear algorithm, since for n large enough,

$$n \leq n \log(n) \leq n\sqrt{n} \leq n^2.$$

Let's go back to our original question of what constitutes a good algorithm. Of course, if we have two different algorithms to solve the same problem, then we shall consider one better if it is more efficient in the sense described above. But we make a global judgment now about what is known in computer science circles as a "good" algorithm.

Definition. Suppose that A is a correct algorithm with complexity function $a(n)$. Then A is called **good** (or **polynomial**) if there exists a polynomial $p(n)$ with $a = O(p)$. A is called **bad** if a is not $O(p)$ for any polynomial p. If $a(n) \geq r^n$ for some constant r (with $r > 1$), then A is called **exponential**. A problem that does not have a good algorithmic solution is called **intractable**.

There are some important ideas here. An algorithm will be called good regardless of what polynomial gives the bound on $a(n)$. It might be that the polynomial is huge, and so the algorithm takes a long time to run. That would not be a very "good" situation from the point of view of efficiency, but in theory at least the situation is not as bad as having $a(n) = 2^n$, an exponential function.

Now we see the relevance of Theorem 9.1. It states that $r^n \neq O(p(n))$ for any polynomial $p(n)$. Thus if we find an algorithm with complexity 2^n or 3^n or even 1.000005^n, this is an exponential algorithm, not a polynomial one.

For example, look at the algorithm SUBSET. Our input to the algorithm is a set of n elements, and we want as output all 2^n subsets. The size of the input data is n. We systematically create and list all subsets of the n-set. Since there are 2^n subsets, our list making will need 2^n steps and the complexity will be at least 2^n, an exponential function. Thus SUBSET is an exponential algorithm, and the problem of listing all subsets of an n-set is intractable.

But why do we make such a harsh judgment about an exponential algorithm? As we all know from the news media, exponential growth is considered to be very fast growth into large numbers, perhaps dangerously out of control when related to, for example, population. Is this also the case for algorithms? Does this mean that exponential algorithms are too large or take too long? Surely, modern computers can handle large amounts of calculation extraordinarily quickly.

Let's do some arithmetic that is specific to an IBM PC but that would be approximately the same for any microcomputer. The PC can perform about 17,800 single-digit multiplications in a minute. Suppose that we compare seven algorithms that do multiplications and whose complexities are given by the functions $\log(n)$,

$\sqrt{n}$, n, n^2, n^3, 2^n, and 10^n. What size problems could we reasonably solve on the PC? Table 2.9 shows various values of n, the seven complexity functions, and roughly the amount of time needed to run the algorithms. (We stop filling in the table once the numbers become inhumanly large!)

Table 2.9

n	$\log(n)$	$\sqrt{n}$	n	n^2	n^3	2^n	10^n
8	0.01 sec	0.01 sec	0.027 sec	0.216 sec	1.73 sec	0.863 sec	3.9 days
16	0.013 sec	0.013 sec	0.054 sec	0.863 sec	13.8 sec	3.7 min	10,689 cent.
24	0.015 sec	0.017 sec	0.081 sec	1.94 sec	46.6 sec	15.7 hr	...
32	0.017 sec	0.019 sec	0.108 sec	3.45 sec	1.84 min	168 days	...
64	0.02 sec	0.027 sec	0.216 sec	13.8 sec	14.7 min	19,717,160 cent.	...
128	0.024 sec	0.038 sec	0.43 sec	55 sec	2 hr		
256	0.027 sec	0.054 sec	0.86 sec	3.7 min	16 hr		

Thus we see that we will be in trouble as soon as we have at least 24 pieces of data on which we must run an exponential algorithm. Of course, we could switch over to minicomputers, which typically run about 100 times as fast, but our problem size still demands too much computing with data of size 32 or larger. Thus there really is a problem when we must do an exponential amount of computing on even a moderate amount of material.

EXERCISES FOR SECTION 10

1. Suppose that algorithm A has complexity function $a(n)$ and algorithm B has complexity function $b(n)$. In each of the following cases decide whether the algorithms are equivalent or if not, which is more efficient.

(i) $a(n) = 36$. $\quad b(n) = 2n - 10$.
(ii) $a(n) = n^2$. $\quad b(n) = n$.
(iii) $a(n) = n^2$. $\quad b(n) = n - 6$.
(iv) $a(n) = 2n^2$. $\quad b(n) = 3n^2$.
(v) $a(n) = n^2 + 2n$. $\quad b(n) = n^2 + n$.

2. Explain why $n^n \neq O(r^n)$ for any constant r.

3. Find a function that is not $O(n^n)$.

4. In Exercise 1.7.6 you designed an algorithm that lists all $n(n-1)/2$ subsets of size 2 of a set A containing n elements. Suppose that you count as one step the formation and output of a single subset. Is your algorithm good or bad?

5. Suppose that A contains n objects and we have an algorithm that outputs all elements of the Cartesian product A^n. Counting one step as the formation and output of one element of the Cartesian product, is this algorithm good or exponential?
6. Show that the algorithm to settle the Satisfiability Problem for a Boolean function (see Section 1.10), which consists of trying all possibilities, is bad.

2:11 ANOTHER LOOK BACK

By far the most important ideas in this chapter have been the proof techniques of induction and contradiction. Their use permeates all of mathematics and computer science. Indeed the practitioners of these disciplines regularly apply these methods without acknowledgment. We have met typical instances of proofs using these techniques.

Induction proofs work well on set theory problems, on summation formulas, and on algorithmic problems, that is, in proving that an algorithm does what it is supposed to do in all cases. The Principle of Induction gives us a three-step format that allows us to set up and attack problems in a straightforward way. This doesn't mean that proof by induction will always be easy and automatic. We shall still need to think carefully and creatively about each instance.

Proofs by contradiction arise naturally for statements of the form, "Object A does not have property P." Often we can begin with the assumption that A has property P, do some algebra or a logical argument, and arrive at a contradiction. Often some experimentation is needed to find a reasonable argument, leading to, say, $0 = 1$. Probably, the best advice for both kinds of proofs is to study the examples and theorems in the text and to try to imitate these in the exercises.

This chapter also presents the central computer science and mathematical ideas about the analysis of algorithms. We now have a sequence of tasks to perform when we look for an algorithmic solution to a problem. Not only must we come up with the algorithm, but we must also prove that it is correct in all cases and we must analyze its time and space requirements. The worst-case analysis is commonly used, since we can often estimate an upper bound on the maximum number of the most time-consuming steps. Notice that these upper bounds may be crude, too large, and it may be that our algorithm works more efficiently.

Of course, even if we know an exact analysis of the number of steps performed, we have no guarantee that there isn't a faster algorithm. For example, how do we know that computing x^n can't be done by an algorithm faster than FASTEXP, that is, by an algorithm that is faster than logarithmic? We haven't discussed this issue at all. There are other ways to analyze algorithms. For example, we might ask for typical or average-case behavior. Or we might be more demanding and count all kinds of steps: multiplication, division, addition, subtraction, comparisons, assignment statements, and so on. Such details are suitable for more advanced courses in theoretical computer science.

Most algorithms presented in the computer science literature include a discussion of worst-case behavior, and this discussion inevitably entails use of the big oh jargon. The big oh definitions are subtle yet important; it is well worth doing lots of exercises on this concept. It is the primary way that computer scientists and mathematicians distinguish between the quality of algorithms. There are a number of important problems (including searching and sorting, which we will consider in Chapter 6, and the fast Fourier transform, which we will not consider) that have naive algorithms that are $O(n^2)$ and better algorithms that are $O(n\log(n))$. The existence of the better algorithms greatly expands the size of the problems that are feasible to solve. One further satisfactory point about the $O(n\log(n))$ sorting algorithms is that we can also prove that every algorithm (of a certain type) that solves a sorting problem must use at least $cn\log(n)$ steps for some constant c. That tells us that we've found essentially the very best algorithm.

In the SUBSET problem and algorithm we realized that every algorithm must form and list all 2^n subsets, and so is necessarily exponential. We have a lower bound on the number of steps needed; often it is difficult to obtain such lower bounds. The distinction between polynomial and exponential algorithms is the most actively researched area within computer science. We shall soon see many problems for which every known algorithm is exponential, but no one has proved that there is no polynomial algorithm. (The Satisfiability Problem of Section 1.10 is one such example.) Thus upper bounds, but not lower bounds, are known, and it is not yet known whether the problems are inherently intractable. This leads to the heart of some fascinating, unsolved problems in computer science.

SUPPLEMENTARY EXERCISES FOR CHAPTER 2

1. Show that if there is a counterexample to Fermat's Last Theorem, then there exists a counterexample in which no pair of the integers x, y, and z has a common factor. Use this fact to show how large z must be in any counterexample.
2. It was claimed that Fermat's Last Theorem is known to be true for $n \leq 125{,}080$. From the previous problem you know that z can't be 1 (for instance). Suppose that you wanted to write down the digits in this hypothesized counterexample. If you could write 100 digits a minute for the rest of your life (with no resting time), would you be able to transcribe this example?
3. (*a*) The distinguished mathematician Gauss was a child prodigy. Legend has it that at the age of 10 he was asked to add the integers $81{,}297 + 81{,}495 + 81{,}693 + \cdots + 100{,}899$. Almost immediately he wrote one number, 9,109,800, on his slate. Was he correct?

(*b*) Let a and b be constants. Show that the sum of the arithmetic progression $a + (a + b) + (a + 2b) + \cdots + (a + nb)$ is given by $(n + 1)(a + nb/2)$.

4. Prove by induction that if b is a number such that $1 \leq b \leq n(n + 1)/2$, then there is a subset of $\{1, 2, \ldots, n\}$ whose sum equals b.

5. Prove by induction that if $a \neq b$, then

$$a^n + a^{n-1} \cdot b + \cdots + a^j \cdot b^{n-j} + \cdots + b^n = \frac{a^{n+1} - b^{n+1}}{a - b}.$$

6. Consider the following algorithm:

STEP 1. Input r, s {r and s positive integers}
STEP 2. While $s > 0$ do
Begin
STEP 3. $r := r - s$
STEP 4. $s := s - 1$
End {step 2}
STEP 5. Output r, then stop.

(*a*) Trace through this algorithm when 12 and 3 are input for r and s, respectively.
(*b*) Show that this algorithm must terminate no matter what positive integer values of r and s are input.
(*c*) What is the output in terms of r and s?

7. Given the sum $t + 2(t - 1) + \cdots + j(t - j) + \cdots + t$, guess a formula in terms of t for this sum. Use induction to prove that your guess is correct.

8. Determine the sums: 1^2, $2^2 - 1^2$, $3^2 - 2^2 + 1^2$, and $4^2 - 3^2 + 2^2 - 1^2$. Then deduce and prove a general formula for the sum

$$n^2 - (n - 1)^2 + (n - 2)^2 - \cdots + (-1)^{n+1} \cdot 1^2.$$

9. Your calculator probably has logarithms to the base 10 or to the base e. Here's how to change from base 10, written $\log_{10}(x)$, to base 2, $\log(x)$:

$$\log(x) = \frac{\log_{10}(x)}{\log_{10}(2)} = 3.3219 \log_{10}(x).$$

Prove the above identity. [*Hint:* Begin with $x = 2^{\log(x)}$.]

10. Find a formula for changing natural logarithms, logs to the base e, to base 2 logarithms. Justify your formula.

11. Count the number of comparisons made in the algorithm MAX of Exercise 4.12 and in BUBBLES in 4.13. Which is more efficient? Count the number of assignment statements in the two algorithms. Which is more efficient from the assignment point of view? Suppose that an assignment can be done in 1 second and a comparison in 2 seconds, then which algorithm is more efficient?

12. Suppose that you have a computer that can only perform addition, subtraction, and multiplication. Write an algorithm that upon input x, a positive real number, calculates and outputs $\lfloor \sqrt{x} \rfloor$.

13. Suppose that you have a computer that can perform addition, subtraction, and multiplication, but not division. Design an algorithm that upon input of real numbers x and y (with $x \neq 0$) calculates and outputs $\lfloor y/x \rfloor$.

14. Show that if $f = O(g)$, then $2^f = O(2^g)$. Is the converse true? Give a proof or counterexample.

15. Here is one of the oldest and most famous results in all of mathematics.

Theorem. There are infinitely many primes.

(*a*) What is the negation of this theorem?
(*b*) If the theorem is false, suppose that t equals one plus the product of all the primes. Is t prime?
(*c*) Is the t you formed in part (b) divisible by 2? 3? 5? Any prime?
(*d*) Prove the Theorem by contradiction.

16. Every positive integer can be expressed in the form $3n$, $3n + 1$, or $3n + 2$ for some integer n.
(*a*) Explain why every prime number greater than 3 is of the form $3n + 1$ or $3n + 2$.
(*b*) Explain why a number of the form $3n + 2$ must have a prime divisor of the form $3n + 2$.
(*c*) Prove that there are infinitely many primes of the form $3n + 2$. [*Hint*: Suppose that $p_1, p_2, \ldots, p_k$ are all the primes of the form $3n + 2$. Then consider $3(p_1 \cdot p_2 \cdot \cdots \cdot p_k) + 2$.]

17. Here we prove that $(x + 1)^2$ does not equal $x^2 + 2x + 1$. The proof is by contradiction, so we begin by assuming the negation:

$$(x + 1)^2 = x^2 + 2x + 1.$$

We subtract $2x + 1$ from both sides to obtain

$$(x + 1)^2 - (2x + 1) = x^2.$$

Next we subtract $x(2x + 1)$ to get

$$(x + 1)^2 - (2x + 1) - x(2x + 1) = x^2 - x(2x + 1),$$

which by factoring becomes

$$(x+1)^2 - (x+1)(2x+1) = x^2 - x(2x+1).$$

Next we add $(2x+1)^2/4$ to both sides:

$$(x+1)^2 - (x+1)(2x+1) + (2x+1)^2/4$$
$$= x^2 - x(2x+1) + (2x+1)^2/4.$$

Since both sides are perfect squares, we can factor them into

$$\left[(x+1) - \frac{(2x+1)}{2}\right]^2 = \left[x - \frac{(2x+1)}{2}\right]^2.$$

Taking square roots of both sides, we obtain

$$(x+1) - \frac{(2x+1)}{2} = x - \frac{(2x+1)}{2}$$
$$(x+1) = x$$
$$1 = 0.$$

This contradiction forces us to conclude that $(x+1)^2$ does not equal $x^2 + 2x + 1$. What is wrong with this proof?

18. Prove Theorem 9.1 for arbitrary $r > 1$.

19. There are functions $f(n)$ such that $f(n) \neq O(p(n))$ for any polynomial $p(n)$ and $f(n) < r^n$ for every positive r and sufficiently large n. Find such a function.

20. Here is another exponentiation algorithm.

STEP 0. Input x and m {m a positive integer}
STEP 1–6. Algorithm DtoB {assume that the output is in string s, a $(j+1)$-bit binary number}
STEP 7. ans := x
STEP 8. For $i := (j-1)$ down to 0 do
  STEP 9. If the ith entry of s is 1,
    then set ans := x * ans * ans
    Otherwise, set ans := ans * ans
STEP 10. Output ans and stop.

[COMMENT: "For $i := (j-1)$ down to 0 do" means that i successively takes on the values $(j-1), (j-2), \ldots, 0$ and for each value performs step 9.] Run this algorithm on some data to check that it correctly calculates x^m for m, a positive integer. Then compare the efficiency of this algorithm with that of FASTEXP.

3

ARITHMETIC OF SETS

3:1 INTRODUCTION

College Hall has decided to automate the internal mail system. Although the administrators communicate easily by electronic mail, they also need to circulate documents, such as memos, reports, and minutes of meetings, among their offices. They have hired an outside consultant who suggests that the flow of paper will be improved by the use of robotlike mail carts, called Mailmobiles. These carts can be programmed to travel through office corridors, to pause at designated points (or if the bumper hits anything!), and to stop at a location where they can be reprogrammed for another journey. Although most college administration buildings are irregularly shaped, let's think about a simple floor plan, the rectangular grid shown in Figure 3.1. We have marked the Mail Room with M and the President's Office with P. Each line segment indicates a corridor, and the corridor intersections are possible stopping points for the mailmobile.

Here are some questions that occur to the consultant while planning for the mailmobile.

1. A shortest trip from M to P requires the traversal of 11 corridors. In how many different ways can the mailmobile make a trip of shortest length from M to P?
2. Is it possible to visit every stopping point exactly once on a trip (necessarily of longer length) from M to P? If so, in how many different ways can such a trip be planned?
3. Is it possible to travel along every corridor exactly once on a trip from M to P? If so, in how many different ways can such a trip be planned?

Figure 3.1 A 6×5 rectangular grid.

Question 1.1. Answer the preceding three questions on the 3×2 rectangular grid in Figure 3.2.

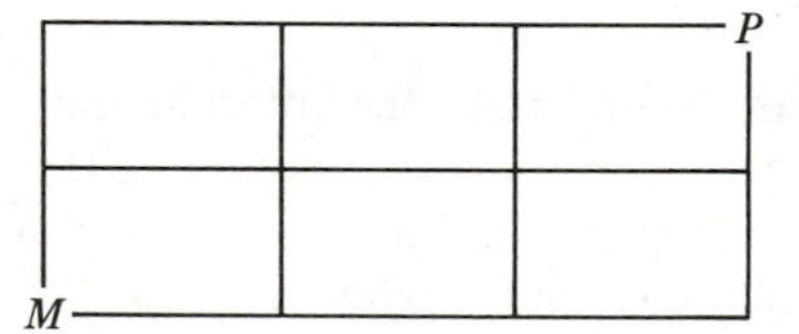

Figure 3.2 A 3×2 rectangular grid.

In this chapter we focus on the first of the three questions posed above. The second and third questions will be studied in depth in Chapter 8.

Let's consider some of the shortest trips between M and P in Figure 3.1. Such a trip is shortest if and only if it covers exactly 11 corridors. Think of Figure 3.1 as being a map, oriented with north at the top of the figure. Then we could describe a shortest trip from M to P, for example, by

$$\text{E, E, E, E, E, E, N, N, N, N, N,}$$

where N stands for moving north along a corridor and E for moving east. Two other shortest trips are

$$\text{E, E, E, E, E, N, E, N, N, N, N,} \qquad \text{and} \qquad \text{N, N, N, N, E, E, E, E, E, E, N.}$$

In fact, to move from M to P in a shortest path we must travel six units east and five units north. Furthermore if we write down any sequence of five Ns and six Es, then these give instructions for a shortest path from M to P.

Question 1.2. Why does a sequence of five Ns and six Es always stay inside the rectangle of Figure 3.1? Why does such a sequence describe a path that always reaches P, starting at M? Describe all sequences of Ns and Es that correspond with a trip from M to P in Figure 3.2.

One way to describe these shortest paths from M to P in Figure 3.1 is as a subset of $\{N, E\}^{11}$ (recall the Cartesian product from Chapter 1). Specifically, the set of all shortest paths corresponds with the subset consisting of all of the 11-tuples that contain exactly five Ns.

Has the introduction of the Ns and Es helped us count the number of shortest paths from M to P or to find these paths? So far, not at all! We have only found another way to look at the problem. We shall construct an algorithm to list all suitable N and E sequences; however, there is a simple formula for the number of shortest paths, to which the work in this chapter will lead.

The diagrams in Figures 3.1 and 3.2, although they may not be representative of building or office floor plans, are ones that arise repeatedly in mathematics and computer science. A grid that has $(m + 1)$ vertical lines and $(n + 1)$ horizontal lines is known as an ***m* × *n* rectangular grid**; a picture of the $m \times n$ grid is given in Figure 3.3. This grid might represent streets in a city (like New York); it might represent the intricate connections on layers of silicon on a computer chip; or it might represent the pixels on a computer monitor. Thus this configuration is studied for a variety of reasons.

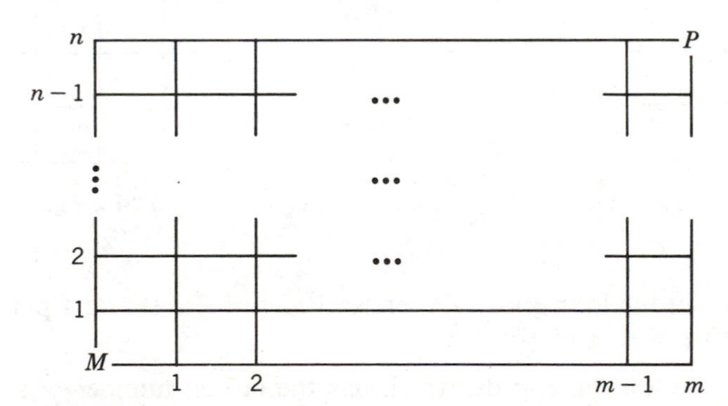

Figure 3.3

Question 1.3. Estimate the number of shortest paths from M to P in Figure 3.1: Choose the interval from the following list that most likely contains the correct number.
(a) Less than 25.
(b) Between 25 and 50.

(c) Between 50 and 100.
(d) Between 100 and 250.
(e) Between 250 and 500.
(f) Between 500 and 1000.
(g) Between 1000 and 1500.
(h) More than 1500.

EXERCISES FOR SECTION 1

1. Here are four rectangular grids. For each draw all possible shortest paths from M to P.

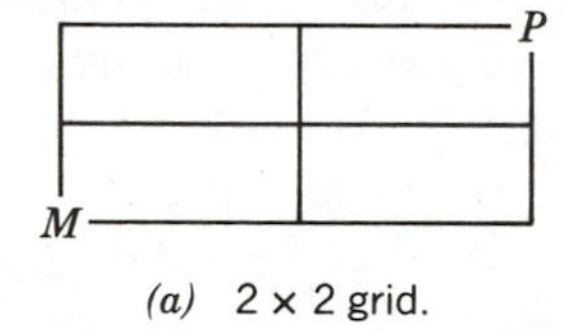

(a) 2×2 grid.

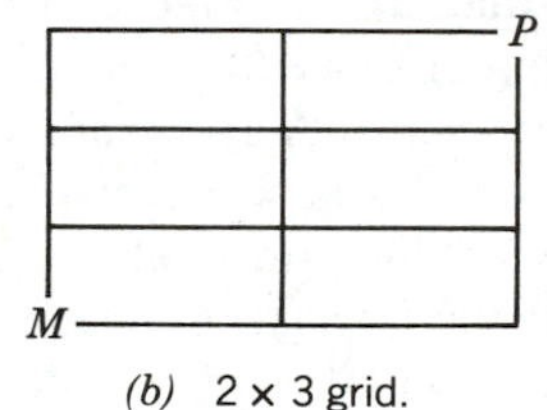

(b) 2×3 grid.

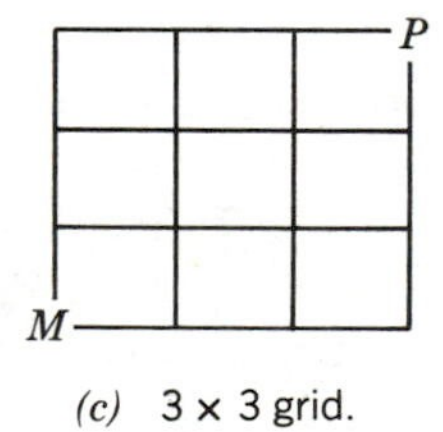

(c) 3×3 grid.

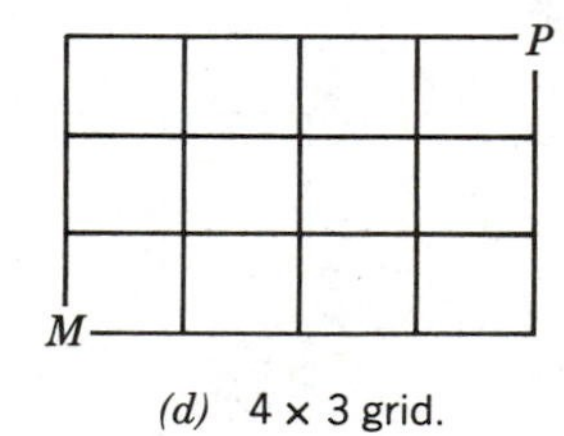

(d) 4×3 grid.

2. For each of the four grids in Exercise 1, describe a shortest path from M to P in terms of Es and Ns.

3. Among the following grids, which has the largest number of shortest paths from M to P? 1×5, 2×4, 3×3, 4×2, and 5×1.

4. Explain why the number of shortest paths from M to P in an $m \times n$ grid is the same as the number of shortest paths from M to P in an $n \times m$ grid.

5. What are the dimensions of the rectangular grid on which the sequence N, N, E, E, N, E, E, N, N, E, N, N, E gives a path from M to P?

6. Check that the number of sequences of Ns and Es of length six containing exactly two Ns equals the number of six-digit binary numbers with exactly two ones. Is the same result true if there are three Ns and if the binary numbers have three ones? Is the same true if the sixes in these statements are all changed to sevens? Explain why.

7. In the four grids of Exercise 1, find a path from M to P that passes through every intersection point exactly once, or else deduce that there is no such path. In which grids is there more than one such path?
8. Make a conjecture about what values of m and n are such that an $m \times n$ grid contains a path from M to P that passes through every intersection point exactly once.
9. Explain why in the grids of Exercise 1 it is impossible to travel from M to P traveling along every corridor (or line segment) exactly once. Are there values of m and n for which such a path exists?
10. Show that an $m \times n$ rectangular grid with M at the lower left corner and P at the upper right corner contains no more than 2^{m+n} shortest paths from M to P. When $n = 1$, determine exactly how many shortest paths there are from M to P.

3:2 BINOMIAL COEFFICIENTS

This section introduces the important factorial function and a counting device from the seventeenth century known as Pascal's triangle. With these we can readily calculate the number of shortest paths from the lower left-hand corner to the upper right-hand corner of any $m \times n$ grid. In addition, the factorial function and Pascal's triangle lead to the study of subsets of sets and related algorithms.

We return to rectangular grids. Figure 3.4 presents the $m \times n$ grid with each point in the grid labeled with its Cartesian coordinates. The point M is placed at the origin $(0,0)$ and P lies at the point (m,n).

Next we define a function f whose domain is the set of points of the grid; let $f(i,j)$ equal the number of shortest paths from $M = (0,0)$ to (i,j). Although we were originally looking only for the value $f(P)$, we'll find this number using the other function values.

Let's figure out some values of f. We say that $f(0,0) = 1$ because there is only one shortest way to travel from M to M, that is, by doing nothing! Now $f(1,0) = f(0,1) = 1$, and $f(1,1) = 2$, since we can reach $(1,1)$ via $(0,1)$ or $(1,0)$.

Question 2.1. (a) Show that $f(0,2) = f(2,0) = 1, f(1,2) = f(2,1) = 3$, and $f(2,2) = 6$. (b) Determine the values of $f(0,3)$, $f(1,3)$, $f(2,3)$, $f(3,2)$, $f(3,1)$, and $f(3,0)$.

Question 2.2. Explain why in the $m \times n$ rectangular grid $f(i,0) = 1$ for $i = 0, 1, \ldots, m$ and $f(0,j) = 1$ for $j = 0, 1, \ldots, n$.

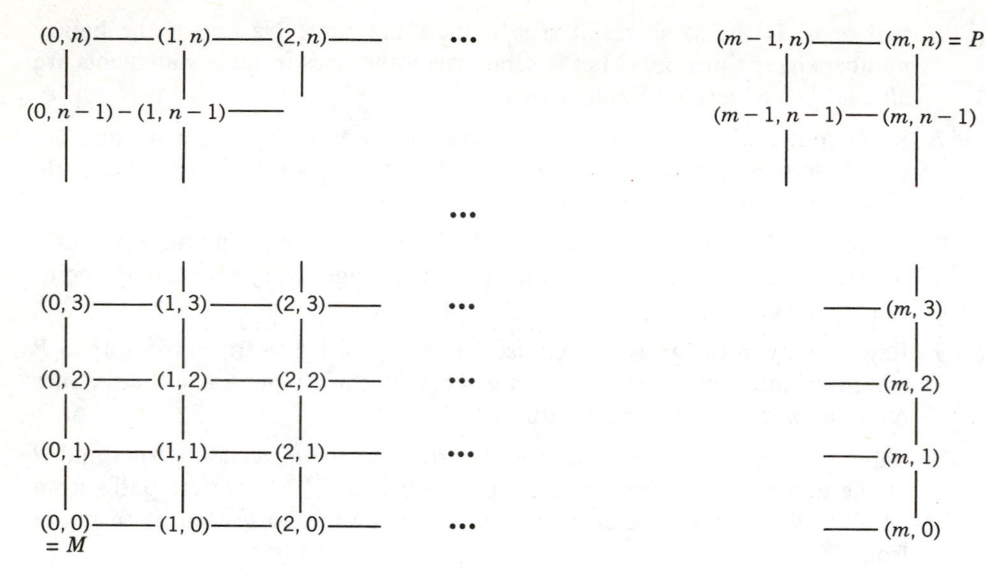

Figure 3.4 An $m \times n$ grid.

Consider a point (i,j) some place out in the middle of the grid shown in Figure 3.5. There are exactly two ways to arrive at (i,j) along a shortest path. Either we approach it from the point to the left $(i-1,j)$ or from the point below $(i,j-1)$. Thus the number of shortest paths from $(0,0)$ to (i,j) is the number of shortest paths to $(i-1,j)$ plus the number of shortest paths to $(i,j-1)$, that is,

$$f(i,j) = f(i-1,j) + f(i,j-1).$$

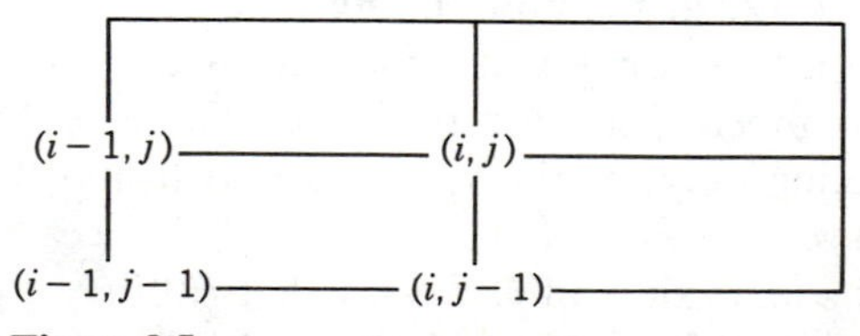

Figure 3.5

This relationship provides a method for calculating all the f values in a given grid. We know from Question 2.2 that $f(i,0) = f(0,j) = 1$ for all i and j. From these initial values we can fill in the remaining values, moving from the lower left up to the upper right:

$$f(1,1) = f(0,1) + f(1,0) = 1 + 1 = 2$$
$$f(1,2) = f(0,2) + f(1,1) = 1 + 2 = 3$$
$$f(2,1) = f(1,1) + f(2,0) = 2 + 1 = 3$$
$$f(1,3) = f(0,3) + f(1,2) = 1 + 3 = 4$$
$$f(2,2) = f(1,2) + f(2,1) = 3 + 3 = 6,$$

and so on. Eventually, $f(m, n)$ can be computed as the sum of $f(m-1, n)$ and $f(m, n-1)$.

Question 2.3. Show that $f(3, 3) = 20$ and that $f(4, 2) = 15$.

This process would be tedious on the 6×5 grid and hopeless for much larger grids. We haven't yet found the promised simple formula for the number of shortest paths.

We now uncover Pascal's famous triangle. Suppose that we redraw Figure 3.4, omitting the coordinate labels and the line segments and writing instead the values of f at each point [see Figure 3.6(*a*)]. Next we rotate the figure 135° ($= 3\pi/4$ radians) clockwise so that M is at the top and P at the bottom. Then Pascal's triangle emerges as shown in Figure 3.6(*b*). A larger version appears in Figure 3.7.

```
1                               1
1  4  10  20                  1   1
1  3   6  10                1   2   1
1  2   3   4              1   3   3   1
1  1   1   1  1         1   4   6   4   1

      (a)                      (b)
```

Figure 3.6 Pascal's triangle

The numbers in Pascal's triangle are organized by rows; numbers on the kth horizontal row (starting with $k = 0$) are said to form the kth row of Pascal's triangle. For example, here are the first rows:

the 0th row: 1,
the 1st row: 1 1,
the 2nd row: 1 2 1,
the 3rd row: 1 3 3 1,

and

$$\text{the 4th row:} \quad 1 \quad 4 \quad 6 \quad 4 \quad 1.$$

Exactly as in the shortest path problem (once we account for the rotation), each row of **Pascal's triangle** begins and ends with a 1, and in the middle the entries are the sum of the two numbers immediately above.

Question 2.4. Calculate the 5th row of Pascal's triangle. Then determine the coordinates of all points in the $m \times n$ grid that end up, after the 135° rotation, on the 5th row of Pascal's triangle. Compare their f values with the entries of the 5th row.

We want to determine explicitly the numbers that lie on the kth row of Pascal's triangle for an arbitrary positive integer k. These in turn will give us the f values needed in the shortest path problem. To do this, we introduce the **factorial function**: For each natural number n, we define $n!$, read "n factorial," by

$$n! = n(n-1)(n-2)\cdots 3\cdot 2\cdot 1 \qquad \text{if } n \text{ is positive,}$$

and

$$0! = 1.$$

The definition of 0! may be surprising, but as you'll see, it's useful to have 0! defined in this way. From the definition we see that

$$1! = 1, \quad 2! = 2\cdot 1 = 2, \quad \text{and } 3! = 3\cdot 2\cdot 1 = 6.$$

Question 2.5. Calculate $n!$ for $n = 4, 5, 6, 7$, and 8. Then find a value of n such that $n!$ is greater than 1,000,000.

Notice that the values of $n!$ grow rapidly as n increases.

Here's a useful property of n factorial that follows immediately from the definition:

$$n! = n[(n-1)!].$$

Now we can solve many mysteries of Pascal's triangle and of shortest paths within a grid. We define the **binomial coefficients** $\binom{k}{i}$, read "k choose i" or "k above i" or "k pick i" for all natural numbers k and i with $0 \le i \le k$ by

$$\binom{k}{i} = \frac{k!}{i!(k-i)!}.$$

1
1 1
1 2 1
1 3 3 1
1 4 6 4 1
1 5 10 10 5 1
1 6 15 20 15 6 1
1 7 21 35 35 21 7 1
1 8 28 56 70 56 28 8 1
1 9 36 84 126 126 84 36 9 1
1 10 45 120 210 252 210 120 45 10 1
1 11 55 165 330 462 462 330 165 55 11 1

Figure 3.7 Pascal's Triangle, Rows 0–11.

Example 2.1. Here are a few binomial coefficient values:

$$\binom{2}{0} = \frac{2!}{0! \cdot 2!} = \frac{2}{1 \cdot 2} = 1 = \frac{2!}{2! \cdot 0!} = \binom{2}{2}$$

and

$$\binom{2}{1} = \frac{2!}{(1! \cdot 1!)} = 2.$$

Question 2.6. Determine the following binomial coefficients

$$\binom{3}{0}, \binom{3}{1}, \binom{3}{2}, \binom{3}{3}.$$

Then determine all binomial coefficients of the form

$$\binom{4}{i} \quad \text{and} \quad \binom{5}{j}$$

for $0 \leq i \leq 4$ and $0 \leq j \leq 5$.

Question 2.7. Explain why the following facts are true for all positive k.

$$\binom{k}{0} = \binom{k}{k} = 1 \quad \text{and} \quad \binom{k}{k-1} = \binom{k}{1} = k.$$

Notice that the assignment of the value 1 to 0! allows $\binom{k}{0}$ and $\binom{k}{k}$ to be defined.

Theorem 2.1. The nth row of Pascal's triangle consists of the binomial coefficients

$$\binom{n}{0}, \binom{n}{1}, \binom{n}{2}, \dots, \binom{n}{i}, \dots, \binom{n}{n-1}, \binom{n}{n}.$$

Proof. The proof is by induction on n. Some of the rows of the Pascal triangle are listed in Figure 3.7 and for $n < 6$ agree with the results of Example 2.1 and Question 2.6. Thus we assume that the kth row of Pascal's triangle consists of the binomial coefficients $\binom{k}{i}$ for $i = 0, 1, \dots, k$. In Figure 3.8 we display this row and label the unknown values in the $(k + 1)$st row with variables $x_1, x_2, \dots, x_k$, whose values we must now determine.

$$\begin{matrix} & \binom{k}{0} & & \binom{k}{1} & & \binom{k}{2} & \cdots & \binom{k}{i-1} & & \binom{k}{i} & \cdots & \binom{k}{k-1} & & \binom{k}{k} & \\ 1 & & x_1 & & x_2 & & x_3 \quad \cdots & & x_i & & x_{k-1} & & x_k & & 1 \end{matrix}$$

Figure 3.8

We know that the zeroth and $(k + 1)$st entries of the $(k + 1)$st row equal 1. We also know that each entry equals the sum of the two numbers above. Thus x_1 must equal the sum

$$\binom{k}{0} + \binom{k}{1} = 1 + k = \binom{k+1}{1}.$$

Question 2.8. Show that $x_2 = \binom{k+1}{2}$.

In general, we see that

$$\begin{aligned} x_i = \binom{k}{i-1} + \binom{k}{i} &= \frac{k!}{(i-1)!(k-i+1)!} + \frac{k!}{i!(k-i)!} \\ &= \frac{k!}{(i-1)!(k-i)!} \cdot \left[\frac{1}{k-i+1} + \frac{1}{i} \right] = \frac{k!}{(i-1)!(k-i)!} \cdot \frac{i + (k-i+1)}{i(k-i+1)} \\ &= \frac{(k+1)!}{i!(k-i+1)!} = \binom{k+1}{i}. \end{aligned}$$

Thus the ith element in the nth row of Pascal's triangle is $\binom{n}{i}$. □

Corollary 2.2. $\binom{k+1}{i} = \binom{k}{i-1} + \binom{k}{i}$ for all $1 \le i \le k$.

Proof. This was exactly the central calculation in the induction proof of Theorem 2.1. □

Why do we want to know all these facts about rows of the Pascal triangle? First of all, they provide the solution to the shortest path problem. We began this section with an effort to calculate $f(P)$, the number of shortest paths from M to P in the 6×5 rectangular grid. The point P lies on the 11th row of Pascal's triangle and is the 6th entry, counting beginning with 0. Thus by Theorem 2.1,

$$f(P) = \binom{11}{6} = \frac{11!}{6! \cdot 5!} = \frac{11 \cdot 10 \cdot 9 \cdot 8 \cdot 7}{5 \cdot 4 \cdot 3 \cdot 2 \cdot 1} = 462.$$

In the case of the general $m \times n$ rectangular grid, the point $P = (m, n)$ lies on the $(m+n)$th row of the Pascal triangle and is the mth entry. Thus in this case

$$f(P) = \binom{m+n}{m} = \frac{(m+n)!}{m!\,n!},$$

an explicit formula for the number of shortest paths.

Question 2.9. How many shortest paths are there from $(0, 0)$ to $(4, 3)$ in a rectangular grid?

Look back at Pascal's triangle in Figure 3.7 and notice that there is a great deal of symmetry in the binomial coefficients. For a fixed value of k the numbers $\binom{k}{0}, \binom{k}{1}, \ldots, \binom{k}{i}, \ldots, \binom{k}{k}$ begin with 1, first increase as i increases, and then decrease through the same values back to 1. In other words, the second half of this sequence is a mirror image of the first half. The largest value occurs in the middle.

Theorem 2.3. $\binom{k}{i} = \binom{k}{k-i}$ for all $i = 0, 1, \ldots, k$.

Proof.

$$\binom{k}{i} = \frac{k!}{i!(k-i)!} = \frac{k!}{(k-i)!i!}$$

$$= \frac{k!}{(k-i)!(k-(k-i))!} = \binom{k}{k-i}.$$ □

The largest value of $\binom{k}{i}$ for fixed k occurs when $i = k/2$ if k is even and when $i = (k-1)/2$ and $(k+1)/2$ if k is odd.

Theorem 2.4. Given natural numbers k and i with $i \leq (k-1)/2$,

$$\binom{k}{i} \leq \binom{k}{i+1}.$$

Proof. Given $i \leq (k-1)/2$, we multiply both sides by 2 to get $2i \leq k-1$. Next we add 1 and subtract i to get $i+1 \leq k-i$. Taking reciprocals, we get

$$\frac{1}{k-i} \leq \frac{1}{i+1}. \tag{*}$$

Now

$$\binom{k}{i} = \frac{k!}{i!(k-i)!} = \frac{1}{(k-i)} \cdot \frac{k!}{i!(k-i-1)!}$$

which after substituting (*) yields

$$\binom{k}{i} \leq \frac{1}{(i+1)} \cdot \frac{k!}{i!(k-i-1)!} = \binom{k}{i+1}. \qquad \square$$

Corollary 2.5. For natural numbers k and j with $j \geq (k-1)/2$

$$\binom{k}{j} \geq \binom{k}{j+1}.$$

Proof. Since $j \geq (k-1)/2$, $(k-j-1) \leq k-(k-1)/2-1 = (k-1)/2$. Applying Theorem 2.4 with $i = k-j-1$, we have

$$\binom{k}{k-j-1} \leq \binom{k}{k-j}.$$

Applying Theorem 2.3, we get

$$\binom{k}{j+1} \leq \binom{k}{j}. \qquad \square$$

Theorem 2.4 and Corollary 2.5 combine to tell us that the largest value among

$$\binom{k}{0}, \binom{k}{1}, \ldots, \binom{k}{i}, \ldots, \binom{k}{k}$$

occurs at

$$\binom{k}{\lfloor k/2 \rfloor}.$$

Theorems 2.3, 2.4, and Corollary 2.5 are attractive and interesting, and they are also useful. We need these results for the analysis of the complexity of the next major algorithm, which generates all k-subsets of an n-set.

EXERCISES FOR SECTION 2

1. Fill in the f value for every point on the 4×4 rectangular grid.
2. In rotating the rectangular grid into Pascal's triangle, the vertical and horizontal rows of the grid become diagonals in the triangle. Even more important, it is the diagonals from the point $(0, i)$ to the point $(i, 0)$ that become the rows of Pascal's triangle. Determine the coordinates of all the points in the grid that end up on the 3rd row of Pascal's triangle. Then do the same for the 4th row.
3. Find the integer coordinates of all the points in the rectangular grid that end up on the kth row of Pascal's triangle. On what row of Pascal's triangle does the point (i, j) end up?
4. Is the following statement true or false: Every number on the kth row of Pascal's triangle, except for the first and last, is divisible by k. If true, explain why. If false, characterize the rows for which it is a true statement.
5. Calculate the sum of the kth row of the Pascal triangle for $k = 3$, 4, 5, and 6. Formulate a conjecture about the sum of the kth row of the triangle.
6. Calculate the following binomial coefficients:

$$\binom{6}{3}, \quad \binom{7}{3}, \quad \binom{18}{2}, \quad \binom{18}{15}, \quad \binom{13}{3}, \quad \binom{97}{1}, \quad \binom{100}{2}.$$

7. Verify Corollary 2.2 by calculating the following sums and checking that the corollary is correct in these cases:

$$\binom{5}{3} + \binom{5}{4}, \quad \binom{6}{2} + \binom{6}{3}, \quad \binom{8}{4} + \binom{8}{5}.$$

8. Find the largest binomial coefficient of the form $\binom{14}{i}$ and $\binom{15}{j}$.
9. The number 10 equals at least four different binomial coefficients:

$$\binom{10}{1}, \quad \binom{10}{9}, \quad \binom{5}{2}, \quad \text{and} \quad \binom{5}{3}.$$

Are there any other values of k and j such that $10 = \binom{k}{j}$? Find a number $n > 1$ that equals at least five different binomial coefficients, and then find a number k that equals only the two binomial coefficients $\binom{k}{1}$ and $\binom{k}{k-1}$.

10. Construct an algorithm FACTORIAL that upon input of a positive integer j, calculates and outputs $j!$.

11. How many zeros does 15! end in? Find the smallest value of $n!$ that is divisible by 1,000,000.

12. Notice that $2 = 2!$, $3 = 2! + 1!$, $4 = 2 \cdot 2!$, and $5 = 2 \cdot 2! + 1!$. Express 6, 7, 8, 9, and 10 as sums of multiples of factorials in the following form:

$$n = a_j \cdot j! + a_{j-1} \cdot (j-1)! + \cdots + a_1 \cdot 1!$$

where $a_j, a_{j-1}, \ldots, a_1$ are integers satisfying $a_i \le i$ for $i = 1, 2, \ldots, j$. This form is called the **factorial representation** of the integer n. (See Supplementary Exercises 8 and 9.)

13. Determine whether the following are true or false. Give reasons for your answers.

(*a*) $n! = O((n+1)!)$
(*b*) $n! + 5 = O(n!)$
(*c*) $n! = O((n-1)!)$
(*d*) $n! + (n-1)! = O(n!)$
(*e*) $1! + 2! + 3! + \cdots + n! = O(n!)$.

14. If p is a prime, show that each of the following numbers is a composite number. (A number that is not prime is called **composite**.)

$$p! + 2,\ p! + 3, \ldots, p! + p.$$

Describe how to obtain 1000 consecutive composite numbers. If p is not a prime, can you use the same construction to obtain m consecutive composites for any integers p and m?

15. Prove Theorem 2.3 using the grid path formulation of binomial coefficients.

16. Each row of Pascal's triangle is called **symmetric** because the ith entry equals the $(k-i)$th. Each row is called **unimodal**, since the numbers increase to a maximum and then decrease. Find an example of a sequence of 10 numbers that is symmetric but not unimodal. Then find an example of a sequence of 10 numbers that is unimodal but not symmetric.

17. Explain why

$$\binom{k}{i} = \binom{k-2}{i} + 2\binom{k-2}{i-1} + \binom{k-2}{i-2}.$$

Then express $\binom{k}{i}$ in terms of binomials coefficients of the form "$k-3$ choose something." Interpret these identities in the grid.

18. How many shortest paths proceed from $(0,0)$ to $(6,9)$ in the rectangular grid? How many of these go through the point $(4,5)$?

19. How many shortest paths proceed from $(0,0)$ to (m,n) through the point (r,s) if $0<r<m$ and $0<s<n$?

20. Suppose that k is odd. Explain why Theorem 2.3 and Corollary 2.4 imply that

$$\binom{k}{\frac{(k-1)}{2}} = \binom{k}{\frac{(k+1)}{2}} \geq \binom{k}{i} \qquad \text{for all } 0 \leq i \leq k.$$

Then when k is even, explain why for all $0 \leq i \leq k$

$$\binom{k}{k/2} \geq \binom{k}{i}.$$

3:3 SUBSETS OF SETS

This section considers the problems of counting and generating all the j-subsets of an n-set.

Example 3.1. Suppose that we want to know how many 3-subsets there are in a 6-set. One method would be to list them all. We now do this for the 6-set $A = \{a_1, a_2, a_3, a_4, a_5, a_6\}$:

$$\begin{array}{lll}
\{a_1,a_2,a_3\}, & \{a_1,a_2,a_4\}, & \{a_1,a_2,a_5\} \\
\{a_1,a_2,a_6\}, & \{a_1,a_3,a_4\}, & \{a_1,a_3,a_5\} \\
\{a_1,a_3,a_6\}, & \{a_1,a_4,a_5\}, & \{a_1,a_4,a_6\} \\
\{a_1,a_5,a_6\}, & \{a_2,a_3,a_4\}, & \{a_2,a_3,a_5\} \\
\{a_2,a_3,a_6\}, & \{a_2,a_4,a_5\}, & \{a_2,a_4,a_6\} \\
\{a_2,a_5,a_6\}, & \{a_3,a_4,a_5\}, & \{a_3,a_4,a_6\} \\
\{a_3,a_5,a_6\}, & \{a_4,a_5,a_6\}. &
\end{array}$$

Thus there are 20 3-subsets of a 6-set. Notice that $20 = \binom{6}{3}$.

In this problem we asked for the number of 3-subsets. We would have been content to have the number 20 without the list of subsets.

Exercises 1.7.2 and 1.7.3 asserted that n equals both the number of 1-subsets of an n-set and the number of $(n-1)$-subsets of an n-set. Notice that

$$n = \binom{n}{1} = \binom{n}{n-1}.$$

Similarly, the number of 2-subsets and the number of $(n-2)$-subsets of an n-set equals $\binom{n}{2}$ as seen in Exercises 1.7.5 and 1.7.7. Thus in these cases the exact subset count is given by a binomial coefficient.

Question 3.1. If $A = \{a_1, a_2, a_3, a_4, a_5, a_6\}$, check that the number of 4-subsets of A is $\binom{6}{4}$ and the number of 5-subsets is $\binom{6}{5}$.

Specific examples all point toward the truth of the following theorem. The idea of the proof is similar to the proof by induction that an n-set has 2^n subsets, given in Section 2.4.

Theorem 3.1. For $j = 0, 1, \ldots, n$, the number of j-subsets of an n-set equals the binomial coefficient $\binom{n}{j}$.

Proof. The proof of this theorem is by induction on n. For $n = 0$ and $j = 0$, there is only one subset of the empty set (itself) and $1 = \binom{0}{0}$. When $n = 1$, there is $\binom{1}{0}$ 0-subset, the empty set, and $\binom{1}{1}$ 1-subset, the entire set.

We assume that the theorem is valid for $n = k$ and try to prove it for $n = k + 1$. Suppose that $A = \{a_1, a_2, \ldots, a_{k+1}\}$, and consider all j-subsets of A. We divide these into two piles, those that contain the last element a_{k+1} and those that don't.

Question 3.2. Divide the 3-subsets of the 6-set given in Example 3.1 into two piles depending on whether a_6 is contained in the 3-subset or not.

Let $A' = A - \{a_{k+1}\}$. See Figure 3.9. If S is a j-subset of A that contains a_{k+1}, then $S - \{a_{k+1}\}$ is a $(j-1)$-subset of the k-set A'. By the inductive hypothesis we know that there are $\binom{k}{j-1}$ of these $(j-1)$-subsets of A'. Thus there are

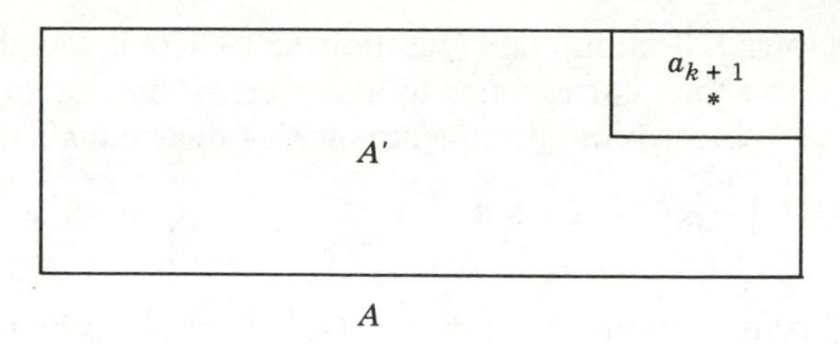

Figure 3.9

$\binom{k}{j-1}$ j-subsets of A that contain a_{k+1}. If S is a j-subset of A that does not contain a_{k+1}, then S is a j-subset of A'. By the inductive hypothesis there are $\binom{k}{j}$ of these subsets. Thus the total number of j-subsets of A is given by $\binom{k}{j-1} + \binom{k}{j}$. By Corollary 2.2 this sum equals $\binom{k+1}{j}$. □

Question 3.3. Use Theorem 3.1 to determine the number of k-subsets of an n-set, where (a) $k = 3$, $n = 6$; (b) $k = 5$, $n = 11$; and $k = 9$, $n = 17$.

Theorem 3.1 will enable us to prove a variety of results. For example, look back at Figure 3.7 and Exercise 2.5 and notice that the sum of the numbers in one row of Pascal's triangle seems to be a power of 2. It appears that for fixed n, the binomial coefficients of the form $\binom{n}{i}$ sum to 2^n. This result is an easy corollary of Theorem 3.1 once we look at the problem in the right way.

Corollary 3.2. For every natural number n

$$\binom{n}{0} + \binom{n}{1} + \cdots + \binom{n}{i} + \cdots + \binom{n}{n} = 2^n.$$

Proof. As we've seen, the number of subsets of an n-set is 2^n. This number of subsets also equals the number of 0-subsets plus the number of 1-subsets plus $\cdots$ plus the number of i-subsets plus $\cdots$ plus the number of n-subsets. By Theorem 3.1 this sum is exactly the left-hand side of the equation and so the two expressions are equal. □

Example 3.2. Consider an n-digit number in binary. We can also think of this as an n-bit sequence of zeros and ones. How many n-bit sequences are there that contain exactly i ones, for i an integer between 0 and n? To answer this question, we need to recall a discussion about subsets and "bit vectors" of zeros and ones presented in Chapter 1, Section 7 (see Question 7.5.) There it was shown that every n-bit sequence with i ones corresponds to a unique i-subset of an n-set, and conversely. Thus the number of n-bit sequences with i ones equals the number of i-subsets of an n-set, $\binom{n}{i}$ for $0 \le i \le n$.

Example 3.3. From Section 3.1 the number of shortest paths from M to P in the 6×5 grid equals the number of sequences of 11 Es and Ns with exactly 6 Es, and we know that this number equals $\binom{11}{6}$. Now we can see directly why there are $\binom{11}{6}$ sequences of 6 Es and 5 Ns. Suppose that we take any sequence of Es and Ns and replace every E by a one and every N by a zero. Then we have an 11-bit sequence with 6 ones; from Example 3.2 we know that there are $\binom{11}{6}$ of these.

Furthermore, every 11-bit sequence with 6 ones can be transformed back into a sequence of 6 Es and 5 Ns. In fact, Es and Ns or ones and zeros are two ways to represent exactly the same idea. Thus there are the same number of these two types of sequences.

Question 3.4. How many 11-letter sequences of Ns and Es are there that contain 3 Es?, 4 Es?, 7 Es? What is the total number of all possible 11-letter sequences of Es and Ns?

Problem. Given positive integers j and n with $j \le n$, list all j-subsets of $\{1, 2, 3, \ldots, n\}$.

We know the number of j-subsets of an n-set without listing them all. However, we may wish to list all subsets of a certain size. We turn to the construction of an algorithm that upon input of natural numbers n and j with $j \le n$ will list all j-subsets of an n-set. Up until now we've used the generic n-set $A = \{a_1, a_2, \ldots, a_n\}$. We simplify the notation and consider the n-set $I_n = \{1, 2, \ldots, n\}$ and its j-subsets. This change merely saves on writing: the j-subsets of A and I_n correspond exactly. We agree to always write the j-subset with its entries in increasing order.

Question 3.5. Describe how to transform a j-subset of A into a j-subset of I_n, and conversely. In particular, find the 4-subset of I_n that corresponds with $\{a_1, a_2, a_4, a_{n-1}\}$ and the 5-subset of A that corresponds with $\{1, 2, 3, 5, 8\}$.

For small, fixed values of j, the problem of creating an algorithm to list all j-subsets is not hard, and we begin with some specific cases. Exercise 14 asks you to design the straightforward algorithms to list all 0-subsets and all 1-subsets of $I_n = \{1, 2, \ldots, n\}$.

Example 3.4. Here is an algorithm that lists all 2-subsets of $I_n = \{1, 2, \ldots, n\}$. (This also was Exercise 1.7.6.)

Algorithm PAIR

STEP 1. Input n, a positive integer
STEP 2. For $i = 1$ to $(n - 1)$ do
  Begin
  STEP 3. For $j = (i + 1)$ to n do
    STEP 4. List $\{i,j\}$
  End
STEP 5. Stop.

COMMENTS. Section 2.4 contained examples of this form of a loop, often called a "**do loop**." Recall that the instructions, "For $i = 1$ to $(n - 1)$ do, Begin, . . . , End," mean that first i should be set equal to 1 and the instructions between the Begin and End statements should be executed. Next, i is set equal to 2 and the Begin . . . End statements are executed. We repeat these steps, increasing i by one each time until i is set equal to $(n - 1)$, the Begin . . . End statements are executed for the last time, and then we move on to the next statement, here step 5. This is also our first example of a loop within a loop; such pairs of loops are called **nested**. For each assignment of a value to i, step 3 is repeated several times, once for each of the values $j = (i + 1), (i + 2), \ldots,$ and n. Then i is assigned its next value and j runs through its sequence of values.

The output of PAIR has the following properties. The subset $\{i,j\}$ is always listed with its smaller element first. The first subsets listed are those that contain 1; next the subsets that contain 2 (but not 1) are listed; then those with 3; and so on. Here is how the output looks:

$$\begin{array}{l} \{1,2\}, \quad \{1,3\}, \quad \{1,4\}, \quad \ldots, \quad \{1,n\}, \\ \{2,3\}, \quad \{2,4\}, \quad \ldots, \quad \{2,n\}, \\ \{3,4\}, \quad \ldots, \quad \{3,n\}, \\ \qquad\qquad \cdots \\ \{n-2, n-1\}, \quad \{n-2, n\}, \\ \{n-1, n\}. \end{array}$$

This ordering is known as **lexicographic** or **dictionary order**.

The idea behind PAIR could be generalized to generate all 3-subsets of an n-set by having 3 nested loops. However, if we want to list all j-subsets, where j is part of the input to the algorithm, then this idea won't work, since we can't have the correct number of loops programmed in advance. Instead we generalize the idea of lexicographic order for arbitrary $j > 2$ and use this order to list all j-subsets.

We approach lexicographic ordering in two ways. First we describe to you, a human, which of any two j-subsets of an n-set should be listed before the other. Next we need to explain this order to a computer, that is, in language sufficiently precise so that it can be implemented as an algorithm. We need to know exactly how, once the algorithm has created and listed a j-subset S, to move on and produce the next j-subset T. At the same time we must explain why lexicographic order as described for humans is the same as the order described for computers.

We begin our description for humans with an example.

Example 3.5. Here are the 3-subsets of $I_5 = \{1,2,3,4,5\}$ listed in lexicographic order: $\{1,2,3\}$, $\{1,2,4\}$, $\{1,2,5\}$, $\{1,3,4\}$, $\{1,3,5\}$, $\{1,4,5\}$, $\{2,3,4\}$, $\{2,3,5\}$, $\{2,4,5\}$, $\{3,4,5\}$.

One way to think about lexicographic order is to disregard all the set notation that accompanies each subset and just list the integers in the order in which they appear. Then we see

$$123 < 124 < 125 < 134 < 135 < 145 < 234 < 235 < 245 < 345,$$

and the ordering is the same as the "less than" order on the natural numbers.

Question 3.6. Place the following sets of I_8 in lexicographic order. $\{1,3,4,7\}$, $\{3,6,7,8\}$, $\{2,3,4,7\}$, $\{4,6,7,8\}$, $\{2,3,5,6\}$, and $\{1,3,5,6\}$.

This description would suffice for humans and computers except for the fact that if $n > 9$, we might see sets like $\{2,3,45,678\}$ and $\{2,34,56,78\}$. This could be salvaged if we were willing to do arithmetic in base n, but since n is part of the input, there are still problems with this approach.

Here is a way for humans and computers to decide which of two j-subsets, S and T, to list first. Suppose that $S = \{s_1,\ldots,s_j\}$ and $T = \{t_1,\ldots,t_j\}$. If $s_1 < t_1$, then S is listed before T. If $s_1 > t_1$, then T is listed before S. The only other alternative is that $s_1 = t_1$ in which case we compare the second element of each set. In general, if $s_1 = t_1, s_2 = t_2, \ldots, s_{i-1} = t_{i-1}$, and $s_i < t_i$, then we list S before T, and if $s_i > t_i$ we list T before S. In English, S appears before T if in the first place the two sets differ, the element of S is smaller than the element of T.

Example 3.6. $\{2,3,4,7,8,9\}$ is listed after $\{2,3,4,6,8,9\}$, since the sets agree until the fourth entry at which point $7 > 6$.

When listing all subsets of a set, if S is listed immediately before T, we should expect that S and T agree as much as possible. Furthermore, we should expect that at the location where S and T first disagree, the entry in T is just one larger than the corresponding entry in S.

Example 3.5 (revisited). A second way to view lexicographic ordering of the 3-subsets of $\{1,2,3,4,5\}$ is to start with the "first" set $\{1,2,3\}$. Then we let the last entry increase by 1 and we get $\{1,2,4\}$. Repeating this, we get $\{1,2,5\}$. So our listing begins

$$\{1,2,3\}, \quad \{1,2,4\}, \quad \{1,2,5\}.$$

Now the last entry is as big as it can be. So we agree to increase the second to last entry by 1 and make every other entry as small as it can be. Thus after $\{1,2,5\}$ we get $\{1,3,4\}$. Now we return our attention to the last entry, which can once again be incremented. Thus the list continues

$$\{1,3,4\}, \quad \{1,3,5\}, \quad \{1,4,5\}.$$

Now our second to last entry is as big as it can be, so we have to increment our third to last (here the same as the first) entry. We finish with

$$\{2,3,4\}, \quad \{2,3,5\}, \quad \{2,4,5\}, \quad \{3,4,5\}.$$

Question 3.7. List all 3-subsets of $I_6 = \{1,2,3,4,5,6\}$ in lexicographic order.

Thus, in general, the idea of lexicographic order is to begin with the j-subset $\{1,2,\ldots,j-1,j\}$ and then to let the last entry increase through all possible values until we reach the subset $\{1,2,\ldots,j-1,n\}$. Next we move back to the second to last entry, increase it by one, and let the last entry take on all larger values. In general, after a subset $S = \{s_1,\ldots,s_h,\ldots,s_j\}$ is listed, we search from right to left in S looking for the first entry that can be increased by one. Suppose that s_h can be increased to $s_h + 1$, but no s_k with $k > h$ can be increased. Then we produce the new subset $T = \{s_1,\ldots,s_{h-1},s_h + 1,s_h + 2,s_h + 3,\ldots\}$.

Example 3.7. Here is how these ideas apply when listing all 4-subsets of $\{1,2,\ldots,8\}$ in lexicographic order.

$$\begin{array}{llll}
\{1,2,3,4\}, & \{1,2,3,5\}, & \ldots, & \{1,2,3,8\}, \\
\{1,2,4,5\}, & \{1,2,4,6\}, & \ldots, & \{1,2,4,8\}, \\
\{1,2,5,6\}, & \{1,2,5,7\}, & \{1,2,5,8\}, & \\
\{1,2,6,7\}, & \{1,2,6,8\}, & \{1,2,7,8\}, & \\
\{1,3,4,5\}, & \{1,3,4,6\}, & \ldots, & \{1,3,4,8\}, \\
\{1,3,5,6\}, & \ldots, \{1,6,7,8\}, & & \\
\{2,3,4,5\}, & \{2,3,4,6\}, & \ldots, & \{5,6,7,8\}.
\end{array}$$

We know that we are finished, since no entry can be increased.

Question 3.8. List all 3-subsets of $I_7 = \{1, 2, \ldots, 7\}$ in lexicographic order.

Notice that in Example 3.1 if the set $I_6 = \{1, 2, 3, 4, 5, 6\}$ is substituted for A and the corresponding 3-subsets of A are replaced by those of I_6, then the subsets are listed in lexicographic order.

This work gives us the ideas for an algorithm to list all j-subsets of an n-set for a general value of j and n. The only imprecision left is how, after a j-subset S is listed, to find the entry s_h that can be increased to $s_h + 1$. Note that the jth entry in a j-subset can be no larger than n, the $(j-1)$st entry can be no larger than $n-1$, and in general the ith entry can be no larger than $n + i - j$. Thus we read from right to left until we find s_h such that $s_h < n + h - j$. We replace s_h by $s_h + 1$ and fill in the remaining subset entries with $s_h + 2$, $s_h + 3, \ldots$. The algorithm stores each subset in an array $\{b_1, b_2, \ldots, b_j\}$.

Algorithm JSET

STEP 1. Input j and n with $1 \le j \le n$
STEP 2. Set $\{b_1, b_2, \ldots, b_j\} := \{1, 2, \ldots, j\}$ and list this set
STEP 3. Set $h := j + 1$ and FOUND := FALSE
{Next we determine which entry to increase by 1. When we find the entry, we set FOUND to be TRUE.}
STEP 4. While $h > 1$ and FOUND = FALSE do
Begin
STEP 5. $h := h - 1$
STEP 6. If $b_h < n + h - j$, set FOUND := TRUE
End
{Right now the value assigned to h is the rightmost entry that can be increased. If FOUND = FALSE, then we are finished listing all subsets.}
STEP 7. If FOUND = FALSE, then stop.
Else
Begin
STEP 8. Set $b_h := b_h + 1$
STEP 9. For $k := h + 1$ to j do
STEP 10. Set $b_k := b_{k-1} + 1$
STEP 11. List $\{b_1, b_2, \ldots, b_j\}$
STEP 12. Go to step 3
End

COMMENTS. We introduce the so-called Boolean variable FOUND. (See also Section 10 of Chapter 1.) A **Boolean variable** can be assigned only two values, either TRUE or FALSE. We initialize FOUND to be FALSE and change its value if the correct entry to increment has been located. Before repeating the steps inside the While do loop at step 4, two conditions must be satisfied. It must be the case that both $h > 1$ and FOUND = FALSE. If either condition is contradicated, then we proceed to step 7. In general, Boolean variables are useful in the design of loops when the number of iterations through the loop is not known in advance.

With $j = 3$ and $n = 5$, JSET lists the subsets as shown in Example 3.5 and with $j = 4$ and $n = 8$ as in Example 3.7.

Question 3.9. Run JSET with $j = 4$ and $n = 6$.

We conclude with an informal analysis of the complexity of JSET. The input is j and n with $j \leq n$; so we use n as a measure of the input size. We let $f(n)$ equal the maximum number of time-consuming operations performed in JSET and carry out a worst-case analysis. We hope to find a simple function $g(n)$ [like $\log(n)$, n^2, or 2^n] such that $f(n) = O(g(n))$. Instead of counting all the additions, assignments, and comparisons in JSET, we choose to define a basic, time-consuming operation to be the creation and listing of a j-subset. (For a more precise analysis see Exercise 23.) Since $\binom{n}{j}$ j-subsets are produced, $f(n) = \binom{n}{j}$. Is JSET good? That depends on the value of j. For example, when $j = 2$, the complexity of JSET is

$$f(n) = \binom{n}{2} = O(n^2)$$

and in this case JSET is a good algorithm.

Question 3.10. Show that JSET is a good algorithm when $j = 3$.

Question 3.11. Show that if j is a fixed constant less than $n/2$, then

$$\binom{n}{j} = O(n^j).$$

However, there are cases in which JSET is an exponential algorithm. We have seen in Section 3.2 that $\binom{n}{j}$ is largest when $j = \lfloor n/2 \rfloor$. We shall show in the next theorem that $\binom{n}{\lfloor n/2 \rfloor}$ is exponential in n [thus not $O(p(n))$ for any polynomial p].

Does that mean that we have failed to be sufficiently clever in our algorithm design? No, because when $j = \lfloor n/2 \rfloor$, there is an exponential number of j-subsets, and any j-subset algorithm will be exponential, since each of these must be listed. Thus although this algorithm is exponential and not good, it is nonetheless more or less as efficient as possible and so we'll use it when needed.

Theorem 3.3 For n even $\binom{n}{n/2} \geq (\sqrt{2})^n$.

Proof

$$\binom{n}{n/2} = \frac{n!}{(n/2)!\,(n-n/2)!} = \frac{n!}{(n/2)!\,(n/2)!}$$
$$= \frac{n(n-1)(n-2)\cdots(n/2+1)}{(n/2)(n/2-1)(n/2-2)\cdot\cdots 2\cdot 1}$$
$$= \frac{n}{n/2}\cdot\frac{n-1}{n/2-1}\cdot\cdots\cdot\frac{n-j}{n/2-j}\cdot\cdots\cdot\frac{n/2+1}{1}.$$

Next we consider an arbitrary factor in the above expression.

$$\frac{n-j}{n/2-j} = \frac{n-j}{\frac{(n-2j)}{2}} = 2\cdot\frac{n-j}{n-2j} \geq 2.$$

Thus each of these factors is at least 2. Since there are $n/2$ factors, the product is at least $2^{n/2} = (\sqrt{2})^n$. □

EXERCISES FOR SECTION 3

1. For $i = 0, 1, \ldots, 6$ calculate the number of i-subsets of a 6-set. Check that these numbers sum to 2^6.
2. For $i = 0, 2, 4, 6, 8$ calculate the number of i-subsets of an 8-set. Check that these numbers sum to 2^7.
3. The number of j-subsets of an n-set equals $\binom{n}{j}$, and the number of $(n-j)$-subsets equals $\binom{n}{n-j}$. Give a set theory argument to show that these two numbers are equal.
4. From a class of 14 students, a committee of 5 students is formed. In how many different ways can this committee be chosen? How many of these committees contain one particular student, Sue, and how many committees do not contain her?
5. Suppose that a class is composed of eight freshmen and six sophomores. How many committees are there that consist of three freshmen and two sophomores?

6. Suppose that two committees with no members in common are to be selected from a class of 14 students. First a committee of 5 students is selected and then a committee of 3 students. In how many ways can these two committees be formed? Is the answer to the last question the same as the number of ways to form first a committee of 3 students and then second a committee of 5 students?

7. The local Pizza Factory produces pizzas topped with tomato sauce and cheese. Optional items are mushrooms, green peppers, olives, onions, anchovies, pepperoni, and hamburger. The prices are $3.00 for plain pizza, $3.50 for one extra item, $4.00 for two items, and $4.50 for three. How many different kinds of pizza cost $3.50? $4.00? $4.50?

8. The ice cream parlor in town offers 10 different kinds of ice cream daily plus toppings of fudge sauce, butterscotch sauce, raspberry sauce, walnuts, M&Ms, Heath Bar Crunch, and whipped cream. How many different ways are there to make a sundae if (by definition) a sundae consists of one kind of ice cream plus at least one topping? How many kinds of sundae are there that have exactly 3 toppings? That have no more than 3 toppings?

9. Prove by induction on n that

$$\binom{n}{0} + \binom{n}{1} + \cdots + \binom{n}{n} = 2^n.$$

10. If n is even, show that

$$\binom{n}{0} + \binom{n}{2} + \binom{n}{4} + \cdots + \binom{n}{n} = 2^{n-1}.$$

Then find and justify a similar formula when n is odd.

11. List all seven-bit vectors with exactly two ones. Then list all sequences of Es and Ns that contain exactly five Ns and two Es.

12. For $j = 2, 3, \ldots, 7$ calculate the binomial coefficients $\binom{j}{2}$ and find their sum. Then find a formula in terms of n that compactly expresses the sum

$$\binom{2}{2} + \binom{3}{2} + \cdots + \binom{n}{2}.$$

13. Design an algorithm to create and list all bit vectors of length n.

14. Describe an algorithm that will list all 0-subsets of $I_n = \{1, 2, \ldots, n\}$. Then describe an algorithm that will list all 1-subsets of I_n.

15. **(a)** Modify the algorithm PAIR to list all bit vectors of length n that contain exactly 2 ones.

(b) Modify PAIR to list all sequences of Es and Ns that contain exactly 2 Es and $n-2$ Ns.

16. How many 2-subsets of an n-set contain the element 1? How many 2-subsets contain the element 2 but not 1? How many contain 3 but neither 1 nor 2? In general, how many 2-subsets contain the element i but no smaller number? Then add up all these answers for $i = 1, 2, \ldots, (n-1)$ and explain why that sum adds up to $\binom{n}{2}$, the total number of 2-subsets of an n-set.

17. List all 2-subsets of $\{1, 2, \ldots, 6\}$ in lexicographic order. Then list all 5-subsets of the set $\{1, 2, 3, 4, 5, 6, 7\}$ in lexicographic order.

18. We defined lexicographic order of subsets of numbers by saying that we list a subset S before a subset T "if in the first place the two sets differ, the element of S is smaller than the element of T." If we consider now subsets of a set $A = \{a, b, c, d, e, f\}$ and use alphabetical ordering of letters, how should we define lexicographic order of subsets of A? List all 2-subsets of A in lexicographic order. Then check that in your list, when a is replaced by 1, b by $2, \ldots,$ and f by 6, that the 2-subsets are still listed in lexicographic order (for numbers).

19. Run the algorithm JSET with
(a) $j = 1$ and $n = 6$.
(b) $j = 3$ and $n = 6$.
(c) $j = 3$ and $n = 7$.

20. Is JSET a good algorithm when $j = n - 2$? When $j = n - 1$?

21. If n is odd, show that

$$\binom{n}{\frac{(n-1)}{2}} \geq r^n \qquad \text{for some } r > 1.$$

22. Show that

$$\binom{n}{j} = O(2^n) \qquad \text{for all } j.$$

23. Give a more detailed, worst-case complexity analysis of JSET by counting the number of additions, subtractions, comparisons, and assignments. Note that the While do loop of step 4 is executed at most j times and the For do loop of step 9 at most $(j-1)$ times. Show that the total number of these operations is $O\left(j\binom{n}{j}\right)$. Then explain why this analysis shows JSET to be a polynomial algorithm for j a constant number, but for j arbitrary JSET is exponential.

3:4 PERMUTATIONS

When we were generating all the j-subsets of the n-set I_n in Section 3, it was crucial to be able to arrange the elements within each subset in increasing order. The reason we could do this is that in sets, the order of presentation of the elements does not matter and so we could choose an algorithmically useful presentation. In contrast we have seen some cases earlier where the given order is important and we cannot rearrange elements. Examples include applications of the Multiplication Principle and Cartesian products. We turn now to another area where order is crucial.

Given a set $A = \{a_1, a_2, \ldots, a_n\}$ containing n distinct elements, an ordered list of these n elements is called a **permutation** of A. Often, but not always, we shall use the set $I_n = \{1, 2, \ldots, n\}$ as the generic set with n elements. We distinguish sets, which will continue to be denoted by $\{\cdots\}$, from permutations, which will be surrounded by $\langle \cdots \rangle$.

Example 4.1. If $I_6 = \{1, 2, 3, 4, 5, 6\}$, then $\langle 1\ 2\ 3\ 4\ 5\ 6 \rangle$, $\langle 2\ 1\ 4\ 3\ 6\ 5 \rangle$, and $\langle 1\ 3\ 5\ 2\ 4\ 6 \rangle$ are three permutations of I_6; however, $\langle 1\ 2\ 4\ 5 \rangle$ is not a permutation of I_6, neither is $\langle 1\ 2\ 3\ 4\ 5\ 5 \rangle$ nor $\langle 1\ 2\ 3\ 4\ 5\ 7 \rangle$.

Within the permutation $\langle 2\ 1\ 3\ 5\ 4\ 6 \rangle$ we say that 2 is in the first position, 1 in the second position, 3 in the third position, and so on.

Example 4.2. Here are all permutations on $\{1, 2, 3\}$:

$$\langle 1\ 2\ 3 \rangle, \quad \langle 1\ 3\ 2 \rangle, \quad \langle 3\ 1\ 2 \rangle, \quad \langle 2\ 1\ 3 \rangle, \quad \langle 2\ 3\ 1 \rangle, \quad \langle 3\ 2\ 1 \rangle.$$

Notice there are $6 = 3!$ such permutations.

Question 4.1. List all 4! permutations on the four numbers $\{1, 2, 3, 4\}$.

The Multiplication Principle shows us why the factorial function counts permutations of an n-set. Since a permutation is an ordered list, we may count the ways to fill in the n blanks:

$$\langle \_\_ \ \_\_ \ \_\_ \cdots \_\_ \rangle.$$

First we see that we have n choices for the element to place in the first position of the permutation. After that we have only $(n - 1)$ choices for the entry in the second position, then $(n - 2)$ for the third position, and so on. At the next to last position we have two choices remaining and then one choice for the last position. Thus the number of permutations on n elements is

$$n(n - 1)(n - 2) \cdots \cdot 2 \cdot 1 = n!.$$

This argument proves the next theorem. One proof is sufficient for any theorem, but we choose to present a second proof using induction because it will cause us to think about generating permutations algorithmically.

Theorem 4.1. There are $n!$ permutations of a set containing n distinct elements.

Proof. First we check the base cases for $n = 1$ and 2: $\langle 1 \rangle$ is the only permutation on one number and $\langle 1\ 2 \rangle$ and $\langle 2\ 1 \rangle$ are the two permutations on two numbers. In Example 4.2 and Question 4.1 we checked the cases for $n = 3$ and $n = 4$.

Our inductive hypothesis is that there are $k!$ permutations of any set with k elements. Using this hypothesis, we need to show that any set with $k + 1$ elements has $(k + 1)!$ permutations. Let $I_{k+1} = \{1, 2, \ldots, k, k + 1\}$. As usual we remove an element, setting $I_k = I_{k+1} - \{k + 1\} = \{1, 2, \ldots, k\}$. Pick any permutation of I_k: for instance,

$$\langle 2\ 1\ 3\ 4 \cdots k \rangle.$$

Each such permutation of I_k generates $k + 1$ different permutations of I_{k+1} because the element $(k + 1)$ can be inserted in $k + 1$ different positions. From the permutation above we create

$$\begin{array}{lccccccl}
\langle & 2 & 1 & 3 & \cdots & k & k+1 & \rangle, \\
\langle & 2 & 1 & 3 & \cdots & k+1 & k & \rangle, \\
 & & & & \cdots & & & \\
\langle & 2 & 1 & 3 & k+1 & \cdots & k & \rangle, \\
\langle & 2 & 1 & k+1 & 3 & \cdots & k & \rangle, \\
\langle & 2 & k+1 & 1 & 3 & \cdots & k & \rangle,
\end{array}$$

and

$$\langle k+1 \quad 2 \quad 1 \quad 3 \quad \cdots \quad k \rangle.$$

Thus the number of permutations of I_{k+1} equals $k + 1$ times the number of permutations of I_k, or $(k + 1)(k!) = (k + 1)!$. □

Remember this proof because it suggests how to write an algorithm to generate all permutations.

Example 4.3. Suppose that we start with the permutation $\langle 2\ 1\ 3\ 4 \rangle$. We can insert the element 5 in any one of five different positions to obtain $\langle 2\ 1\ 3\ 4\ 5 \rangle$, $\langle 2\ 1\ 3\ 5\ 4 \rangle$, $\langle 2\ 1\ 5\ 3\ 4 \rangle$, $\langle 2\ 5\ 1\ 3\ 4 \rangle$, and $\langle 5\ 2\ 1\ 3\ 4 \rangle$.

Permutations give us another way to see that an n-set has $\binom{n}{j}$ j-subsets for $0 \leq j \leq n$. A 2-subset of the set $I_n = \{1, 2, \ldots, n\}$ can be formed by filling in two blanks: $\{\_\_, \_\_\}$. By the Multiplication Principle there are n choices for the first blank and $(n - 1)$ choices for the second; however, once the set is filled in, every permutation of the elements in the set will give the same set. On two elements there are $2! = 2$ permutations and so in the $n(n - 1)$ choices, each set is listed twice. Thus there are $n(n - 1)/2 = \binom{n}{2}$ 2-subsets.

Question 4.2. Use the Multiplication Principle and permutations to show that there are $n(n - 1)(n - 2)/3!$ different 3-subsets of an n-set.

Next consider the case of counting the number of j-subsets of an n-set for an arbitrary value of j between 0 and n. To create a j-subset, begin by filling in j empty blanks: $\{\_\_, \_\_, \ldots, \_\_\}$. There are n choices for the first blank, then $(n - 1)$ choices for the second blank, and eventually for the jth blank, there are $n - (j - 1) = n - j + 1$ choices remaining. Thus there are

$$n(n-1)\cdots(n-j+1) = \frac{n!}{(n-j)!}$$

choices for filling in j empty blanks. But each subset with j elements can arise in $j!$ different orders, one for each of the permutations on j elements. Thus the total number of subsets of size j is given by

$$\frac{n!}{(n-j)!j!} = \binom{n}{j}.$$

Now we turn to the algorithmic question of generating all permutations. We displayed all permutations on three elements in Example 4.2, and you were asked to do the same for four elements in Question 4.1. For $n > 4$, $n!$ is larger than 100, and listing all permutations is clearly a task suitable for computers rather than humans.

Problem. Given an integer $n > 0$, list all permutations of $\{1, \ldots, n\}$.

Look back at the proof of Theorem 4.1. It tells us that from each permutation on $(n - 1)$ elements, we get n permutations on n elements by moving the element n through all possible positions in the $(n - 1)$ permutation. To use this idea, we need to begin at the base case of our inductive proof. We first write down all permutations on one element: $\langle 1 \rangle$. Then we use the idea of the inductive step to generate all permutations on two elements: $\langle 1\ 2 \rangle$ and $\langle 2\ 1 \rangle$. How about three elements? First we expand $\langle 1\ 2 \rangle$ into the three permutations $\langle 1\ 2\ 3 \rangle$, $\langle 1\ 3\ 2 \rangle$ and $\langle 3\ 1\ 2 \rangle$.

Then we expand the permutation ⟨2 1⟩ into ⟨2 1 3⟩, ⟨2 3 1⟩, and ⟨3 2 1⟩. Here is an algorithm that works exactly in this way.

Algorithm PERM

STEP 1. Set $j := 1$ and write down the permutation ⟨1⟩
STEP 2. Set $j := j + 1$
STEP 3. For each permutation $\langle a_1\, a_2 \ldots a_{j-1} \rangle$ on $j - 1$ elements do
    Begin
    STEP 4. Create and list $P := \langle a_1 a_2 \ldots a_{j-1} j \rangle$
    STEP 5. For $i = j - 1$ down to 1 do
        STEP 6. Set $P := P$ with the values assigned to positions i and $(i + 1)$ switched, and list P
    End {step 3}
STEP 7. If $j < n$, then go to step 2; otherwise, stop.

COMMENTS. There are some new features in this algorithm and some old. The first new feature is the form of step 5, "**For** $i = j - 1$ **down to** 1 **do**." This indicates a do-loop; here the variable i will decrease from $j - 1$. Thus initially i equals $j - 1$ and step 6 is executed; then i equals $j - 2$ for step 6, . . . , until finally i equals 1 for the last execution of step 6, at least for this round.

We have previously seen an example (in PAIR) of a "nested loop," a loop within a loop. Step 3 is an instruction that causes steps 4 and 5 to be executed for each permutation on $j - 1$ elements. Also, step 5 instructs us to carry out step 6 $j - 1$ times for each time step 5 is encountered. Actually, the algorithm also has a third loop that sends us back to step 2 repeatedly. We do not specify step 6 precisely, but notice that it requires the switching of the values of two variables, exactly the task that we learned to do in Section 2.1.

Example 4.4. In Table 3.1 we trace through algorithm PERM in the case of $n = 3$.

Table 3.1

Values of j	*Permutations*
1	⟨1⟩
2	⟨1 2⟩
	⟨2 1⟩
3	⟨1 2 3⟩
	⟨1 3 2⟩
	⟨3 1 2⟩
	⟨2 1 3⟩
	⟨2 3 1⟩
	⟨3 2 1⟩

Question 4.3. Use Algorithm PERM to generate all permutations on $I_4 = \{1, 2, 3, 4\}$.

Our next task is to analyze the complexity of PERM. We do even fewer kinds of time-consuming operations than in JSET. The algorithm mainly assigns, reassigns, and lists permutations, $n!$ of these, in fact.

To analyze PERM, we need to study the factorial function and relate it to our hierarchy of functions: we know that for sufficiently large n,

$$1 \leq \log(n) \leq \sqrt{n} \leq n \leq n^2 \leq n^3 \leq \cdots \leq 2^n \leq \cdots \leq 10^n.$$

We shall find that not only is $n! \neq O(p(n))$ for any polynomial n, but even more, in Theorem 4.2 we prove that for any positive integer $r > 0$ and any constant C, $n! > C \cdot r^n$ for n sufficiently large. Thus the correct position for $n!$ in the hierarchy is at the end (so far):

$$\cdots \leq 10^n \leq \cdots \leq n!.$$

We count the formation and listing of each permutation as the fundamental time-consuming operation. PERM performs $n!$ of these steps and is thus not a polynomial algorithm. However, PERM accomplishes even more than it was asked to do. Besides generating all permutations on $\{1, 2, \ldots, n\}$, it also generates all permutations on the smaller sets $\{1, 2, \ldots, i\}$ for each $i = 1, 2, \ldots, n - 1$. Thus PERM really performs $1! + 2! + \cdots + n!$ steps and its complexity is consequently even larger than $n!$.

Also notice that in PERM we must store all $(i - 1)!$ permutations on $i - 1$ numbers before creating the permutations on i numbers. Hence this algorithm requires more than an exponential amount of storage as well as more than an exponential amount of time, and so it is bad with respect to both time and space, the worst of both worlds.

Generating permutations is always time-consuming; however, there are ways to generate permutations on n elements without first generating all permutations on smaller sets. Also there are ways to create a "next" permutation from a given permutation (as in the algorithm JSET) so that in an application the permutations can be produced one after the other without storing previous permutations. Some of the ideas behind these algorithms are presented in the exercises.

But why is $n!$ such a fast growing function? Why does it belong at the end of the function hierarchy? Suppose that we add the function $n!$ to Table 2.9 in Chapter 2. That is, suppose that we perform $n!$ single digit multiplications on an IBM PC. The time involved for differing values of n is listed in Table 3.2.

The numerical evidence is clear that the function $n!$ grows even faster than 10^n. Suppose as with multiplication on the PC, we could figure out and write

Table 3.2

	$n = 8$	$n = 11$	$n = 16$	$n = 24$	$n = 32$
$n!$	2.27 min	37.4 hr	2236 yr	663,178,306,400 cent	...

down 17,800 permutations per minute (and this would be working pretty fast!) Then Table 3.2 tells us that using PERM in, say, a 24-hour run we could only hope to list all permutations for $n \leq 10$.

Here is theoretical evidence that also points to the rapid growth of $n!$.

Theorem 4.2. $n! \neq O(r^n)$ for any positive integer r.

Proof. The theorem states that it is not the case that given a positive integer r, there is a constant C such that $n! \leq C \cdot r^n$ for all sufficiently large integers n. Thus given an integer r we must show that for every constant C, $n! > C \cdot r^n$ for n sufficiently large. So pick a constant C and suppose that $n > C \cdot r^r$. (This number is cooked up because we found that it made the proof work easily!) We know that $r = r$ and $r + 1 > r$, and $r + 2 > r$, and $\ldots$, $n - 1 > r$. Thus we have

$$
\begin{aligned}
n! &= n(n-1)\cdots(r+1)r\cdots 2\cdot 1 && \text{by definition}\\
&\geq n(n-1)\cdots(r+1)r && \text{by ignoring the last factors}\\
&> n\cdot r\cdot \cdots \cdot r\cdot r && \text{(a product of } (n-r)\ r\text{'s)}\\
&> (C\cdot r^r)\cdot r\cdot \cdots \cdot r\cdot r && \text{by choice of } n\\
&= C\cdot r^n.
\end{aligned}
$$

□

Question 4.4. Theorem 4.2 says that $n! \neq O(2^n)$. Find an integer N so that for $n \geq N$, $n! > 10 \cdot 2^n$.

We state without proof a result that gives some feeling for the size of $n!$.

Stirling's Formula. $n!$ is approximately equal to the function

$$\sqrt{2\pi n}\left(\frac{n}{e}\right)^n$$

where $e = 2.71828\ldots$ is the base of the natural logarithm.

We do not define what "approximately equal" means, but a consequence of Stirling's formula is that both

$$n! = O\left(\left(\frac{n}{e}\right)^n \sqrt{n}\right) \quad \text{and} \quad \left(\frac{n}{e}\right)^n \sqrt{n} = O(n!).$$

By all accounts, $n!$ quickly becomes humongous.

We have used the word permutation previously in Chapter 1 in the study of functions. We said that a one-to-one and onto function f, $f:D \to D$, is called a permutation. When D is a finite set, such as $I_n = \{1, 2, \ldots, n\}$, then the two definitions are really the same. If $f:I_n \to I_n$ is a one-to-one function (which is necessarily onto by Theorem 9.1 from Chapter 1), then we get a permutation, in the sense of this section, from the listing $\langle f(1)\ f(2)\ \ldots\ f(n)\rangle$. Conversely, given a permutation of I_n $\langle a_1\ a_2\ \ldots,\ a_n\rangle$, this defines a one-to-one function $f:I_n \to I_n$ by setting $f(i) = a_i$ for $i = 1, \ldots, n$.

Example 4.5. Corresponding to the permutation $\langle 1\ 3\ 2\ 5\ 4\ 6\rangle$ is the one-to-one function f defined by $f(1) = 1$, $f(2) = 3$, $f(3) = 2$, $f(4) = 5$, $f(5) = 4$, and $f(6) = 6$. Corresponding to the one-to-one function f defined by $f(i) = 7 - i$ for $i = 1, 2, \ldots, 6$, is the permutation $\langle 6\ 5\ 4\ 3\ 2\ 1\rangle$.

Permutations occur in many mathematical and computer science contexts. In this course we shall meet them again in a game in the next section as well as in later chapters. In other courses they occur in diverse settings such as linear algebra, sorting and searching problems, in a variety of games and path traversing questions, and even in English bell ringing. The goal in English bell ringing, called change ringing, is to ring all permutations on however many bells are in the tower. Fortunately, most towers have at most 8 or 10 bells.

EXERCISES FOR SECTION 4

1. List all permutations of $\{x, y, z\}$ and of $\{a, b, c, d\}$.
2. Find the least integer n such that $n! > 3^n$. Then find any integer n such that $n! > 4^n$.
3. English bell ringers, called change ringers, ring tower and hand bells following sequences of permutations; however, one of the requirements of change ringing is that a permutation p may be followed by a permutation p' only if for $i = 2, 3, \ldots, n-1$, the number in position i of p is in position $i-1$, i or $i+1$ of p'. The number in position 1 of p may stay in position 1 or move to position 2, and the number in position n may stay in position n or move to position $n-1$ of p'. Thus the following sequence of three permutations is legal from

the perspective of bell ringing: $\langle 1\ 2\ 3\rangle$, $\langle 1\ 3\ 2\rangle$, $\langle 3\ 1\ 2\rangle$, but the next permutation (as produced by PERM) $\langle 2\ 1\ 3\rangle$ is not a legal follow-up to $\langle 3\ 1\ 2\rangle$, since number 3 jumped from the front to the back. Find a sequence of all 3! permutation on $\{1, 2, 3\}$ that is suitable for bell ringing. Do the same for the 4! permutations on $\{1, 2, 3, 4\}$.

4. Which is larger $n^{\sqrt{n}}$ or $(\sqrt{n})^n$?

5. Prove that $n^{\sqrt{n}} = O(2^n)$. Either prove or disprove the statement $(\sqrt{n})^n = O(2^n)$.

6. Prove that $n^{\sqrt{n}} \neq O(p(n))$ for any polynomial $p(n)$.

7. Prove that $r^n = O(n!)$.

8. Prove the following corollary of Theorem 4.2: For every real number r, $n! \neq O(r^n)$.

9. Check the accuracy of Stirling's formula: For $n = 5, 6, \ldots, 10$ use a calculator to evaluate $\sqrt{2\pi n}(n/e)^n$ using $e = 2.71828$ and compare these values with $n!$.

10. Let f be the one-to-one function defined on $\{1, 2, \ldots, 6\}$ by $f(1) = 2$, $f(2) = 1$, $f(3) = 4$, $f(4) = 3$, $f(5) = 6$, and $f(6) = 5$. Show that $f^2 = f \circ f$ is the identity mapping. Find a one-to-one function g on the same domain for which g^2 is not the identity. For this g find the least integer i such that g^i (g composed with itself i times) is the identity. Then write down the i permutations: $\langle g(1) \ldots g(6)\rangle$, $\langle (g(1))^2 \ldots (g(6))^2\rangle, \ldots$, and $\langle (g(1))^i \ldots (g(6))^i\rangle$. Are these all of the possible permutations on six numbers?

11. Given a permutation $p = \langle a_1\ a_2\ \ldots\ a_n\rangle$, an **adjacent transposition** is the switching of two items in positions i and $(i + 1)$ for some value of $i = 1, 2, \ldots, n - 1$. For example, $\langle a_2\ a_1\ \ldots\ a_n\rangle$ is the result of applying an adjacent transposition to p as is $\langle a_1\ a_2\ \ldots\ a_n\ a_{n-1}\rangle$. Given permutations p and p' we define the distance between them to be the minimum number of adjacent transpositions needed to transform one into the other; we denote this by $d(p, p')$. Determine the following values of $d(p, p')$:
(*a*) $p = \langle 1\ 2\ 3\ 4\rangle$ $\quad p' = \langle 4\ 3\ 2\ 1\rangle$
(*b*) $p = \langle 1\ 2\ 3\ 4\rangle$ $\quad p' = \langle 3\ 1\ 4\ 2\rangle$
(*c*) $p = \langle 3\ 2\ 1\ 4\rangle$ $\quad p' = \langle 1\ 3\ 2\ 4\rangle$.

12. We define a permutation p to be either **even** or **odd** according as $d(p, p^*)$ is an even or an odd number when p^* is the identity permutation $\langle 1\ 2\ 3\ \ldots\ n\rangle$. Determine which of the four nonidentity permutations of the previous problem are even and which are odd.

13. For which permutation p on $\{1, 2, \ldots, n\}$ is $d(p, p^*)$ the largest when p^* is the identity permutation? Also, what is the value of $d(p, p^*)$ for this permutation?

14. How many even and how many odd permutations are there on $\{1, 2, \ldots, n\}$? Prove your answer.

15. Modify the algorithm PERM so that it outputs only even permutations.

16. Design an algorithm that upon input n, a positive integer, lists all permutations of $\{1,2,\ldots,n\}$ without first listing all permutations on $\{1,2,\ldots,n-1\}$. (*Hint:* Look at the order of the output of PERM and see if there is a way to move directly from one permutation to the next.)

17. Prove that $n! \le n^n$ and thus show that n^n is a new candidate for the biggest function in the hierarchy.

18. Find a function $f(n)$ such that $f(n) \ne O(n^n)$.

19. Let P be the positive integers and suppose that we define a permutation of P to be a reordering of P. If f is a function with domain and target P, what properties must f have so that a listing of its range $\langle f(1)\ f(2)\ \ldots\rangle$ is a permutation (or reordering) of P?

20. There are $n!$ different ways to make a straight line of n people. Suppose instead that n people are seated at a circular table and two seating arrangements are considered the same if everyone has the same person on their left side and the same person on their right side. How many different seating arrangements are there? Suppose that two arrangements are considered the same if everyone has the same set of two people on their left and right, but which side doesn't matter. Then how many different seating arrangements are there?

3:5 AN APPLICATION OF PERMUTATIONS: THE GAME OF MASTERMIND

We turn to some colorful "near" permutations that arise in the game of Mastermind. Mastermind is a two-person guessing game that is usually played without special knowledge of permutations and set theory. As you read this section, you might consider whether mathematical training helps in playing this game.

Mastermind is played as follows. The first player secretly writes down a list or code of four colors, chosen from red, yellow, green, blue, purple, and white. We denote the set of colors by

$$C = \{r, y, g, b, p, w\}$$

where each initial stands for the color beginning with the same letter. Repetitions of colors in the secret list are allowed. Then the second player tries to guess the colors and their order in the secret list.

For example, suppose that I choose

$$p\ b\ p\ g$$

as my secret list. Then when you guess a list of four colors, I must tell you two pieces of information. First I must tell you how many of your colors are correct and in the right position. Next I compare the remaining colors in my secret code

Table 3.3

Guess	*Position and Color Correct*	*Color Correct; Position Incorrect*
$r\ b\ r\ b$	1	0
$b\ g\ b\ g$	1	1

and in your guess and tell you how many of these colors match correctly but are in the wrong position. Two guesses and responses are listed in Table 3.3.

To use the language of set theory, both the secret code and all guesses come from the Mastermind universe, which is the Cartesian product C^4. When player 2 makes a guess, that is, selects an element from C^4, player 1 produces $f(\text{guess})$, where f is a function whose domain is C^4 and whose target is $\{0, 1, 2, 3, 4\}^2$. Thus the image of any particular guess is an ordered pair of integers as described above.

Example 5.1. The answers to some possible guesses of a secret code are listed in Table 3.4.

Table 3.4

Guess	*Position and Color Correct*	*Color Correct; Position Incorrect*
$r\ r\ r\ r$	1	0
$y\ y\ y\ y$	1	0
$g\ g\ g\ g$	0	0
$b\ b\ b\ b$	1	0
$p\ p\ p\ p$	0	0
$w\ w\ w\ w$	1	0

Question 5.1. Figure out the four colors involved in the secret code of Example 5.1. Can you tell what their order is?

Example 5.1 (continued). Some more guesses and answers about the secret code in Example 5.1 are shown in Table 3.5.

Table 3.5

$r\ y\ b\ w$	1	3
$r\ y\ w\ b$	0	4
$r\ w\ y\ b$	1	3
$w\ r\ y\ b$	0	4

Question 5.2. Can you determine from the information in Example 5.1 what the secret code is? If not, try to figure out as many of the correct color positions as possible from these answers.

Question 5.3. Find a classmate, roommate, or any willing soul, and play six games of Mastermind in three of which you pick the secret code and in three of which you try to guess the secret code. How many guesses were needed in each game?

Experienced Mastermind players can figure out the secret code in four or five guesses. Their strategy combines a knowledge of good initial guesses plus close logical analysis of the responses to these. We develop some strategies of our own now. One strategy would be to repeatedly pick elements of C^4 at random until an answer of $(4, 0)$ results. Since there are six colors and four positions to fill with repetitions possible, there are $6^4 = 1296$ different possible Mastermind codes. Thus random guessing will not be an effective playing strategy! We shall describe a method that can determine the list of colors with only six guesses.

An effective playing strategy for this game (and any game) is just an algorithm that tells us what to do in any possible situation. We shall develop three different strategies for Mastermind. We begin with one that comes from the idea behind the guesses used in Example 5.1.

Problem. To determine the secret list of colors in a Mastermined game.

Algorithm 1

STEP 1. For each color x in $\{r, y, g, b, p, w\}$ guess $x\ x\ x\ x$.

STEP 2. (Now that you know the colors in the list) guess all possible lists using the four (or fewer) colors involved.

Notice that both steps 1 and 2 really involve several steps. Step 1 is shorthand for six steps or six guesses.

Question 5.4. Suppose that you learn in step 1 that the colors involved are green, blue, purple, and white. How many different secret codes are there involving these colors? How could you go about writing down all of these possible lists?

In fact, when we discover that the code involves four different colors, then the code is one of the permutations of these colors. But what happens when there are repeated colors?

Example 5.2. Suppose that the colors involved in the secret code are green, blue, and purple and that green occurs twice. Then the code will be a **"near" permutation** of $\{g, g, b, p\}$, that is, an ordered list of these letters including two g's. (A "real" permutation is an ordered list of distinct elements.) A set with repeated elements is known as a **multiset**. However, our counting and algorithmic techniques will solve this case too. In each code word the g's will occupy two positions, for example, the first and the third position. We associate their positions with the 2-subset (in this instance $\{1,3\}$) of the four possible positions $P = \{1,2,3,4\}$. Conversely, every 2-subset of P gives us a prescription for where to place the g's. Thus there are $\binom{4}{2}$ ways to fill in the g's into a color code. The remaining two places can be filled with first b and then p or with first p, then b. Thus the total number of codes using $\{g, g, b, p\}$ is $\binom{4}{2} \cdot 2 = 12$.

Furthermore, PAIR and JSET, with $n = 4$ and $j = 2$, provide a list of all 2-subsets of the 4-set P, and from these we position the g's. Then each of these leads to two code words by filling in b and g in the two possible orders. For example $\{1,2\}$ leads to $g\ g\ b\ p$ and $g\ g\ p\ b$.

Question 5.5. List all possible codes formed from $\{g, g, b, p\}$.

Question 5.6. Look back at Algorithm 1. What is the maximum number of steps or guesses that you will need in step 2?

The total number of guesses needed in Algorithm 1 is rather large, and you probably can think of better ways to play. Here is another approach that may more closely resemble how you and most people play Mastermind. The idea is to focus on the set of all possible codes, then on a subset of this set, and on a subset of the subset until the subset is pared down to one code.

Algorithm 2

STEP 1. Let $L = \{$all possible Mastermind codes$\}$
STEP 2. Repeat
  Begin
    STEP 3. Pick an element from L to be your guess
    STEP 4. Update L {remove from L all those codes that are inconsistent with the response to your guess}
  End
  Until your guess is correct
STEP 5. Stop.

COMMENT. The form of step 2, "**Repeat** . . . **Until** . . ." is like the While . . . do loop, only the condition for ending the loop is checked at the end of the execution of the loop.

Algorithm 2 requires a lot of work. Initially, the list L in step 1 has 1296 codes in it, and the checking and updating will be time-consuming. Neither this approach nor that of Algorithm 1 would be so bad for a computer to implement, since we can use the computer's large memory and quick access to it to play the game effectively.

Here is a third approach, possibly effective for use by both humans and computers.

Algorithm 3

STEP 1. Make the following four guesses:

$$r\ r\ y\ y$$
$$r\ g\ r\ g$$
$$b\ b\ p\ p$$
$$b\ w\ b\ w$$

STEP 2. Create L, a list of all possible Mastermind codes that are consistent with the answers you receive from the first four guesses

STEP 3. Repeat

Begin

STEP 4. Pick an element from L to be your guess

STEP 5. Update L {remove from L all those codes that are inconsistent with the response to your guess}

End

Until your guess is correct

STEP 6. Stop.

Try playing a few games using this algorithm. You will be surprised at how small the list L is that you create in step 2. In fact, we claim that after at most one more guess you can always figure out the secret list; so, in a total of six guesses you will have the answer nailed down. (See Supplementary Exercise 15.)

There are a few ideas that can be derived from thinking about and playing Mastermind. At first you might wonder why we bother to create a formal algorithm to play the game. We all can quickly learn effective guesses to make, and so why not play just using these hunches and logical deductions? In fact, that's how most of us do play this and many other games. But as soon as we turn the process around so that the computer is making the guesses, then it becomes essential to have an algorithm so that the computer has a way to proceed. If we start to analyze how we play the game, we soon see that we have lots of different strategies that we adapt according to the responses. It would be quite a feat to design an algorithm that would accommodate all the different possibilities that the mind thinks up. In fact, it is quite impressive how clever and logical the human mind is.

Instead of trying to simulate the logical working of a human brain, we have chosen very straightforward algorithms that could be programmed easily and that will lead to success with a relatively small number of guesses. We have also exploited the fact that the computer can do lots of checking rather quickly. You are invited to write (or find a friendly computer programmer to write) a program using Algorithms 2 or 3, and then to race the computer to see who is faster!

EXERCISES FOR SECTION 5

1. Find a solution to each of the following Mastermind games given the indicated guesses and answers:

	Guess	Position and Color Correct	Color Correct; Position Incorrect
(a)	r b r b	0	0
	w y w y	0	0
	g p g p	3	0
	g g p g	0	2
(b)	r b r b	2	0
	w w y y	0	2
	r b y w	1	3
	y b r w	0	4
(c)	r y w b	1	1
	g b p w	1	0
	r y p w	2	0
	y g b p	0	0
	w r w r	1	2
(d)	r r y y	0	0
	r g r g	0	0
	b b p p	2	0
	b w b w	1	1
	w p w p	2	2
(e)	b g p r	0	4
	r p g b	2	2
	r g p b	1	3
	r p b g	1	3
	r b g p	0	4

2. Explain why there are no solutions to the following guesses and answers:

	Guess	Position and Color Correct	Color Correct; Position Incorrect
(a)	r b r b	0	1
	y g y g	1	0
	p w p w	0	1
	r y p w	1	3

(*b*)	$r\ g\ b\ b$	2	2
	$b\ g\ r\ b$	1	3
	$g\ r\ b\ b$	0	4
(*c*)	$b\ r\ g\ b$	0	3
	$y\ y\ p\ w$	1	0
	$y\ r\ g\ b$	1	2
	$r\ b\ y\ b$	2	1

3. Suppose you learn that there are two colors, green and blue, in the code word. How many code words are possible? List them all.
4. Write a list of four Mastermind guesses and responses for which there are **(a)** no solutions, **(b)** one solution, and **(c)** two or more solutions.
5. How many Mastermind codes begin with r? How many Mastermind codes begin with r and contain no other r? How many Mastermind codes begin with r and contain another r?
6. How many possible codes are there for which the response to the guess $r\ b\ b\ b$ is 0 (color and position correct) and 1 (color but not position correct)?
7. List all Mastermind codes formed from $\{p, p, y, b\}$. Then do the same for $\{p, p, y, y\}$.
8. Associated with the game of Mastermind is a function $f: C^4 \rightarrow \{0, 1, 2, 3, 4\}^2$ as described in this section. What is the range of f?
9. Answer the following questions (without listing all possibilities.) Suppose that we play a version of Mastermind in which we pick a code of length 5 from the same set of six colors. Otherwise, all the rules are the same.
   (*a*) How many possible code words are there?
   (*b*) How many code words are there in which the five colors $\{r, b, g, y, p\}$ each appear once?
   (*c*) How many code words are there if the colors $\{r, b, g\}$ each appear once and w appears twice?
   (*d*) How many code words are there if the colors $\{r, b\}$ each appear once and p appears three times?
10. How many code words are consistent with the following set of guesses and responses?

$r\ r\ y\ y$	1	0
$r\ g\ r\ g$	0	0
$b\ b\ p\ p$	1	0
$b\ w\ b\ w$	2	1

11. Define a new game called Trivialmind in which only three colors are used (red, blue, and white) and the secret code consists of a list of only two colors. How many different secret color codes are there in Trivialmind?

12. Describe a variation on Algorithm 1 that will work for Trivialmind. What is the maximum number of guesses that will be made using this algorithm?

13. Describe a variation on Algorithm 3 that will work for Trivialmind. Using this new version of Algorithm 3, how many guesses will player 2 need to guarantee a correct guess?

14. Show that no algorithm will in all cases correctly solve Trivialmind with three or fewer guesses.

15. The point of this exercise is to improve Algorithm 1. In step 1 we guessed $x\ x\ x\ x$ for each x in $\{r, y, g, b, p, w\}$. Figure out a way to reduce these six questions to five. Now suppose that we know the colors present in the secret code and their frequency. Suppose that the color x is present (at least once) and that the color z is missing. Notice that at least two colors must be missing. What can you learn from the following four questions:

$$z\ z\ z\ x$$

$$z\ z\ x\ z$$

$$z\ x\ z\ z$$

$$x\ z\ z\ z?$$

Can you learn this same information with only three guesses? Using these ideas, try to devise an algorithm along these lines with as few guesses as possible. How many guesses do you make in general?

16. Pick a mastermind code and apply Algorithm 3. How large is your initial list L?

17. Suppose that you guess $r\ b\ r\ b$. How many different responses are possible? If you guess $r\ b\ r\ w$, how many different responses are there? What is the minimum number of different responses to any Mastermind guess? What are the guesses that produce this minimum? What is the maximum number of different responses to any Mastermind guess? What are the guesses that produce this maximum?

3:6 THE BINOMIAL THEOREM

This chapter concludes with a result that unifies a variety of facts about binomial coefficients. We know, for example, that the sum of the binomial coefficients $\binom{n}{i}$ for $i = 0, 1, \ldots, n$ equals 2^n, and from Exercise 3.10 we know that the sum of the "even" binomial coefficients $\binom{n}{2j}$ for $j = 0, 1, \ldots, \lfloor n/2 \rfloor$ equals 2^{n-1}. There are many other similar identities lurking around, waiting to be discovered.

Question 6.1. For $n = 3, 4$, and 5 verify that

$$\binom{n}{0} - \binom{n}{1} + \binom{n}{2} - \binom{n}{3} + \cdots + (-1)^i \binom{n}{i} + \cdots + (-1)^n \binom{n}{n} = 0.$$

Question 6.2. Verify that

$$\binom{2}{2} + \binom{3}{2} + \binom{4}{2} = \binom{5}{3}.$$

In general, the sum of the binomial coefficients

$$\binom{2}{2} + \binom{3}{2} + \cdots + \binom{i}{2} + \cdots + \binom{n}{2}$$

can be expressed as a single binomial coefficient $\binom{k}{j}$. For $n = 5$ and 6 find such a binomial coefficient $\binom{k}{j}$. Then for arbitrary n find k and j so that

$$\binom{2}{2} + \binom{3}{2} + \cdots + \binom{i}{2} + \cdots + \binom{n}{2} = \binom{k}{j}.$$

(Here j will be a constant and k will depend upon n.)

Example 6.1. We review some polynomial arithmetic:

$$(1 + x)^2 = 1 + 2x + x^2$$
$$(1 + x)^3 = 1 + 3x + 3x^2 + x^3$$
$$(1 + x)^4 = 1 + 4x + 6x^2 + 4x^3 + x^4.$$

Notice that the coefficients in the expansion above of $(1 + x)^k$ are exactly the numbers in the kth row of Pascal's triangle. Thus the coefficient of x^i in $(1 + x)^n$ appears to be the binomial coefficient $\binom{n}{i}$ for $0 \leq i \leq n$.

Example 6.1 sets us on the right track. We prove the suggested result and find that the answers to Questions 6.1 and 6.2 and much more follow from the so-called Binomial Theorem.

Theorem 6.1. The Binomial Theorem. For n, a natural number,

$$(1 + x)^n = \binom{n}{0} + \binom{n}{1}x + \binom{n}{2}x^2 + \cdots + \binom{n}{i}x^i + \cdots + \binom{n}{n}x^n.$$

Proof. We prove this theorem by induction on n. When $n = 0$,

$$(1 + x)^0 = 1 = \binom{0}{0}.$$

When $n = 1$,

$$(1 + x)^1 = (1 + x) = \binom{1}{0} + \binom{1}{1}x.$$

(You might check the cases of $n = 2$ and 3 to make sure that the statement of the theorem is clear.)

The inductive hypothesis is that

$$(1 + x)^k = \binom{k}{0} + \binom{k}{1}x + \cdots + \binom{k}{i}x^i + \cdots + \binom{k}{k}x^k.$$

From this we must determine $(1 + x)^{k+1}$. By factoring,

$$\begin{aligned}(1 + x)^{k+1} &= (1 + x)^k(1 + x) \\ &= \left[\binom{k}{0} + \binom{k}{1}x + \cdots + \binom{k}{i}x^i + \cdots + \binom{k}{k}x^k\right](1 + x)\end{aligned}$$

by the inductive hypothesis,

$$\begin{aligned}&= \left[\binom{k}{0} + \binom{k}{1}x + \cdots + \binom{k}{i}x^i + \cdots + \binom{k}{k}x^k\right] \\ &\quad + \left[\binom{k}{0}x + \cdots + \binom{k}{i-1}x^i + \cdots + \binom{k}{k-1}x^k + \binom{k}{k}x^{k+1}\right]\end{aligned}$$

by multiplying first by 1 and then by x.

Now we must collect like terms together. That is, first we find the constant term, the coefficient of x^0. It is just $\binom{k}{0} = 1 = \binom{k+1}{0}$, since all terms in the second bracket involve x. What are the terms involving $x^1 = x$? We have $\binom{k}{1}x$ and $\binom{k}{0}x$, from the first and second bracket, respectively, or

$$\binom{k}{1}x + \binom{k}{0}x = (k + 1)x = \binom{k+1}{1}x.$$

In general, what are the coefficients of x^i? In the first bracket we have $\binom{k}{i}x^i$ and in the second bracket we have $\binom{k}{i-1}x^i$. Thus in the sum we have

$$\begin{aligned}\binom{k}{i}x^i + \binom{k}{i-1}x^i &= \left[\binom{k}{i} + \binom{k}{i-1}\right]x^i \\ &= \binom{k+1}{i}x^i \qquad \text{by Corollary 2.2.}\end{aligned}$$

Finally, the term x^{k+1} appears only in the second bracket and we have

$$\binom{k}{k}x^{k+1} = x^{k+1} = \binom{k+1}{k+1}x^{k+1}. \qquad \square$$

For comparison we offer the following.

Second proof of Theorem 6.1. We want to determine the expansion of $(1 + x)^n$ as a polynomial:

$$(1 + x)^n = a_0 + a_1x + \cdots + a_ix^i + \cdots + a_nx^n.$$

That is, we want to determine the coefficients $a_0, a_1, \ldots, a_n$. Now

$$(1 + x)^n = (1 + x)(1 + x)\cdots(1 + x),$$

a product of n identical factors. A term in the product results from every way of selecting one element (either 1 or x) from each pair of parentheses and multiplying them together. For example, if we select 1 from the first parenthesis, x from the second and third, and 1 from all the rest, we multiply these to get the product x^2. Thus this selection contributes 1 to a_2, the coefficient of x^2. Suppose that we designate this choice by an n-vector with a zero in the entries where we select a $1\ (= x^0)$, and a one in the entries where we select an $x\ (= x^1)$. Thus in our example we designate our selection by the vector $(0, 1, 1, 0, 0, \ldots, 0)$. Then the number of ways to choose elements one from each parenthesis, to multiply together and get x^2 is the same as the number of bit vectors with two ones and $(n - 2)$ zeros. From Example 3.2 we know that the number of such vectors equals $\binom{n}{2}$, and so $a_2 = \binom{n}{2}$.

In general, a_i, the coefficient of x^i in the expansion of $(1 + x)^n$ equals the number of ways to select i xs and $(n - i)$ 1s, one from each of the n sets of parentheses. This number equals the number of bit vectors with i ones and $(n - i)$ zeros or $\binom{n}{i}$ as we saw in Example 3.2. $\square$

Notice that Theorem 6.1 gives us a polynomial identity, that is, an equation that is true upon substitution of any real value of x. We now substitute different values for x and see what happens.

Example 6.2. Substituting $x = 1$ in the Binomial Theorem shows that

$$\begin{aligned} 2^n &= (1+1)^n \\ &= \binom{n}{0} + \binom{n}{1}1 + \binom{n}{2}1^2 + \cdots + \binom{n}{n}1^n \\ &= \binom{n}{0} + \binom{n}{1} + \cdots + \binom{n}{n}, \end{aligned}$$

the same result as in Corollary 3.2.

Example 6.3. Substituting $x = -1$ in the Binomial Theorem leads to

$$\begin{aligned} 0 &= (1-1)^n \qquad \text{if } n > 0 \\ &= \binom{n}{0} + \binom{n}{1}(-1) + \binom{n}{2}(-1)^2 + \cdots + \binom{n}{n}(-1)^n \\ &= \binom{n}{0} - \binom{n}{1} + \binom{n}{2} - \binom{n}{3} + \cdots + (-1)^n\binom{n}{n}, \end{aligned}$$

a phenomenon we observed in Question 6.1. Notice that we may rearrange the last equation to read

$$\binom{n}{0} + \binom{n}{2} + \binom{n}{4} + \cdots = \binom{n}{1} + \binom{n}{3} + \binom{n}{5} + \cdots.$$

Since the sum of the left-hand side plus the right-hand side equals 2^n, we see that each side separately equals $(\frac{1}{2})(2^n) = 2^{n-1}$, a result that we noted in the first paragraph of this section.

How does Question 6.2 relate to the Binomial Theorem? We can't obtain the result suggested there by merely substituting in a number for x, but we can use the Binomial Theorem to derive it. We have previously seen in Chapter 2, Example 3.3, another polynomial identity that gives the sum of a (finite) geometric series:

$$1 + x + x^2 + \cdots + x^i + \cdots + x^n = \frac{1 - x^{n+1}}{1 - x}.$$

Since this is an identity that holds true for all x, except for $x = 1$, we can substitute both values and expressions for x, and the equation still holds true. For example, if we substitute $(1 + x)$ for x everywhere in the geometric series we get the result that

$$\begin{aligned} 1 + (1 + x) + (1 + x)^2 + \cdots + (1 + x)^i + \cdots + (1 + x)^n \\ = \frac{1 - (1 + x)^{n+1}}{1 - (1 + x)} \qquad \text{provided } x \neq 0 \\ = \frac{1 - (1 + x)^{n+1}}{-x} \\ = \frac{(1 + x)^{n+1} - 1}{x} \qquad \text{for } x \neq 0. \qquad (*) \end{aligned}$$

This result tells us that if we multiply out both the right-hand side and the left-hand side of equation (*), then the resulting polynomials are the same.

In particular, let's look at the x^2 term on both sides of equation (*). On the left-hand side at first there are no x^2 terms, but then x^2 appears in the expansion of all the terms from $(1 + x)^2$ on up to $(1 + x)^n$. By the Binomial Theorem, here are the x^2 terms:

$$\begin{aligned} \binom{2}{2}x^2 + \binom{3}{2}x^2 + \cdots + \binom{i}{2}x^2 + \cdots + \binom{n}{2}x^2 \\ = \left[\binom{2}{2} + \binom{3}{2} + \cdots + \binom{i}{2} + \cdots + \binom{n}{2}\right]x^2. \end{aligned}$$

The x^2 term on the right-hand side of (*) will result from dividing the x^3 term of the numerator by the x in the denominator. By the Binomial Theorem the x^3 term of $[(1 + x)^{n+1} - 1]$ is $\binom{n+1}{3}x^3$. Thus

$$\binom{n+1}{3}\frac{x^3}{x} = \binom{n+1}{3}x^2.$$

Since (*) gives two expressions for the same underlying polynomial, the coefficient of x^2 on the left and on the right must be the same, that is,

$$\left[\binom{2}{2} + \binom{3}{2} + \cdots + \binom{i}{2} + \cdots + \binom{n}{2}\right]x^2 = \binom{n+1}{3}x^2,$$

or

$$\binom{2}{2} + \binom{3}{2} + \cdots + \binom{i}{2} + \cdots + \binom{n}{2} = \binom{n+1}{3}.$$

There is another more general form of the Binomial Theorem that gives the expansion of a polynomial with two variables:

$$(y + x)^n = \binom{n}{0} y^n x^0 + \binom{n}{1} y^{n-1} x^1 + \cdots + \binom{n}{i} y^{n-i} x^i + \cdots + \binom{n}{n} y^0 x^n.$$

This result can be proved in essentially the same way as Theorem 6.1.

A common problem in combinatorial mathematics is to count objects with specified properties, for example, to count the number of j-subsets of an n-set. Sometimes a neat formula can be obtained; other times the answer can be derived from a polynomial identity. One often can deduce that the answer to the problem is, say, the coefficient of x^j in a certain polynomial. We have seen that the number of j-subsets of an n-set is the coefficient of x^j in the polynomial $(1 + x)^n$. Such polynomial solutions are known as **generating functions**. The use of generating functions is a powerful technique with wide application. In this section we have seen an introduction to the methods and use of generating functions.

EXERCISES FOR SECTION 6

1. Expand $(1 + x)^5$ and $(1 + x)^6$ and check that the coefficients are the binomial coefficients promised by Theorem 6.1. [Reminder: Do you really have to multiply $(1 + x)$ by itself four and five times, respectively?]

2. ***(a)*** Find a simple formula for

$$\binom{n}{0} + \binom{n}{1} 2 + \binom{n}{2} 4 + \cdots + \binom{n}{i} 2^i + \cdots + \binom{n}{n} 2^n,$$

and for

$$\binom{n}{0} - \binom{n}{1} 2 + \binom{n}{2} 4 - \cdots + (-1)^i \binom{n}{i} 2^i + \cdots + (-1)^n \binom{n}{n} 2^n.$$

(b) Prove that if n is even, then

$$\binom{n}{0} + \binom{n}{2} 4 + \binom{n}{4} 16 + \cdots + \binom{n}{n} 4^{n/2} = \frac{3^{n+1}}{2}.$$

3. Use the Binomial Theorem to derive the expansion of $(1 + 1/x)^n$.

4. In each of the following, first verify the equation by selecting a value for r, n, and m and checking the equation for these values. Then prove the general result about sums of binomial coefficients.

(*a*) $\binom{r}{0} + \binom{r+1}{1} + \cdots + \binom{r+i}{i} + \cdots + \binom{r+n}{n} = \binom{r+n+1}{n}$.

(*b*) For $m \le n$, $\binom{m}{m} + \binom{m+1}{m} + \cdots + \binom{i}{m} + \cdots + \binom{n}{m} = \binom{n+1}{m+1}$.

(*c*) For $n < r$, $\binom{r}{0} - \binom{r}{1} + \cdots + (-1)^i\binom{r}{i} + \cdots + (-1)^n\binom{r}{n} = (-1)^n\binom{r-1}{n}$.

5. (*a*) For $r \le 5$, verify that

$$\binom{5}{0}\binom{6}{r} + \binom{5}{1}\binom{6}{r-1} + \binom{5}{2}\binom{6}{r-2} + \cdots + \binom{5}{r}\binom{6}{0}$$
$$= \binom{5+6}{r} = \binom{11}{r}.$$

(*b*) Now we want to show, in general, that for r less than or equal to both m and n,

$$\binom{m}{0}\binom{n}{r} + \binom{m}{1}\binom{n}{r-1} + \binom{m}{2}\binom{n}{r-2} + \cdots + \binom{m}{r}\binom{n}{0} = \binom{m+n}{r}.$$

It is clear that $(1+x)^m(1+x)^n = (1+x)^{m+n}$. Use the Binomial Theorem to find the coefficient of x^r in $(1+x)^{m+n}$. Then use the Binomial Theorem to expand $(1+x)^m$ and $(1+x)^n$ and determine the coefficient of x^r in the product

$$(1+x)^m(1+x)^n.$$

From these results, conclude that the previous equation holds.

6. Explain why

$$(1+x)^n\left(1+\frac{1}{x}\right)^n = \frac{(1+x)^{2n}}{x^n}.$$

Then find the constant term (the coefficient of x^0) in $(1+x)^n\left(1+\frac{1}{x}\right)^n$ and find the constant term in the expansion of $\frac{(1+x)^{2n}}{x^n}$. See Supplementary Exercise 17 for an application.

7. Prove the general form of the Binomial Theorem that gives the expansion of $(y+x)^n$.

8. Find a simple formula for the sum of the binomial coefficients

$$\binom{3}{3} + \binom{4}{3} + \cdots + \binom{i}{3} + \cdots + \binom{n}{3}.$$

Next prove the formula using the ideas of the solution to Question 6.2. Then find and prove a formula for the sum

$$\binom{k}{k} + \binom{k+1}{k} + \cdots + \binom{i}{k} + \cdots + \binom{n}{k} \qquad \text{for } k \leq n.$$

3:7 IMPORTANT SUBSETS

The most important ideas in this chapter are those about sets: j-subsets, permutations, and the counting of these sets. It is worth repeating that this set theory along with that in Chapter 1 is a crucial element in all further study of mathematics, computer science, and algorithms. The new counting functions, the factorial and the binomial coefficients, will appear frequently. Already the latter have given us the number of j-subsets of an n-set, the number of shortest paths in a rectangular grid, the number of n-bit sequences with a specified number of ones, and the coefficients in the expansion of $(1 + x)^n$. We shall meet these functions in every future chapter; we shall use the subset and permutation ideas repeatedly and without further ado. To be honest, Mailmobiles haven't caught on with quite the popularity initially expected, but the path counting problem is a classic.

The algorithms of this chapter, JSET and PERM, are of more than illustrative value. They are used repeatedly, for example, in Mastermind and in more serious applications in Chapter 5. In these algorithms we introduced nested loops and Boolean variables. Although these algorithms and algorithmic techniques are more complex than those of the previous chapter, they were selected for their relative simplicity. The "state of the art" algorithms for generating subsets and permutations are more sophisticated and a little more efficient. However, no amount of trickery can get around the fact that generating all permutations of an n-set and all subsets of size $\lfloor n/2 \rfloor$ requires an exponential amount of work.

This study of counting, generating, and listing objects of a certain size provides an introduction to the areas of combinatorial analysis and combinatorial algorithms. In these modern and fast-growing fields, a typical problem involves counting all objects with a particular structure and listing them in a certain order. The counting task is often solved using functions like the binomial coefficients and techniques like generating functions. The listing is frequently done by imposing an order on the structure such as lexicographic order.

In Chapter 5 we shall focus on finite sets and their 2-subsets, an area known as graph theory. The study of orderings, both constructing them and searching them, is the subject of Chapter 6.

SUPPLEMENTARY EXERCISES FOR CHAPTER 3

1. Given the rectangular grid shown here how many different shortest paths join the vertices labeled R and S? How many different shortest paths are there from R to S that do not go through the vertex labeled P? How many shortest paths are there from R to S that do not go through either the vertex labeled P or the vertex labeled Q?

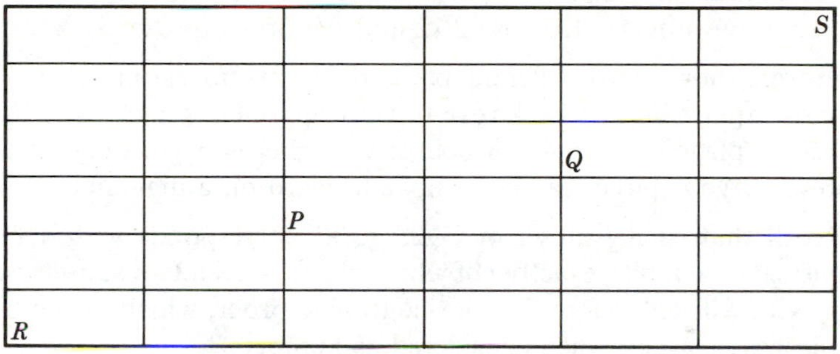

2. How many 4-subsets of $\{1, \ldots, 8\}$ contain the number 3? How many contain 5? How many 4-subsets contain either 3 or 5?

3. Count the number of 10-digit ternary numbers (i.e., numbers that use only the digits 0, 1, and 2). How many of these contain exactly four zeros? How many contain exactly four ones? How many contain three zeros and three ones?

4. Find a formula for the number of n-digit ternary numbers that contain i zeros and $(n - i)$ ones and twos (combined). How many contain exactly i zeros and j ones when i and j are positive integers with $i + j \leq n$?

5. The number of n-letter "words" made up from the letters $a, b,$ and c is the same as the number of n-digit ternary numbers—why? How many n-letter words are there with i a's, j b's and $(n - i - j)$ c's?

6. ***(a)*** Is the following true or false? $(2n)! = 2^n \cdot n!$ Explain.

(b) For an odd number $(2n - 1)$ we define the odd factorial

$$(2n - 1)O! = (2n - 1)(2n - 3) \cdot \cdots \cdot 3 \cdot 1.$$

Thus $(3)O! = 3 \cdot 1 = 3$, and $(5)O! = 5 \cdot 3 \cdot 1 = 15$. Calculate $(7)O!$ and $(9)O!$.

(*c*) Explain why the following new formula is correct:

$$\binom{2n}{n} = \frac{2^n \cdot (2n-1)O!}{n!}.$$

(*d*) Show that $(2n-1)O!/n! < 2n$ and then deduce that $\binom{2n}{n} = O(4^n)$.

7. Define the even factorial $(2n)E!$ in a way analogous to the odd factorial $(2n-1)O!$. Then find a formula that relates $(2n)E!$ and $n!$.

8. Reread Exercise 2.12 where the factorial representation of a number is defined. Then prove that every positive integer has a unique factorial representation.

9. Find an algorithm that with input n, a positive integer, outputs the factorial representation of n (see Exercise 2.12 and the previous exercise).

10. In checkers, how many different paths (of legal moves) are there from the rightmost square in your back row, to your opponent's back row? Recall that checkers is played on an 8×8 board; your pieces begin only on the black squares and your pieces can only move forward on a diagonal.

11. (*a*) Recall that each j-subset of $\{1, 2, \ldots, n\}$ corresponds with a binary sequence with n bits, exactly j of which are 1s. Given two subsets, say S and T, with S listed before T in lexicographic order, which of the binary sequences is larger (when considered as numbers)?

(*b*) Design an algorithm that will produce all the j-subsets of an n-set in lexicographic order by manipulating the bit vectors.

12. Let $I_n = \{1, \ldots, n\}$ and j be fixed. Define

$$U = \{B\colon B \text{ is a } j\text{-subset of } I_n \text{ that contains } 1\}$$

and

$$V = \{B\colon B \text{ is a } j\text{-subset of } I_n \text{ that does not contain } 1\}.$$

Which of U and V is the larger set?

13. Let $I_6 = \{1, 2, 3, 4, 5, 6\}$. How many permutations of I_6 have the even numbers in their correct positions (i.e., 2 is in the second position, etc.)? How many permutations of I_6 have the even numbers in even-numbered positions?

14. How many numbers from 1000 to 3000 have their digits all in the set $\{1, 2, 3, 4, 5\}$? How many such numbers have no repetitions in their digits?

15. Show that Algorithm 3 from Section 5 will always work in at most six guesses. (*Hint:* One way to verify this is using a computer search.)

16. Find an algorithm for Mastermind that uses at most five questions in all cases.

(*Warning:* The solution to this problem is hard enough so that it has been written up in a journal.)*

17. Guess a formula for $\binom{n}{0}^2 + \binom{n}{1}^2 + \cdots + \binom{n}{n}^2$. Prove your guess.

18. How many multiplications are needed to calculate $n!$? How many multiplications and divisions are needed to calculate $\binom{n}{j}$?

19. For $j \le \lfloor n/2 \rfloor$, find a way to calculate $\binom{n}{j}$ using at most $2j - 2$ multiplications and divisions.

20. Prove the following identity: $\binom{n}{i} = (n/i)\binom{n-1}{i-1}$. Here is another way to calculate $\binom{n}{i}$:

$\binom{n-i+1}{1} = n - i + 1,$	using no multiplications or divisions
$\binom{n-i+2}{2} = \frac{n-i+2}{2}\binom{n-i+1}{1}$	by the identity above
$\binom{n-i+3}{3} = \frac{n-i+3}{3}\binom{n-i+2}{2}$	by the same identity
$\cdots$	
$\binom{n-i+i}{i} = \binom{n}{i} = (n/i)\binom{n-1}{i-1}$	by repeated application of the identity

Using this approach, determine the number of multiplications and divisions used to calculate $\binom{n}{i}$. Compare this result with your answers in the previous two exercises.

21. Use the results of Exercise 4 to find the expansion of $(1 + x + y)^n$. That is, write this as a polynomial of two variables:

$$a_{0,0} + a_{1,0}x + a_{0,1}y + a_{1,1}xy \\ + a_{2,0}x^2 + a_{0,2}y^2 + \cdots + a_{j,k}x^j y^k + \cdots$$

and find a formula for each coefficient $a_{j,k}$, where $j + k \le n$.

* Donald Knuth, "The Computer as Master Mind", *Journal of Recreational Mathematics*, Vol. 9(1), 1976–77, pp. 1–6.

22. If $(1 + x + x^2)^n$ is expanded as

$$b_0 + b_1 x + \cdots + b_i x^i + \cdots + b_{2n} x^{2n},$$

then find an expression for each coefficient b_i.

23. Reread the definition of even and odd permutations in Exercises 4.11 and 4.12. Prove that for a permutation p, $d(p, p^*)$, where p^* is the identity permutation, is even if and only if every number of transpositions that transform p into p^* is even.

24. If $p = \langle s_1 \ s_2 \ \cdots \ s_n \rangle$ is a permutation, we define $Inv(p)$, called the number of **inversions** of p, to be the number of pairs (s_i, s_j) such that $i < j$ and $s_i > s_j$ for $1 \leq i < j \leq n$. Determine $Inv(p)$ for each of the following:

(*a*) $\langle 1 \ 2 \ 3 \ 4 \ 5 \ 6 \rangle$
(*b*) $\langle 2 \ 1 \ 3 \ 4 \ 5 \ 6 \rangle$
(*c*) $\langle 2 \ 1 \ 4 \ 3 \ 6 \ 5 \rangle$
(*d*) $\langle 6 \ 5 \ 4 \ 3 \ 2 \ 1 \rangle$

25. Prove that $Inv(p)$ is an even number if and only if p is an even permutation.

4

NUMBER THEORY

4:1 GREATEST COMMON DIVISORS

In this chapter seemingly elementary questions from integer arithmetic lead to surprising and elegant mathematics. We shall look at divisibility properties of integers, the greatest common divisor of two integers, and the Fibonacci numbers. These topics have aesthetic appeal and are applicable, as we shall see, in cryptography.

Here are two problems on which we spent many (dull?) hours in elementary school. Recall that a fraction a/b is **simplified** (or reduced) if a and b have no common factor greater than 1.

Problem 1. Is the fraction a/b simplified? If not, simplify it.

Problem 2. Compute $a/b + c/d$ and leave the answer simplified.

Question 1.1. Simplify, if possible, the following: $\frac{3}{12}$, $\frac{13}{121}$, $\frac{65}{130}$, $\frac{34,567}{891,011}$. Add and simplify the following: $\frac{1}{3} + \frac{1}{2}$, $\frac{1}{4} + \frac{1}{3}$, $\frac{1}{15} + \frac{1}{65}$.

You might wonder why we did these exercises in elementary school as well as how we did them. Probably being dutiful and bright students, we just did them. But why bother? Certainly, calculators remove the need to simplify fractions.

Try an experiment. Add $\frac{1}{3}$ to itself three times on a calculator. You might get 1 or you might get .99999999 (depending on your calculator). In either case subtract 1 from your total. Surprisingly enough you won't get zero (unless your calculator is fancy or broken). There are instances (you will see one in Section 7) when we know quantities to be integers and want to retain the accuracy and precision of integer arithmetic. Most computer languages give us the option of exact arithmetic with integers, provided that the integers are not too large.

How did we do Problems 1 and 2? To find the sum of two fractions, most of us would compute

$$\frac{a}{b} + \frac{c}{d} = \frac{ad + bc}{bd}$$

and then simplify this fraction. Both problems require the ability to simplify fractions. As a practical technique, most people would simplify the fraction a/b by searching for integers that are divisors of both a and b. When such an integer, say c, is found, they cancel c from both the numerator and the denominator to obtain the smaller problem of reducing $(a/c)/(b/c)$. This is fine if the numbers a and b are small or have common divisors that are easy to find, for instance, if both a and b are even or both end in 0 or 5.

A slightly more sophisticated approach is to look for common divisors among the primes, for if two numbers have a common divisor, then they have a common prime divisor. An even better description of how to proceed is to find the greatest common divisor of a and b and then cancel that number. Although this is better as a description, if the numbers a and b are at all large, we might be at a loss in finding the greatest common divisor or, for that matter, any common divisor.

Question 1.2. Find the greatest common divisor of the pairs (a) (65, 130), (b) (48, 88), and (c) (34567, 891011).

In this section we work out a straightforward, although slow, procedure for finding the greatest common divisor of two integers. A more efficient algorithm will be presented in a later section.

We begin with some precise definitions pertaining to integer arithmetic. If b and c are integers, we say that b **divides** c (b is a **divisor** of c, and c is a **multiple** of b) if c/b is an integer. Then as the name implies, the **greatest common divisor** of two positive integers b and c is the largest integer that is a divisor of both b and c. We denote the greatest common divisor of b and c by **gcd (*b*, *c*)**.

Does every pair have a greatest common divisor? Any pair of positive integers has 1 as a common divisor, and the largest number that could possibly be a common divisor of b and c is the minimum of b and c. Thus the greatest common divisor always exists and lies somewhere between 1 and the minimum of b and c.

Question 1.3. Find b and c (with $b \leq c$) such that (i) $\gcd(b, c) = 1$, (ii) $1 < \gcd(b, c) < b$, and (iii) $\gcd(b, c) = b$. Why is it impossible for $\gcd(b, c)$ to be larger than the minimum of b and c?

Our first gcd algorithm, a brute force search, looks for $\gcd(b, c)$ starting with the largest possibility, the minimum of b and c, and then checks each smaller integer in turn until a common divisor is found. The first common divisor found will be the greatest. The algorithm must stop, since 1 is a common divisor.

Algorithm GCD1

STEP 1. Input b, c; set $g := $ minimum of b and c
STEP 2. While $g > 1$ do
 Begin
 STEP 3. If b/g and c/g are both integers, then output g and stop.
 STEP 4. Set $g := g - 1$
 End
STEP 5. Output gcd = 1 and stop.

Question 1.4. Carry out GCD1 on the pairs (3, 4), (3, 12), and (6, 20).

We judge the efficiency of this algorithm by the number of divisions (which occur only in step 3). The exact number will depend upon b and c, and so we carry out a worst-case analysis to obtain an upper bound. Our input to GCD1 is two integers b and c; suppose that $b \leq c$. We measure the size of the input by c and let the complexity function $f(c)$ count the maximum number of divisions carried out for any pair of numbers $b \leq c$. Two divisions are performed every time step 3 is encountered. Step 3 will be executed with $g = b$, then $g = (b - 1)$, then $g = (b - 2)$, and so on, until g has decreased down to the real gcd. Thus step 3 will happen most often when the gcd is 1. In this event we would encounter step 3 a total of $b - 1$ times, performing $2(b - 1)$ divisions. Then

$$f(c) \leq 2(b-1) \leq 2(c-1) < 2c \qquad \text{so } f(c) = O(c).$$

We see that the number of divisions in GCD1 is linear in the size of the input, and thus it seems to be an efficient algorithm.

Question 1.5. Find two positive integers b and c such that when GCD1 is applied to them we find the following.
(a) The number of divisions is exactly $2(b - 1)$.
(b) The number of divisions is less than $2(b - 1)$.
(c) The number of divisions is as small as possible.

With GCD1 we can respond precisely to Problems 1 and 2. With a more efficient gcd algorithm, we could upgrade our responses by replacing GCD1. Here is a solution to Problem 1.

Algorithm SIMPLIFY

STEP 1. Input a and b {The fraction a/b is to be simplified.}

STEP 2. Use GCD1 and set $g := \gcd(a, b)$

STEP 3. Set $a' := a/g$ and $b' := b/g$

STEP 4. Output the fraction a'/b' and stop.

Question 1.6. Write an algorithm ADDFRACT1 that solves Problem 2. Upon the input of fractions a/b and c/d, it should calculate their sum and output that sum as a simplified fraction. You may use the algorithm SIMPLIFY within ADDFRACT1.

Question 1.7. Count the number of multiplications and divisions performed by SIMPLIFY and by ADDFRACT1, including those in GCD1.

Previously, we have called linear algorithms fast and claimed that they were more efficient than, say, quadratic algorithms. Although GCD1 performs at most $O(c)$ divisions, it seems slow and inefficient on hand calculations. In fact, it is not the approach that many humans would take to find the gcd of two integers, and it doesn't use any properties of integers that might speed up the process. In the next sections we shall reexamine the complexity of GCD1 and the way we perform complexity analyses. We shall find that GCD1 is not an efficient algorithm, but we shall develop a good gcd algorithm, one that performs $O(\log(c))$ divisions in the worst case upon input of integers b and c with $b \le c$.

EXERCISES FOR SECTION 1

1. Simplify the following fractions: ***(a)*** $\frac{138}{240}$, ***(b)*** $\frac{75}{615}$, ***(c)*** $\frac{357}{189}$, and ***(d)*** $\frac{164}{644}$.
2. Combine the following into one simplified fraction: ***(a)*** $\frac{138}{15} - \frac{138}{16}$ and ***(b)*** $\frac{1}{15} + \frac{34}{615}$.
3. If both a/b and c/d are simplified, is $(ad + bc)/(bd)$ simplified?
4. If a/b is simplified, is a^2/b^2 simplified?
5. If a^2/b^2 is simplified, is a/b simplified?
6. Suppose that we find the lowest common denominator of $a/b + c/d$ to be e, and with this denominator we get $a/b + c/d = f/e$ for some integer f. Is f/e always a simplified fraction?
7. Trace GCD1 on the following pairs: ***(a)*** (4, 7), ***(b)*** (4, 6), ***(c)*** (8, 10), ***(d)*** (8, 12), ***(e)*** (15, 35), and ***(f)*** (18, 42).

8. Algorithm GCD1 begins with g equal to the minimum of b and c and then decreases g, searching for a common divisor of b and c. Design an algorithm that instead begins with $g = 1$ and then increases g until the gcd is found. How does the efficiency of this algorithm compare with that of GCD1?

9. Suppose that a, b, and c are three positive integers with $a \leq b \leq c$. We define **gcd (a, b, c)** to be the largest integer that divides all three numbers, a, b, and c. Explain why $\gcd(a, b, c) \leq a$. Design an algorithm that upon the input of a, b, and c finds $\gcd(a, b, c)$. Find $\gcd(24, 68, 128)$, $\gcd(28, 70, 98)$, and $\gcd(112, 148, 192)$.

10. Find pairs (b, c) such that when GCD1 is applied, the number of divisions is exactly ***(a)*** 12, ***(b)*** 16, and ***(c)*** $b/2$.

11. Given two integers b and c, the **least common multiple** of b and c, denoted by **lcm (b, c)**, is the smallest integer that is a multiple of both b and c. Find a pair of integers b and c with $b \leq c$ such that **(i)** $\text{lcm}(b, c) = bc$ and **(ii)** $\text{lcm}(b, c) = c$. Then explain why in all cases $c \leq \text{lcm}(b, c) \leq bc$.

12. Find the following: $\text{lcm}(2, 3)$, $\text{lcm}(3, 4)$, and $\text{lcm}(6, 8)$. Then add and simplify the fractions: $\frac{1}{2} + \frac{1}{3}$, $\frac{1}{3} + \frac{1}{4}$, and $\frac{1}{6} + \frac{3}{8}$.

13. Calculate the following:
(a) $\gcd(5, 7)$ and $\text{lcm}(5, 7)$.
(b) $\gcd(4, 9)$ and $\text{lcm}(4, 9)$.
(c) $\gcd(6, 10)$ and $\text{lcm}(6, 10)$.
(d) $\gcd(6, 9)$ and $\text{lcm}(6, 9)$.
(e) $\gcd(8, 12)$ and $\text{lcm}(8, 12)$.
(f) $\gcd(5, 10)$ and $\text{lcm}(5, 10)$.

14. Here is a proof that $\text{lcm}(b, c) \cdot \gcd(b, c) = bc$. Give reasons for each step. $\{$Let $g = \gcd(b, c)$, $b' = b/g$, $c' = c/g$, and $m = \text{lcm}(b, c)$.$\}$
1. bc/g is a multiple of b and a multiple of c
2. $\text{lcm}(b, c) \leq bc/g$
3. $\gcd(b, c) \cdot \text{lcm}(b, c) \leq bc$
4. bc/m divides both b and c
5. $\gcd(b, c) \geq bc/m$
6. $\gcd(b, c) \cdot \text{lcm}(b, c) \geq bc$
7. $\gcd(b, c) \cdot \text{lcm}(b, c) = bc$.

15. Given the following pairs of integers b and c, find $g = \gcd(b, c)$, $b' = b/g$, $c' = c/g$, and $\text{lcm}(b, c)$. Then check that $\text{lcm}(b, c) = b'c'g$. ***(a)*** 3 and 4, ***(b)*** 6 and 8, ***(c)*** 4 and 6, ***(d)*** 3 and 9, and ***(e)*** 8 and 20.

16. Prove that $\text{lcm}(b, c) = b'c'g$, where b', c', and g are as defined in Exercise 15.

17. Find pairs (b, c) such that $\gcd(b, c)$ equals **(a)** 3, **(b)** 8, **(c)** $b/2$, **(d)** $b/3$, and ***(e)*** $\sqrt{b}$. Find pairs (b, c) such that $\text{lcm}(b, c)$ equals ***(a)*** 14, ***(b)*** 29, **(c)** $2b$, ***(d)*** $3b$, and ***(e)*** b^2.

18. What can be said about the relation between $\gcd(a, b)$ and $\gcd(at, bt)$ where t is any positive integer?

19. Prove that if a and b are positive integers and x and y are nonzero integers such that $ax + by = 1$, then

$$\gcd(a, b) = \gcd(a, y) = \gcd(x, b) = \gcd(x, y) = 1.$$

Show that exactly one of the numbers x and y must be negative. [We can define $\gcd(c, d)$, where one or both of c and d are negative with exactly the same definition as for positive integers.]

20. If a, b, x, and y are nonzero integers such that $ax + by = 2$, is it true that $\gcd(a, b) = 2$?

21. Prove that if $\gcd(a, b) = 1$ and if c divides b, then $\gcd(a, c) = 1$.

22. Suppose that $a = qb + r$, where a, b, q, and r are integers. Is it true that $\gcd(a, b) = \gcd(a, r)$? Is $\gcd(a, r) = \gcd(b, r)$? Explain your answers.

23. Here is the idea for another algorithm to add the fractions a/b and c/d. Set $g := \gcd(b, d)$, $b' := b/g$, $d' := d/g$, and $m := \text{lcm}(b, d)$. First calculate m by $m = bd/g$. Then $a/b = ad'/m$ and $c/d = cb'/m$ (Why?) and $a/b + c/d = (ad' + cb')/m$. Finally, simplify this last fraction. Implement these ideas as an algorithm ADDFRACT2. How many variables does ADDFRACT2 use? Count the number of multiplications and divisions performed, including those of GCD1.

24. Compare the algorithms ADDFRACT1 and ADDFRACT2 with respect to number of variables used and number of multiplications and divisions performed. Which uses less space and which is quicker?

4:2 ANOTHER LOOK AT COMPLEXITIES

We want to reexamine the complexity of algorithms, especially those from number theory. In a formal analysis of an algorithm the size of the input should be measured by the number of bits (zeros and ones) needed to represent the input. For number theory algorithms whose input is typically one or more positive integers, the size of the input should be the total number of zeros and ones needed to represent the input integers in binary notation. As before, we count the number of time-consuming operations performed in the worst case of the algorithm (usually multiplications and divisions for number theory algorithms) and express the resulting upper bound as a function of the number of input bits. In this section we discuss the effects of this change of perspective on complexity analysis.

Why the change? There is a certain (bureaucratic-style) inefficiency built into our previous approach to the analysis of algorithms. We measured how efficient an algorithm was by estimating the number of steps it required as a function of the input size. The problem with this is that if we are careless about measuring the size of the input, that is, if we let it be artificially large, then the algorithm might appear to take a correspondingly small number of steps. This is just what happened in our study of GCD1 and the exponentiation algorithms of Chapter 2. Measuring input size in terms of bits leads to complexities that reflect actual running times.

Changing the input measure, to bit size, is not hard. Suppose that an integer n is the input to an algorithm. As we saw in Section 2.6 the number of bits needed to represent n is precisely

$$B = \lfloor \log(n) \rfloor + 1.$$

This formula gives the translation from n to B, and it implies the following useful relationships.

$$\begin{aligned} \log(n) < B &\leq \log(n) + 1 \\ &\leq 2\log(n) \qquad \text{for } n \geq 2. \end{aligned} \tag{1}$$

Example 2.1. Suppose that algorithm A performs at most $C\log(n)$ time-consuming operations upon input of an integer n for some constant C. Then what can be said about the complexity function as a function of B, the number of bits needed to represent n? By (1)

$$C\log(n) \leq CB = O(B).$$

Thus in terms of the variable B, the number of time-consuming operations is a linear function.

Look back in Section 2.6 at the complexity analysis of FASTEXP. There we found that no more than $3\log(n) + 3$ multiplications and divisions are needed to compute x^n. Using (1), we see that

$$3\log(n) + 3 \leq 3B + 3 = O(B).$$

In terms of input bits FASTEXP is a linear algorithm and so deserving of its name.

Question 2.1. Suppose that algorithms R, S, and T each have an integer n as input, and their complexity functions are, respectively, $(\log(n))^2$, $\log(n^2)$, and $\log(\log(n))$. Find an upper bound on their complexity functions in terms of B, the number of bits needed to represent n.

Example 2.2. Suppose that algorithm A′ performs at most Cn time-consuming operations upon input of an integer n for some constant C. Then what can be said about the complexity function as a function of B, the number of bits needed to represent n?

$$\begin{aligned} Cn &= C\,2^{\log(n)} && \text{by properties of log} \\ &\le C\,2^{B} && \text{using (1)} \\ &= O(2^{B}). \end{aligned}$$

Thus in terms of the variable B, the number of time-consuming operations is big oh of an exponential function. Furthermore, if there are instances when A' uses all Cn operations, then

$$\begin{aligned} Cn &= C\,2^{\log(n)} && \text{by properties of log} \\ &\ge C\,2^{(B/2)} && \text{using (1)} \\ &= C(\sqrt{2})^{B} \\ &\ge C(1.414)^{B}. \end{aligned}$$

Thus A' is an exponential algorithm.

The analysis in Example 2.2 shows why both the algorithms GCD1 and EXPONENT of Chapter 2 are bad algorithms. Since GCD1 has integers b and c input, the number of bits needed to express b and c in binary is given by

$$\begin{aligned} \log(c) \le B &= \lfloor \log(b) \rfloor + 1 + \lfloor \log(c) \rfloor + 1 \\ &\le 2\log(b) + 2\log(c) \qquad \text{for } b \ge 2 \\ &\le 4\log(c). \end{aligned} \tag{2}$$

We know that GCD1 performs at most $2c$ divisions. From Example 2.2 we know that $2c \le 2(2^{B})$, giving an exponential upper bound. In addition, when $b = c - 1$, $\gcd(b, c) = 1$ (see Exercise 2). In that case GCD1 performs exactly $2(b - 1) = 2c - 4$ divisions.

$$\begin{aligned} 2c - 4 &= 2(2^{\log(c)}) - 4 && \text{by properties of log} \\ &\ge 2(2^{(B/4)} - 2) && \text{from (2)} \\ &\ge 2(2^{(B/4 - 1)}) && \text{when } B \ge 8 \\ &= 2^{(B/4)} \\ &= (2^{(1/4)})^{B} \\ &> (1.189)^{B}. \end{aligned}$$

Thus in the worst case GCD1 performs an exponential number of divisions in terms of the input bit size.

Question 2.2. In Section 2.5 it was observed that EXPONENT always performs n multiplications. If B is the number of bits needed to represent n in binary, explain why EXPONENT is an exponential algorithm.

Since GCD1 is now recognized to be bad, it is clear why we continue to search for a faster algorithm. From now on we shall measure the input size by the number of bits needed. This approach is standard in the study of algorithms using Turing Machines.

EXERCISES FOR SECTION 2

1. Comment on the following statement: "Most of the time $\lfloor \log(n) \rfloor = \lceil \log(n) \rceil - 1$."
2. Explain why $\gcd(c-1, c) = 1$ for all integers $c > 1$.
3. Let $B = \lfloor \log(n) \rfloor + 1$. For each function f listed in the table find the smallest function g such that $f(n) \le g(B)$.

$f(n)$	$g(B)$
$2\log(n) - n$	$\sqrt{B}$
$\sqrt{\log(n)}$	B
$(\log(n))^2 + 2\log(n) + 1$	$2B$
$2^{\log(n)}$	B^2
$3^{\log(n)}$	$10B^2$
$\sqrt{n}$	$\sqrt{2}^B$
$3n + 3$	2^B
$n\log(n)$	$2^{(B+3)}$
n^2	$B(2^B)$
$n^3 - n$	2^{2B}
2^n	6^B
	8^B
	$2^{(B^2)}$
	$2^{(2^B)}$

4. Let the input to algorithm A be an integer n. Thus the number of bits needed is $B = \lfloor \log(n) \rfloor + 1$. Suppose that the complexity function for algorithm A is $a(n) = g(B)$.

(*a*) Show that if $g(B) = O(p(B))$, then $a(n) = O(p(n))$.
(*b*) Show that if $a(n) \neq O(p(n))$ for any polynomial p, then $g(B) \neq O(q(B))$ for any polynomial q.
(*c*) If $a(n) = O(p(n))$ for some polynomial p, is it true that $g(B) = O(q(B))$ for some polynomial q?

5. In the algorithms SUBSET, JSET, and PERM we measured the input by the integer variable n. If we translate now to the number of bits input, $B = \lfloor \log(n) \rfloor + 1$, do these algorithms remain exponential in the variable **B** using the worst-case analysis? (See Exercise 4.)

6. Suppose that the input to an algorithm A is an integer n and suppose the size of the input is measured by the number of decimal digits needed to express n. Would this change of measure of input size change whether or not A is a good algorithm?

4:3 THE EUCLIDEAN ALGORITHM

We have developed the simplistic (but bad) algorithm GCD1 to determine gcd (b, c). Fortunately, there is a much more efficient algorithm that appeared in 300 B.C. in Euclid's *Elements*. This Euclidean algorithm is probably the oldest algorithm still in use today. The Babylonians wrote down some precise arithmetic procedures about 1500 years before Euclid, but these have all been replaced by more efficient methods. The amazing fact about the Euclidean algorithm is that, except for minor variations, it is the best (most efficient) algorithm for calculating the greatest common divisor of two integers. In this section we'll learn the algorithm and in subsequent sections the mathematics needed to determine its complexity.

Here is the idea behind the algorithm. Suppose that we are given positive integers $b \leq c$ and want to calculate gcd (b, c). If d divides both b and c [i.e., d is a candidate for gcd (b, c)], then d divides $c - b$. Indeed if d divides $c - b$ and b, then it divides c also. What is the advantage of working with b and $c - b$ instead of b and c? Very simply, $c - b$ is smaller than c.

Question 3.1. Find gcd (18, 30), gcd (18, 48), and gcd (18, 66).

If $c - b$ is better than c, then $c - 2b$ should be better still. While we're at it, there is $c - 3b$, $c - 4b$, and so on, to consider. Indeed why not subtract off as many bs as possible subject to the condition that the remaining value is not negative?

Question 3.2. For each pair (b, c), find the maximum integer q such that $c - qb \geq 0$. (a) (24, 36), (b) (36, 120), and (c) (34, 170).

This question illustrates the general rule that the right number of bs to subtract from c is the floor function of the quotient c/b. Thus we divide c by b to obtain an integer quotient q_1 and a remainder r_1, where

$$q_1 = \left\lfloor \frac{c}{b} \right\rfloor \qquad \text{and} \qquad \frac{c}{b} = q_1 + \frac{r_1}{b}.$$

We rewrite the previous equation in the form

$$c = q_1 b + r_1, \tag{A}$$

and note that the remainder r_1 must satisfy $0 \leq r_1 < b$. We call q_1 the **quotient** and r_1 the **remainder** of the division c/b.

Here is an important fact about the numbers in (A).

Lemma 3.1. If b, c, q, and r are integers such that $c = qb + r$, then $\gcd(b, c) = \gcd(b, r)$.

Proof. Since an integer that divides b and c also divides b and r, $\gcd(b, c)$ divides both b and r and so is at most $\gcd(b, r)$. Thus

$$\gcd(b, c) \leq \gcd(b, r).$$

An integer that divides b and r also divides c. Thus

$$\gcd(b, r) \leq \gcd(b, c),$$

and the lemma follows. □

Applying the lemma to line (A) gives $\gcd(b, c) = \gcd(r_1, b)$.

Question 3.3. For each of the following pairs of numbers, determine q_1 and r_1. Check that (A) holds and that $0 \leq r_1 < b$. Finally, compute $\gcd(b, c)$ and $\gcd(r_1, b)$. (a) (3, 12), (b) (13, 121), (c) (233, 377), and (d) (34567, 891011).

Notice that if in (A) $r_1 = 0$, then $c = q_1 b$ and $\gcd(b, c) = b$. But if $r_1 > 0$, then we don't have the gcd at our fingertips and consequently must do more work. The problem is simpler now because we have smaller numbers. This technique of replacing c by a smaller number, the remainder, worked once. Let's do it again. Thus we divide b by r_1, a number smaller than b, to obtain a new integer quotient q_2 and a new remainder r_2:

$$b = q_2 r_1 + r_2 \qquad \text{with } 0 \leq r_2 < r_1.$$

If $r_2 = 0$, then r_1 divides b and so $r_1 = \gcd(r_1, b) = \gcd(b, c)$ by Lemma 3.1. More generally (even when $r_2 \neq 0$), we have by Lemma 3.1 that

$$\gcd(b, c) = \gcd(r_1, b) = \gcd(r_2, r_1).$$

Next we divide r_1 by r_2, then r_2 by r_3, and keep dividing each remainder by the next until we reach a remainder of zero. Here is the sequence of divisions spelled out precisely; for future reference we call these the **Euclidean equations**. Note that every variable assumes only integer values.

The Euclidean Equations

$$\begin{aligned}
c &= q_1 b + r_1 &\quad& \text{with } 0 \le r_1 < b \\
b &= q_2 r_1 + r_2 && \text{with } 0 \le r_2 < r_1 \\
r_1 &= q_3 r_2 + r_3 && \text{with } 0 \le r_3 < r_2 \\
&\cdots && \\
r_{i-2} &= q_i r_{i-1} + r_i && \text{with } 0 \le r_i < r_{i-1} \\
&\cdots && \\
r_{k-3} &= q_{k-1} r_{k-2} + r_{k-1} && \text{with } 0 \le r_{k-1} < r_{k-2} \\
r_{k-2} &= q_k r_{k-1} + 0 && \text{with } r_k = 0.
\end{aligned}$$

The claim made by Euclid is that r_{k-1}, the last nonzero remainder, equals $\gcd(b, c)$. Before we verify this, how do we know that this algorithm stops? That is, how do we know that eventually we shall find a remainder of zero? Notice that the remainders satisfy

$$b > r_1 > r_2 > r_3 > \cdots > r_{i-1} > r_i > \cdots > r_{k-1},$$

and all the remainders are nonnegative integers. Eventually, a remainder must equal zero, certainly after no more than b remainders.

Example 3.1. Let's carry out the Euclidean algorithm on the numbers 26 and 32:

$$\begin{aligned}
32 &= 1 \cdot 26 + 6 \\
26 &= 4 \cdot 6 + 2 \\
6 &= 3 \cdot 2 + 0.
\end{aligned}$$

We know that $\gcd(26, 32) = 2$, the last nonzero remainder.

Next we try 233 and 377:

$$
\begin{aligned}
377 &= 1 \cdot 233 + 144\\
233 &= 1 \cdot 144 + 89\\
144 &= 1 \cdot 89 + 55\\
89 &= 1 \cdot 55 + 34\\
55 &= 1 \cdot 34 + 21\\
34 &= 1 \cdot 21 + 13\\
21 &= 1 \cdot 13 + 8\\
13 &= 1 \cdot 8 + 5\\
8 &= 1 \cdot 5 + 3\\
5 &= 1 \cdot 3 + 2\\
3 &= 1 \cdot 2 + 1\\
2 &= 2 \cdot 1 + 0
\end{aligned}
$$

(That took a while!) This calculation implies that $\gcd(233, 377) = 1$. To check this, note that 233 is a prime while $377 = 13 \cdot 29$.

Question 3.4. Use the Euclidean algorithm to calculate the following: (a) $\gcd(12, 20)$, (b) $\gcd(5, 15)$, (c) $\gcd(377, 610)$, and (d) $\gcd(34567, 891011)$. Check that the gcd divides each remainder in the Euclidean equations. In each instance count the number of divisions needed to find the gcd.

In our development of the Euclidean algorithm the concurrent explanation can readily be turned into a proof that the algorithm is correct. We now give such a proof.

Theorem 3.2. Given positive integers b and c, the last nonzero remainder produced by the Euclidean algorithm equals $\gcd(b, c)$.

Proof. Suppose that b and c produce the Euclidean equations as listed above. We must prove that $r_{k-1} = \gcd(b, c)$. The last equation tells us that r_{k-1} is a divisor of r_{k-2}, since r_{k-2}/r_{k-1} is the integer q_k. Thus

$$r_{k-1} = \gcd(r_{k-2}, r_{k-1}). \tag{B}$$

Applying Lemma 3.1 to the next to last equation, we get

$$\gcd(r_{k-3}, r_{k-2}) = \gcd(r_{k-2}, r_{k-1}) = r_{k-1} \qquad \text{by (B).}$$

Continuing and repeatedly applying Lemma 3.1, we get

$$\begin{aligned}\gcd(b,c) &= \gcd(b,r_1)\\ &= \gcd(r_1,r_2)\\ &= \gcd(r_2,r_3)\\ &\;\;\cdots\\ &= \gcd(r_{k-2},r_{k-1})\\ &= r_{k-1} \qquad \text{by (B).}\end{aligned}$$

□

Corollary 3.3. If $g = \gcd(b,c)$, then there are integers x and y such that $g = xb + yc$.

Proof. Look at the Euclidean equations. Notice that r_1 can be expressed as $r_1 = c - q_1 b$. If $g = r_1$, then we have demonstrated this result. If not, we can use the second Euclidean equation to express

$$\begin{aligned} r_2 &= b - q_2 r_1 \\ &= b - q_2(c - q_1 b) && \text{by substitution} \\ &= (1 + q_1 q_2)b - q_2 c && \text{simplifying and factoring.}\end{aligned}$$

We continue this process until we reach

$$g = r_{k-1} = r_{k-3} - q_{k-1} r_{k-2}$$

and can substitute in expressions for r_{k-3} and r_{k-2} found earlier, to express $g = r_{k-1}$ in the form $xb + yc$. □

We say that the resulting equation expresses the gcd as a **linear combination** of b and c. This result will be useful in Section 7; other applications are explored in the exercises.

Example 3.1 (continued). We found that $\gcd(26,32) = 2$. Now we use the Euclidean equations to express 2 as a linear combination of 26 and 32. From the first Euclidean equation we have

$$6 = 1 \cdot 32 - 1 \cdot 26.$$

From the second equation

$$2 = 1 \cdot 26 - 4 \cdot 6.$$

We substitute the first equation into the second to get

$$2 = 1 \cdot 26 - 4(1 \cdot 32 - 1 \cdot 26) = 5 \cdot 26 - 4 \cdot 32.$$

The same procedure applied to 233 and 377 yields

$$1 = (-144) \cdot 233 + 89 \cdot 377,$$

but we spare you the 11 equations needed to derive this. Notice that once derived, it is easy to check that the values of x and y work.

Now we write the Euclidean algorithm in pseudocode. Note that the Euclidean equations all are in the same form.

Algorithm EUCLID

STEP 1. Input b and c $\{0 < b \leq c\}$; set $r := b$
STEP 2. While $r > 0$ do
  Begin
  STEP 3. Set $q := \lfloor c/b \rfloor$
  STEP 4. Set $r := c - q * b$
  STEP 5. If $r = 0$, then output gcd $= b$
    else
   set $c := b$ and $b := r$
  End {step 2}
STEP 6. Stop.

Question 3.5. Run EUCLID on the following pairs of integers and express the gcd as a linear combination of the pair of numbers. (a) (6, 20), (b) (3, 4), and (c) (55, 89).

What can we say about the complexity of EUCLID? We begin as we did with GCD1. Let $e(c)$ count the maximum number of divisions and multiplications performed in the algorithm upon input of numbers $b \leq c$. Not surprisingly, there are lots of these operations. One division occurs in step 3 and one multiplication in step 4. Every time we execute step 3 we immediately execute step 4. Thus $e(c) = 2m$, where m is the number of times that step 3 is executed. Another way to count this is to notice that $e(c)$ equals twice the number of Euclidean equations needed to calculate gcd (b, c). This is so, since we do one division to get the quotient and one multiplication to get the remainder in each new equation.

Thus $e(c) = 2k$, where k is the number of Euclidean equations used upon the pair b and c. We search for an upper bound on k that will give us an upper bound on $e(c)$. Since the remainders in the equations decrease, we know that in the worst case we can have no more than b equations. For this to occur, the remainders must be precisely $(b-1), (b-2), \ldots, 1$, and 0. Then

$$e(c) \leq 2b \leq 2c = O(c),$$

a complexity result no better than that of GCD1.

We shall see in the next sections that the remainders cannot behave in such a perverse manner and that EUCLID is considerably more efficient than GCD1. In fact, we shall see that as a function of the size of the bit input, EUCLID is a linear algorithm.

EXERCISES FOR SECTION 3

1. Use EUCLID to find the gcd of the following pairs: **(*a*)** (10, 14), **(*b*)** (14, 35), **(*c*)** (24, 42), **(*d*)** (128, 232), **(*e*)** (98, 210).
2. For each of the pairs in Exercise 1, express the greatest common divisor as a linear combination of the given numbers.
3. Suppose that you EUCLID the pair (b, c) and then the pair (tb, tc) for some integer constant t. What is the relationship between the two sets of Euclidean equations? What is the relationship between the pairs of integers x and y that express b and c and tb and tc as linear combinations of their gcd's?
4. Suppose that $a = bc + d$. Which of the following are true and which false? Explain.
   (i) If e divides a and b, then e divides d.
   (ii) If e divides a and c, then e divides d.
   (iii) If e divides a and d, then e divides b.
   (iv) If e divides c and d, then e divides a.
   (v) If e divides b and d, then e divides a.
   (vi) If e divides b and c, then e divides a.
   (vii) $\gcd(a, c) = \gcd(c, d)$.
   (viii) $\gcd(a, c) = \gcd(b, d)$.
   (ix) $\gcd(a, b) = \gcd(b, d)$.
   (x) $\gcd(a, c) = \gcd(b, c)$.
5. Find a number c such that with $b = 3 \leq c$, the remainders in the Euclidean equations are precisely the numbers 2, 1, and 0. Is there a number c such that with $b = 4 \leq c$ the remainders are (all) the numbers 3, 2, 1, and 0? Can you find a pair of numbers b and c with $4 \leq b \leq c$ such that the remainders in the Euclidean algorithm are all the numbers $(b-1), (b-2), \ldots, 1$, and 0?
6. Suppose that d divides b and $c - sb$, where s is an integer such that $c - sb < 0$. Is it still true that d divides c?
7. What is the maximum number of Euclidean equations you can have if **(*a*)** $b = 4$, **(*b*)** $b = 5$, and **(*c*)** $b = 6$?

8. What is the maximum number of Euclidean equations you can have if **(*a*)** $c = 7$, **(*b*)** $c = 9$, and **(*c*)** $c = 10$?

9. Rewrite the Euclidean algorithm so that all qs and rs are stored in arrays as they are calculated. Then extend this algorithm so that it also calculates x and y such that $g = xb + yc$.

10. Construct a modified Euclidean algorithm incorporating the following idea. Given the Euclidean equation $c = qb + r$, if $r < b/2$, set $c := b - r$ and $b := r$. Otherwise, set $c := r$ and $b := b - r$. Show that the gcd of the new b and c is equal to the gcd of the old b and c. Call the resulting algorithm MODEUCLID.

11. Use MODEUCLID to find the gcd of the following pairs: **(*a*)** (42, 136), **(*b*)** (18, 324), **(*c*)** (148, 268), **(*d*)** (233, 377), and **(*e*)** (324, 432).

12. Discuss the efficiency of MODEUCLID.

13. For each pair (b, c) below characterize $IC(b, c)$, the set of integer combinations of b and c, defined by

$$IC(b, c) = \{mb + nc\colon m, n \text{ are integers}\}.$$

In each case determine the smallest positive integer in $IC(b, c)$. [Note that in the definition of $IC(b, c)$ m and n do not have to be positive integers.]
(*a*) (2, 4) **(*b*)** (6, 8) **(*c*)** (6, 9) **(*d*)** (12, 15)
(*e*) (9, 14) **(*f*)** (5, 7) **(*g*)** (13, 18) **(*h*)** (21, 54).

14. What is the relationship between the Euclidean equations with input (b, c) and those with input $(c - b, c)$?

15. Prove that given integers b and c, there are integers x and y such that $1 = xb + yc$ if and only if $\gcd(b, c) = 1$.

16. Find integers a and b such that $\gcd(a, b) = 3$. Then explain why for these values of a and b there are no integers x and y such that $2 = ax + by$. Comment on the following statement: "If $h \neq \gcd(a, b)$, then there are no integers x and y such that $h = ax + by$."

17. Is the following true or false? Given integers b and c, there are integers x and y such that $d = xb + yc$ if and only if $\gcd(b, c) = d$. Explain your answer.

18. Write a formal induction proof of Corollary 3.3.

4:4 FIBONACCI NUMBERS

We digress to a seemingly unrelated topic, the Fibonacci numbers, because the mathematics associated with them is interesting and because (surprisingly) they are intimately related with the complexity analysis of the Euclidean algorithm.

Here are the first 16 **Fibonacci numbers:**

0 1 1 2 3 5 8 13 21 34 55 89 144 233 377 610.

The convention is to start numbering at zero, so that we have listed the 0th, the 1st, . . . , and the 15th Fibonacci number. We denote the nth Fibonacci number by F_n for each nonnegative integer n.

Question 4.1. Compute $F_{n-1} + F_{n-2}$ for $n = 2, 3, 4, 5, 6, 11$, and 13.

Your answer to the previous question should suggest that there is an easy method for obtaining the Fibonacci numbers. First, $F_0 = 0$, $F_1 = 1$, and then for all $n \geq 2$,

$$F_n = F_{n-1} + F_{n-2}.$$

In fact, this is an inductive sort of a definition. Once you know the two base cases, F_0 and F_1, then you can find all the others, one at a time, by adding successive values.

Question 4.2. Calculate $F_{16}, F_{17}, F_{18}, F_{19}$, and F_{20}. Then compare F_n with 2^n. Which seems to be (or is) larger?

Since the nth Fibonacci number is defined in terms of smaller Fibonacci numbers, it is natural to try to build proofs about these numbers using induction. However, the nth Fibonacci number is not defined solely in terms of its immediate predecessor, but rather in terms of two predecessors. Consequently, we need a strengthening of our induction machine.

Mathematical Induction Revisited. First we repeat the form of induction that we have used so far.

Algorithm INDUCTION

STEP 1. Verify the base case.

STEP 2. Assume that P_k is true for an arbitrary value of k.

STEP 3. Verify that P_{k+1} is true, using the assumption that P_k is true.

Sometimes the truth of P_{k+1} depends on the truth of more than one of the preceding P_j's or depends on the truth of P_j, where $j < k$. There is still hope for the method of induction if we use the following principle, which is known as Complete Induction.

The Principle of Complete Induction. Suppose that P_n is a proposition that depends upon the positive integer n. Then P_n is true for all $n \geq N$, (where N is some fixed integer) provided that

(i) P_N is true,

and

(ii) if $P_N, P_{N+1}, \ldots,$ and P_k are all true, then so is P_{k+1}.

There are two changes here. First we've introduced an unspecified constant N. In the original version of induction we always mentioned the base case P_1 although we admitted that the base case might start off at P_0. Lemma 7.1 from Chapter 2 was only true for $n > 5$; we used this to show that $\sqrt{n}$ is bigger than $\log(n)$ if n is bigger than 64. For situations like this we would like to have the flexibility to begin proofs by induction at different starting points. The variable N allows us this flexibility and tells us what the starting point for the proposition P_n should be. It also tells us the first value of n for which we should check the base case, namely $n = N$.

The second difference between induction and complete induction is in the inductive hypothesis. In this second version we assume that $P_N, P_{N+1}, \ldots,$ and P_k are all true. Since we assume more, it should be easier to use this form.

Question 4.3. Look back at the informal explanation of the Principle of Induction in Section 2.3. Write out a similar argument to explain why the Principle of Complete Induction is valid.

Here is the algorithmic version of complete induction.

Algorithm INDUCTIONC

STEP 1. (The base cases). Verify that $P_N, P_{N+1}, \ldots, P_{N+j}$ are valid for some constants N and j (depending upon the problem).

STEP 2. (The inductive hypothesis). Assume that $P_N, P_{N+1}, \ldots,$ and P_k are all true for an arbitrary value of k.

STEP 3. (The inductive step). Verify that P_{k+1} is true, using the inductive hypothesis.

There is one more change, introduced in this algorithm, namely the constant j in Step 1. At times we shall need to check more than one base case, depending on the proof we construct, to show that P_{k+1} is true. The value of j depends upon the number of $P_N, P_{N+1}, \ldots, P_k$ that we refer back to in our verification of P_{k+1}.

We shall point out explicitly the values of j in each case, but as a rule of thumb you should get in the habit of checking at least two base cases.

Here is an initial example of the use of complete induction.

Example 4.1. **Theorem.** Every integer $n > 1$ has a prime divisor.

Proof. The statement gives the starting point of the proposition: $N = 2$. This statement is true for $n = 2$, since 2 is a prime number and divides itself. Similarly, the statement is valid for $n = 3$, since 3 is a prime. We check that it is also true for $n = 4 = 2 \cdot 2$.

The inductive hypothesis tells us to assume the truth of the statement for $n = 2, 3, \ldots, k$, for some arbitrary value of k. To accomplish the inductive step, we must prove the result that the integer $(k + 1)$ has a prime divisor. Now either $(k + 1)$ is a prime or it isn't. If $(k + 1)$ is a prime, then it has itself as a prime divisor. If $(k + 1)$ is not a prime, then $k + 1 = bc$, where b and c satisfy $1 < b < (k + 1)$ and $1 < c < (k + 1)$. Consider b. By the inductive hypothesis, since $1 < b < (k + 1)$, we may assume that b has a prime divisor, say p. Then

$$\frac{k+1}{p} = \frac{bc}{p} = \frac{b}{p}c.$$

We see that $(k + 1)/p$ is an integer, since it is expressed as the product of two integers. This means that p is a divisor of $(k + 1)$ and so $(k + 1)$ has a prime divisor, namely p. □

Note that we could not have used the standard form of induction in this problem because the truth of the assertion P_{k+1} depends not on the truth of P_k, but on the truth of P_b, where b is less than k. Since our proof depends upon only one earlier case, P_b with $1 < b < (k + 1)$, our base case needed only one value, namely $N = 2$ and $j = 0$, although to get a feel for the problem we checked three base cases.

We now use complete induction to establish some facts about the Fibonacci numbers.

Example 4.2. From the examples calculated in Question 4.2, the following was observed. **Theorem.** $F_n < 2^n$.

We prove this for all nonnegative integers n using complete induction. For the base cases we notice that $F_0 = 0 < 1 = 2^0$, and that $F_1 = 1 < 2 = 2^1$. We require base cases with two consecutive integers because in our proof we use the fact that F_{k+1} can be written in terms of its two immediate predecessors. (Thus $j = 1$ in this example). We shall use complete induction and so assume that $F_i < 2^i$ for all $0 \leq i \leq k$. Then we must prove that $F_{k+1} < 2^{k+1}$. But we know exactly what F_{k+1}

equals:

$$\begin{aligned}
F_{k+1} &= F_k + F_{k-1} && \text{by definition}\\
&< 2^k + 2^{k-1} && \text{by the inductive hypothesis}\\
&= 2^{k-1} \cdot 2 + 2^{k-1} && \text{by algebra}\\
&= 2^{k-1} \cdot (2+1) && \text{by factoring}\\
&< 2^{k-1} \cdot 4 && \text{since } (2+1) < 4\\
&= 2^{k-1} \cdot 2^2 &&\\
&= 2^{k+1} && \text{by laws of exponents.}
\end{aligned}$$ □

Example 4.2 might cause us to ask whether $F_n = O(p)$ for some polynomial p; however, this is not the case. The result of the next question shows that the Fibonacci numbers grow exponentially.

Question 4.4. Find an integer N such that $F_N > (\frac{3}{2})^N$. Prove by induction that $F_n > (\frac{3}{2})^n$ for all $n \geq N$.

Before we do more magic, rabbit-out-of-the-hat tricks with the Fibonacci numbers, let's learn where they come from and why. The Fibonacci numbers first appeared in the book *Liber Abaci* published in 1202 by Leonardo of Pisa (also known as Leonardo Fibonacci, since he was the son of Bonacci). Although Leonardo was mainly interested in their mathematical properties, he also noted the following application.

Example 4.3. A pair of rabbits requires one month to mature to the age when it can reproduce. Suppose that the rabbits then mate and produce another pair in every subsequent month, and that the pair of offspring is always conveniently one male and one female, who then form a new breeding pair. If in the first month we have one pair of rabbits, how many pairs do we have at the beginning of the nth month? For simplicity, we assume no death or loss of fertility. We call the resulting number R_n.

At the beginning of the first month we have one pair, so $R_1 = 1$. At the beginning of the second month we still have one pair, but during the second month they produce a pair of bunnies. Thus $R_2 = 1$ and $R_3 = 2$. During the third month the original pair produces another pair of bunnies, but the new pair of bunnies doesn't reproduce yet. So $R_4 = 3$. Then $R_5 = 5$.

We might as well argue the general case: That is, let's determine R_n in terms of previous values of R. At the beginning of the nth month we have all the rabbit pairs that we had at the beginning of the $(n-1)$st month, R_{n-1}, plus some new bunny pairs. The number of new bunny pairs is the number of rabbit pairs that are at least one month old. The rabbit pairs that are this old are precisely those that were around in the $(n - 2)$nd month, R_{n-2}. In symbols then

$$R_n = R_{n-1} + R_{n-2}.$$

Now we see that the R_n are exactly the same as the Fibonacci numbers and that $R_n = F_n$ for all positive n.

Fibonacci numbers arise in other natural settings. For example, the spacing of leaves on some plants and some arrangements of flower petals and seeds are closely related to the Fibonacci numbers. Mollusk shells spiral in curves derived from Fibonacci numbers. Ratios of successive Fibonacci numbers, like $\frac{5}{3}$, $\frac{8}{5}$, and $\frac{13}{8}$ are considered aesthetically pleasing. The squares in Figure 4.1 each have sides equal to a Fibonacci number. They combine to make rectangles with sides in ratios of 34 to 21, 21 to 13, 13 to 8, 8 to 5, 5 to 3, 3 to 2, and 2 to 1.

In fact, for large values of n, F_n/F_{n-1} gets arbitrarily close to a constant

$$\phi = \frac{1 + \sqrt{5}}{2},$$

known as the golden ratio; ϕ is approximately equal to 1.618. The Fibonacci numbers are even thought to be useful in predicting highs and lows on the stock market. These numbers have so many interesting and varied properties that there is a mathematics research journal, the *Fibonacci Quarterly*, dedicated to results about Fibonacci numbers.

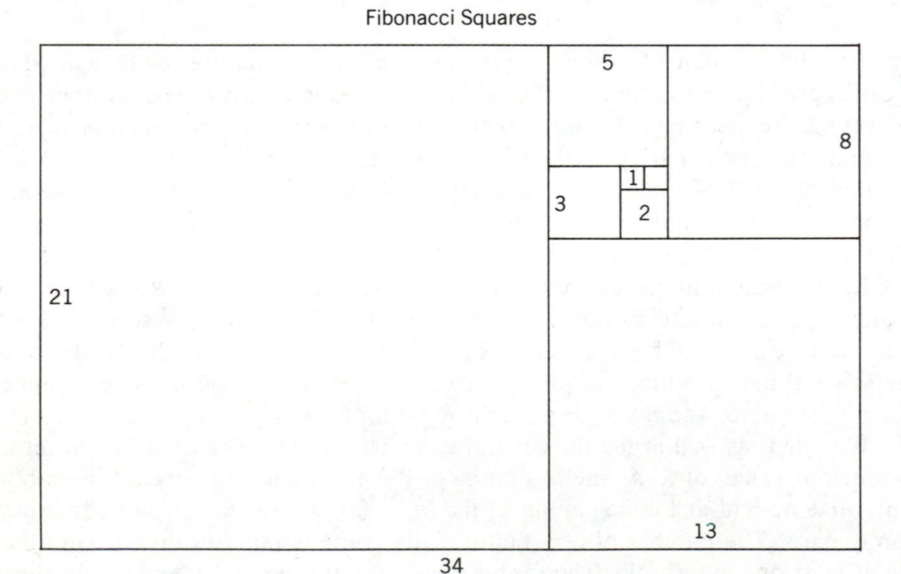

Figure 4.1.

The defining property is useful for proving results about Fibonacci numbers by induction. But one thing seems missing from our knowledge. Is there a formula for F_n? Or to calculate, say, F_{17} must we determine all the smaller Fibonacci numbers? Yes and maybe no, respectively. We shall write down a formula for F_n, but we stress that in most situations the inductive definition that $F_n = F_{n-1} + F_{n-2}$ is the most helpful fact to know. In Chapter 7 we shall do a more systematic study of sequences of numbers and their formulas.

Question 4.5. Show that $\phi = (1 + \sqrt{5})/2$ has the property that its reciprocal is itself minus one. Find all solutions to the equation $x - 1 = 1/x$.

The two solutions to the equation in Question 4.5 are $\phi = (1 + \sqrt{5})/2$ and the closely related $\phi' = (1 - \sqrt{5})/2$. These can be found by rewriting $x - 1 = 1/x$ as $x^2 - x - 1 = 0$ and then solving using the quadratic formula. The relationship between ϕ and ϕ' and the Fibonacci numbers is given in our next result.

Theorem 4.1. For nonnegative integers n, $F_n = (\phi^n - \phi'^n)/\sqrt{5}$.

Proof. The proof will be by complete induction. First we check the base cases. As above we need to verify the truth of the theorem for two consecutive integers, since we shall use the crucial fact that $F_{k+1} = F_k + F_{k-1}$. First we substitute $n = 0$, to obtain

$$\frac{\phi^0 - \phi'^0}{\sqrt{5}} = \frac{1 - 1}{\sqrt{5}} = 0 = F_0.$$

Next, for $n = 1$ we get

$$\frac{\phi^1 - \phi'^1}{\sqrt{5}} = \frac{(1 + \sqrt{5})/2 - (1 - \sqrt{5})/2}{\sqrt{5}} = 1 = F_1.$$

Using complete induction, we assume that the given formula is correct for $F_0, F_1, \ldots, F_{k-1}$ and F_k. We must prove that the formula is correct for F_{k+1}. We write F_{k+1} using smaller values:

$$\begin{aligned} F_{k+1} &= F_k + F_{k-1} \\ &= \frac{\phi^k - \phi'^k}{\sqrt{5}} + \frac{\phi^{k-1} - \phi'^{k-1}}{\sqrt{5}} \\ &= \frac{\phi^{k-1}(\phi + 1)}{\sqrt{5}} - \frac{\phi'^{k-1}(\phi' + 1)}{\sqrt{5}}. \end{aligned}$$

Since ϕ is a root of the equation $x^2 - x - 1 = 0$, we get $\phi^2 = \phi + 1$. Similarly, $\phi'^2 = \phi' + 1$. We substitute these into the equation above to get

$$\begin{aligned} F_{k+1} &= \frac{\phi^{k-1}\phi^2}{\sqrt{5}} - \frac{\phi'^{k-1}\phi'^2}{\sqrt{5}} \\ &= \frac{\phi^{k+1} - \phi'^{k+1}}{\sqrt{5}}, \end{aligned}$$

and that's exactly what we wanted to show. □

Question 4.6. Check the formula for F_2 given in Theorem 4.1.

Corollary 4.2. F_n is approximately equal to $\phi^n/\sqrt{5}$. Specifically,

$$\frac{\phi^n}{\sqrt{5}} - 1 < F_n < \frac{\phi^n}{\sqrt{5}} + 1$$

Proof. We begin by noting that ϕ' is approximately equal to -0.618. What we need is not its exact value but the fact that its absolute value is less than 1. Consequently, ϕ'^n will be less than 1 in absolute value for all positive integers n and $\phi'^n/\sqrt{5}$ will be less than 1 in absolute value for all nonnegative integers n. Thus

$$F_n = \frac{\phi^n}{\sqrt{5}} - \frac{\phi'^n}{\sqrt{5}} < \frac{\phi^n}{\sqrt{5}} + 1.$$ □

The other inequality is proved similarly; see Exercise 17.

We now have two ways to calculate F_n for any fixed n. One involves many additions:

$$\begin{aligned} F_2 &= F_1 + F_0 = 1 + 0 = 1 \\ F_3 &= F_2 + F_1 = 1 + 1 = 2 \\ &\dots \\ F_n &= F_{n-1} + F_{n-2}. \end{aligned}$$

Thus F_n could be calculated with $(n - 1)$ additions.

Question 4.7. To calculate F_n by adding as shown above appears to require that we store all of $F_0, F_1, \ldots, F_{n-1}$. Is it possible to calculate F_n by addition without storing all the previous values in n different memory locations? What is the minimum number of memory locations that you need to calculate F_n in this way?

The second way we now have to calculate F_n is using the formula proved in Theorem 4.1. This requires two exponentiations, one division and one subtraction as well as an approximation of the square root of 5. There are a variety of additional methods known for calculating F_n, including an addition method that is analogous to the FASTEXP algorithm developed in Chapter 2, that is, one that does not require the determination of all intermediate Fibonacci numbers. (See Exercises 13 and 14.)

For years applications of Fibonacci numbers have been found throughout mathematics. For example, a very famous open problem posed by David Hilbert in 1900, known as Hilbert's tenth problem, was finally solved in 1970 when the mathematicians Martin Davis, Yuri Matiasevic, Hilary Putnam, and Julia Robinson thought to examine the Fibonacci numbers carefully. Applications of Fibonacci numbers are also pervasive in computer science. Efficient ways to approximate the local maximum and the local minimum of a function or to merge files can use Fibonacci numbers. In Chapter 8 we shall study problems concerning shortest paths. Recent results have shown "Fibonacci heaps" and "Fibonacci trees" (whatever they are!) to be crucial in developing fast algorithms to solve these problems.

Our interest is to turn now to the complexity analysis of the Euclidean algorithm, where we shall encounter Fibonacci numbers.

EXERCISES FOR SECTION 4

1. Find a sequence of numbers G_n that satisfies the equation $G_n = G_{n-1} + G_{n-2}$ for all n but differs from the sequence of Fibonacci numbers.

2. Let $H_0 = 0$ and $H_1 = 1$. For $n > 1$ define H_n by $H_n = H_{n-1} + 2H_{n-2}$. List the first 11 terms of the H sequence. What happens to the quotient H_n/H_{n-1} as n gets big? Prove that $H_n = [2^n + (-1)^{n-1}]/3$.

3. Show that $F_1 + F_2 + \cdots + F_n = F_{n+2} - 1$.

4. Show that $F_1 + F_3 + \cdots + F_{2n-1} = F_{2n}$.

5. Show that $F_0 + F_2 + \cdots + F_{2n} = F_{2n+1} - 1$.

6. Show that every positive integer can be written as the sum of distinct, positive Fibonacci numbers. Is the choice of numbers for a given sum unique?

7. Suppose that $G_0 = 0$, $G_1 = 1$, and for all $n \geq 2$, $G_n = 2G_{n-1} + G_{n-2}$. Find $G_2, G_3, \ldots, G_8$. Determine which of the following assertions are true.
(a) $G_n = O(n^2)$.
(b) $G_n > 2^{n-1}$ if n is large enough.
(c) $G_n < 3^n$ if n is large enough.
(d) G_n is even if and only if n is even.

8. Knowing F_0 and F_1, one might believe for consistency's sake that F_{-1} should be that number with the property that $F_{-1} + F_0 = F_1$. Since $F_0 = 0$ and

$F_1 = 1$, F_{-1} ought to equal 1. Determine $F_{-2}, F_{-3}, \ldots, F_{-7}$. What is F_{-n} in terms of F_n?

9. Show that $F_{n+1}F_{n-1} - F_n^2 = (-1)^n$.

10. Show that $F_1F_2 + F_2F_3 + \cdots + F_{2n-1}F_{2n} = F_{2n}^2$.

11. Show that $F_1F_2 + F_2F_3 + \cdots + F_{2n}F_{2n+1} = F_{2n+1}^2 - 1$.

12. Find all natural numbers n such that $F_n = n$. Prove that you have found all such numbers.

13. (*a*) Show that $F_{2n} = F_n(F_n + 2F_{n-1})$.
(*b*) Find a similar formula for F_{2n+1} in terms of F_{n+1} and smaller Fibonacci numbers.

14. Using the results of Exercise 13, design an algorithm to determine F_n. Then count the number of multiplications and additions needed in this approach and compare with those discussed at the end of this section.

15. In Example 4.2 and Question 4.4, induction was used to show that $(\frac{3}{2})^n < F_n < 2^n$ provided that n is large enough.
(*a*) Find a number $b < 2$ so that a similar argument will show that $F_n < b^n$ provided that n is large enough. What is the smallest b your argument will support?
(*b*) Find a number $c > \frac{3}{2}$ so that $c^n < F_n$ provided that n is large enough. What is the largest c your argument will support?

16. Are the following equations true or false?

$$F_n = \lceil \phi^n/\sqrt{5} \rceil.$$
$$F_n = \lfloor \phi^n/\sqrt{5} \rfloor.$$

17. Finish the proof of Corollary 4.2 by showing that $F_n > \phi/\sqrt{5} - 1$.

18. Find all pairs of numbers (x, y) such that $x + y = 1$ and $xy = -1$.

4:5 THE COMPLEXITY OF THE EUCLIDEAN ALGORITHM

The question before us is to determine the complexity of the Euclidean algorithm when, given $b \leq c$, it computes $\gcd(b, c)$. To begin with, we let $e(c)$ denote the maximum number of multiplications and divisions performed in EUCLID when c is the larger of the two input integers. Later we shall convert $e(c)$ to a function of the number of bits needed to represent b and c in binary. We shall show that $e(c) = O(\log(c))$; that is, for all pairs of integers $b \leq c$, the Euclidean algorithm requires at most $O(\log(c))$ divisions and multiplications. Consequently, if B denotes the number of bits of input needed to encode the integers b and c, then EUCLID will be linear in B.

Look back in Section 3 and notice that we determined that $e(c) = 2k$, where k is the number of Euclidean equations produced by b and c. Thus we shall search for an upper bound for k in terms of b and c. Before we find such an upper bound, we investigate some pairs for which EUCLID seems to take a long time. Example 3.1, Question 3.4, and some of the exercises point toward the Fibonacci numbers. So let's see what happens if we let b and c be consecutive Fibonacci numbers.

Lemma 5.1. In the Euclidean equations if $c = F_n$ and $b = F_{n-1}$ with $n \geq 4$, then $r_1 = F_{n-2}$.

Proof. The defining relation for the nth Fibonacci number is

$$F_n = F_{n-1} + F_{n-2}.$$

Since $0 \leq F_{n-2} < F_{n-1}$ for $n \geq 4$, this gives the first Euclidean equation with $q_1 = 1$. Thus $r_1 = F_{n-2}$. □

Theorem 5.2. If EUCLID is run with $c = F_{k+2}$ and $b = F_{k+1}$ with $k \geq 1$, then there are exactly k Euclidean equations.

Proof. The proof is by (ordinary) induction on k. If $k = 1$, then $c = F_{k+2} = F_3 = 2$ and $b = F_{k+1} = F_2 = 1$. Given $c = 2$ and $b = 1$, EUCLID produces just one equation, specifically,

$$2 = 2 \cdot 1 + 0.$$

Now we assume that if $c = F_{k+2}$ and $b = F_{k+1}$ with $k \geq 1$, then there are exactly k equations, and try to show that if $c = F_{k+3}$ and $b = F_{k+2}$, then there are exactly $(k + 1)$ equations. Now, if $c = F_{k+3}$ and $b = F_{k+2}$, $k + 3 \geq 4$ and by Lemma 5.1 our first Euclidean equation is

$$F_{k+3} = F_{k+2} + F_{k+1}. \tag{A}$$

Next we divide F_{k+2} by F_{k+1}, but this is the same as if we began the Euclidean algorithm with $c = F_{k+2}$ and $b = F_{k+1}$. By the inductive hypothesis we get k Euclidean equations from this starting point. Hence, in total, (A) is the first of $(k + 1)$ Euclidean equations. □

Question 5.1. Construct the Euclidean equations for (i) $c = F_8$ and $b = F_7$ and (ii) $c = F_{10}$ and $b = F_9$.

Question 5.2. For $0 < b \leq c \leq 5$, what is the maximum number of equations and for what integers does that maximum occur?

What we have done is analyze the complexity of EUCLID for a restricted set of inputs. Specifically, if $b = F_{k+1}$ and $c = F_{k+2}$, then $e(c) = 2k$. Indeed a much stronger statement is true. The smallest integer c that produces as many as k Euclidean equations is $c = F_{k+2}$ and given this c, the only integer b that will (together with c) produce as many as k Euclidean equations is $b = F_{k+1}$. So the Fibonacci numbers provide the worst possible input to EUCLID. We shall not prove this theorem, due to Lamé in about 1845 (but you may prove it in Supplementary Exercise 14). However, we shall show that the worst-case behavior is of the same order of magnitude as that on the Fibonacci numbers.

We begin by investigating the relationship between $c = F_{k+2}$ and k, where $k \geq 1$. Thus $c \geq 2$. Corollary 4.2 and some algebra imply that

$$\phi^{(k+2)} < (c+1)\sqrt{5}.$$

Taking logs (and noting that $\sqrt{5} < 4$), we see that

$$(k+2)\log(\phi) < \log(c+1) + \log(\sqrt{5}) < \log(c+1) + 2.$$

Next we solve for k by dividing by $\log(\phi)$ and subtracting 2 to get

$$k < -2 + \frac{\log(c+1) + 2}{\log(\phi)}.$$

If we estimate $\log(\phi)$ by $\log(1.618) > \frac{1}{2}$ we get

$$\begin{aligned} k < -2 + 2(\log(c+1) + 2) &= 2\log(c+1) + 2 \\ &\leq 2\log(2c) + 2 && \text{since } c \geq 2 \\ &= 2\log(c) + 4 \leq 6\log(c) && \text{since } c \geq 2. \end{aligned}$$

In short, we have that when $c = F_{k+2}$ and $b = F_{k+1}$, the number of Euclidean equations that occur, k, is $O(\log(c))$. In Exercise 11 we ask you to use Corollary 4.2 to bound k from below.

We shall now show that for any inputs b and c, the number of Euclidean equations is no larger than logarithmic in c. To accomplish this, we need a closer look at the Euclidean equations.

Theorem 5.3. If $e(c)$ counts the maximum number of multiplications and divisions in the Euclidean algorithm with input $b \leq c$, then

$$e(c) = O(\log(c)).$$

Proof. Suppose that we consider the first two Euclidean equations

$$c = q_1 b + r_1$$

and

$$b = q_2 r_1 + r_2.$$

We know that $b > r_1 > r_2$. From these inequalities r_2 might be as large as $b - 2$. In fact, we shall obtain the much better estimate that $r_2 < b/2$. This estimate is better in the sense that it allows us to conclude that r_2 is smaller than we otherwise knew. If r_2 is smaller, then we expect to use fewer Euclidean equations.

Example 5.1. In the Euclidean equations of Example 3.1,

$$\begin{aligned} b = 233 \text{ and } r_2 &= 89 < b/2, \\ r_4 &= 34 < r_2/2, \\ r_6 &= 13 < r_4/2, \\ r_8 &= 5 \ \ < r_6/2, \\ r_{10} &= 2 \ \ < r_8/2, \\ \text{and} \qquad r_{12} &= 0 \ \ < r_{10}/2. \end{aligned}$$

Question 5.3. Show that the Euclidean equations with $b = 77$ and $c = 185$ have $r_2 < \frac{77}{2}$.

To show that r_2 is less than $b/2$ in general, we look first at r_1. If $r_1 \le b/2$, then $r_2 < r_1$ implies that $r_2 < b/2$. If, on the other hand, $r_1 > b/2$, then $2r_1 > b$. Thus in the second Euclidean equation $q_2 = 1$ and we have that

$$b = r_1 + r_2.$$

By solving for r_2, we get

$$\begin{aligned} r_2 &= b - r_1 \\ &< b - \frac{b}{2} = \frac{b}{2} \qquad \text{since } r_1 > \frac{b}{2}. \end{aligned}$$

So in either case we have r_2 less than $b/2$. By doing the same thing to the next two Euclidean equations, we can show that

$$r_4 < \frac{r_2}{2} < \frac{b}{4}.$$

Let's be careful at this next step so that the pattern becomes clear. As above we argue that

$$r_6 < \frac{r_4}{2} < \frac{b}{8}.$$

Thus, in general, we have

$$r_{2t} < \frac{b}{2^t}. \tag{B}$$

How long can this go on? If $b/(2^t) \leq 1$, then by the preceding equation the remainder r_{2t} equals zero. But $b/(2^t) \leq 1$ implies that $\log(b) \leq t$. Thus once $t \geq \log(b)$ or equivalently once $t = \lceil \log(b) \rceil$, then

$$\frac{b}{2^t} \leq \frac{b}{2^{\log(b)}} = \frac{b}{b} = 1,$$

and from (B) we get $r_{2t} = 0$. Thus k, the number of Euclidean equations, is at most $2t$, where $t = \lceil \log(b) \rceil$. Thus

$$e(c) = 2k \leq 2(2t) = 2(2\lceil \log(b) \rceil) \leq 4\lceil \log(c) \rceil = O(\log(c)).$$

Question 5.4. Look at the Euclidean equations from Example 3.1. For each integer t compute r_{2t+2}/r_{2t}.

In conclusion we have that the Euclidean algorithm is a good algorithm. Look back in Section 2 at the inequalities of line (2):

$$\log(c) \leq B = \lfloor \log(b) \rfloor + 1 + \lfloor \log(c) \rfloor + 1.$$

Since it requires B bits to input b and c in binary, the number of multiplications and divisions is bounded by

$$e(c) = O(\log(c)) = O(B).$$

Thus EUCLID is a linear algorithm.

EXERCISES FOR SECTION 5

1. Show that if $c = F_{k+2}$ and $b > F_{k+1}$ in the Euclidean equations, then $r_1 < F_k$.

2. In the Euclidean equations, if $c = F_{k+2}$ and $b < F_{k+1}$, is $r_1 < F_k$?
3. Construct the Euclidean equations if $c = F_9$ and $b = F_7$.
4. If $c = F_{k+3}$ and $b = F_{k+1}$, what can you say about the number of Euclidean equations?
5. Suppose that $G_0 = 4$, $G_1 = 7$, and for $n > 1$, $G_n = G_{n-1} + G_{n-2}$. Exhibit G_n for $n = 2, 3, \ldots, 8$.
6. Exhibit the Euclidean equations with $c = G_8$ and $b = G_7$. (G_n is defined in Exercise 5.)
7. How many Euclidean equations are there if $c = G_{k+2}$ and $b = G_{k+1}$? (G_n is defined as above.)
8. In the Euclidean equations we know that $r_4 < b/4$. Is $r_3 < b/4$?
9. Choose a value of t so that in the Euclidean equations $r_8 < r_4/t$.
10. Suppose that C_n is a sequence of nonnegative integers with the property that $C_n < C_{n-1}/4$. If $C_1 = M$, for what value z can you guarantee that $C_z = 0$?
11. Use Corollary 4.2 to find a constant D such that if $c = F_{k+2}$, then $k > D \log(c)$.
12. The complexity of EUCLID was shown to be $O(B)$, where B equals the number of bits needed to represent the integers b and c. Thus the number of multiplications and divisions performed is at most $sB + t$ for some integers s and t. Find integers s and t that give an upper bound on the number of these arithmetic operations that is as small as possible based on the analysis in this section.

4:6 CONGRUENCES AND EQUIVALENCE RELATIONS

Integer arithmetic is fundamental to the mathematical field of number theory and to the computer science field of cryptography. The particular kind of arithmetic used in these fields is known as **modular** or **congruence arithmetic**. In this section we introduce the basics of **arithmetic modulo *n*** and develop simultaneously the concept of an equivalence relation. In the next section we apply this work to encryption schemes.

Definition. If n is a positive integer and a and b any two integers, we define

$$a \equiv b \pmod{n}$$

(read "***a* is congruent to *b* modulo *n***") if $(a - b)$ is divisible by n. We let $[a]$ denote the set of all integers congruent to a modulo n,

$$[a] = \{x\colon a \equiv x \pmod{n}\}.$$

This is called the **equivalence class** containing a.

Example 6.1. Let $n = 12$. Then $1 \equiv 13 \pmod{12}$, $1 \equiv 25 \pmod{12}$, $13 \equiv 25 \pmod{12}$, $1 \equiv -11 \pmod{12}$, and

$$\begin{aligned}[1] &= \{\ldots, -23, -11, 1, 13, 25, \ldots\} \\ &= \{1 + 12k\colon k \text{ an integer}\}.\end{aligned}$$

If "$\not\equiv$" means "not congruent to," then $1 \not\equiv 0 \pmod{12}$, $1 \equiv 1 \pmod{12}$, $1 \not\equiv 2 \pmod{12}$, and $1 \not\equiv i \pmod{12}$ for $i = 3, 4, \ldots, 12$.

We are used to working "modulo 12," since that is how our clocks and some of our measurements work. If it is 11 A.M. and I have an appointment in 3 hours, then since $11 + 3 = 14 \equiv 2 \pmod{12}$, the appointment is for 2 P.M.

Question 6.1. Determine which of the following are true and which false: (a) $2 \equiv 3 \pmod{12}$, (b) $2 \equiv 4 \pmod{12}$, (c) $2 \equiv 10 \pmod{12}$, (d) $2 \equiv 14 \pmod{12}$, (e) $2 \equiv -10 \pmod{12}$, and (f) $-10 \equiv -22 \pmod{12}$. Describe all integers x that are congruent modulo 2 to 0. Working modulo 3, list six elements of $[1]$ and then describe the entire set precisely.

Question 6.2. Let n be a positive integer and i an integer such that $0 \le i < n$. What is the least integer $j > i$ such that $i \equiv j \pmod{n}$? What is the largest negative number m such that $i \equiv m \pmod{n}$?

Congruences modulo n behave like equalities.

Lemma 6.1. Let n be a positive integer and a, b, and c arbitrary integers. Then
(i) $a \equiv a \pmod{n}$,
(ii) If $a \equiv b \pmod{n}$, then $b \equiv a \pmod{n}$, and
(iii) If $a \equiv b \pmod{n}$ and $b \equiv c \pmod{n}$, then $a \equiv c \pmod{n}$.

Proof. We prove part (iii). If $a \equiv b \pmod{n}$, then n divides $(a - b)$ and so there is an integer i such that $a - b = in$. If $b \equiv c \pmod{n}$, then there is an integer j such that $b - c = jn$. Thus

$$a - c = (a - b) + (b - c) = in + jn = (i + j)n,$$

and $(a - c)$ is divisible by n, that is, $a \equiv c \pmod{n}$. □

Question 6.3. Prove parts (i) and (ii) of Lemma 6.1.

Here's a vocabulary to highlight the similarities between relationships like "$=$," "$\equiv \pmod{n}$," and others. We say that a **relation** "$\sim$" is defined on a set S

(finite or infinite) if for each pair of elements (a, b) in S, $a \sim b$ is either true or false. Colloquially, if $a \sim b$ is true, then we say a is related to b. A more formal way to describe a relation on S is to say that $\sim$ corresponds to a function

$$T: S \times S \to \{\text{True, False}\}$$

such that $T(a, b) = \text{True}$ if and only if $a \sim b$ is true (or equivalently a is related to b).

Example 6.2. Let S be the set Z of all integers. Then equality gives us a relation on Z by defining $i \sim j$ to be true for integers i and j if and only if $i = j$. Similarly, for a fixed positive integer n, congruence modulo n is a relation on Z if we define $i \sim j$ to mean that $i \equiv j \pmod n$.

Question 6.4. Which of the following defines a relation on the given set S? (a) $S = Z$ and $\sim$ stands for $\leq$; (b) $S =$ all subsets of Z and $\sim$ stands for $\subseteq$; and (c) $S =$ all real numbers and $r \sim s$ means that $(r - s)$ is even.

Definition. A relation $\sim$ defined on a set S is said to be an **equivalence relation** if it satisfies the following three properties. If a, b, and c are arbitrary elements of S, then

(i) $a \sim a$ (**reflexive property**)
(ii) If $a \sim b$, then $b \sim a$ (**symmetric property**)
(iii) If $a \sim b$ and $b \sim c$, then $a \sim c$ (**transitive property**).

Lemma 6.1. says that the relation "congruence modulo n" defined on Z is an equivalence relation.

Example 6.3. Let $U = \{1, 2, \ldots, n\}$ and let P be the set of all subsets of U. If $A, B, C \subseteq U$, then (i) $A \subseteq A$ is true for every subset A, and (iii) if $A \subseteq B$ and $B \subseteq C$, then $A \subseteq C$, but it is not true that (ii) if $A \subseteq B$, then $B \subseteq A$. Thus the relation of containment is not an equivalence relation on P.

Example 6.1 (continued). Working modulo 12, we saw that 1 is not congruent modulo 12 to $0, 2, 3, \ldots,$ or 11. Let us look at the equivalence classes $[0], [1], [2], \ldots [11]$.

$$[0] = \{\ldots -24, -12, 0, 12, 24, 36 \ldots\}$$
$$[1] = \{\ldots -23, -11, 1, 13, 25, 37 \ldots\}$$
$$[2] = \{\ldots -22, -10, 2, 14, 26, 38 \ldots\}$$
$$\ldots$$
$$[11] = \{\ldots -25, -13, -1, 11, 23, 35 \ldots\}.$$

Notice that no two of these sets intersect and that every integer is in precisely one of these sets.

Definition. If $\sim$ is an equivalence relation on the set S, then we define for a in S, the **equivalence class** containing a to be

$$[a] = \{x \text{ in } S\colon a \sim x\}.$$

Then just as in the case of congruence of integers modulo n, the collection of all distinct equivalence classes of S divides up S into disjoint subsets. Such a division is called a **partition** of S.

Lemma 6.2. If $\sim$ is an equivalence relation on a set S, then
(i) a is in $[a]$ for all a in S,
(ii) $[a] = [b]$ if and only if $a \sim b$, and
(iii) if $[a] \neq [b]$, then $[a] \cap [b] = \varnothing$.

Proof. (i) a is in $[a]$, since $a \sim a$ by the reflexive property of equivalence relations.

(ii) If $[a] = [b]$, then a in $[a]$ implies that a is in $[b]$ and so $b \sim a$ by definition. By the symmetric property, $a \sim b$. Conversely, suppose that $a \sim b$ and let x be in $[a]$. Then $a \sim x$ and $x \sim a$, and by the transitive property $x \sim b$ and so $b \sim x$. Thus x is in $[b]$ and $[a] \subseteq [b]$. The proof that $[b] \subseteq [a]$ is carried out in the same way.

(iii) Suppose that $[a] \cap [b] \neq \varnothing$ so that there is an element x in $[a] \cap [b]$. Then $a \sim x$ and $b \sim x$. By the symmetric property $x \sim b$ and by the transitive property $a \sim b$. Using part (ii), we have that $[a] = [b]$. We have proved the contrapositive of (iii). □

Fix a positive integer n. When working with the integers modulo n, there are many ways to express the same equivalence class. For example, $[0] = [n] = [-n] = [17n]$. It is often convenient to represent an equivalence class $[i]$ using the least nonnegative integer to which i is congruent modulo n. We can find that integer by dividing i by n:

$$i = qn + r \qquad \text{with } 0 \leq r < n$$

and so $[i] = [r]$.

Definition. If $i = qn + r$ with $0 \leq r < n$, then r is called the **least nonnegative residue** of i modulo n.

This process also shows that every integer i is in one of the classes $[0], [1], \ldots, [n-1]$ modulo n. Furthermore, no two of the equivalence classes $[0], [1], \ldots,$

$[n-1]$ are equal, for if $[i] = [j]$, then $i \equiv j \pmod{n}$ by Lemma 6.2 (part ii). But since $0 \leq i, j < n$, then n cannot divide $(i - j)$ unless $i = j$. Thus it is not the case that $[i] = [j]$ when $0 \leq i < j < n$. The equivalence classes $[0], [1], \ldots, [n-1]$ are a complete (and useful) set of equivalence classes of the integers modulo n.

Example 6.1 (continued). When working modulo 12, we use the equivalence classes $[0], [1], \ldots, [11]$.

We can also do arithmetic with the equivalence classes modulo n: addition, subtraction, multiplication, exponentiation, and sometimes division.

Definition. The equivalence classes $\{[0], [1], \ldots, [n-1]\}$ are called the **integers modulo *n*** and are denoted by $\mathbf{Z}_n$.

Lemma 6.3. If $a \equiv b \pmod{n}$ and $c \equiv d \pmod{n}$, then
(i) $a + c \equiv b + d \pmod{n}$,
(ii) $a - c \equiv b - d \pmod{n}$, and
(iii) $ac \equiv bd \pmod{n}$.

Proof. We prove part (iii). Since $a \equiv b \pmod{n}$, there is an integer i such that $a - b = in$ or equivalently $a = b + in$. Since $c \equiv d \pmod{n}$, there is an integer j such that $c = d + jn$. Thus

$$\begin{aligned} ac &= (b + in)(d + jn) \\ &= bd + bjn + din + ijn^2 \\ &= bd + (bj + di + ijn)n. \end{aligned}$$

Thus $(ac - bd)$ is divisible by n, and $ac \equiv bd \pmod{n}$. □

Question 6.5. Verify the first two parts of Lemma 6.3.

Thus we can define arithmetic modulo n on equivalence classes as follows.

$$[a] + [b] = [a + b], \quad [a] - [b] = [a - b],$$
$$[a] \cdot [b] = [ab], \quad \text{and } [a]^k = [a^k] \text{ for } k \text{ a positive integer.}$$

These definitions look sensible, but there are some important points to be checked. We must check that addition and multiplication are **"well defined"** by these equations. (Subtraction is just addition of negative numbers, and exponentiation is just repeated multiplication, so we concentrate on the other two operations.) What this means is that if x is any element of $[a]$ and y is any element of

$[b]$, then

$$[x + y] = [a + b] \qquad \text{and} \qquad [xy] = [ab].$$

We now show that addition of equivalence classes modulo n is well defined. If x is in $[a]$ and y is in $[b]$, then

$$a \equiv x \pmod{n} \qquad \text{and} \qquad b \equiv y \pmod{n}.$$

By Lemma 6.3

$$a + b \equiv x + y \pmod{n} \qquad \text{and} \qquad [x + y] = [a + b].$$

Question 6.6. Show that multiplication of equivalence classes is well defined.

Example 6.1 (once more). Working with the integers modulo 12, we want to add and multiply in the following way:

$$[1] + [1] = [2], \quad [1] + [2] = [3], \quad [8] + [9] = [17] = [5],$$
$$[-5] + [10] = [5], \quad [3] \cdot [0] = [0], \quad \text{and} \quad [5] \cdot [6] = [30] = [6].$$

But is this consistent? We know that $[1] = [25]$, since $1 \equiv 25 \pmod{12}$. Thus it should be the case that

$$[1] + [25] = [1] + [1] = [2].$$

Fortunately, $[1] + [25] = [26] = [2]$, since $2 \equiv 26 \pmod{12}$. Similarly, $[17] = [5]$ and $[-6] = [6]$. Thus

$$[17] \cdot [-6] = [-102] = [6] = [5] \cdot [6].$$

When can we do division or cancel modulo n?

Lemma 6.4. If $ab \equiv cd \pmod{n}$ and $a \equiv c \pmod{n}$, then $b \equiv d \pmod{n}$ provided that $\gcd(a, n) = 1$.

Proof. By assumption there are integers i and j such that

$$ab - cd = in \qquad \text{and} \qquad a - c = jn.$$

Substituting $c = a - jn$ in the first equation yields

$$ab - (a - jn)d = in.$$

Thus

$$ab - ad = in - jdn$$

and

$$a(b - d) = (i - jd)n. \tag{A}$$

Since n divides the right-hand side of (A), n also divides the left-hand side, $a(b - d)$. Since a and n have gcd 1 and thus no factors in common, n must divide $(b - d)$. Thus $b \equiv d \pmod{n}$. □

Question 6.7. Pick five distinct integers a, b, c, d, and n such that $\gcd(a, n) = 1$, $a \equiv c \pmod{n}$, and $ab \equiv cd \pmod{n}$. Verify that $b \equiv d \pmod{n}$. Then find integers a, b, c, d, and n such that $\gcd(a, n) \neq 1$, $a \equiv c \pmod{n}$, $ab \equiv cd \pmod{n}$, but $b \not\equiv d \pmod{n}$.

Now what would it mean to say that we can do division with the integers modulo n? Division by a number x is the same as multiplying by $1/x$, and $1/x$ has the property that $x(1/x) = 1$.

Definition. Given $[a]$ in the integers modulo n, we say that $[a]$ has a **multiplicative inverse** if there is another equivalence class $[b]$ such that

$$[a] \cdot [b] = [1].$$

Thus $[b]$ is playing the role of "$1/[a]$" and is called the multiplicative inverse of $[a]$. If $[a] \cdot [b] = [1]$, then $[b] \cdot [a] = [1]$ and so $[a]$ is also the multiplicative inverse of $[b]$.

Similarly, if a and b are two integers with $0 < a, b < n$ such that $ab \equiv 1 \pmod{n}$, then we say that a and b are each other's **multiplicative inverses**.

Corollary 6.5. Let n be a positive integer. Then the equivalence class $[a]$ has a multiplicative inverse if and only if $\gcd(a, n) = 1$.

Proof. If $1 = \gcd(a, n)$, Corollary 3.3 says there are integers x and y such that

$$1 = xa + yn.$$

Thus

$$1 \equiv xa \pmod{n},$$

and so

$$[1] = [xa] = [x] \cdot [a].$$

Thus $[x]$ is the multiplicative inverse of $[a]$.

Conversely, if

$$[a] \cdot [x] = 1,$$

then

$$ax = 1 + kn \qquad \text{for some integer } k.$$

Thus any common divisor of a and n must also divide 1, and so $\gcd(a, n) = 1$. □

Two integers a and b are said to be **relatively prime** if $\gcd(a, b) = 1$. Thus an integer a has a multiplicative inverse modulo n if and only if a and n are relatively prime.

Example 6.1 (again). In Z_{12}, only 1, 5, 7, and 11 are relatively prime to 12. Here are their multiplicative inverses:

$$[1] \cdot [1] = [1], \quad [5] \cdot [5] = [25] = [1],$$
$$[7] \cdot [7] = [49] = [1], \quad \text{and } [11] \cdot [11] = [121] = [1].$$

In other words, each of 1, 5, 7, and 11 is its own multiplicative inverse. Here is a brute force check that $[2]$ does not have a multiplicative inverse:

$$\begin{array}{lll} [2] \cdot [0] = [0], & [2] \cdot [1] = [2], & [2] \cdot [2] = [4], \\ [2] \cdot [3] = [6], & [2] \cdot [4] = [8], & [2] \cdot [5] = [10], \\ [2] \cdot [6] = [0], & [2] \cdot [7] = [2], & [2] \cdot [8] = [4], \\ [2] \cdot [9] = [6], & [2] \cdot [10] = [8], & [2] \cdot [11] = [10]. \end{array}$$

Question 6.8. Find multiplicative inverses for all elements of Z_5 and of Z_{10} that have inverses. Which elements of Z_{18} have multiplicative inverses?

Finding inverses will be important in the application presented in Section 7, as will a variation on the next theorem, known as Fermat's little theorem. This one he really did prove.

Theorem 6.6. If p is a prime number and $\gcd(a, p) = 1$, then

$$a^{p-1} \equiv 1 \pmod{p}.$$

Proof. Notice that for any integer i, $\gcd(i, p)$ is either 1 or p, the only divisors of p. Thus $\gcd(i, p) = 1$ if and only if p does not divide i, that is, if and only if $i \not\equiv 0 \pmod{p}$. Thus by assumption $a \not\equiv 0 \pmod{p}$. Consider the equivalence classes

$$[a], [2a], [3a], \ldots, [(p-1)a]. \qquad \textbf{(B)}$$

We claim that none of these is $[0]$ and that no two of them are equal.

First if it were the case that

$$[ia] = [0] \qquad \text{where } 1 \leq i < p,$$

then

$$ia \equiv 0 \pmod{p}.$$

Thus p divides ia, and since $\gcd(a, p) = 1$, p divides i, a contradiction, since $1 \leq i \leq (p-1)$. Thus none of the equivalence classes in (B) equals $[0]$.

Next suppose that

$$[ia] = [ja] \qquad \text{where } 1 \leq i, j < p.$$

Then

$$ia \equiv ja \pmod{p} \qquad \text{by Lemma 6.2 (part ii), and}$$
$$i \equiv j \pmod{p} \qquad \text{by Lemma 6.4,}$$

a contradiction, since both i and j are positive integers less than p. Thus the $(p-1)$ equivalence classes listed in (B) are the same as the equivalence classes $[1], [2], [3], \ldots, [p-1]$, although probably listed in a different order. Then

$$[a] \cdot [2a] \cdots [(p-1)a] = [1] \cdot [2] \cdots [p-1]$$
$$[a(2a) \cdots ((p-1)a)] = [(p-1)!] \qquad \text{multiplying equivalence classes}$$
$$a(2a) \cdots ((p-1)a) \equiv (p-1)! \pmod{p} \qquad \text{by Lemma 6.2 (part ii)}$$
$$a^{p-1}(p-1)! \equiv (p-1)! \pmod{p} \qquad \text{simplifying}$$
$$a^{p-1} \equiv 1 \pmod{p} \qquad \text{by Lemma 6.4}$$

since $\gcd(p, (p-1)!) = 1$. □

Question 6.9. Pick a prime p and an integer b such that $\gcd(b, p) = 1$, write down the equivalence classes $[b], [2b], \ldots, [(p-1)b]$ modulo p, and verify that they are the same as the classes $[1], [2], \ldots, [p-1]$. Check that $b^{p-1} \equiv 1 \pmod{p}$. Find c, an integer with $\gcd(c, p) \neq 1$, and show that $c^{p-1} \not\equiv 1 \pmod{p}$.

This has been a brief introduction to arithmetic modulo n and to the ideas of equivalence relations. We shall use this in an application to cryptography (the art of secret messages) in the next section.

EXERCISES FOR SECTION 6

1. In each of the following find the least nonnegative integer i such that
   (*a*) $4^{30} \equiv i \pmod{19}$.
   (*b*) $2^{31} \equiv i \pmod{377}$.
   (*c*) $2^{i} \equiv 1 \pmod{17}$.
   (*d*) $2^{11}3^{13} \equiv i \pmod{7}$.
   (*Hint:* After each multiplication replace the result by its least nonnegative residue modulo n.)
2. Explain why it is always true that $n^5 \equiv n \pmod{10}$ or, in other words, why n^5 and n always have the same last digit.
3. Prove that for every integer n, either $n^2 \equiv 0 \pmod{4}$ or $n^2 \equiv 1 \pmod{4}$. Use this to show that there are no integers x and y such that $x^2 + y^2 = 1987$.
4. The set $\{0, 1, \ldots, n-1\}$ is called a **complete residue system** modulo n because every integer is congruent modulo n to exactly one of these numbers.
   (*a*) Find a complete residue system modulo n in which all numbers are negative.
   (*b*) Find a complete residue system modulo n in which all numbers have absolute value at most $n/2$.
5. Is either the relation "$<$" or "$\leq$" an equivalence relation on the integers?
6. Which of the following define an equivalence relation on the integers? Explain your answer.
   (*a*) $a \sim b$ if a divides b.
   (*b*) $a \sim b$ if $a < b$.
   (*c*) $a \sim b$ if $|a| \leq |b|$.
   (*d*) $a \sim b$ if a and b begin with the same (decimal) digit.
   (*e*) $a \sim b$ if when a and b are expressed as $a = 2^i s$ and $b = 2^j t$ with i and j nonnegative integers and s and t odd integers, then $s = t$.
   (*f*) $a \sim b$ if when a and b are expressed as in part (e), then $i = j$.

7. Prove that the following is an equivalence relation defined on the integers: $a \sim b$ if a and b have the same number of prime divisors, counting multiplicity (e.g., $18 = 2 \cdot 3^2$ has three prime divisors). For this equivalence relation are addition and multiplication of equivalence classes well defined by $[a] + [b] = [a + b]$ and $[a] \cdot [b] = [ab]$? Explain.

8. Give an example of a relation on a set that has the following properties.
(*a*) Reflexive and symmetric, but not transitive.
(*b*) Reflexive and transitive, but not symmetric.
(*c*) Symmetric and transitive, but not reflexive.

9. **(*a*)** Write down the elements of Z_2. Then write down an addition and multiplication table for Z_2; that is, write down all possible sums $[a] + [b]$ and all possible products $[a] \cdot [b]$.
(*b*) Do the same for Z_3.
(*c*) Do the same for Z_4.

10. Rewrite the Euclidean equations using congruences.

11. Prove that if $[i]$ and $[j]$ are equivalence classes modulo n such that $[i] = [j]$, then $\gcd(i, n) = \gcd(j, n)$.

12. A relation $\sim$ on a set S is called a **total** (or **linear**) **ordering** if
(i) for all a and b in S, exactly one of the following holds:
$a \sim b$, $a = b$, or $b \sim a$, and
(ii) for elements a, b, and c in S, if $a \sim b$, and $b \sim c$, then $a \sim c$.
Do either $<$ or $\leq$ define a total ordering on the integers? Explain.

13. Give an example of an equivalence relation on the integers that is not a total ordering. Explain which properties of a total order hold for your example and which don't.

14. Explain why, in general, if $\sim$ is a total ordering on a set S, then $\sim$ is not an equivalence relation on S. Conversely, explain why if $\sim$ is an equivalence relation on S, then $\sim$ is not a total ordering.

15. Prove that if p is a prime number, then for every integer n,

$$n^p \equiv n \pmod{p}.$$

16. Investigate whether or not the following is true: If $\gcd(a, n) = 1$, then $a^{n-1} \equiv 1 \pmod{n}$.

17. Suppose that p is a prime number and $\gcd(a, p) = 1$. Then explain why $[a^{p-2}]$ is the multiplicative inverse of $[a]$ modulo p. For each $i = 2, 3, \ldots, p - 1$, find an expression for the multiplicative inverse of $[a^i]$ modulo p.

18. We define an equivalence relation on ordered pairs of integers, $Z \times (Z - \{0\})$, (i.e., on all ordered pairs with the second entry nonzero) by $(a, b) \sim (c, d)$ if $ad = bc$.

(i) Prove that this is an equivalence relation.
(ii) Describe the equivalence classes $[(1,2)]$, $[(1,1)]$, $[(4,2)]$, and $[(2,3)]$.
(iii) In general, describe $[(r, s)]$ and compare this with the rational number r/s.
(iv) Which equivalence classes $[(r,s)]$ have multiplicative inverses? If $[(r,s)]$ has a multiplicative inverse, what is it?

4:7 AN APPLICATION: PUBLIC KEY ENCRYPTION SCHEMES

Although our discussion of the greatest common divisor problem has been couched in modern terminology, most of what we have presented in this chapter is ancient. It was developed without any thought of computing machines like those we now possess and with no anticipation of future applications. It is a truism in mathematics that the purest (i.e., most theoretical and seemingly least applicable) ideas from one generation of mathematicians frequently become indispensable tools of the applied mathematicians of subsequent generations. The application of the Euclidean algorithm and related number theory that we are about to present exemplifies this phenomenon.

The problem that we confront is that Bob wants to send Alice a message, the content of which is to remain a secret from Eve. The message will be a sequence of integers, $M_1, M_2, \ldots, M_k$ with $0 < M_i < N$ for $i = 1, 2, \ldots, k$, where N is a number chosen with which to work modulo N. We'll see how Alice chooses N later. Such a representation of a message by numbers is no restriction. For example, this book has been prepared electronically, and each character of the keyboard of the computer terminal has associated with it a unique decimal number, in this instance called its ASCII code. For use in this chapter we include the ASCII code for capital letters in Table 4.1.

Table 4.1

Letter:	A,	B,	C,	D,	E,	F,	G,	H,	I,	J,	K,	L,	M,	N	O,	P,
Code:	65	66	67	68	69	70	71	72	73	74	75	76	77	78	79	80
	Q,	R,	S,	T,	U,	V,	W,	X,	Y,	Z,	blank					
	81	82	83	84	85	86	87	88	89	90	32					

Thus the assumption that Bob's message is a sequence of decimal numbers is not restrictive. Sending a message might be by telephone or electronic mail or almost anything else. Crucial to the model is that Eve has the technology to intercept the message. This turns out to be surprisingly realistic. Thus if Bob just sends Alice the ASCII code of the real message, we assume that Eve can intercept and correctly interpret its content.

So what can Bob do? Very simply he must devise a way to disguise his message, to **encrypt** it so that after Eve, or anyone else, intercepts the disguised message, she will not be able to figure it out. Of course, if Bob does too good a job encrypting the message, maybe Alice won't be able to figure it out either. So Bob and Alice agree on an encryption scheme. When Bob has a message to send Alice, he pulls out his encryption book (or maybe calls his encryption computer program) and encrypts his message. Alice, having the appropriate decryption book (or computer program), can unscramble or **decrypt** the message. This is fine unless Eve obtains a copy of Bob's encryption procedure. Then it may be that the message is no longer secure. (If Eve obtains Alice's decrypting procedure, presumably Eve can decrypt any message that Alice can decrypt.) We shall describe here a method that will tell Bob exactly how to encrypt his message. It will tell Alice exactly how to decrypt the received message. Finally, (and this is truly magical), even if Eve knows Bob's method of encryption, Eve will not be able to decipher the message.

There is a family of related methods that will accomplish the above goals. These are known as **public key encryption schemes**, and they use so-called **trapdoor functions**. (The analogy is that encrypting information like opening a trapdoor from above is easy, but decrypting like opening a trapdoor from below when one is stuck in the trap is hard.) The scheme we present uses the Euclidean algorithm and modular arithmetic and is known as the **RSA scheme** for Rivest, Shamir, and Adleman, the inventors of the scheme. There are other schemes based on a wide variety of mathematical ideas, and there is a great deal of research being done on the question of just how secure these trapdoor schemes are.

Suppose that Bob wants to send a message with j letters, including blanks between words. Using Table 4.1, this becomes a decimal number with $2j$ digits when we replace each letter by the corresponding two digits of the ASCII code. If Bob simply transmits the ASCII code equivalent, Eve will be able to look up the ASCII code in a table and understand the message.

Example 7.1. The message "HELLO" becomes 7269767679.

Question 7.1. Translate the message "HOWDY" into its ASCII code equivalent. Decipher the message 83858270327383328580.

Actually, most messages, like those just mentioned, will turn into numbers that are far too large to work with. Thus we agree in advance to break the $2j$ digits of the message up into blocks of length B and then send k messages $M_1, M_2, \ldots, M_k$, each of length at most B.

Example 7.1 (continued). Let $B = 4$. Then we send the encryption of HELLO as three messages: 7269, 7676, and 7932, with a blank added at the end to fill out the last block.

Question 7.2. With $B = 4$, the largest code that can be sent using capital letters is 9090. What letters produce this code? What is the smallest possible decimal number that we can transmit with $B = 4$?

Now it's time for Alice to get sneaky. She picks an integer N and announces that all work will be done modulo N. In particular, the transmitted messages will lie between 0 and N. (Then a convenient choice of block length B is one less than the number of decimal digits in N.) The sneaky part is that Alice picks N to be the product of two nearby, large prime numbers p and q. So $N = pq$ with $p \neq q$. Now it is easy and quick to multiply two prime numbers or any two numbers, even if they are very large. What is very difficult to do, given an integer N, is to determine its prime factors.

(*Note:* All steps in this process are summarized at the end of the section in the algorithm RSA.)

Exercise 14 gives a simple algorithm DIVISORSEARCH that searches for the divisors of an integer N. (A more sophisticated algorithm for finding prime divisors is presented in Supplementary Exercise 4.) DIVISORSEARCH finds divisors by checking whether the integers $2, 3, \ldots$, up to $\sqrt{N}$ divide N. If $N = pq$, then either p or q must be at most $\lfloor \sqrt{N} \rfloor$. If, say, p is discovered to be a divisor, then q is found as N/p.

So why not use this algorithm? DIVISORSEARCH is slow. (In Exercise 15 you are asked to verify that DIVISORSEARCH is exponential.) Faster ones have been derived, using very deep mathematics, but all the known algorithms for factoring a number have nonpolynomial running time. For example, whereas we can easily multiply together two 30-digit numbers to get a 60-digit number, if we are given N with 60 digits it takes much longer to unscramble it into its prime factors. If N is either a prime or the product of two large primes that are near one another, then an algorithm as in Supplementary Exercise 4 would have to run about a year before this fact is discovered. That's no problem for this application: Alice will choose new values of p and q with $N = pq$ every 6 months before Eve is able to find (or to run a computer program to find) the factors of N.

Question 7.3. Each of the following are of the form pq for primes p and q. Try to factor each: 323, 4087, and 8633.

However, Alice has more tricks up her sleeve. After selecting $N = pq$, she selects an integer $e > 1$, known as the **exponent**, with the property that $\gcd(e, (p-1)(q-1)) = 1$. Remember that the gcd of two numbers is easy and quick to calculate. Alice can just try random numbers between 0 and N and run the Euclidean algorithm on them to find an e relatively prime to $(p-1)(q-1)$.

Question 7.4. Let $N = 7 \cdot 11 = 77$. Then search through $2, 3, 4, \ldots$ to find four numbers e that are relatively prime to $6 \cdot 10 = 60$.

Example 7.2. Suppose that $N = 9991 = 97 \cdot 103$, where 97 and 103 are both prime. (These are not particularly large prime numbers but will keep us occupied with calculations by hand.) Let us check that $e = 11$ meets the requirements for an exponent by calculating $\gcd(11, 96 \cdot 102) = \gcd(11, 9792)$.

$$9792 = 890 \cdot 11 + 2$$
$$11 = 5 \cdot 2 + 1$$
$$2 = 2 \cdot 1 + 0.$$

Once Alice has determined e, she will perform one more calculation, described later, but then she may destroy the factors of N or else she must guard them closely as they are the key to the security of this system.

However, Alice can be quite open with the numbers e and N. In fact, she sends them to Bob without any secrecy. Maybe she even lists them in a phone book or publishes them in the newspaper. These numbers will tell Bob (and for that matter anyone who cares to send a message to Alice) how to securely encrypt their message. The procedure goes as follows.

Here's what Bob does to encrypt the message. First the original message is turned into ASCII code using Table 4.1 and then the resulting huge number is broken into blocks of length B. Each block is one of the messages $M_1, M_2, \ldots, M_k$ to be sent. Next Bob must check that each message M_i, for $i = 1, 2, \ldots, k$, is relatively prime to N. If not, in the gcd calculation he will discover that their common divisor is either p or q. In that case he announces to Alice and to the world that he has found a factor of N and it is time to change their protocol (i.e., to change the values of e and N). However, most messages and numbers are relatively prime to N as shown in Exercise 9.

Then for each message M_i, $i = 1, 2, \ldots, k$, Bob calculates R_i, where

$$R_i \equiv M_i^e \pmod{N} \qquad \text{and} \qquad 0 < R_i < N.$$

Precisely, he can divide by N and find the remainder R_i

$$M_i^e = QN + R_i \qquad \text{with } 0 < R_i < N.$$

Then he will transmit the encrypted message $R_1, R_2, \ldots, R_k$.

Now M_i^e will often be a large number, but one that can be determined quickly using FASTEXP; however, there are additional ways in modular arithmetic to keep the numbers relatively small. Recall that by Lemma 6.3, if $a \equiv b \pmod{N}$, then $a^e \equiv b^e \pmod{N}$. Thus when we need M_i, M_i^2, M_i^4, and so on for FASTEXP, we can repeatedly replace the numbers by their least nonnegative residues modulo N, as shown in the next example. This replacing process is also known as **reducing modulo N**.

Examples 7.1 and 7.2 (continued). With $N = 9991$ and $e = 11$, we begin the encryption of the message "HELLO" from its ASCII code 726976767932:

$$M_1 = 7269 \qquad M_2 = 7676 \qquad M_3 = 7932.$$

First we check that $\gcd(7269, 9991) = 1$. Then since $M_1^{11} = M_1^8 M_1^2 M_1$, we calculate

$$\begin{aligned} M_1^2 &\equiv 52838361 \equiv 5953 \pmod{9991} \\ M_1^4 &\equiv (5953)^2 \pmod{9991} \\ &\equiv 35438209 \pmod{9991} \\ &\equiv 132 \pmod{9991} \\ M_1^8 &\equiv (132)^2 \pmod{9991} \\ &\equiv 17424 \pmod{9991} \\ &\equiv 7433 \pmod{9991}. \end{aligned}$$

Thus the first message R_1 that we want to send is

$$\begin{aligned} M_1^{11} &\equiv 7433 \cdot 5953 \cdot 7269 \pmod{9991} \\ &\equiv 44248649 \cdot 7269 \pmod{9991} \\ &\equiv 8501 \cdot 7269 \pmod{9991} \\ &\equiv 61793769 \pmod{9991} \\ &\equiv 9425 \pmod{9991}. \end{aligned}$$

Question 7.5. Show that $M_2 = 7676$ and $M_3 = 7932$ are relatively prime to 9991 and then determine either R_2 or R_3 for these messages.

This procedure to encrypt a message could be implemented easily in a computer program; see algorithm RSA.

There are two questions that require an answer. First, how is Alice to recover the content of the original message? Second, assuming that Eve receives the encrypted message and possesses the numbers e and N, why can't she discover the hidden message?

We answer the second question first. We assume that Eve intercepts the message $R_1, R_2, \ldots, R_k$ with $0 < R_i < N$ for $i = 1, 2, \ldots, k$. How might she find $M_1, M_2, \ldots, M_k$? Why can't she just take the eth root of R_i to get M_i? Or why not try all possible messages M, raise each to the eth power and reduce modulo N until the correct messages are found?

Examples 7.1 and 7.2 (continued). The 11th root of 9425 is 2.2977..., and so this is not much help. The problem is that we took the 11th power of 7269 modulo 9991 and now we would need the 11th root modulo 9991, whatever that means.

The straightforward approach of trying everything would tell us to calculate $1^e(\text{mod } 9991)$, $2^e(\text{mod } 9991)$, $3^e(\text{mod } 9991)$, and so on. But checking with the ASCII code Table 4.1 we can be more clever. The encrypted word is a four-digit number of the form:

$$3232, \qquad 32wz\ (65 \leq wz \leq 90), \quad wz32\ (65 \leq wz \leq 90),$$

or

$$uvwz\ (65 \leq uv, wz \leq 90),$$

a total of 729 possible ASCII codes. Now for any integer i, i^{11} requires five multiplications: three to form i^2, i^4, and i^8, and two more to combine these into i^{11}. If at each stage one reduces modulo 9991, then five more divisions and five more multiplications are needed. Thus in the worst case, after 10,935 multiplications and divisions Eve can uncover which message M_i produced the transmitted message R_i. In Chapter 2, Table 2.9, we assumed that a personal computer can perform 17,800 single-digit multiplications or divisions in a minute. Since multiplying two 4-digit numbers requires at most 16 single-digit multiplications (plus some additions), it will take Eve $10{,}935 \cdot 16/17{,}800$ or about 10 minutes to recover the messages (once she's written the appropriate computer program on her PC). That's not so bad, but as we mentioned earlier, in this example we are really working with small numbers compared with those used in real life.

More generally, one block of a transmitted message may have $B < N$ decimal digits, not just 4. Let's figure out how long it will take Eve to decrypt a B-digit number if she tries all possibilities. If we allow the ASCII code for all characters, not just for capital letters, then the B-digit number will lie between 0 and $10^{B+1} - 1$. Suppose that the modulus N is roughly 10^{B+1} and so exponentiation by e, where $1 < e < N$, might use as many as $O(\log(N))$ multiplications and divisions. Thus a systematic search will involve

$$O(\log(10^{B+1})10^{B+1}) = O((B+1)10^{B+1})$$

operations, clearly an exponential amount of work for Eve.

Question 7.6. Suppose that the modulus is $N = 10^{B+1}$ and the B-digit numbers can be any number between 0 and N and also $e = 11$. Then using the figures from Examples 7.1 and 7.2, find the minimum value of B such that Eve must calculate for a month before she can figure out all possible messages.

There is an interesting sidelight to the above phenomenon. When we say that Eve has an exponential amount of work to do, in general, we are stating an empirical fact about the worst-case scenario. At present there is no theorem that says that Eve will need to examine all or even a large fraction of all numbers. Thus it is conceivable that a clever idea would enable Eve to break this encryption scheme with an efficient decryption scheme. There are other variations on this scheme with the same uncertainty, namely that there is no theorem that says decryption must be exponential in the worst case, and yet no one has determined how to "crack" these schemes with polynomial-time algorithms. This state of uncertainty has prompted a great deal of research on the mathematics behind public key encryption. The state of the art seems to be that encryptors have the upper hand at the moment; however, the decryptors have made some progress that has resulted in the encryptors having to work harder.

If Eve has such a difficult time decrypting the message, then how can Alice successfully decrypt the message? Remember that Alice calculated one additional piece of information about e and $N = pq$ before she destroyed or hid the values of p and q. Since e and $(p-1)(q-1)$ have gcd one, she used Corollaries 6.5 and 3.3 and secretly found the multiplicative inverse d of e modulo $(p-1)(q-1)$, that is,

$$ed \equiv 1 \pmod{(p-1)(q-1)} \qquad \text{with } 0 < d < (p-1)(q-1).$$

The pair (d, N) is called the **decrypting key**.

Example 7.1 and 7.2 (yet again). With $e = 11$ and $N = 9991$, we find the multiplicative inverse d using the Euclidean algorithm. In one part of Example 7.2 we checked that gcd $(11, 9792) = 1$ and we use these Euclidean equations as in Corollary 3.3:

$$\begin{aligned} 1 &= 11 - 5 \cdot 2 \\ &= 11 - 5 \cdot (9792 - 890 \cdot 11) \\ &= 4451 \cdot 11 - 5 \cdot 9792. \end{aligned}$$

Thus 4451 is the multiplicative inverse of 11 modulo 9991.

Decryption now is easy for Alice because she knows a theorem that implies that

$$R_i^d \equiv M_i \pmod{N}$$

for $i = 1, 2, \ldots, k$. Thus all she has to do is to calculate R_i^d, replace it by the least nonnegative residue modulo N and that's the message M_i. And again she pulls out a computer program that can quickly perform this exponentiation.

Examples 7.1 and 7.2 (concluded). The message 7269 was encrypted as 9425. Here is a summary of the calculation of $(9425)^{4451}$:

$$9425^{4451} = 9425^{4096} 9425^{256} 9425^{64} 9425^{32} 9425^{2} 9425.$$

With 12 multiplications we find

$$\begin{aligned}
9425^2 &\equiv 644 \pmod{9991}\\
9425^{32} &\equiv 1975 \pmod{9991}\\
9425^{64} &\equiv 4135 \pmod{9991}\\
9425^{256} &\equiv 5202 \pmod{9991}\\
9425^{4096} &\equiv 1225 \pmod{9991}.
\end{aligned}$$

Then with 5 more multiplications we find

$$\begin{aligned}
9425^{4451} &\equiv 1225 \cdot 5202 \cdot 4135 \cdot 1975 \cdot 644 \cdot 9425 \pmod{9991}\\
&\equiv 7269 \pmod{9991},
\end{aligned}$$

just as we claimed.

Question 7.7. Suppose that $N = 15 = 3 \cdot 5$ and let $e = 7$. Find d such that $ed \equiv 1 \pmod{2 \cdot 4}$. Then encrypt each of the messages 2 and 7 using the exponent e and then decrypt them using d.

Why does Alice's decryption scheme work? The reason is the following theorem; notice its similarity with Fermat's little theorem, Theorem 6.6. They are both cases of a more general result due to Euler; see Supplementary Exercises 23 and 24.

Theorem 7.1. If p and q are distinct primes, $n = pq$, and $\gcd(a, n) = 1$, then

$$a^{(p-1)(q-1)} \equiv 1 \pmod{n}.$$

Why does this explain Alice's decryption procedure? Alice knows that

$$R_i \equiv M_i^e \pmod{N}.$$

Thus

$$\begin{aligned}
R_i^d &\equiv (M_i^e)^d \pmod{N}\\
&\equiv M_i^{ed} \pmod{N}\\
&\equiv M_i^{1+k(p-1)(q-1)} \pmod{N}
\end{aligned}$$

for some integer k, since $ed \equiv 1 \pmod{(p-1)(q-1)}$

$$\begin{aligned}
&\equiv M_i(M_i^{(p-1)(q-1)})^k \pmod{N}\\
&\equiv M_i 1^k \pmod{N}
\end{aligned}$$

by Theorem 7.1, since $\gcd(M_i, N) = 1$

$$= M_i.$$

Remember that Bob checked that $\gcd(M_i, N) = 1$ for $i = 1, 2, \ldots, k$ and if not, announced the need for a change of modulus N and exponent e. Now we see that it is vital that M_i and N be relatively prime for the decryption scheme to work.

Proof of Theorem 7.1. (Notice the similarities between this proof and that of Theorem 6.6.) Let Z_n be the integers modulo n:

$$Z_n = \{[0], [1], [2], \ldots, [n-1]\}$$

and let A be the subset defined by

$$A = \{[x] \text{ in } Z_n\text{: } \gcd(x, n) = 1\}.$$

First we count the number of elements in A by specifying and counting the elements in $Z_n - A$. Now for any $[i]$ in Z_n, $\gcd(i, n)$ is 1, p, q, or pq. The only element $[x]$ for which $\gcd(x, n) = pq$ is $x = 0$. Which elements $[x]$ have $\gcd(x, n) = p$? Exactly $x = p, 2p, \ldots, (q-1)p$, and that's all since $n = pq$. Which ones have $\gcd(x, n) = q$? Exactly $x = q, 2q, \ldots, (p-1)q$. Notice that no two of these numbers x are equal. For example, if $ip = jq$ with $1 \leq i \leq (q-1)$, $1 \leq j \leq (p-1)$, then p divides jq. Since p and q are distinct primes, p divides j, a contradiction since $j \leq (p-1)$.

Thus the equivalence classes $[p], \ldots, [(q-1)p], [q], \ldots, [(p-1)q]$ are all distinct, and so there are $1 + (q-1) + (p-1)$ elements in $Z_n - A$. Then A contains

$$n - 1 - (p-1) - (q-1) = pq - p - q + 1 = (p-1)(q-1)$$

elements. We list the elements of A as

$$A = \{[r_1], [r_2], \ldots, [r_s]\},$$

where $0 < r_1 < r_2 < \cdots < r_s < n$, $s = (p-1)(q-1)$.

Lemma 7.2. If $\gcd(a, n) = \gcd(b, n) = 1$, then $\gcd(ab, n) = 1$.

Proof (of lemma). We prove the contrapositive. Suppose that

$$\gcd(ab, n) = d > 1.$$

By Example 4.1, d has a prime divisor, say p. Thus p divides both n and ab and thus at least one of a and b. So either a or b share a common prime divisor with n contradicting our hypothesis. □

We now use this lemma. Take any number a such that $\gcd(a, n) = 1$ and look at the equivalence classes

$$S = \{[ar_1], [ar_2], \ldots, [ar_s]\}.$$

By Lemma 7.2 $\gcd(ar_i, n) = 1$ for $i = 1, 2, \ldots, s$. In addition,

$$\gcd(r_1 r_2 \ldots r_s, n) = 1. \qquad (*)$$

We claim that the equivalence classes of S are the same as those of A, only perhaps listed in a different order. Since $\gcd(ar_i, n) = 1$, $[ar_i]$ is in A and so $[ar_i] = [r_j]$ for some value of j. Furthermore, no two of the classes in S are equal: if

$$[ar_i] = [ar_k]$$

for some values of i and k with $r_i \neq r_k$, then

$$\begin{aligned} ar_i &\equiv ar_k \pmod{n} \\ r_i &\equiv r_k \pmod{n} && \text{by Lemma 6.4} \\ r_i &= r_k && \text{since } r_i, r_k < n, \end{aligned}$$

a contradiction. Thus $S = A$, and

$$[r_1] \cdot [r_2] \cdot \cdots \cdot [r_s] = [ar_1] \cdot [ar_2] \cdot \cdots \cdot [ar_s]$$

since multiplication is well defined. Thus,

$$\begin{aligned} r_1 r_2 \cdots r_s &\equiv ar_1 ar_2 \cdots ar_s \pmod{n} && \text{by Lemma 6.2} \\ &\equiv a^s r_1 r_2 \cdots r_s \pmod{n} \\ 1 &\equiv a^s \pmod{n} && \text{by Lemma 6.4 and (*)} \\ &\equiv a^{(p-1)(q-1)} \pmod{n}. \end{aligned}$$

□

Why can't Eve find the decrypting pair (d, N)? Precisely because d is the multiplicative inverse of e modulo $(p - 1)(q - 1)$, and she doesn't know the values of p and q. As we saw before, factoring N to obtain p and q would require an exponential amount of work for her, unless she can think of something new and clever. Perhaps Eve's best bet is to study number theory and cryptography and to search for an efficient decrypting algorithm. However, she should be aware that

Alice could do the same. Alice might even someday come up with a provably secure system, that is, a system for which one can prove there is no polynomial-time algorithm to decrypt messages.

Question 7.8. For either R_2 or R_3, calculated in Question 7.5, check that $R_i^{4451} = M_i$.

We conclude with a summary of the steps needed in the RSA encryption and decryption scheme.

Algorithm RSA

STEP 1. (Numerical calculations by receiver)
- (a) Pick primes p and q, and let $N = pq$.
- (b) Find e such that $\gcd(e, (p-1)(q-1)) = 1$.
- (c) Find d such that $ed \equiv 1 \pmod{(p-1)(q-1)}$ with $0 < d < (p-1)(q-1)$.
- (d) Throw away p and q.
- (e) Announce N and e to the world.

STEP 2. (Encryption)
- (a) Translate the message into ASCII code using Table 4.1.
- (b) Pick an integer B less than the number of digits in N.
- (c) Break the ASCII coded message into blocks of B digits each; call these $M_1, M_2, \ldots, M_k$.
- (d) For $i = 1, 2, \ldots, k$ make sure that $\gcd(M_i, N) = 1$; if not, announce that the code is "broken" and return to step 1.
- (e) For $i = 1, 2, \ldots, k$ let $R_i \equiv M_i^e \pmod{N}$ with $0 < R_i < N$.
- (f) Transmit the encrypted messages $R_1, R_2, \ldots, R_k$.

STEP 3. (Decryption)
- (a) For $i = 1, 2, \ldots, k$ calculate $M_i \equiv R_i^d \pmod{N}$ with $0 < M_i < N$.
- (b) For $i = 1, 2, \ldots, k$ translate M_i from ASCII code using Table 4.1.

EXERCISES FOR SECTION 7

1. Using Table 4.1 give the ASCII code for the following: ***(a)*** RIGHT ON, ***(b)*** THE TRUTH, ***(c)*** ENCRYPT ME, and ***(d)*** FOREVER.

2. What do the following ASCII codes stand for in English?

(a) 7279 3272 8577.

(b) 7079 8287 6582 6832.

(c) 7879 3287 6589.

(d) 8479 3266 6932 7982 3278 7984 3284 7932 6669.

3. Determine which of the following are the product of two distinct primes: **(*a*)** 801, **(*b*)** 803, **(*c*)** 807, **(*d*)** 809, **(*e*)** 161, **(*f*)** 1631, and **(*g*)** 17,947.

4. For each of the following values of $N = pq$ (from Question 7.3), find an integer e such that $\gcd(e, (p-1)(q-1)) = 1$ and find the multiplicative inverse of e: **(*a*)** 323, **(*b*)** 4087, and **(*c*)** 8633.

5. If $N = 77$, $e = 7$, and $B = 4$, explain why Bob cannot send the message PEACE to Alice.

6. Using blocks of four digits ($B = 4$), $N = 8633 = 89 \cdot 97$, and $e = 5$ encrypt the message CHEERS.

7. Using $N = 95$ and $e = 29$, decrypt the message (with $B = 2$)

$$53 \quad 29 \quad 02 \quad 51 \quad 29.$$

8. Let p be an odd prime and e an integer such that $\gcd(e, p - 1) = 1$. Suppose that a message M is encrypted as C, where

$$C \equiv M^e \pmod{p} \qquad \text{where } 0 \leq C < p.$$

If d is the multiplicative inverse of e modulo p, then prove that

$$C^d \equiv M \pmod{p}.$$

9. Show that the number of numbers i such that $0 \leq i < n = pq$ and $\gcd(i, n) \neq 1$ is $q + p - 1$. Then deduce that the probability of picking such an i is

$$\frac{1}{p} + \frac{1}{q} - \frac{1}{pq}.$$

If p, $q > 10^{30}$, then show that the probability of choosing, at random, an integer not relatively prime to n is less than 10^{-29}.

10. Write down in pseudocode an algorithm ENCRYPT that upon input of a message $M_1, M_2, \ldots, M_k$ and N and e, encrypts the message using the RSA scheme.

11. Write down in pseudocode an algorithm DECRYPT that upon input of a received message $R_1, R_2, \ldots, R_k$, two primes p and q (where $N = pq$) and the exponent e, decrypts this message.

12. Determine the number of multiplications and divisions performed in the worst case of ENCRYPT and DECRYPT. (You may count each multiplication and division as one, regardless of the number of digits.)

13. Prove the converse of Lemma 7.2.

14. Here is an algorithm to find divisors of an integer N.

Algorithm DIVISORSEARCH

STEP 1. Input N
STEP 2. For $i := 2$ to $\lfloor\sqrt{N}\rfloor$ do
 STEP 3. If i divides N, then output "i is a divisor of N"
STEP 4. If no divisors have been output, then output "N is a prime"
STEP 5. Stop.

Explain why this algorithm correctly determines when N is a prime. Explain why, if N is not a prime, this algorithm finds all, except possibly one, prime divisors of N.

15. Explain why, in the worst case, there is a number $r > 1$ such that the algorithm DIVISORSEARCH performs at least r^D divisions, where D equals the number of bits needed to express N in binary. Find as large a value of r as is possible with your argument.

16. Modify the algorithm DIVISORSEARCH so that its output includes all prime divisors of N. How many divisions does this perform in the worst case?

4:8 THE DIVIDENDS

The overall aim of this chapter has been to introduce the counting and algorithmic ideas of discrete mathematics within number theory. In addition, this chapter introduced specific results from number theory with indications of their applicability in mathematics and computer science.

The chapter has focused on algorithms to determine the greatest common divisor of two integers. In the text and exercises we found straightforward algorithms to solve the gcd problem and then developed the less obvious Euclidean algorithm. From the point of view of bit input, the straightforward algorithms are bad and exponential, but EUCLID is a good and linear algorithm. The worst-case complexity analysis of the latter algorithm is different from that of previous algorithms in that it comes in two stages. First we show that if we use the Euclidean algorithm on two successive Fibonacci numbers, then the number of multiplications and divisions is logarithmic in the input numbers. Next we show that in the worst case of the Euclidean algorithm with arbitrary input $b \leq c$, $O(\log(c))$ operations are performed. Thus the Fibonacci numbers exhibit this worst-case behavior and so the worst-case analysis really does reflect what may happen. What is also true, but we do not prove it, is that the Fibonacci numbers are actually the worst-case input for the Euclidean algorithm. In conclusion, we observe that $O(\log(c)) = O(B)$, where B is the number of bits needed for the input.

Two important general ideas were introduced in this chapter. The first is Complete Induction, which gives us more flexibility at the cost of more checking of base cases. Also we presented the idea that the size of the input to an algorithm ought to be measured in terms of bits. Thus an integer n requires B [roughly $\log(n)$] bits, and it is in terms of this parameter B that we should be determining and analyzing the complexity functions of algorithms. With this perspective we look back to EXPONENT and FASTEXP of Chapter 2 and see that they are exponential and linear algorithms, respectively.

A substantial amount of elementary number theory appears in Section 6. Modular arithmetic and equivalence relations are central to much of mathematics and computer science. For example, the theory of groups and rings involves generalizations of Z_n, the integers modulo n. Many computer languages come with the ability to do arithmetic modulo n; this arithmetic is important in, for instance, random number generation. Equivalence relations will be crucial in further courses in theoretical computer science and mathematics. Thus the lemmas, theorems, and corollaries of Section 6 are worth studying because they will come up again both in applications and in other branches of mathematics and computer science.

There is a variety of different encryption schemes in use today; each uses different aspects of number theory. The approach we pursue relies on the Euclidean algorithm and Fermat's little theorem, but its effectiveness comes from the fact that it is apparently difficult to factor a number into its prime factors. In fact, it has recently been shown that the difficulty of "cracking" a variation of the RSA scheme is computationally equivalent to factoring a number n into two primes. However, there is no known theorem that says it is hard to decrypt a message sent using the RSA scheme or that it is hard to factor a number. A closely related algorithmic problem is that of determining whether a given number is prime. Recent fast, so-called random algorithms have been developed that can test whether "most" numbers are prime, and there is a primality-testing algorithm that has been shown to run in polynomial time on all integers, provided that a famous open problem, the extended Riemann hypothesis, is true. No one has proved the latter result, but most mathematicians believe it is true. Thus if you are tempted to set up an encryption service along the lines of this chapter, take heed. It may be that soon a mathematical or algorithmic breakthrough will occur and destroy the effectiveness of the RSA encryption scheme.

Number theory is an excellent training ground for logical analysis and deduction. It is accessible: Small examples can be explored numerically, general patterns deduced, and proofs constructed by induction and contradiction. The Fibonacci numbers are a sample of the kinds of intriguing problems in the field. Others include prime numbers, modular arithmetic, and solutions of equations. Number theory also gives an introduction to the mathematical discipline of abstract algebra and the computer science discipline of arithmetic and algebraic computations. Especially if the ideas in this chapter interest you, these are fields worthy of further study.

SUPPLEMENTARY EXERCISES FOR CHAPTER 4

1. Design a gcd algorithm called GCD2 that is based on the following idea. If 2 divides b and c, then 2 is a factor of gcd (b, c). Furthermore, we may carry out the division and consider the smaller problem of finding the gcd $(b/2, c/2)$. If 2 does not divide b or c, try 3, ..., try j. Note that the maximum value of j that you need to check is no more than b or $\sqrt{c}$. Why?
2. Use GCD2 to find the gcd of the following pairs: **(*a*)** (8, 12), **(*b*)** (24, 32), and **(*c*)** (72, 96).
3. In the worst case how many divisions will GCD2 need?
4. Design an algorithm that upon input m will find all prime numbers between 1 and m. (*Hint:* Use the idea behind GCD2. This idea is attributed to the Greek mathematician Eratosthenes of the third century B.C. The method is known as the Sieve of Eratosthenes: First cross out all multiples of 2 except for 2 itself. Next cross out all multiples of 3 except for 3 itself, ..., and so on. How far do you have to keep going with this crossing out process?)
5. Use your Sieve of Eratosthenes algorithm to find all prime numbers between 1 and 200.
6. Let U be the set of positive integers less than 49. Set $A = \{x \in U\colon x \text{ is divisible by } 2\}$, $B = \{x \in U\colon x \text{ is divisible by } 3\}$, and $C = \{x \in U\colon x \text{ is divisible by } 5\}$. Find $|A|$, $|B|$, and $|C|$. Find $|A \cup B \cup C|$. (*Hint:* Look at PIE from Chapter 1.) Use the results of this problem to calculate the number of primes less than 50.
7. Two couples are camping in Hawaii with a pet parrot. They collect a pile of macadamia nuts, but during the night one woman gets up, divides the pile of nuts into four equal piles and finds one nut left over, which she gives to the parrot to keep it quiet. She hides one pile, combines the other three piles into one, and goes back to sleep. Then her husband wakes up, looks suspiciously at the pile, divides the (remaining) nuts equally into four with one extra nut for the parrot, hides one pile, and goes back to sleep. The same thing happens two more times; each time the remaining nuts divide evenly into four equal piles with one nut left over, which is given to the parrot, and one pile is hidden. In the morning the four graciously divide the remaining nuts into four equal piles and find they have one macadamia nut left over for the parrot. What is the minimum number of nuts that they could have had at the start of the evening?
8. Look back at the definition of $IC(b, c)$ in Exercise 3.13. Prove that $gcd\,(b, c) = \min\{IC(b, c)\}$.
9. Design an algorithm EXTENDEDEUCLID that first finds the gcd of b and c as in the algorithm EUCLID and then expresses the gcd as a linear combination of b and c. The algorithm should use only a constant number of variables, say 10 at most.

10. **(*a*)** Suppose that n is even. Then the following sum equals a Fibonacci number. Which one is it?

$$\binom{n}{0}+\binom{n-1}{1}+\binom{n-2}{2}+\binom{n-3}{3}+\cdots+\binom{n/2}{n/2}.$$

(*b*) Find a similar sum of binomial coefficients that equals a Fibonacci number when n is odd.

11. Prove the results you obtained in Exercise 10. *Hint:* Use induction and the fact that

$$\binom{n-k}{k}=\binom{(n-1)-k}{k}+\binom{(n-2)-(k-1)}{k-1}.$$

12. If the Euclidean algorithm is applied to $c = F_{k+4}$ and $b = F_{k+1}$, what can you say about the number of Euclidean equations?

13. Suppose that the algorithm EUCLID is modified so that in step 3 the variable q is set equal to the nearest integer to c/b. Run some examples of this algorithm, including some Fibonacci numbers. Then analyze the complexity of the algorithm in terms of c. Is this version more efficient than EUCLID?

14. Prove Lamé's Theorem: In the Euclidean algorithm the smallest values of c that produces k Euclidean equations is $c = F_{k+2}$.

15. Prove that if $a^s \equiv 1 \pmod{n}$ and $a^t \equiv 1 \pmod{n}$, then $a^{\gcd(s,t)} \equiv 1 \pmod{n}$.

16. For ordinary integers, $xy = 0$ if and only if either x or y equals 0. Give examples to show that this is false in Z_n, that is, for equivalence classes modulo n it is not true that

$$[x]\cdot[y]=[0] \quad \text{if and only if} \quad [x]=[0] \quad \text{or} \quad [y]=[0].$$

17. Prove the following about equivalence classes modulo n. Given $[x] \neq [0]$, there is a $[y] \neq [0]$ such that $[x]\cdot[y] = [0]$ if and only if $\gcd(x, n) \neq 1$.

18. Consider Z_n, the integers modulo n, where $n = st$ with $\gcd(s, t) = 1$. Show that there are at least four different equivalence classes $[i]$ modulo n such that $[i]^2 = [i]$. For example, $[0]^2 = [0^2] = [0]$.

19. If $n = pq$, then explain why the following are true:

$$p+q=n-(p-1)(q-1)+1$$
$$p-q=\sqrt{(p+q)^2-4n}.$$

Suppose that an algorithm were discovered that given an integer $n = pq$, a product of two primes, could quickly calculate $(p-1)(q-1)$. Then use the facts that

$$p = \tfrac{1}{2}((p + q) + (p - q))$$
$$q = \tfrac{1}{2}((p + q) - (p - q))$$

and the results of the previous equations to argue that there would be a fast algorithm to factor n into its two prime divisors.

20. We define $\phi(m)$ to be the number of integers i in $\{1, 2, \ldots, m\}$ such that $\gcd(i, m) = 1$. Determine $\phi(6)$, $\phi(7)$, $\phi(9)$, $\phi(10)$, $\phi(p)$, and $\phi(p^2)$, where p is a prime, and $\phi(pq)$, where p and q are distinct primes.

21. Prove Wilson's theorem: If p is a prime, then p divides $((p - 1)! + 1)$. [*Hint:* Show that $(p - 2)! \equiv 1 \pmod{p}$ by pairing numbers with their multiplicative inverses.]

22. Here are some ideas for an alternative proof of Fermat's little theorem, which states that $b^{p-1} \equiv 1 \pmod{p}$ if $\gcd(b, p) = 1$. First show that if strings of beads of length p are formed using b different colors of beads, then the number of such strings that are not all one color is $b^p - b$. (You should assume that there is an unlimited supply of beads of each color.) If the ends of each string are tied together to form a bracelet, explain why the number of different colored bracelets is $(b^p - b)/p$. (For example, the string of red, blue, and green beads forms the same bracelet as the string of blue, green, and red beads.)

23. A theorem due to Euler states that if $\gcd(a, m) = 1$, then

$$a^{\phi(m)} \equiv 1 \pmod{m},$$

where the function ϕ is defined in Exercise 20. Verify that this theorem is true for m a prime or a product of two primes.

24. Prove Euler's theorem (of Exercise 23) using the following hints: Let $A = \{[s_1], [s_2], \ldots, [s_{\phi(m)}]\}$, where the s_i are all the integers in $\{1, 2, \ldots, m\}$ that are relatively prime to m. Let $S = \{[as_1], [as_2], \ldots, [as_{\phi(m)}]\}$. Then proceed as in the proof of Theorem 7.1.

5

GRAPH THEORY

5:1 BUILDING THE LAN

A college's minicomputers, terminals, and microcomputers are joined in a **Local Area Network (LAN** for short). The advantages include the potential to connect terminals and microcomputers with any minicomputer, the capacity to support more terminals on campus, and rapid transmission of data between terminals and minicomputers. The first step in the installation of the LAN was to link the basement of every building on campus using coaxial cable. This does not mean that each pair of buildings is joined by a cable. What it does mean is that it is possible to send electrical signals via the cable from any building on campus to any other building perhaps using one or more intermediate buildings. How should we, or the LAN designers, decide which pairs of buildings to join directly by coaxial cable?

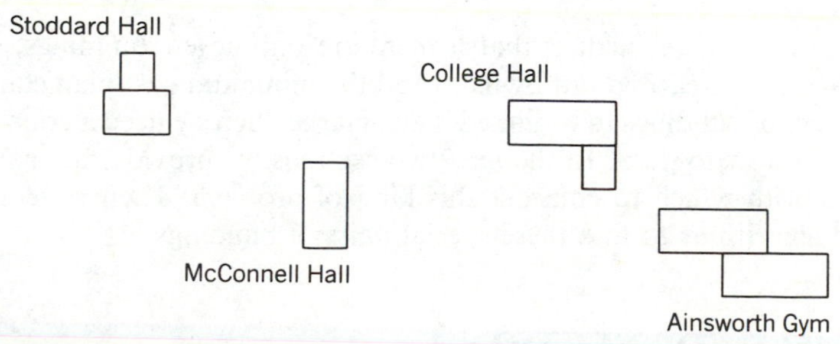

Figure 5.1 Campus map.

Example 1.1. Suppose that we want to connect College Hall (C), Ainsworth Gymnasium (A), and McConnell Hall (M) by coaxial cable. Which pairs of buildings should be directly connected? See Figure 5.1. If we connect C directly to M and M directly to A, then there is no need to connect C directly to A, since it is already possible to send electronic signals from C to A through M. There are three possible direct connections to make. Choosing any two makes the third unnecessary.

Question 1.1. Suppose that Ainsworth Gym (A), College Hall (C), McConnell Hall (M), and Stoddard Hall (S) are buildings on campus to be connected by the LAN. See Figure 5.1. How many pairs of possible direct connections are there among the four buildings? How many direct connections do you need to install so that communication (though not necessarily direct) is possible between each pair of buildings? Will any set of this many direct connections work? Answer the same questions if five buildings are to be joined.

In Example 1.1 why should any particular pair of connections be selected in preference to any other pair? In practice, such decisions are made to minimize the total installation cost.

Example 1.2. Suppose that it costs \$85,000 to install a cable between McConnell Hall and College Hall, \$78,000 to install a cable between Ainsworth and McConnell, and \$87,000 to install a cable between Ainsworth and College Hall. Which two direct cable links have minimum cost? The MC and AM links have a total cost of \$163,000; the MC and AC links have a total cost of \$172,000; and the AC and AM links have a total cost of \$165,000.

Question 1.2. Suppose that the cost of joining S with C is \$30,000, the cost of joining S with M is \$51,000, and the cost of joining S with A is \$67,000. Using the data from Example 1.2, determine which pairs of A, C, M, and S should be directly linked in order to minimize the total cost. Compare your answer with that of the preceding example.

What should be evident is that if there are only a few buildings, then these calculations can be carried out by hand and the minimum cost plan can be found. If the number of buildings to be linked is at all large, then we need a good algorithm to figure out how to do it. In the next two sections we provide the mathematical framework with which to consider this kind of problem. Then we contrast bad and good algorithms to find these special pairs of buildings.

EXERCISES FOR SECTION 1

1. Neilson Library (N) is to be included on the LAN with McConnell, Ainsworth, Stoddard, and College Hall. The cost to join Neilson with the other buildings is given by \$20,000 to join N with S, \$27,000 to join N with C, \$45,000 to join

N with M, and \$75,000 to join N with A. The other costs are the same as in Example 1.2 and Question 1.2. Which pairs of these five academic buildings should be directly linked in order to minimize the total cost?

2. Suppose that there are six buildings, say A, B, C, D, E, and F, that are to be joined in a LAN system. How many pairs of buildings are there that might be joined by cable? How many pairs will you need to join in order to establish the LAN? The estimated costs (in tens of thousands of dollars) are given in the following table. What should be the pairs of the LAN system?

	A	B	C	D	E	F
A	0	2.7	3.8	2.9	7.8	9.3
B	2.7	0	4.8	5.2	8.4	6.9
C	3.8	4.8	0	3.5	4.6	5.7
D	2.9	5.2	3.5	0	6.4	7.1
E	7.8	8.4	4.6	6.4	0	3.9
F	9.3	6.9	5.7	7.1	3.9	0

3. Suppose that you wanted to connect four buildings in a communications system so that there are two different ways to send messages from any building to any other building. How many pairwise connections would you need? Answer the same question for five buildings.

4. Suppose in Question 1.2 that we require a direct link between McConnell Hall and College Hall. Otherwise, we still wish to find the least expensive LAN connections. Which additional direct connections should we add?

5. Repeat Exercise 1 with the two constraints that McConnell Hall and College Hall must be directly linked, as must Neilson Library and Stoddard Hall.

6. Draw the 3×2 and 6×5 rectangular grid (see Section 3.1). Suppose that each line represents a street and at each street intersection there is a fire hydrant. In each grid find a subset of the streets that connects up all fire hydrants to the lower left-hand corner and whose total length is as short as possible.

5:2 GRAPHS

The LAN of the preceding section can be modeled using what mathematicians and computer scientists call a graph (not to be confused with the graph of a function). For this problem the sets that we consider are the set of buildings and the set of pairs of buildings.

Graph Theory Definitions. A **graph** consists of a finite set of **vertices** together with a finite set of edges. Each **edge** consists of a distinct pair of distinct vertices. If the edge e consists of the vertices $\{u, v\}$, we often write $e = (u, v)$ and say that u

is **joined** to v (also that v is joined to u) and that u and v are **adjacent**. We also say that both u and v are **incident** with the edge e.

Although graphs are frequently stored in a computer as lists of vertices and edges, humans have a more picturesque way to think about graphs. Typically, we shall represent the vertex set of a graph as a set of points in the plane. An edge will be represented by a line segment or an arc (not necessarily straight) joining the two vertices incident with it.

Example 2.1. Figure 5.2 exhibits two typical graphs. The vertices and edges of the graph in Figure 5.2(b) are labeled so that we can reinforce the above definitions. In this graph x and z are adjacent as are y and r, and so on. The edge $f = (z, y)$ is incident with both z and y, and y and w are incident with the edge h. Note that r is not adjacent to w. However, if the edges of this graph represent the direct cable connections in a Local Area Network, then w and r are connected in the sense that an electronic message could be sent from w to r going through y.

Look back at Chapter 3 and notice that we were really doing a graph theory problem there, only we didn't call it by that name. We now call the diagrams of that chapter by their usual names: **grid graphs**. Figure 5.3 shows the 3×2 grid graph.

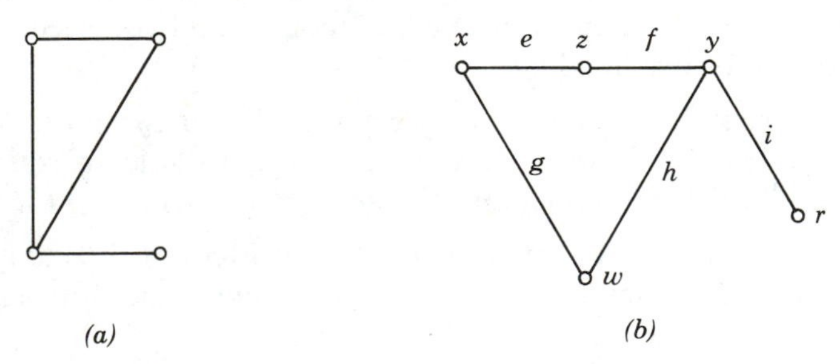

Figure 5.2

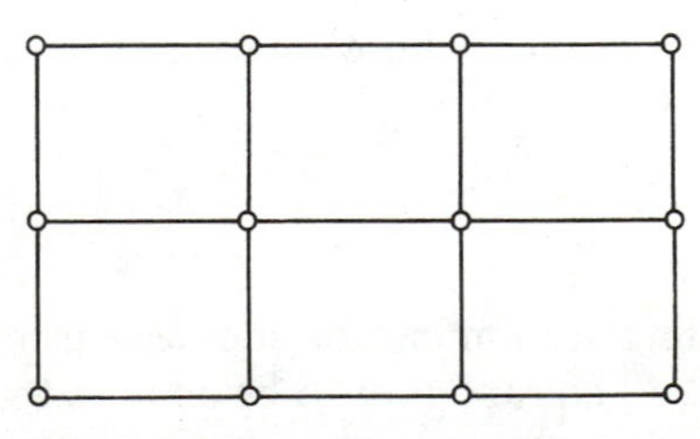

Figure 5.3

Some Notation. In a graph G the vertex set will be denoted by $\boldsymbol{V(G)}$ and the number of vertices by $|V(G)| = V$ (if there is no ambiguity as to which graph we are talking about). Similarly, the edge set will be denoted by $\boldsymbol{E(G)}$ and the number of edges by $|E(G)| = E$. Often the vertices of a graph will be labeled so that they can be distinguished or named according to the intended application.

Example 2.1 (again). In Figure 5.2(*b*), $V(G) = \{x, y, z, w, r\}$ and $V = 5$. Also, $E(G) = \{e, f, g, h, i\} = \{(x, z), (y, z), (x, w), (y, w), (y, r)\}$ and $E = 5$.

Question 2.1. Draw a graph with $V = 4$ and $E = 3$. Is there more than one such graph?

Another Definition. The **degree** of a vertex x is the number of vertices adjacent to x (or equivalently, the number of edges incident with x). Within a graph G we denote the degree of x by **deg(*x*, *G*)** or **deg(*x*)** if there is no confusion as to which graph G we refer to.

One way to think about a graph is as if it were constructed of buttons and thread: The buttons represent the vertices and the thread represents the edges. In this model the degree of a vertex is the number of strands of thread emanating from the corresponding button.

Example 2.1 (once more). In Figure 5.2(*b*), $\deg(x) = 2 = \deg(z) = \deg(w)$, while $\deg(y) = 3$ and $\deg(r) = 1$.

Question 2.2. For the graph shown in Figure 5.4 determine V, E, and the degree of each vertex. Find the sum of the degrees of all the vertices.

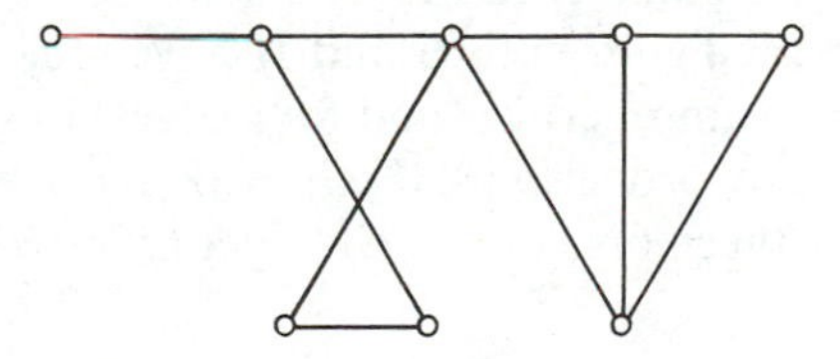

Figure 5.4

In the preceding question you might have noticed that the sum of the degrees was exactly twice the number of edges. That this happens in general is our first result.

Theorem 2.1. If $V(G) = \{x_1, \ldots, x_V\}$, then

$$\deg(x_1) + \cdots + \deg(x_V) = 2E.$$

Proof. Since the degree of a vertex is the number of edges incident with that vertex, the sum of the degrees counts the total number of times an edge is incident with a vertex. Since every edge is incident with exactly two vertices, each edge gets counted twice, once at each end. Thus the sum of the degrees equals twice the number of edges. □

Question 2.3. A vertex whose degree is odd is called **odd**. Show that in any graph there is an even number of odd vertices.

To answer the second part of Question 2.1 with assurance you need to know how to decide if two graphs are the same or different.

Definition. Two graphs, say G and H, are said to be **isomorphic** if there exists a function $f: V(G) \rightarrow V(H)$ such that
(i) f is both one-to-one and onto,
(ii) f preserves adjacencies, and
(iii) f preserves nonadjacencies.

Isomorphic is a fancy mathematical word meaning fundamentally the same. Two graphs that are not isomorphic are also called **different**. Properties (ii) and (iii) can be formalized as
(ii′) If (x, y) is in $E(G)$, then $(f(x), f(y))$ is in $E(H)$ and
(iii′) If (x, y) is not in $E(G)$, then $(f(x), f(y))$ is not in $E(H)$.

The function f that shows the correspondence between the vertices of G and the vertices of H is called an **isomorphism**. In practice, it is displayed by labeling the vertices of G and H with letters or numbers and then explicitly writing out the function values $f(x)$ for each x in $V(G)$. Or the vertices of G and H can be labeled so that if x is labeled A in G, then $f(x)$ is labeled A in H.

Example 2.2. The graphs G and H in Figure 5.5 are isomorphic as are the graphs K and M. Note that in the figure both G and H have four vertices and five edges while K and M both have four vertices and four edges. In general, if G and H are isomorphic graphs, then by property (i), $|V(G)| = |V(H)|$ and by properties (ii) and (iii), $|E(G)| = |E(H)|$. Furthermore, $\deg(x, G) = \deg(f(x), H)$ for every x in $V(G)$.

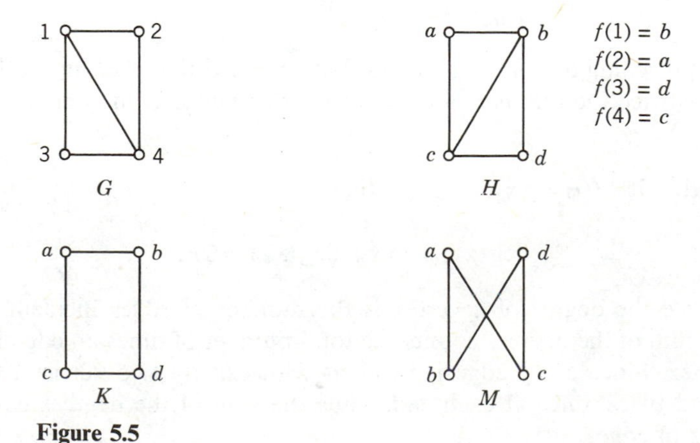

Figure 5.5

Example 2.3. Figure 5.6 shows the six different graphs on two and three vertices. (Note that a graph need not be connected in the sense of the LAN.)

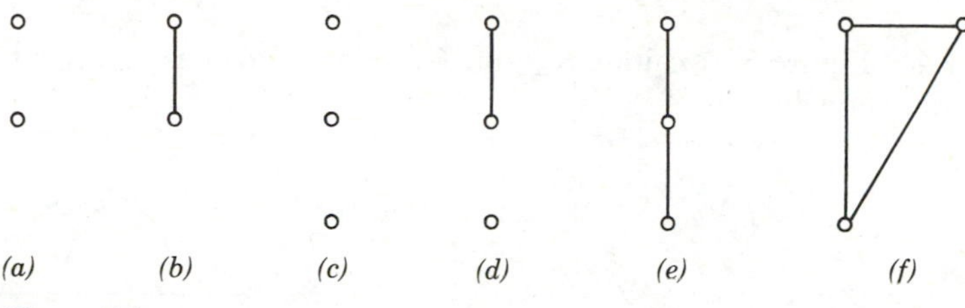

Figure 5.6

Question 2.4. Find the 11 different graphs on 4 vertices.

Question 2.5. Note that the graphs G and H shown in Figure 5.7 have identical degrees. Show that they are not isomorphic.

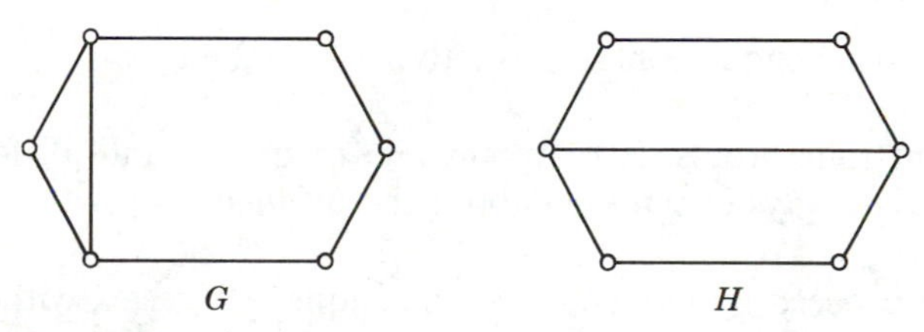

Figure 5.7

It is an open question whether there exists a good algorithm to determine if two graphs are isomorphic. This question is of considerable interest to, for instance, chemists. They model chemical molecules using graphs, where vertices represent the individual atoms and edges represent the chemical bonds. It is sometimes difficult given two molecules with identical atomic constituents to determine whether they are the same. If chemists synthesize a particular molecule in the laboratory and want to find out what is known about it, they need some method of recognizing when the chemical is discussed in the literature, that is, they want to be able to test graph isomorphism. Supplementary Exercises 6 to 9 ask you to construct and analyze straightforward graph isomorphism algorithms.

There is a notable connection between graphs and relations, as defined in Section 4.6. Given a symmetric relation $\sim$ on a finite set S, we may create a graph G with $V(G) = S$ and for $a, b \in S$ an edge $(a, b) \in E(G)$ if and only if $a \sim b$ is true. Conversely, a graph G and its edges define a relation on the elements of $V(G)$. See Exercises 24 to 27 and Supplementary Exercises 32 to 34.

We close this introductory section on graphs by defining and presenting some special classes of graphs.

Definition. A graph in which every pair of distinct vertices is joined by an edge is called **complete**. A complete graph with r vertices is also called an ***r*-clique** and is denoted by $\boldsymbol{K_r}$.

Example 2.4. Figure 5.8 exhibits K_4 and K_5. Note that these graphs have 6 and 10 edges, respectively.

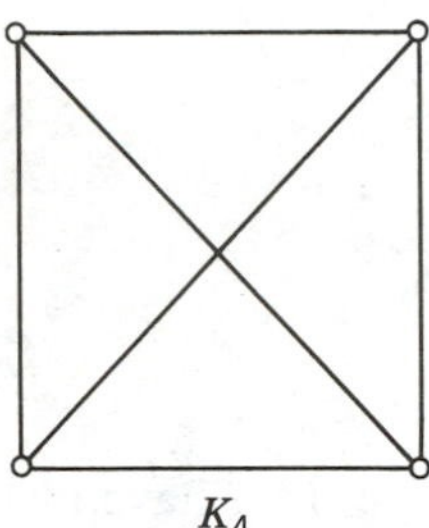

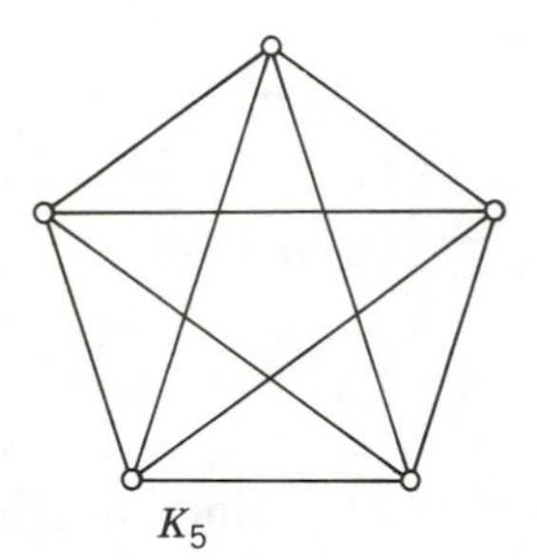

Figure 5.8

Theorem 2.2. An r-clique contains exactly $r(r-1)/2$ edges.

We present four proofs. It is worthwhile to understand all four, since they represent different ways of thinking about the problem.

Proof 1. We proceed by induction. A one-clique is a vertex without any edges, which satisfies the formula. In Figures 5.6 and 5.8 we see that $K_2, \ldots, K_5$ have the correct number of edges. Thus the base case is safely accounted for. Suppose that the theorem is true for $r = k$. We must show that it is then true for $r = k + 1$. Given a $(k+1)$-clique, pick a vertex, say x. If we erase x and all edges incident with x from our graph, we are left with a k-clique, which has $k(k-1)/2$ edges by the inductive hypothesis. In our original graph the vertex x was incident with k edges, one to each of the other vertices. Thus the total number of edges in the $(k+1)$-clique equals

$$\frac{k(k-1)}{2} + k = k\left(\frac{k-1}{2} + 1\right) = \frac{k(k+1)}{2} = \frac{(k+1)\{(k+1)-1\}}{2},$$

which is what we needed to prove. $\square$

Proof 2. Let's draw K_r, adding one vertex at a time. When we draw the first vertex, we need no edges. When we add the second vertex, we need to draw one edge to connect the two vertices. When we add the third vertex, we must join it to the two previously created vertices, so we need to draw two more edges. In general, when we add the kth vertex, we need to draw $k - 1$ new edges. Thus the total number of edges we need to draw is

$$1 + 2 + 3 + \cdots + (r - 1).$$

We have seen that this sum equals $(r - 1)\{(r - 1) + 1\}/2$ in Section 2.3. □

Proof 3. In K_r each vertex has degree $r - 1$. Thus the sum of the degrees equals $r(r - 1)$. By Theorem 2.1, this sum also equals $2E$. Thus $2E = r(r - 1)$ and $E = r(r - 1)/2$. □

Proof 4. The number of edges in K_r equals the number of 2-subsets of an r-set, which equals $\binom{r}{2}$ as we saw in Chapter 3. □

Question 2.6. How many edges are there in K_7?

Corollary 2.3. For any graph, $E \leq V(V - 1)/2$.

A graph is said to be **bipartite** if its vertex set can be partitioned into two sets, say R and B, with the property that every edge joins a vertex in R with a vertex in B. Such graphs are also called **2-colorable**, since you can think of the vertices in R as being painted red while the vertices in B are painted blue. With this painting no vertex is joined by an edge to a vertex with the same color. A bipartite graph is said to be a **complete bipartite graph** if every red vertex is joined by an edge to every blue vertex. The complete bipartite graph with p red vertices and q blue vertices is often denoted by $\boldsymbol{K_{p,q}}$. (Notice that this definition cries out for an algorithm to decide if a graph is bipartite, and if it is, to construct the sets R and B; see Supplementary Exercise 10.)

Example 2.5. Each of the graphs in Figure 5.9 is bipartite. The graph in Figure 5.9(*c*) is the complete bipartite graph $K_{3,3}$.

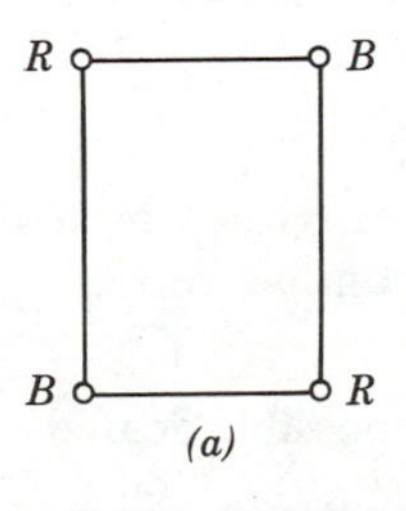

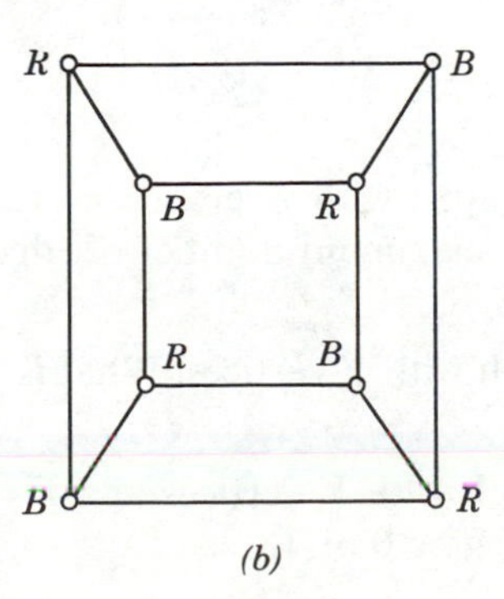

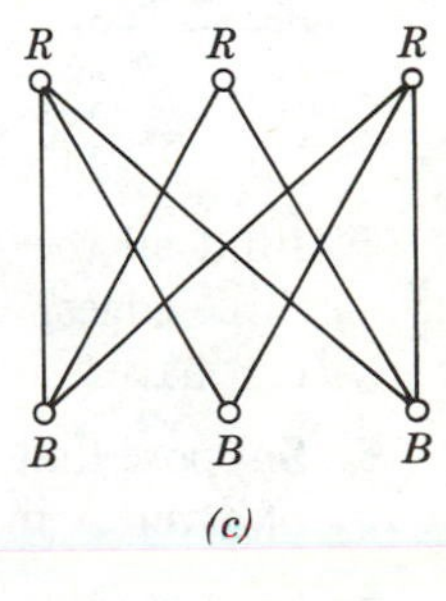

Figure 5.9

Question 2.7. Which of the graphs in Figure 5.10 are bipartite? If a graph is bipartite, color its vertices red and blue so that no edge joins two vertices of the same color. If it is not bipartite, explain why its vertices cannot be so colored.

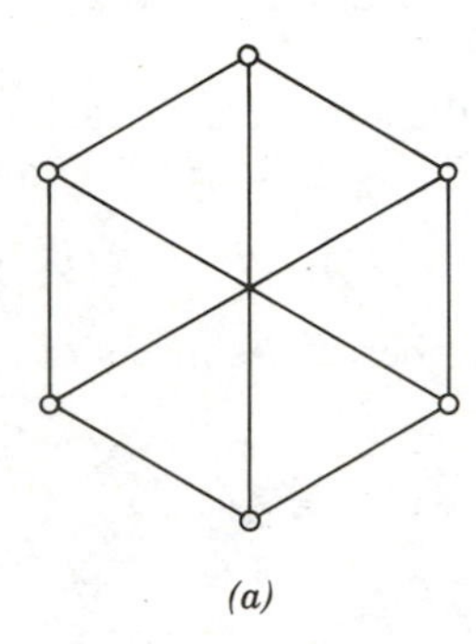
(a)

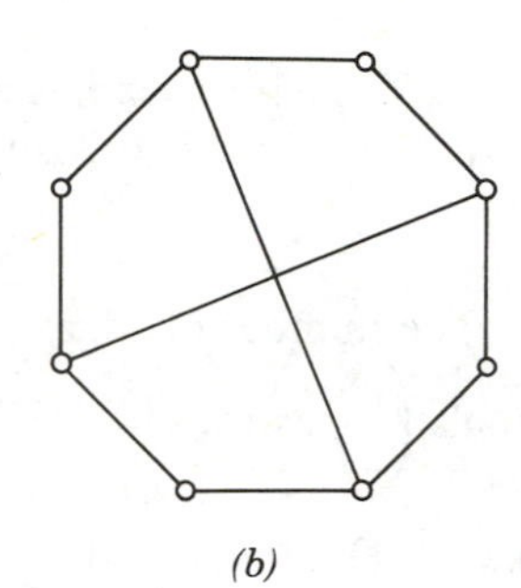
(b)

Figure 5.10

EXERCISES FOR SECTION 2

1. Draw a graph with five vertices that illustrates your LAN connections from Exercise 1.1.
2. Draw a graph with 6 vertices and 10 edges.
3. For the following graphs find V, E, and the degree of each vertex.

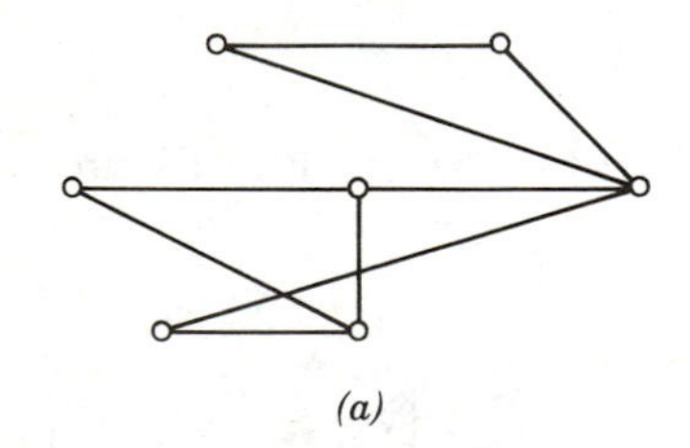
(a)

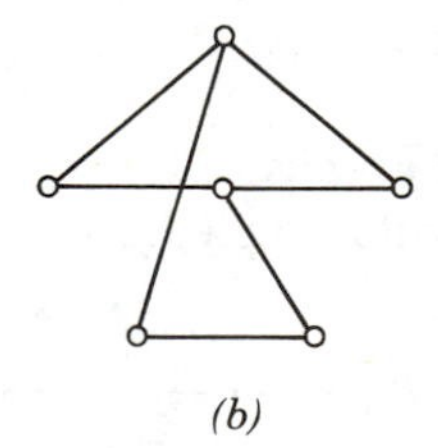
(b)

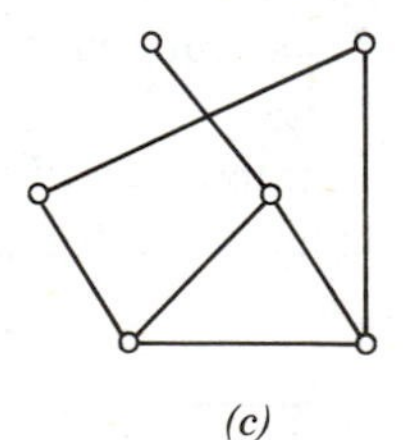
(c)

4. Find all the different graphs with five vertices and two edges. How about three edges? What is the maximum number of edges a graph on five vertices can have?
5. Suppose that G is a graph with V vertices. What is the largest possible degree of a vertex in G?
6. Suppose that G is a graph with V vertices and $E = V - 1$ edges. Prove that G contains a vertex of degree 0 or 1.

7. Prove Theorem 2.1. using induction on E.

8. At the beginning of a business meeting some of the participants are introduced to each other. In an introduction A is introduced to B and B is introduced to A for some pair A and B. Show that the number of individuals who have been introduced to an odd number of other individuals is even.

9. A graph in which every vertex has degree r is called **regular** of degree r. Find examples of graphs that are not cliques but are regular of (a) degree 1, (b) degree 2, and (c) degree 3.

10. For each of the following sequences either find a graph whose vertices have exactly these degrees or show that such a graph cannot exist.

(*a*) 3, 3, 1, 1. (*b*) 3, 2, 2, 1. (*c*) 5, 4, 4, 2, 2, 2.
(*d*) 3, 3, 2, 2, 1, 1. (*e*) 4, 1, 1, 1, 1. (*f*) 2, 2, 2, 1, 1.
(*g*) 7, 3, 3, 3, 2, 2. (*h*) 5, 5, 5, 2, 2, 2, 2, 1.

11. (*a*) Find all different graphs with 6 vertices and 15 edges.
(*b*) Find all different graphs with 6 vertices and 14 edges.
(*c*) Find all different graphs with 6 vertices and 13 edges.

12. Which of the following pairs of graphs are isomorphic? For each isomorphic pair, exhibit the isomorphism.

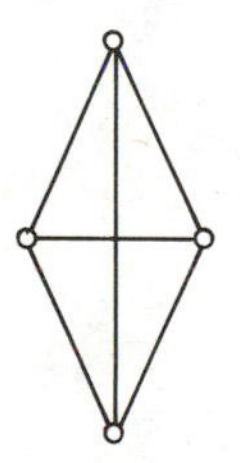
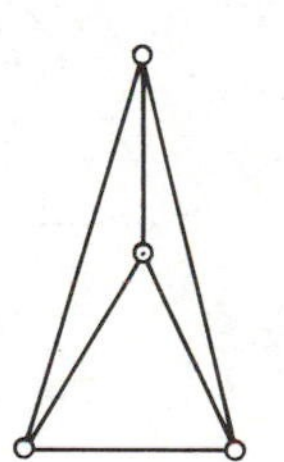

(a)

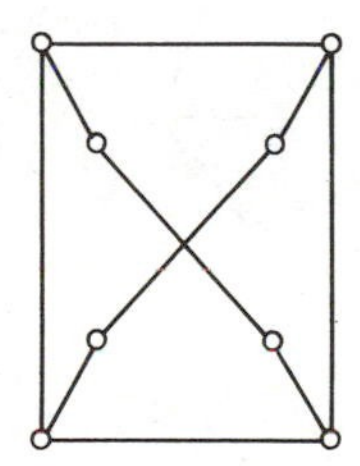
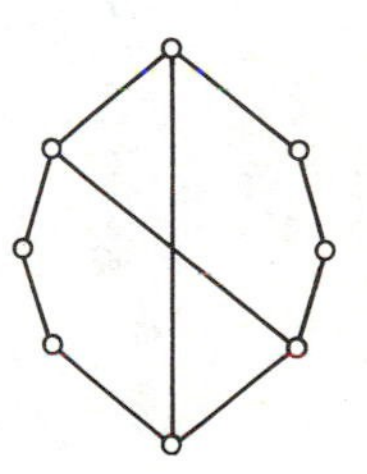

(b)

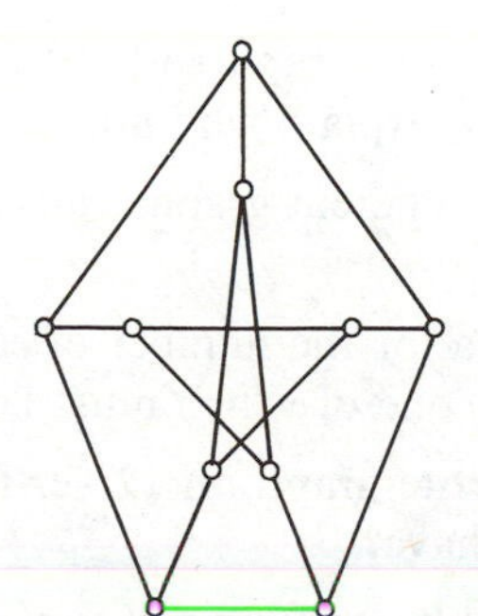

(c)

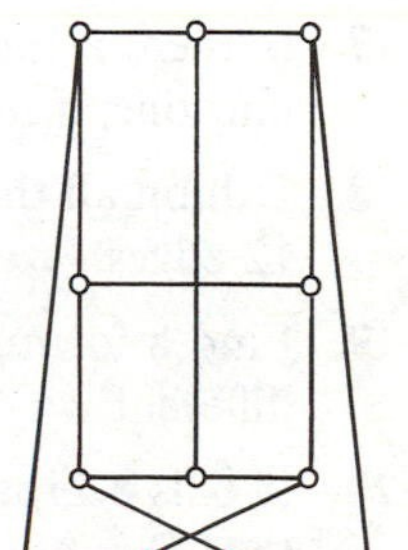

(d)

13. Find two nonisomorphic graphs with six vertices both of which are regular of degree 3.

14. Explain why a regular graph with V vertices and E edges must have all vertices of degree $2E/V$. (See Exercise 9.)

15. Which of the following graphs are bipartite? Label the vertices of the bipartite graphs with R and B so that no edge joins two vertices with the same label, and explain why the others cannot be so labeled.

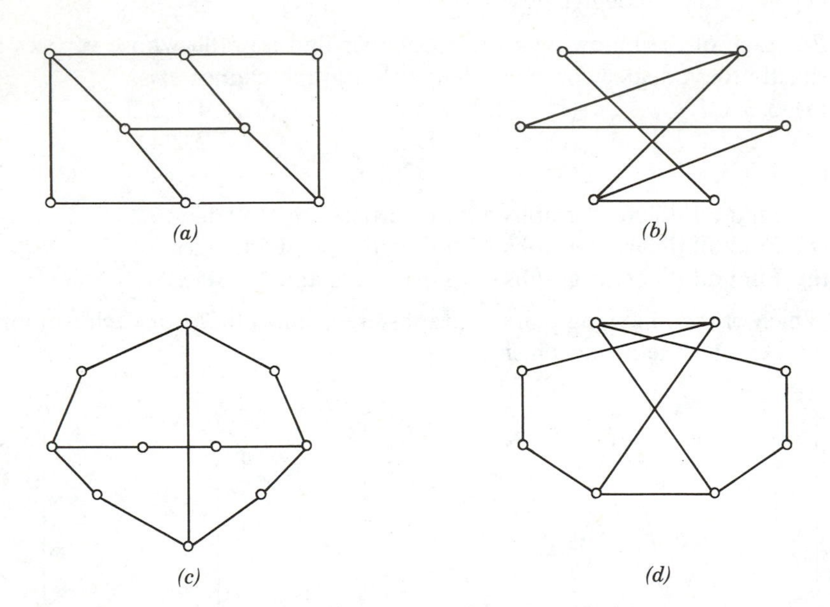

16. Draw K_6 and K_7. What is the smallest value of n such that K_n has at least 1000 edges?

17. Is there a bipartite graph with 10 vertices that is regular of degree 3? If so, find one; if not, explain why not.

18. Exhibit all the different graphs with 6 vertices and 3 edges, and 6 vertices and 12 edges.

19. Find a formula for the number of edges in $K_{p,q}$. Prove your formula in two different ways, one of which must be by induction.

20. If G is a bipartite graph on 12 vertices, what is the largest number of edges that G might have?

21. If G is a bipartite graph on V vertices, what is the largest number of edges that G might have?

22. The **generalized cubes Q_n** are graphs defined as follows: The vertices of Q_n consist of all binary sequences of length n, that is, $V(Q_n) = \{0, 1\}^n$. Two vertices in Q_n are joined by an edge precisely if the corresponding binary sequences differ in exactly one entry. Thus the vertices of Q_1 are 0 and 1 and there is an edge joining them. The graph Q_2 follows. Draw Q_3 and Q_4. How many vertices and how many edges does Q_n have?

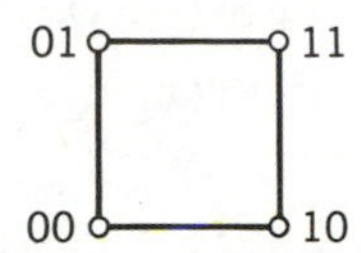

23. Show that for all n, Q_n {defined in Exercise 22} is bipartite.
24. Let $S = \{0, 1, 2, \ldots, 9\}$ and suppose that for $x, y \in S$, $x \sim y$ is true if and only if x and y are both even or both odd. Draw the corresponding graph.
25. Let $G = K_{3,3}$. Specify the corresponding relation on $V(G) = \{1, 2, 3, 4, 5, 6\}$. Do the same for $G = K_6$.
26. Are the relations in Exercise 25 reflexive or transitive?
27. If G is a graph, explain why the corresponding relation $\sim$ is necessarily a symmetric relation.

5:3 TREES AND THE LAN

In our model of cable connections for the LAN system we wanted to be able to send an electrical signal from any building to any other building. This property corresponds with the notion of connectivity. Here is an informal description. With the button and thread image from the previous section, a graph is connected if whenever you pick up a button and walk out of the room, the entire graph comes with you. If only some of the graph comes with you, the part that comes with you is called a (connected) component.

Example 3.1. In Figure 5.11, G is connected while H has three components.

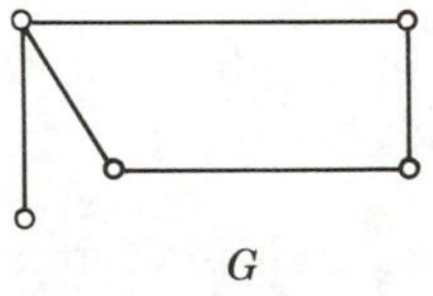

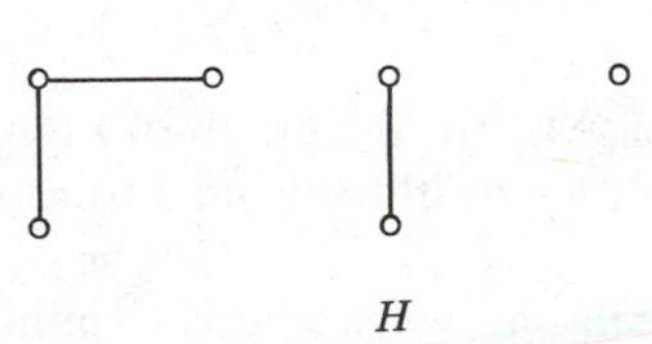

G $\qquad$ H

Figure 5.11

Even though this informal definition of connectivity captures the spirit of the concept correctly, we need a less "seamy" definition to enable us to prove theorems about connected graphs.

Definition. Given a graph G with vertices x and y, a **path** from x to y of length k is a sequence of k distinct edges, $e_1, e_2, \ldots, e_k$, such that

$$e_1 = (x, x_1),$$
$$e_2 = (x_1, x_2),$$
$$\ldots$$
$$e_j = (x_{j-1}, x_j),$$
$$\ldots$$

and finally,

$$e_k = (x_{k-1}, y).$$

Frequently, we just list the vertices that are incident with the edges of the path, like $x, x_1, x_2, \ldots, x_{j-1}, x_j, \ldots, x_{k-1}, y$. We may say that there is a path of length zero from a vertex x to itself. A path from x to itself of length k is called a ***k*-cycle**.

Example 3.2. In the graph of Figure 5.12 the vertices a, b, c, d, and e form a 5-cycle. There is a path from a to z of length four using the vertices b, c, and r as well as a path from a to z of length three using the vertices e and d. In the latter path $e_1 = (a, e)$, $e_2 = (e, d)$, and $e_3 = (d, z)$.

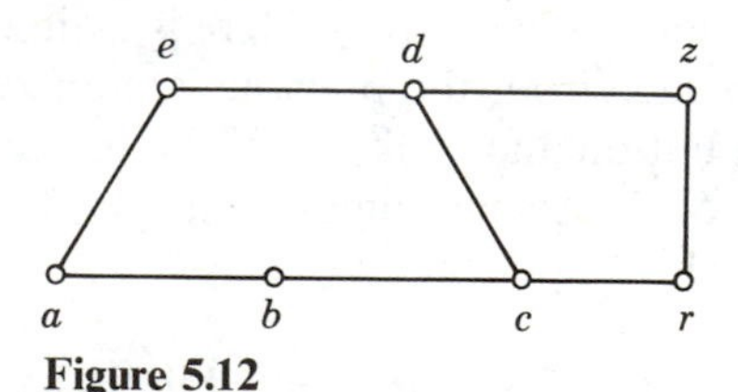

Figure 5.12

Question 3.1. In the graph of Figure 5.13, find a path of length five from a to b, a path of length three from z to r, and a 4-cycle through b.

Recall that we used this terminology when discussing grids in Chapter 3. For example, in the 6×5 grid graph we were searching for paths of length 11 from (the vertex) M to (the vertex) P.

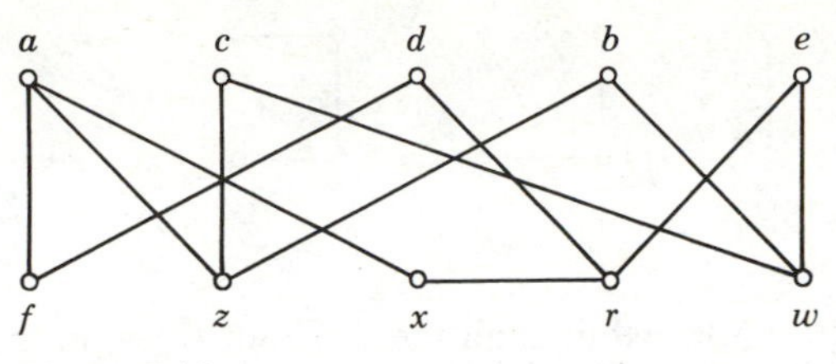

Figure 5.13

Definition. The graph that consists of a path of length k from one vertex to another vertex with no repeated vertices is called $\boldsymbol{P_k}$. The graph that is a k-cycle with no repeated vertices is called $\boldsymbol{C_k}$.

Example 3.3. The graphs P_3 and C_5 are shown in Figure 5.14.

Figure 5.14

We use the concept of path to define connectivity precisely.

Definition. A graph G is said to be **connected** if for every pair of vertices x and y, there exists a path from x to y. The **distance** between x and y, denoted $\boldsymbol{d(x, y)}$, is the smallest number of edges in a path from x to y. Given a vertex x within a graph G, the **component** of G containing x consists of the set of vertices and the set of edges in G that are a part of some path beginning at x.

We emphasize that this does not mean that in a connected graph every pair of vertices is joined by an edge.

If we construct a graph to model the LAN, the vertices will represent the buildings of the campus and the edges will represent the pairs of buildings that are directly joined by a coaxial cable. We want this graph to be connected. On the other hand, we don't want to include unnecessary connections. These properties suggest the following definitions.

Definition. A **forest** is a graph with no cycles. (Such a graph is also called **acyclic**.) A connected graph with no cycles is called a **tree**.

Trees are the most widely applicable type of graph. We shall explore some of their properties in the remainder of this section and use them to settle the LAN question.

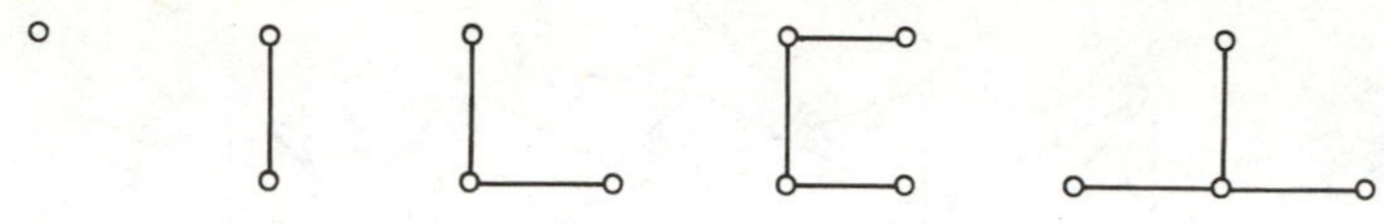

Figure 5.15

Example 3.4. Figure 5.15 exhibits all the different trees on fewer than 5 vertices. The union of these graphs forms a forest with 5 components and 14 vertices. Figure 5.16 shows a tree whose bark is worse than its bite.

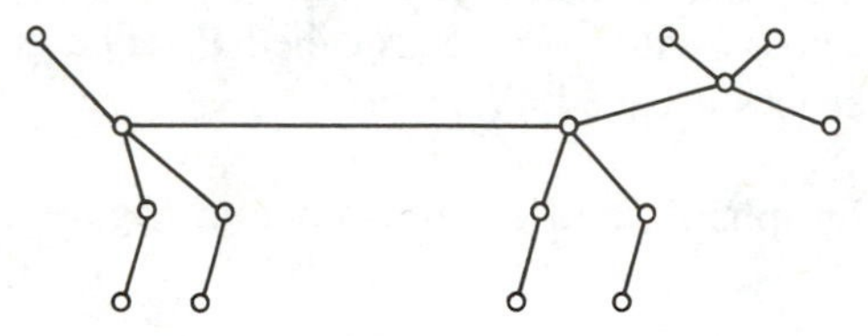

Figure 5.16

Question 3.2. Find the three different trees on five vertices.

Question 3.3. How many edges does a tree on six vertices contain? Does the answer depend upon the tree or is the answer the same for all trees on six vertices?

Theorem 3.1. If T is a tree with V vertices and E edges, then

$$E = V - 1.$$

Proof. We proceed by induction on the number of vertices. You can consult Figure 5.15 to see that the result is true for $V = 1$, 2, 3, and 4. We assume that the result is true for all trees with fewer than V vertices. Consider an arbitrary tree T with V vertices and an edge $e = (x, y)$. What happens if we remove the edge e from T? We illustrate in Figure 5.17.

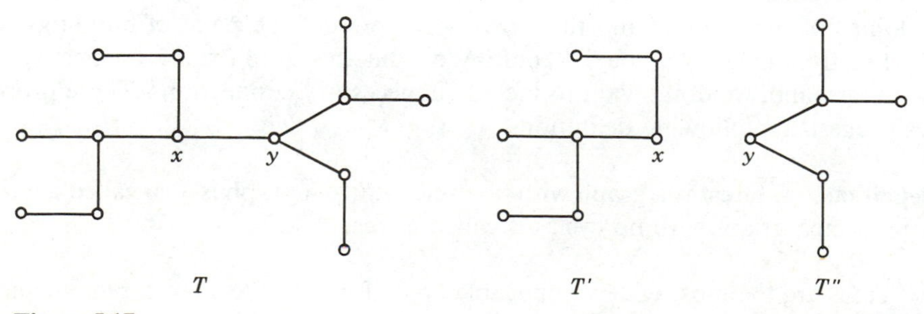

Figure 5.17

The resulting graph is a forest made up of two trees, one containing the vertex x and the other containing the vertex y. Let T' denote the tree containing x and T'' denote the tree containing y. Suppose that the number of vertices in T' is V' and the number of edges in T' is E'. Similarly, the number of vertices in T'' is V'' and the number of edges in T'' is E''. Since both T' and T'' are trees that have fewer vertices than T, we can assume by the inductive hypothesis that

$$E' = V' - 1 \qquad \text{and} \qquad E'' = V'' - 1.$$

The number of edges in T is one more than $E' + E''$ (since e is in T) while the number of vertices in T equals $V' + V''$. Thus

$$\begin{aligned} E &= E' + E'' + 1 \\ &= (V' - 1) + (V'' - 1) + 1 \\ &= V' + V'' - 1 \\ &= V - 1. \end{aligned}$$

□

Question 3.4. If you want to join every pair of buildings on a campus with 40 buildings, how many different coaxial cables do you need? What is the minimum number of cables that must be installed to sustain a LAN system connecting these 40 buildings?

Question 3.5. (a) Let C denote the number of component trees in the forest shown in Figure 5.15. Show that $E = V - C$. (b) More generally, show that if F is any forest with V vertices, E edges, and C component trees, then $E = V - C$. [*Hint:* Suppose that for $i = 1, \ldots, C$, the ith component tree contains V_i vertices and E_i edges.] (See also Exercises 24 and 25.)

That a tree on V vertices has exactly $V - 1$ edges is fundamental. Indeed this property can be exchanged with either of the two defining properties of a tree. This is shown in the next theorem and the questions that follow it.

Theorem 3.2. If G is an acyclic graph with V vertices and $V - 1$ edges, then G is a tree.

Proof. If G is acyclic, then G is by definition a forest. By the result in Question 3.5 a forest with V vertices, E edges, and C component trees necessarily has

$$E = V - C.$$

Solving for C and substituting for E yields

$$\begin{aligned} C &= V - E \\ &= V - (V - 1) = 1. \end{aligned}$$

Thus $C = 1$ and G is a tree. □

Question 3.6. Show that if G is a connected graph containing a cycle and if e is any edge of the cycle, then $G - e$ is connected. (By $G - e$ we mean the graph obtained from G by erasing e, that is, removing e from the edge set while leaving e's incident vertices in the vertex set.) (*Hint:* Pick a pair of vertices in G, say x and y, and describe a path from x to y in $G - e$.)

Question 3.7. Show that a connected graph with V vertices and $V - 1$ edges is a tree. (*Hint:* Use the result of Question 3.6.)

We now have most of the mathematical machinery necessary to construct an algorithm to pick the pairs of buildings that ought to be joined directly by coaxial cables in the LAN system. In our model of the campus buildings we wanted enough cables to be installed so that the graph that represents the cable connections is a tree. Recall the two defining properties of trees: They are connected and acyclic. We need the connectivity because that is just the property that mimics the "real world requirement" that electrical signals can be sent between any pair of buildings. If the cable graph had a cycle, then by Question 3.6, it would remain connected if some cable in the cycle were removed. Consequently, in seeking a minimum cost connected graph we are inexorably led to a tree (provided that no edge has a negative cost, i.e., no one is willing to pay us to install an extra cable connection). We need just a few more definitions to be able to finish the task.

Definitions. Suppose that $V(G)$ and $E(G)$ denote the vertex and edge sets of a graph G. If H is a graph with the properties that
(1) $V(H) \subseteq V(G)$,
(2) $E(H) \subseteq E(G)$, and
(3) every edge of $E(H)$ has both its incident vertices in $V(H)$,
then H is called a **subgraph** of G. If $V(H) = V(G)$, then H is called a **spanning subgraph** of G. If in addition H is a tree, then H is called a **spanning tree** of G.

Example 3.5. Let G be the graph shown in Figure 5.18. If $V(H) = \{1, 2, 3\}$ and $E(H) = \{(1, 2)\}$, then H is a subgraph. If $V(H) = \{1, 3, 4\}$ and $E(H) = \{(1, 2)\}$, then H is not a subgraph. If $V(H) = \{1, 2, 3, 4, 5\}$ and $E(H) = \{(1, 2), (2, 3), (3, 4), (4, 5), (1, 5)\}$, then H is a spanning subgraph. If $V(H) = \{1, 2, 3, 4, 5\}$ and $E(H) = \{(1, 2), (2, 3), (3, 4), (4, 5)\}$, then H is a spanning tree.

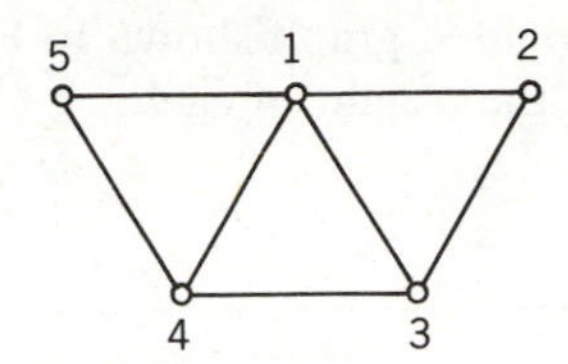

Figure 5.18

Question 3.8. Find a spanning tree of the graph in Figure 5.19.

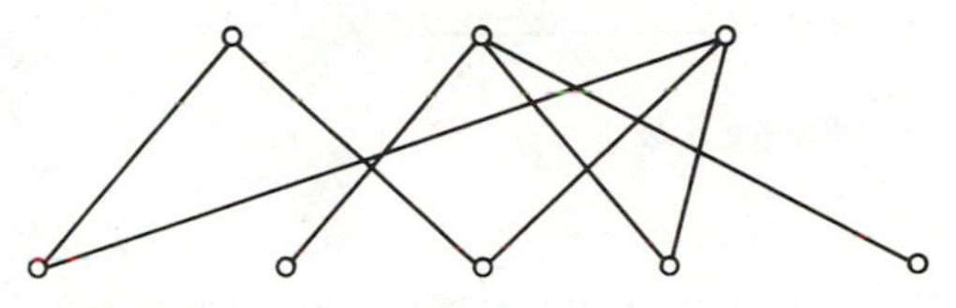

Figure 5.19

Does every graph have a spanning tree and how can we find a spanning tree in a graph with one? In Exercise 13 you are asked to prove that a graph contains a spanning tree if and only if it is connected. If a graph is disconnected, it contains a spanning tree for each connected component; such a collection of trees is called a **spanning forest**. In Exercise 19 we present an algorithm SPTREE that produces a spanning tree of a connected graph. The algorithm SPTREE actually does more than promised and produces a spanning forest of a disconnected graph.

We return to the LAN problem. We improve our graph theory model to incorporate the costs of installing the cables.

Weighty Definitions. Let R^+ denote the set of positive real numbers. A graph G together with a function $w: E(G) \to R^+$ is called a **weighted** (or an **edge-weighted**) **graph**. If e is in $E(G)$, then $w(e)$ will be the weight of e. If $F \subseteq E(G)$, then $\boldsymbol{w(F)}$, the **weight of** $\boldsymbol{F}$, is defined to be the sum of the weights of the edges in F.

Typically, the numbers assigned to the edges will represent costs, capacities, lengths, or some parameter of real-world interest.

Example 3.6. Given the weighted graph G shown in Figure 5.20, the graph H is a (weighted) spanning tree of G whose total weight is 37.

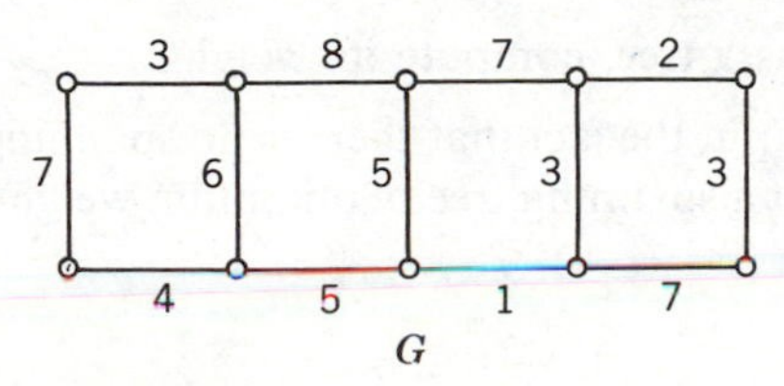

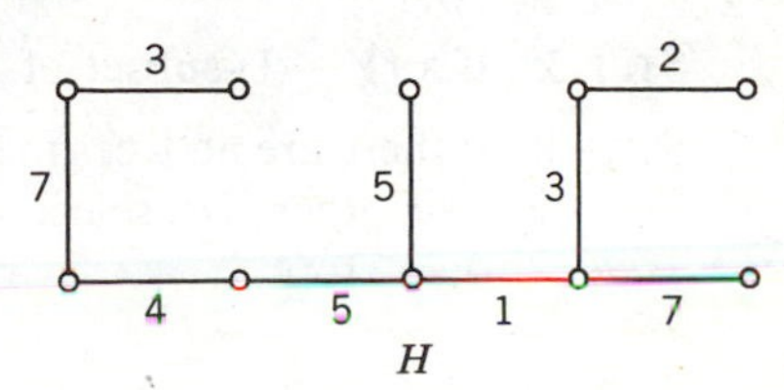

Figure 5.20

Question 3.9. Given the weighted graph shown in Figure 5.21, find all the spanning trees of this graph and the weight of each.

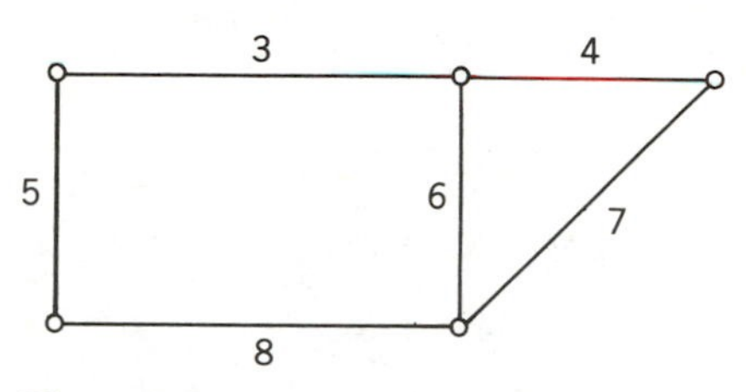

Figure 5.21

Our LAN problem can now be formalized as follows. Given the weighted graph G with vertices representing the buildings, edges representing the possible direct cable connections, and edge weights representing the installation costs of the corresponding cables: find H, a minimum weight spanning tree of the graph G. It is important to distinguish the edges of G that represent the possible direct-cable connections from those of H that represent the actual direct-cable connections. We shall examine two different algorithmic solutions to this question. First we state the problem succinctly.

Problem. Given a weighted graph G, find a spanning tree H with minimum total weight.

The idea of our initial, naive algorithm is that every subset of $V - 1$ edges of a graph that forms a connected or an acyclic subgraph gives a spanning tree by Theorem 3.2 and Question 3.6.

Algorithm BADMINTREE

STEP 1. Input the weighted graph G

STEP 2. Use algorithm JSET from Chapter 3 to find all subsets of the edges of G that contain exactly $V - 1$ edges

STEP 3. If a $(V - 1)$-subset of $E(G)$ forms a tree, compute its weight

STEP 4. If there are no trees in step 3, output the fact that there is no spanning tree; otherwise, select and output a spanning tree of minimum weight; then stop.

Question 3.10. Run the algorithm BADMINTREE on the weighted graph in Figure 5.22.

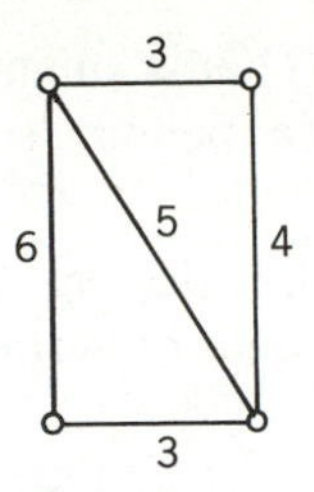

Figure 5.22

The name of this algorithm gives away the quality of its performance. We know that JSET performs at least $\binom{E}{j}$ steps to find all j-subsets of a set with E elements; here it's the set of E edges whose $(V-1)$-subsets we list. Each edge represents the potential installation of a cable. Since, in general, we cannot eliminate any of these possibilities, by Theorem 2.3, $E = V(V-1)/2$. Thus the number of steps is at least

$$\binom{V(V-1)/2}{(V-1)}. \qquad (*)$$

Question 3.11. Evaluate (*) for $V = 3, 4, 5, 6, 7$.

This binomial coefficient evidently grows rapidly. Exercise 23 asks you to find lower bounds. In the next section we shall find a much more efficient way to solve the LAN Problem.

EXERCISES FOR SECTION 3

1. Find all trees with six vertices. Find a graph with six vertices and five edges that is not a tree. Can you find such a graph that is also acyclic?
2. For what values of n is C_n bipartite?
3. Show that trees are bipartite.
4. How many different 5-cycles are there in the following graph? Before answering, specify what it means for two 5-cycles to be different.

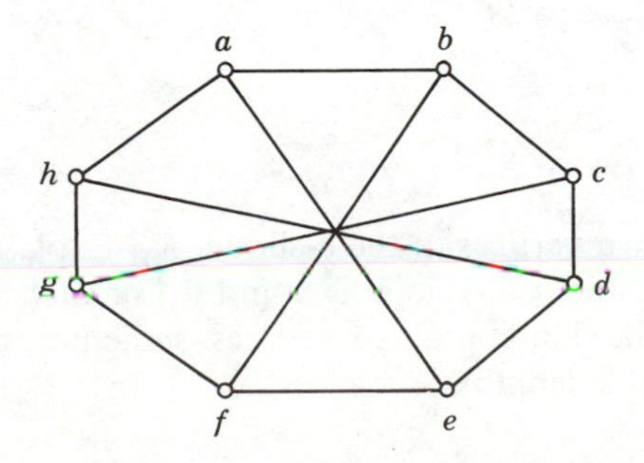

5. Give an example of a graph G and a subgraph H of G that is not a spanning subgraph. Give an example of a spanning subgraph of G that is not a spanning tree.

6. The **average degree** of a graph is the sum of the degrees of all the vertices of the graph divided by the number of vertices. Show that the average degree of a forest is less than 2.

7. Which of the following pairs of trees are isomorphic?

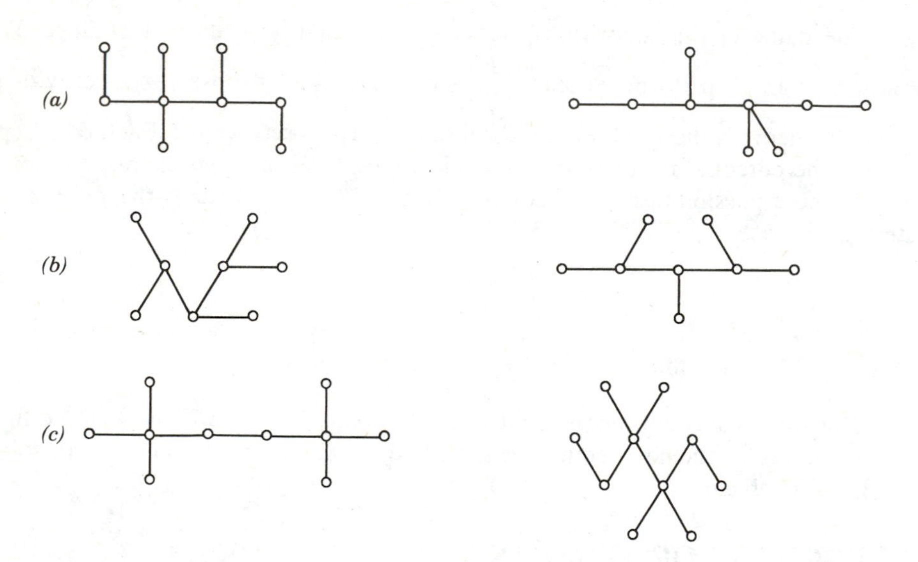

8. Find the longest path and the longest cycle that is a subgraph of the following graphs and give their lengths. Then find the largest value of j and k such that P_j and C_k are subgraphs of the given graph.

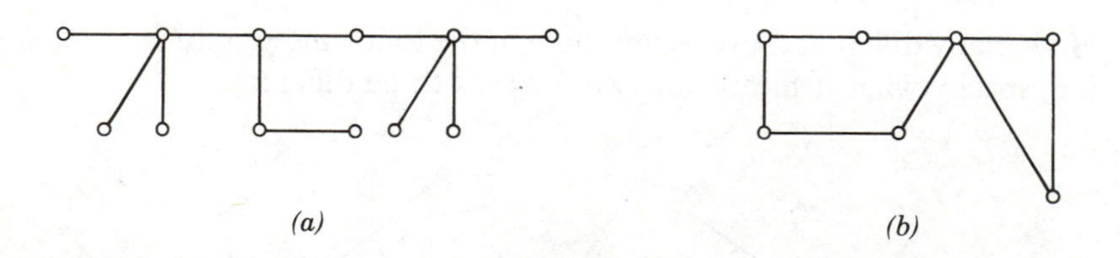

9. Let x and y be two vertices in the r-clique, K_r. Explain why there are paths of length $1, 2, \ldots,$ and $(r - 1)$ joining x and y. For each $r > 3$, describe a graph other than a clique that contains r vertices, some pair of which are joined by paths of all possible lengths.

10. Identify the components in the following graphs.

(*a*)

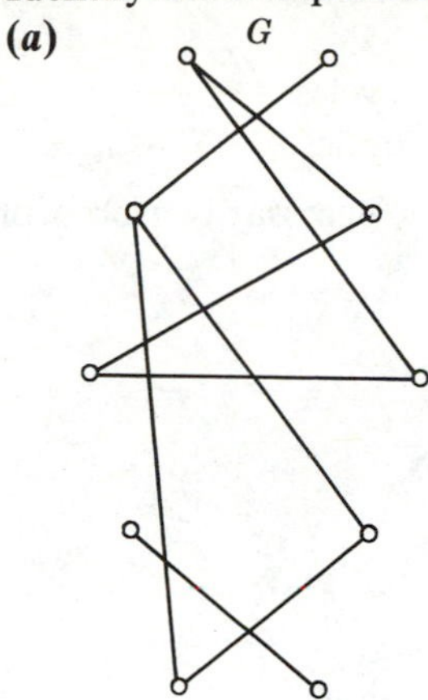

(*b*) H where $V(H) = \{1, \ldots, 10\}$ and $E(H) = \{(1,2), (3,4), (5,6), (7,8), (9,10), (3,5), (5,8), (1,9), (4,10), (6,9)\}$

11. If G is a connected graph and x, y, and z are vertices, is it always true that $d(x, y) + d(y, z) \geq d(x, z)$? Give a proof or counterexample.

12. Given the weighted graph shown here, find all of the spanning trees of this graph and the weight of each.

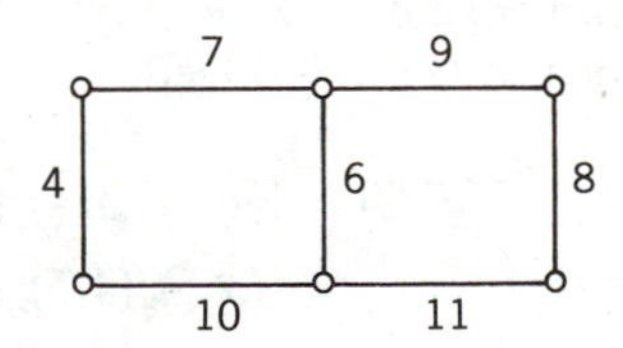

13. (*a*) Prove that a graph is connected if and only if it has a spanning tree.

(*b*) Prove that every graph contains a spanning forest that consists of a spanning tree of each component.

14. Determine whether the following are true or false. Prove each true statement and give a counterexample for each false statement.

(*a*) A graph with $E \geq V$ is connected.

(*b*) A graph with $E \geq V$ contains a cycle.

(*c*) A graph with $E \leq V - 2$ is not connected.

(*d*) A graph with $E \leq V - 2$ is acyclic.

(*e*) A graph with two components has at most $V - 2$ edges.

(*f*) A graph with $E = V - 2$ has at least two components.

(*g*) A graph with $E = V - 2$ has exactly two components.

(*h*) If G is a connected graph containing a cycle, then the removal of any edge of the graph leaves the graph connected.

(*i*) Every spanning subgraph H of a connected graph G has $|E(H)| = |V(H)| - 1$.
(*j*) A graph with $E = V + 1$ contains at least two cycles.
(*k*) A graph with $E = V + 1$ contains exactly two cycles.

15. Are two trees with the same sequence of degrees necessarily isomorphic? Give a proof or a counterexample.

16. Find a spanning forest of the following graph.

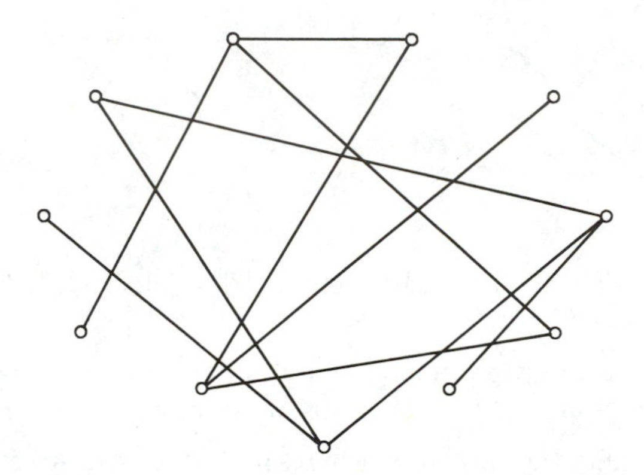

17. A subgraph H of a graph G is called an **induced subgraph** if whenever x and y in $V(H)$ are joined by an edge e in G, then e is also in H. Determine which of the following subgraphs of G are induced:

$$V(G) = \{1,2,3,4,5,6\} \qquad E(G) = \{(1,2), (1,4), (1,5), (2,3), (2,5), (2,6), (3,6), (4,5), (5,6)\}$$

(*a*) $V(H) = \{1,2,3\}$ $\quad E(H) = \{(1,2), (2,3)\}$
(*b*) $V(H) = \{1,2,4,5\}$ $\quad E(H) = \{(1,2), (1,4), (2,5), (4,5)\}$
(*c*) $V(H) = \{1,3,4,6\}$ $\quad E(H) = \{(1,4), (3,6)\}$

18. Suppose that G is a connected graph and T is a spanning tree of G. When is it the case that T is an induced subgraph of G?

19. Here is an algorithm that finds (if possible) a spanning tree of the input graph.

Algorithm SPTREE

STEP 1. Input the graph G with V vertices and edge list $e_1, e_2, \ldots, e_E$
STEP 2. For $j = 1$ to E do
    STEP 3. Add edge e_j to the spanning tree T if it creates no cycle with the edges already in T

STEP 4. If T contains $V - 1$ edges, then output T as the desired spanning tree and stop; otherwise, declare that G contains no spanning tree and stop.

Run the algorithm SPTREE on the following graph.

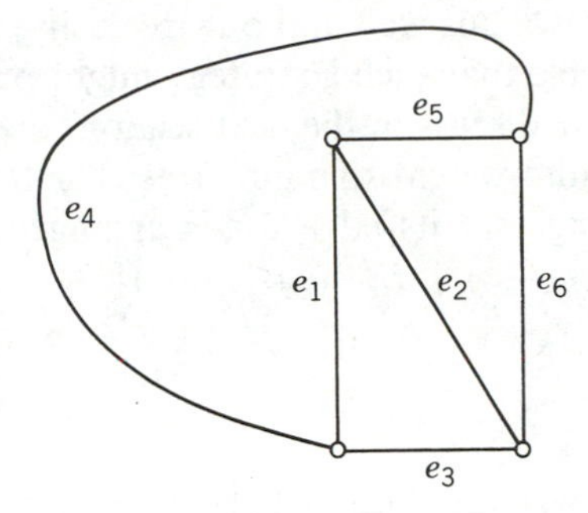

20. Prove that SPTREE stores a spanning tree in T if and only if G is connected.

21. Prove that SPTREE stores a spanning forest in T in all cases. Modify SPTREE so that it always outputs T and determines the number of connected components of G. (*Hint:* See Question 3.5.)

22. Show that in a tree, every pair of vertices is joined by a unique path. Is the converse true, that is, if G is a graph in which every pair of vertices is joined by a unique path, then is G a tree? Give a proof or a counterexample.

23. Find N such that if $V > N$, then the binomial coefficient given in (*) is greater than 2^V. Show that, in any case, the binomial coefficient is greater than $\left(\frac{V-1}{2}\right)^{V-1}$.

24. (An alternate solution to Question 3.5). Using induction on the number of components, show that if F is a forest with V vertices, E edges, and C component trees, then $E = V - C$.

25. (Another alternate solution to Question 3.5). Given a forest F with V vertices, E edges, and C component trees labeled $T_1, \ldots, T_C$, for $i = 1, \ldots, C - 1$, add an edge from some vertex of T_i to some vertex of T_{i+1}. Use the resulting graph to prove that $E = V - C$.

5:4 A GOOD MINIMUM-WEIGHT SPANNING TREE ALGORITHM

We now present an algorithm that is dramatically better than BADMINTREE. It is universally known as Kruskal's algorithm. Kruskal did write the first paper developing this particular algorithm in 1956. However, there are earlier algorithms that correctly and efficiently find minimum-weight spanning trees. The earliest known such algorithm is due to Otakar Borůvka, who as an electrical engineer, was working on the problem of the electrification of Southern Moravia about 60 years ago.

We are seeking a spanning tree of small weight. A lightweight spanning tree contains lightweight edges. Thus we build our tree using the lightest possible edge at each stage. It is plausible that such a strategy might produce a reasonably light tree. It is surprising (as we discuss in the next section) that this strategy is guaranteed to produce a minimum-weight spanning tree. The following algorithm works on the graph G assuming that its edge list is arranged so that the weights are **increasing** [i.e., for all $1 \leq i < j \leq E$, $w(e_i) \leq w(e_j)$].

Algorithm KRUSKAL

STEP 1. Input the weighted graph G {Assume that G has V vertices and E edges and that the edge list of G is in increasing order, by weight.}
STEP 2. Set $j := 1$ {j will index the edges of G.}
STEP 3. Set T to be empty {T will contain the edges of the minimum-weight spanning tree.}
STEP 4. Set $k := 0$ {k records $|E(T)|$.}
STEP 5. Repeat
    Begin
    STEP 6. If $T + e_j$ is acyclic, then do {add e_j to tree}
        Begin
        STEP 7. $T := T + e_j$
        STEP 8. $k := k + 1$
        End {step 6}
    STEP 9. $j := j + 1$
    End {step 5}
    Until either $k = V - 1$ or $j > E$
STEP 10. If $k = V - 1$ report success, output T, and stop. Otherwise, report failure and stop.

COMMENT. We use "+," as in $T + e_j$, to denote set theoretic union.

Theorem 4.1. If G is weighted graph, then KRUSKAL outputs a minimum-weight spanning tree if and only if G is connected.

Example 4.1. We run KRUSKAL on G shown in Figure 5.23 to obtain the minimum-weight spanning tree T.

Question 4.1. Run KRUSKAL on the weighted graphs shown in Figure 5.24.

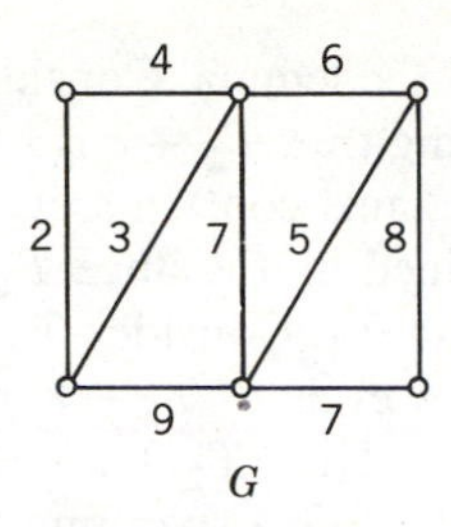

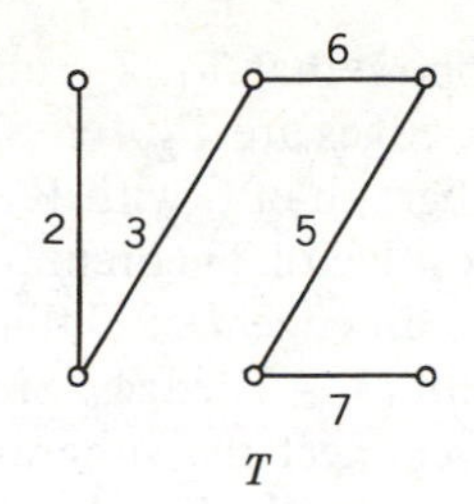

Figure 5.23

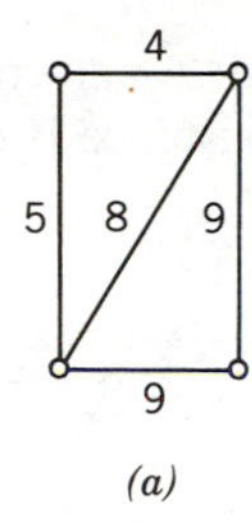

(a)

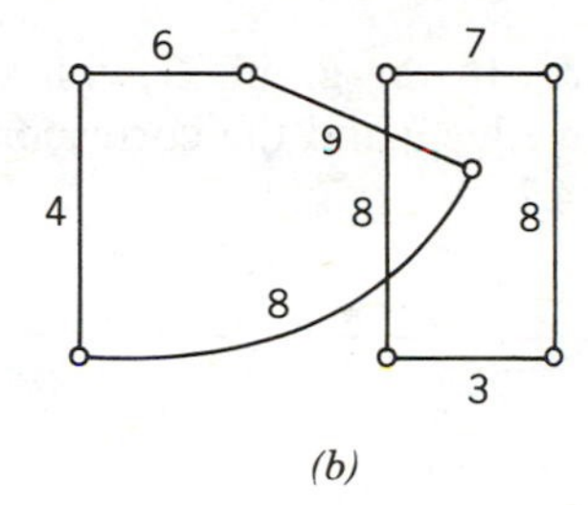

(b)

Figure 5.24

Proof of Theorem 4.1. First let's see that if KRUSKAL reports success, then the graph in T is a tree. In step 7 the edge e_j is added to the set T precisely when $T + e_j$ is acyclic. Thus the subgraph formed by T must always be acyclic. If KRUSKAL returns success, then T has $V - 1$ edges. By Theorem 3.2 we know that an acyclic graph with $V - 1$ edges is necessarily a tree. Consequently, if KRUSKAL returns success, then the edges stored in T are the edges of a tree.

Next let's see that if KRUSKAL returns failure, then G does not contain a spanning tree. For KRUSKAL to return failure, after examining all the edges of the graph, the set T does not contain $V - 1$ edges but still forms an acyclic subgraph. Thus T is a forest. By Question 3.5 if the number of edges in T is $V - C$, then C equals the number of components of T. Since T contains fewer than $V - 1$ edges, T contains more than one component. Suppose that Q is a component of T. If the whole graph G were connected, then there would be an edge in G, say e, that joins a vertex of Q with a vertex of another component of T. When KRUSKAL examined the edge e, it would have found that $T + e$ was acyclic and thus T would have contained e. Since e did not make it into T, G must not have been connected. Thus KRUSKAL reports a failure when G is disconnected.

A more difficult thing to verify is that the tree returned by KRUSKAL is a minimum-weight spanning tree. We begin with a lemma.

Lemma 4.2. Suppose that T_1 and T_2 are two different spanning trees of a connected graph G and that c is an edge of T_1 but not of T_2. Then there is an edge d of T_2 but not of T_1 such that $T_2 + c - d$ is a spanning tree of G.

Proof. Suppose that T_1, T_2, and c are as given in the lemma. Consider the subgraph whose edges are $T_2 + c$. Since T_2 is a spanning tree, $T_2 + c$ is a connected, spanning subgraph of G with V edges. Thus $T_2 + c$ must contain a cycle C (using the contrapositive of Theorem 3.1). C is not contained in T_1, since T_1 is acyclic. Thus there is an edge d of C that is in T_2 but not in T_1. Consider $T_2 + c - d$, a subgraph with $(V - 1)$ edges. Since d is an edge in a cycle of $T_2 + c$, its removal does not disconnect the subgraph by Question 3.6. Thus $T_2 + c - d$ is a connected subgraph with $(V - 1)$ edges. By Question 3.7 it is a spanning tree of G. □

Question 4.2. In the graph G with spanning trees T_1 and T_2 given in Figure 5.25, find an edge c and the corresponding edge d whose existence is guaranteed by Lemma 4.2.

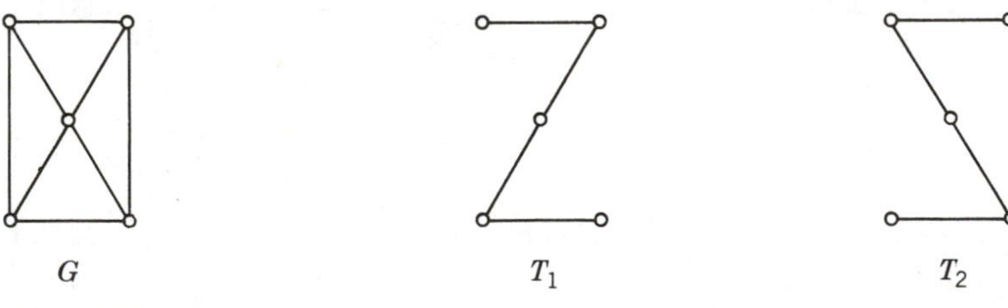

Figure 5.25

To finish the proof that if KRUSKAL reports success, then the spanning tree in T is of minimum weight, we assume that the edge weights of G are all distinct. (You can complete the general proof in Supplementary Exercise 31.) The proof will be by contradiction. We negate the conclusion we are seeking to prove and assume that T is not a minimum-weight spanning tree. Thus there exists a minimum-weight spanning tree, say F, with $w(F) < w(T)$. Since $T \neq F$, there exists an edge in T that is not in F. Let e denote the lightest-weight such edge. We apply Lemma 4.2 with $T_1 = T$, $T_2 = F$, and $c = e$. Thus there exists an edge, say f, such that f is in F but not in T. By the lemma $F' = F + e - f$ is a spanning tree. If $w(e) < w(f)$, then $w(F') < w(F)$, contradicting the assumption that F is a minimum-weight spanning tree. On the other hand, suppose that $w(f) < w(e)$. Since e was the lightest-weight edge in T, but not in F, all the edges selected by KRUSKAL before e are also in F. Thus f would have been added to T instead of e, another contradiction. We conclude that T is a minimum-weight spanning tree. □

There are two steps within KRUSKAL that should provoke comment. The first is the requirement that the edges of G be input in increasing order. There is a straightforward way to do this. Specifically, we could examine the edge list of

G to find an edge with the smallest weight and list it first. We could then find a second smallest edge and list it second, and so on. We shall see in the next chapter that this procedure would perform $O(E^2)$ comparisons to list the E edges of G in increasing order. This method is analogous to the algorithm MAX, presented in Exercise 2.4.12. We shall also see that there are more efficient ways to do this sorting.

The second difficulty with KRUSKAL as presented above occurs in step 6. Specifically, how should we decide, given an acyclic set of edges T and an edge e_j, whether $T + e_j$ is acyclic? In small examples we can obviously "eyeball" the set of edges for cycles. One way to check for cycles in larger and more general examples is to keep track of the connected components of T at each stage of its creation. If e_j joins the two vertices x and y, then the addition of e_j creates a cycle if and only if x and y are in the same component of T.

Example 4.1 (continued). Initially, T is empty, and we consider every vertex to be a separate component of T. First we added the edges of weights 2 and 3 because not only are they the lightest-weight edges, but also they join vertices in different components of T. Now T has a component consisting of these two edges and three vertices as well as three additional components, each consisting of an isolated vertex. The next edge of weight 4 is rejected because it joins two vertices in a component, whereas the edge of weight 5 joins two vertices in different components and is accepted in T.

It is not hard to estimate the complexity of KRUSKAL. The principal operations in this and most graph theory algorithms are comparisons. In KRUSKAL we first compare edge weights so that the edges are rearranged in increasing order. As mentioned previously, ordering the edges might take $O(E^2)$ comparisons. The loop at step 5 is repeated no more than E times. Testing $T + e_j$ for cycles requires that we keep track of the components of T at each stage, so with one comparison we can tell whether e_j joins two vertices in the same component. However when e_j does not form a cycle and is added to T, we need to update the components of T because the addition of e_j causes two components to be joined into one. This updating can be done with at most V comparisons. (For more details, see Exercises 11 to 13.) Thus there are $O(V)$ comparisons done within the loop beginning at step 5 and no more than $O(E\,V)$ comparisons after the ordering of the edges. Thus the total number of comparisons and assignments is $O(E^2) + O(E\,V)$.

A careful analysis of a more efficient implementation could achieve the result that the algorithm including an efficient sorting routine in step 1 is $O(E \log(E))$. Notice that this bound and the previous one can also be expressed in terms of (only) V, since by Corollary 2.3, $E \leq V(V-1)/2 = O(V^2)$.

Question 4.3. Express $O(E^2) + O(E\,V)$ and $O(E \log(E))$ as $O(f(V))$, where f is a function of V but not of E.

A formal analysis of any graph algorithm must consider how to input the graph G as a string of zeros and ones. One convenient method uses what is called the adjacency matrix of a graph. A **matrix** is a rectangular array of entries, usually numbers; an $r \times s$ matrix consists of $r \cdot s$ entries arranged in r rows and s columns. There are exactly s entries in every row and exactly r entries in every column.

Example 4.2. Here is a 2×3 matrix:

$$\begin{bmatrix} 1 & 2 & 3 \\ 4 & 5 & 6 \end{bmatrix}.$$

Suppose that G is a graph with V vertices that are labeled $1, \ldots, V$. We define $A(G)$, the **adjacency matrix of G**, to be the $V \times V$ matrix that has a one in the ith row and jth column if the vertex labeled i is adjacent to the vertex labeled j. All other entries of $A(G)$ equal zero.

Example 4.3. Here is a graph given first by its edge list and then by its adjacency matrix:

Edge List	*Adjacency Matrix*
$\{1,2\}$	0 1 1 1 1 1
$\{1,3\}$	1 0 1 0 0 0
$\{1,4\}$	1 1 0 0 0 0
$\{1,5\}$	1 0 0 0 0 0
$\{1,6\}$	1 0 0 0 0 0
$\{2,3\}$	1 0 0 0 0 0

Question 4.4. Find the adjacency matrix of each graph in Figure 5.26.

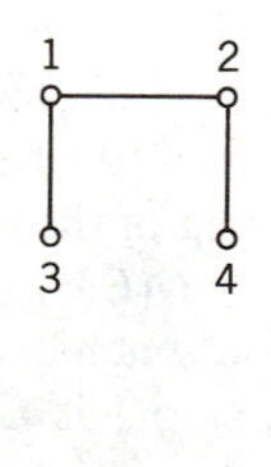

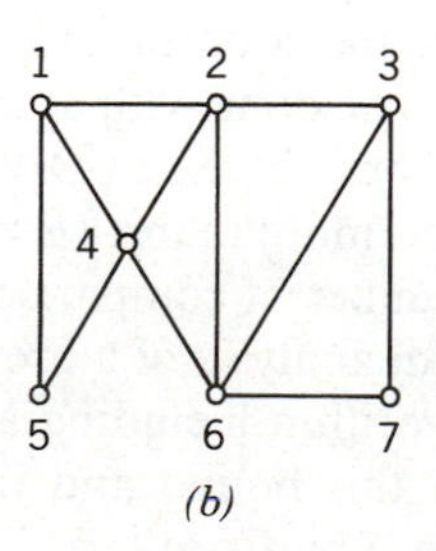

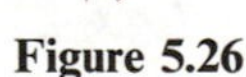
Figure 5.26

Question 4.5. Draw the graphs whose adjacency matrices are as follows.

(a) $\begin{bmatrix} 0 & 1 & 0 & 1 \\ 1 & 0 & 1 & 0 \\ 0 & 1 & 0 & 1 \\ 1 & 0 & 1 & 0 \end{bmatrix}$ (b) $\begin{bmatrix} 0 & 1 & 0 & 0 & 0 \\ 1 & 0 & 1 & 0 & 0 \\ 0 & 1 & 0 & 1 & 0 \\ 0 & 0 & 1 & 0 & 1 \\ 0 & 0 & 0 & 1 & 0 \end{bmatrix}$

This form of representing a graph has advantages and disadvantages. Notice that it gives us a way to input a graph into an algorithm as a string of V^2 zeros and ones obtained by laying out the matrix, row by row, as one long string.

Example 4.4. Here is the string of $V^2 = 36$ zeros and ones that represents the graph of Example 4.3.

011111010001100001000001000001000000

Thus the number of bits needed to input the adjacency matrix of a graph with V vertices is given by $B = V^2$.

Question 4.6. Suppose that $f(n)$ counts the number of comparisons made in the worst case of a graph algorithm and suppose that $f(n) = 0(n^k)$ for some positive integer k. If $B = n^2$, find a big oh bound on the number of comparisons made in terms of B.

The result of Question 4.6 indicates that if we determine the complexity of a graph theory algorithm to be bounded by a polynomial in V, then it is also bounded by a polynomial in B and hence is a good algorithm. The converse is also true, that if an algorithm requires an exponential number of steps in terms of V, then it also requires a nonpolynomial number in terms of B and is a bad algorithm. (See Supplementary Exercise 30.) In particular, KRUSKAL is a good algorithm and BADMINTREE is not.

EXERCISES FOR SECTION 4

1. Run KRUSKAL on the following weighted graphs.

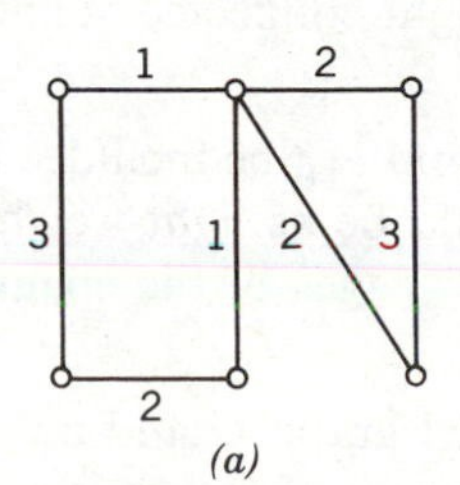

(a)

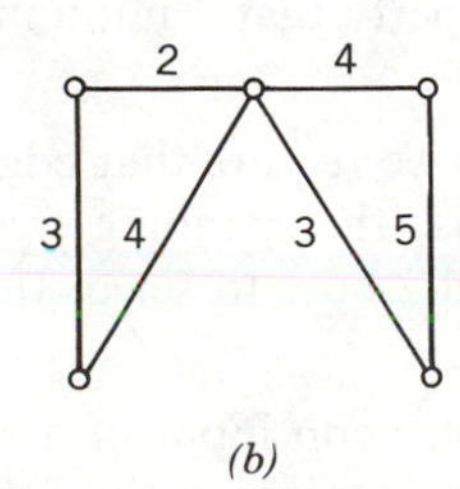

(b)

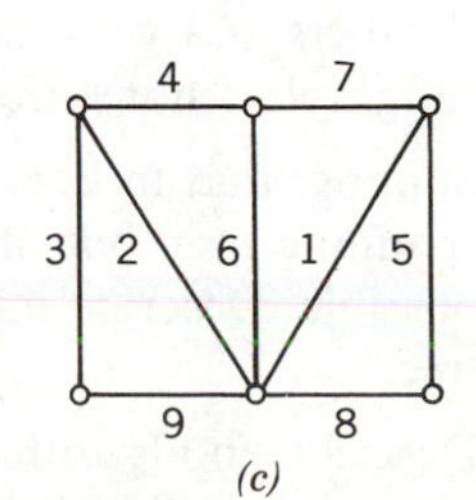

(c)

2. Suppose that G is the weighted graph with $V = 7$ and $E = 10$ whose edges are (in lexicographic order) $(1, 2)$, $(1, 5)$, $(1, 6)$, $(2, 3)$, $(2, 6)$, $(2, 7)$, $(3, 4)$, $(4, 5)$, $(5, 6)$ and $(6, 7)$. The weights are given by 1, 1, 2, 2, 1, 2, 3, 2, 1, and 3, respectively. Run KRUSKAL on this graph. Find all minimum-weight spanning trees of this graph.

3. Is the following variation on Lemma 4.2 true or false? Suppose that H_1 and H_2 are two different spanning subgraphs of G that are themselves connected graphs. Suppose that e is an edge of H_1 but not of H_2. Then there exists an edge f of H_2 but not H_1 such that $H_2 + e - f$ is a connected spanning subgraph of G. Explain.

4. Find adjacency matrices for the graphs in Exercises 1 and 2. You will have to label the vertices of the graphs from 1 to V.

5. Suppose that a graph G has adjacency matrix

$$\begin{bmatrix} 0 & 1 & 0 & 1 & 1 \\ 1 & 0 & 1 & 0 & 1 \\ 0 & 1 & 0 & 1 & 0 \\ 1 & 0 & 1 & 0 & 1 \\ 1 & 1 & 0 & 1 & 0 \end{bmatrix}.$$

Without drawing G determine the number of edges of G and the degree of each vertex. Describe, in general, how to obtain the degrees of the vertices from the adjacency matrix.

6. Suppose that G is a weighted graph with V vertices. Find a way to describe G including the weights as a $V \times V$ matrix.

7. Which of the following is the adjacency matrix of a graph? Explain.

(*a*) $\begin{bmatrix} 0 & 1 & 0 & 0 & 1 \\ 1 & 0 & 1 & 0 & 0 \\ 0 & 0 & 0 & 1 & 0 \\ 0 & 0 & 1 & 0 & 1 \\ 1 & 0 & 0 & 1 & 0 \end{bmatrix}$ (*b*) $\begin{bmatrix} 0 & 1 & 1 & 1 & 1 \\ 1 & 0 & 1 & 1 & 1 \\ 1 & 1 & 0 & 1 & 1 \\ 1 & 1 & 1 & 0 & 1 \\ 1 & 1 & 1 & 1 & 0 \end{bmatrix}$ (*c*) $\begin{bmatrix} 0 & 1 & 0 & 1 & 0 \\ 1 & 0 & 1 & 0 & 1 \\ 0 & 1 & 0 & 1 & 0 \\ 1 & 0 & 1 & 1 & 0 \\ 0 & 1 & 0 & 0 & 0 \end{bmatrix}$

8. Find an example of a weighted graph G whose edge weights include negative numbers and with the property that a minimum-weight connected spanning subgraph is not a tree.

9. Suppose that in Exercise 2 we require that edges e_7 and e_{10} be included in a spanning tree, but otherwise the spanning tree should be as light-weight as possible. Describe informally how to select the other edges of the spanning tree.

10. Describe an algorithm that, upon input of a weighted graph G and a designated subset S of $E(G)$, finds a minimum-weight subgraph of G that is a connected and spanning subgraph that contains all the edges of S.

11. Here are more details on how to test algorithmically for cycles in KRUSKAL. Initially, in step 4.5, we define the component number of vertex i, denoted $cn(i)$, to be equal to itself, i.

STEP 4.5. For $i = 1$ to V do
$cn(i) := i$

Then in the new step 6, to test an edge $e_j = (x_j, y_j)$ we compare $cn(x_j)$ and $cn(y_j)$. The edge e_j forms a cycle if and only if these component numbers are equal. If they are not equal, we add e_j to T and reset the component numbers of the new component. This is accomplished by the Procedure Renumber.

STEP 6. {Suppose $e_j = (x_j, y_j)$.}
If $cn(x_j) \neq cn(y_j)$, then do
Begin
STEP 7. $T := T + e_j$
STEP 8. $k := k + 1$
STEP 8.5. Call Procedure Renumber (x_j, y_j)
End {step 6}

Here is the procedure:
Procedure Renumber (a, b)

STEP 1. Set $bigcn := \max(cn(a), cn(b))$,
set $smallcn := \min(cn(a), cn(b))$
STEP 2. For $i = 1$ to V do
STEP 3. If $cn(i) = bigcn$, then
$cn(i) := smallcn$
STEP 4. Return

Run KRUSKAL with these additional steps and this procedure on the example in Exercise 2. Keep track of the component numbers at each vertex. Do you get the same spanning tree?

12. Run the extended version of KRUSKAL as given in Exercise 11 on the following graph. Keep track of the component numbers at each vertex.

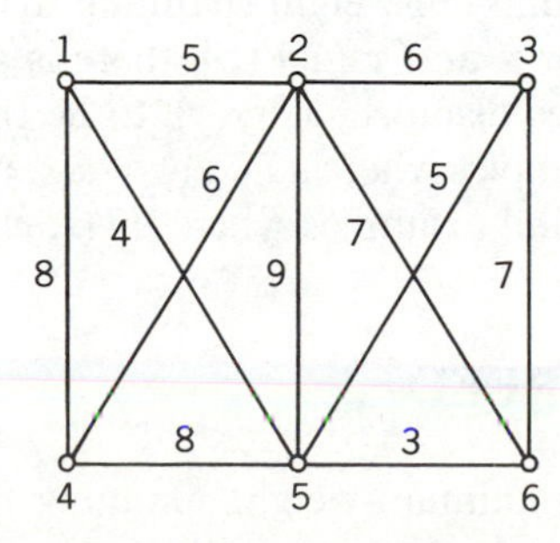

13. Explain why step 6 of the extended KRUSKAL now performs $V + 3 = O(V)$ comparisons and why step 5 performs $O(E\,V)$ comparisons.

14. Within a tree, the **eccentricity** of a vertex x is defined to be the number of edges in a longest path that begins at x. (Or equivalently, it is the maximum value of $d(x, v)$ taken over all vertices v.) For each of the following trees find the eccentricity of each vertex.

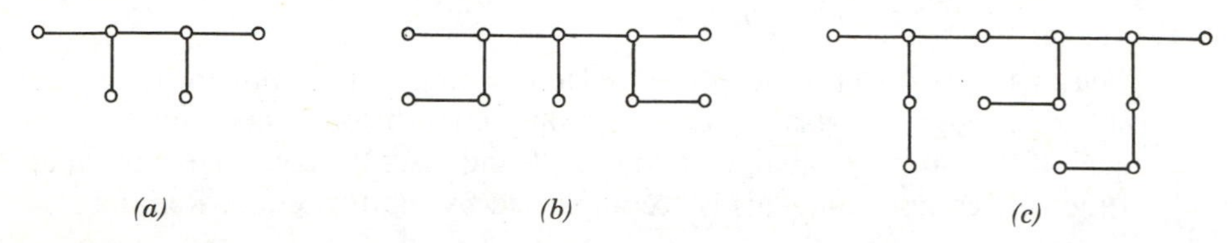

(a) (b) (c)

15. The **center** of a tree is the set of vertices whose eccentricities are as small as possible. Find the center of each of the trees in Exercise 14.

16. Given a tree T with more than one edge, let $p(T)$ denote the tree obtained from T by erasing all of the leaves of T and their incident edges. For each tree from Exercise 14 find $p(T)$. {Within a tree a vertex of degree one is called a **leaf**. We could say that $p(T)$ is obtained from T by **pruning** all of T's leaves.}

17. If you know the eccentricities of every vertex in a tree T, what can you say (with proof) about the eccentricities of the vertices in $p(T)$ [where $p(T)$ is as defined in Exercise 16]?

18. If T is a tree with more than one edge, show that the center of T equals the center of $p(T)$.

19. Prove that every tree has a center that consists of either one or two vertices.

20. Construct an algorithm that will, upon input of a tree T, find the center of T.

21. Let G be a weighted and connected graph. For x and y in $V(G)$ we define the distance from x to y, $d(x, y)$, to be the length of the shortest path from x to y, where by shortest path we mean that the sum of the edge weights along that path is a minimum among all paths from x to y. Find $d(x, y)$ for each pair of distinct vertices in the graph in Exercise 2.

22. Suppose that T is a minimum-weight spanning tree in G, a weighted and connected graph. Then for x and y in $V(G)$, there is a unique path in T from x to y. We define the tree distance, $dT(x, y)$, to be the sum of the edge weights on that path. Find examples where $dT(x, y) = d(x, y)$, where $d(x, y)$ is as defined in Exercise 21. Then find examples where $dT(x, y) \neq d(x, y)$.

5:5 AN ODE TO GREED

The problem of finding a minimum-weight spanning tree of a graph is typical of a large number of problems in discrete mathematics. In a more general context there is a set of objects with positive numbers assigned to them. The subsets of these objects are partitioned into **desirable subsets** and **undesirable subsets**. We assume that if S is a desirable subset and T is a subset of S, then T is desirable. This property is known as the **hereditary** property. In the particular tree problem of this chapter the objects are the weighted edges. The desirable subsets are those that when considered as subgraphs are acyclic. The undesirable subsets are those that contain cycles. The property of being acyclic is hereditary.

The problem is to find a maximal desirable subset of the objects whose total value is a minimum (or in some cases a maximum). The word **maximal** means that the subset cannot be extended to a larger desirable subset. So a maximal desirable subset of a given set S is first of all a subset of S, second of all it is desirable, and finally it is not properly contained in any desirable subset of S. In the context of the tree problem, a maximal desirable subset of the edges of a graph G is a subset of the edges that is acyclic and not contained in any larger acyclic subgraph. If G is connected, a maximal desirable subset is just a spanning tree. If G is not connected, a maximal desirable subset is a spanning forest, composed of spanning trees of each connected component.

Problem. Given a set of weighted objects E and a partition of the subsets of E into desirable subsets and undesirable subsets such that the property of being desirable is hereditary, find a minimum-weight maximal desirable subset of E.

Algorithm GREEDYMIN

STEP 1. Order the objects of E in order of increasing weight; assume E contains m objects $e_1, \ldots, e_m$
STEP 2. Set $j := 1$ {j will index the objects.}
STEP 3. Set T to be empty {T will contain the desirable subset being created.}
STEP 4. Repeat
Begin
STEP 5. If $T + e_j$ is desirable, set $T := T + e_j$
STEP 6. $j := j + 1$
End
Until $j > m$
STEP 7. Output T and stop.

This algorithm is called **greedy** because at each stage it tries to do as well as it can without regard to what will happen at future steps. Notice that if E is the set of weighted edges in a graph and if desirability is defined as being acyclic, then GREEDYMIN is identical with KRUSKAL. {Actually, KRUSKAL contains an additional stopping criterion that was possible because we knew exactly how many edges a tree has.}

Question 5.1. Construct GREEDYMAX, a greedy algorithm to find a maximum-weight, maximal desirable subset of E. If E is the set of weighted edges in a graph and if desirability is defined as being acyclic, does GREEDYMAX produce a maximum-weight spanning tree?

Example 5.1. Suppose that we greedily attempt to find not a minimum-weight spanning tree but a minimum-weight spanning path. Specifically, we implement GREEDYMIN with desirability defined as follows. S is said to be desirable if S is contained in some path P_{V-1} within the graph G on V vertices. We show in Figure 5.27 a weighted graph G whose greedily chosen path is heavier than the minimum-weight spanning path.

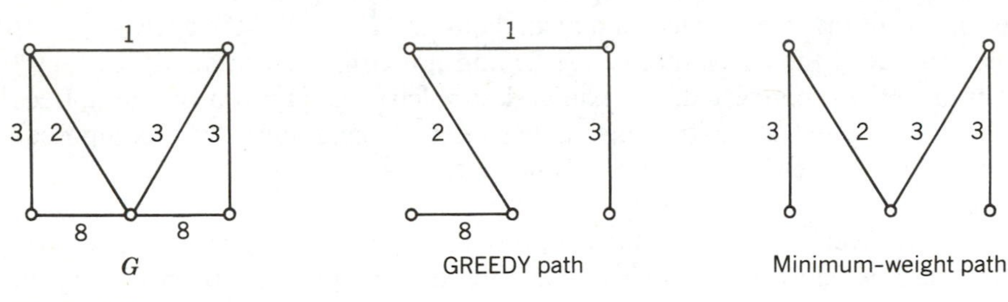

Figure 5.27

This example illustrates the important fact that being greedy does not always produce a best answer, that is, greed does not always pay. In fact, researchers in the field of combinatorial optimization are actively seeking an understanding of just how bad an answer GREEDY will produce for specific applications.

Question 5.2. Suppose that G is a complete weighted graph on V vertices. Further suppose that you wanted to find a minimum-weight cycle C_V as a subgraph of G. Formulate a greedy algorithm to "solve" this problem. Find an example where your algorithm fails to produce the minimum-weight cycle.

This last question is not just whimsy. A variation of this is known as the **Traveling Salesrepresentative Problem**. Suppose that the vertices of G represent a collection of cities and the weight on each edge represents the cost of flying between the two cities. Then an economy-minded salesrepresentative might wish to visit all the cities in a cyclic tour but wants a tour of minimum cost. In fact, no good algorithm is known to solve this problem, or the graph theoretical version in Question 5.2. It is also not known that the problem requires an exponential algorithm. In fact, the problem is computationally equivalent to the Satisfiability Problem introduced in Section 1.10. This area is an active and important one in computer science, operations research, and combinatorics. In Chapter 8 we shall use KRUSKAL to give an approximate solution to this problem.

EXERCISES FOR SECTION 5

1. Here is a new algorithm:

STEP 1. Input the weighted graph G with edges $e_1, e_2, \ldots, e_E$ with $w(e_1) \geq w(e_2) \geq \cdots \geq w(e_E) > 0$
STEP 2. For $j = 1$ to E do
 STEP 3. If $G - e_j$ is connected, set $G := G - e_j$
STEP 4. Output G and stop.

Run this algorithm on the graphs of Exercises 4.1 and 4.2.

2. Describe, in general, for any weighted graph G, the output of the algorithm in Exercise 1. Is this a greedy algorithm?

3. A graph is said to be **unicyclic** if it contains exactly one cycle. Suppose that, given a weighted graph G, we wanted to find a minimum-weight, connected, unicyclic subgraph of G. Does greed pay?

4. Suppose you attempt to find a minimum-weight path using a greedy algorithm with the following criterion of desirability: S is said to be desirable if its edges form a path. Is this GREEDYMIN different from the GREEDYMIN in Example 5.1? If not prove that the two are the same. If they are different, determine whether this GREEDYMIN produces a minimum-weight path.

5. Recall that a property P is called hereditary if whenever S has property P and T is a subset of S, then T has property P. Decide which of the following properties are hereditary:

(*a*) P is the property that the subset is nonempty.
(*b*) P is the property that the subset contains an even number of elements.
(*c*) P is the property that the subset S satisfies $|S| = \lfloor n/2 \rfloor$, where the universe has n elements.
(*d*) P is the property that S is such that $|S| < n/2$, where the universe has n elements.
(*e*) P is the property that S does not contain a fixed element x.
(*f*) P is the property that S contains a fixed element x.
(*g*) P is the property that S contains at most one of the two elements x and y.

6. Call a subgraph of G desirable if by itself it is a connected graph. In this instance is desirability hereditary? Describe the maximal desirable subgraphs of G.

7. We list some properties that a graph G might have. In each instance if H is a subgraph of G, does H necessarily have the specified property? (That is, is the property hereditary?)

(*a*) The maximum degree of a vertex in G is less than 7.
(*b*) G is bipartite.

(*c*) G is a forest.
(*d*) G contains a cycle.
(*e*) G is a complete graph.

8. A graph G is said to be **triangle-free** if G does not contain a 3-clique as a subgraph. Show that being triangle-free is hereditary.

9. Some subgraphs of a graph are induced subgraphs (see Exercise 3.17 for the definition.) Is this a hereditary property?

10. Suppose that you wanted to find a maximum-weight, triangle-free subgraph. Does greed pay?

11. Let G be a graph with $V(G) = \{A, B, C, D, E, F, H\}$. Suppose that G is complete and its edges (in lexicographic order) have weights 1, 4, 14, 4, 15, 21, 2, 3, 2, 3, 3, 1, 3, 5, 2, 2, 2, 5, 2, 17, 1. Find a minimum-weight spanning cycle C_7 that begins and ends at A. What is the cycle that the greedy algorithm produces?

12. Call a subset of $E(G)$ desirable if it is contained in a spanning cycle of G. Show that with this definition of desirable GREEDY will not produce a minimum-weight spanning cycle.

5:6 GRAPHICAL HIGHLIGHTS

Graph theory is a rapidly expanding mathematical discipline. It is important in its own right, as the mathematical basis of many applications, and as a fertile ground for logical and algorithmic thinking. Like the number theory of the previous chapter, graph theory is accessible and concrete. Pictures of graphs make small examples workable; computer programs make large examples tractable. Examples lead to conjectures and ideas for proofs and counterexamples. In fact, this is the effective learning process for both students and research mathematicians.

It may seem as if this chapter contains an overwhelming number of definitions. Each is there for a reason relevant to our work. Most definitions are needed immediately to understand Kruskal's algorithm. Others are needed for wide-ranging applications that mathematics and computer science students will meet. A tree may be thought of as the basic underlying structure on which the rest of a connected graph hangs. Trees also arise as a structure used for information storage. These so-called data structures, when formed as a tree, allow for quick retrieval of stored information. For example, most computer operating systems allow for directories, subdirectories, and so on, that are organized by the vertices of a tree.

Our modeling of the Local Area Network with graphs is a true-to-life depiction of how the problems of linking computers and terminals are now being attacked. In fact, computer scientists and electrical engineers regularly look to graphs and their properties to aid them in network design. Telephone companies use graphs, for example, to design systems of switching stations. Their goals are to have short distances between vertices while still having each vertex of small degree. Kruskal's algorithm and other minimum-weight spanning tree algorithms are used in a whole spectrum of applications dealing with transportation systems, commodity flows, and efficient robot manufacturing, as well as the cable connection problem we've seen here. The tree minimizing problem of this chapter provides an introduction to the area of combinatorial optimization. In this field it is now well understood when greedy algorithms work. On the other hand, the search for effective methods to find a minimum-weight spanning cycle in a weighted graph is one of the central problems of mathematics and computer science. Of seemingly intermediate difficulty, the graph isomorphism problem has so far resisted satisfactory solution, yet workers in the field expect this problem to be solved in the near future.

In Chapter 8 we consider more graph theory, both abstract and applied to optimization problems. We also present some approximation algorithms, that is, algorithms that work efficiently but only produce a near-optimal answer.

SUPPLEMENTARY EXERCISES FOR CHAPTER 5

1. Given a graph G, define G^c, the **complement** of G, to be the graph that has $V(G^c) = V(G)$ and $E(G^c) = \{(x, y): (x, y) \notin E(G)\}$. Find the complement of each of the following graphs.

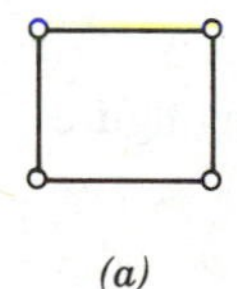
(a)

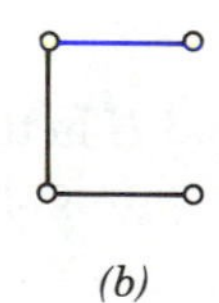
(b)

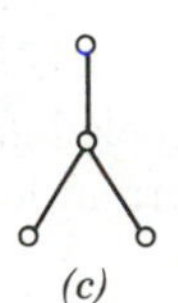
(c)

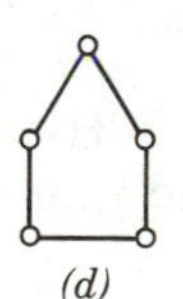
(d)

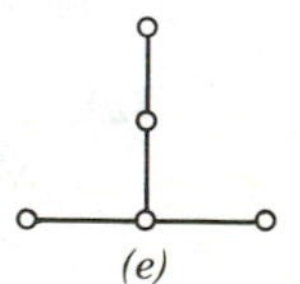
(e)

2. If G is isomorphic to H, is G^c isomorphic to H^c? Give a proof or counterexample.

3. If $d_1, d_2, \ldots, d_V$ are the degrees of G, a graph on V vertices, what are the degrees of G^c?

4. What is the largest clique contained in the complement of Q_3? Q_4? (For a definition of Q_n see Exercise 2.19.)

5. We define the **diameter** of a graph to be the maximum value of $d(x, y)$ among all pairs of vertices x and y. Show that if G has diameter 4 or more, then G^c has diameter 2 or less.

6. Suppose that G is a graph with V vertices and E edges and with vertices labeled $\{1, 2, \ldots, V\}$. Then we can list the edges in lexicographic order, as defined in Section 3.3: If each edge e_i is given as a pair of vertices (x_i, y_i), then the edges are numbered and listed in the order $e_1, e_2, \ldots, e_E$ subject to the restrictions that for all i and j with $1 \le i, j \le E$,
(1) $x_i \le y_i$;
(2) $i < j$ implies that $x_i \le x_j$; and
(3) $i < j$ and $x_i = x_j$ implies that $y_i \le y_j$.

(*a*) Explain why the edge list $e_1 = (1,2)$, $e_2 = (1,3)$ and $e_3 = (3,4)$ is in lexicographic order, but the list $f_1 = (1,2)$, $f_2 = (2,4)$, and $f_3 = (2,3)$ is not in lexicographic order.

(*b*) The following edge list is not in lexicographic order. Rearrange it so that conditions (1), (2), and (3) are met:

$$\begin{aligned} e_1 &= (1,2) \qquad & e_4 &= (5,4) \\ e_2 &= (1,3) \qquad & e_5 &= (5,1) \\ e_3 &= (2,3) \qquad & e_6 &= (4,1). \end{aligned}$$

7. Here is an algorithm to solve the so-called Labeled Graph Isomorphism Problem.

Algorithm LABGPHISO

STEP 1. Input G and H with edges in lexicographic order; let e_i denote the ith edge of G and f_i the ith edge of H {Assume $|V(G)| = |V(H)| = V$; the vertices are labeled with $1, 2, \ldots, V$, and $|E(G)| = |E(H)| = E$.}
STEP 2. For $j = 1$ to E do {j indexes the edges}
  STEP 3. If $e_j \neq f_j$ {as ordered pairs}, then output "no" and stop.
STEP 4. Output "yes" and stop.

(*a*) Run LABGPHISO on the labeled graphs G and H in the following figure. (Make sure your edge lists are in lexicographic order.)

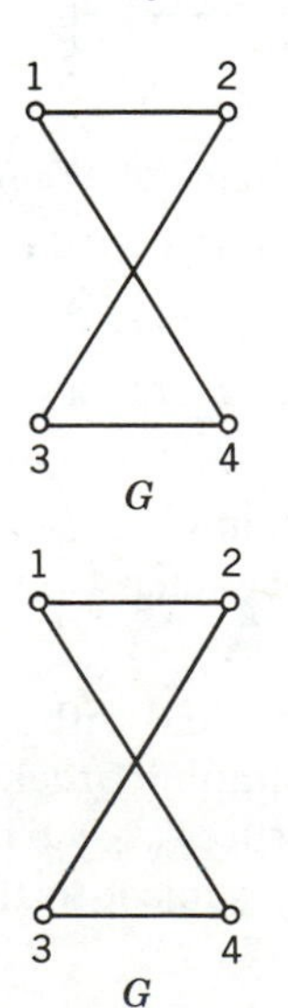

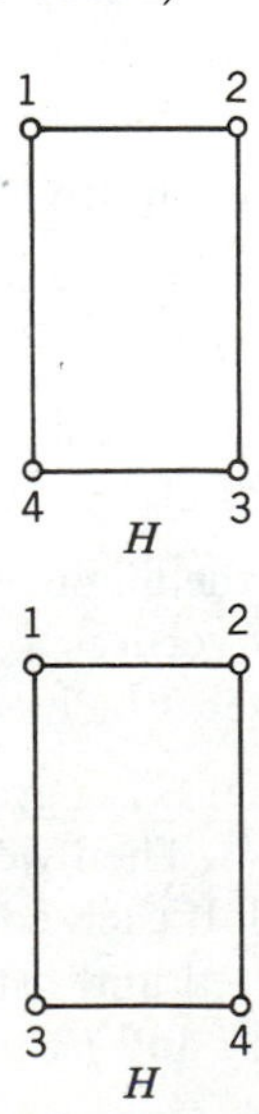

(*b*) Explain why the number of comparisons made in LABGPHISO is at most $2E = O(E)$.

8. Here is an idea for solving the more difficult problem of determining graph isomorphism for unlabeled graphs: Fix a labeling of the vertices of G with $1, 2, \ldots, V$. Then run through all permutations of labels of H, and for each permutation run the algorithm LABGPHISO. Design an algorithm GPHISO that uses these ideas. (You may use the algorithms PERM from Chapter 3 and LABGPHISO within your algorithm, without repeating it in its entirety.) Is your algorithm good or bad?

9. Here is an idea to try to speed up the algorithm GPHISO. As seen in Example 2.2 vertices with the same labels must have the same degrees. Thus, for example, if S is the set of all vertices of degree 3 of G and T all vertices of degree 3 of H, then we need to check only $|T|!$ permutations of the labels of T. (The same is true for each degree of vertices in G and H.) Use this idea to redesign GPHISO and then analyze whether this speeds up the algorithm in some or all cases.

10. Suppose that we are given a graph G with $V(G) = \{x_1, \ldots, x_V\}$ and edge list $E(G) = \{(x_1, x_2), \ldots\}$. Suppose that we want to decide if G is bipartite. Begin by placing x_1 in R. Next place each of x_1's neighbors in B, and so on. Construct a precise algorithm BIPARTITE. How many comparisons does BIPARTITE make?

11. What is the maximum number of edges a graph on V vertices can have and still not be connected?

12. For each of the following sequences, either draw a tree whose vertices have these degrees or show that such a tree cannot exist.

(*a*) $\langle 4, 1, 1, 1, 1 \rangle$.
(*b*) $\langle 6, 2, 2, 1, 1, 1, 1, 1 \rangle$.
(*c*) $\langle 5, 2, 2, 1, 1, 1, 1, 1 \rangle$.
(*d*) $\langle 4, 3, 2, 1, 1, 1, 1, 1 \rangle$.
(*e*) $\langle 3, 2, 2, 2, 2, 2, 2, 2, 1 \rangle$.
(*f*) $\langle 3, 3, 3, 1, 1, 1, 1, 1 \rangle$.

13. Give an algorithm that will, given a sequence of positive integers $d_1, \ldots, d_V$ with $d_1 + \cdots + d_V = 2V - 2$, construct a tree whose vertices have the given sequence as its sequence of degrees.

14. Show that in any gathering of people, some pair of people have the same number of acquaintances. (*Hint:* Assume that if A knows B, then B knows A. Think of the graph that could represent acquaintances and try a proof by contradiction.)

15. Prove that if d equals the maximum degree of a vertex in a tree T, then T contains at least d vertices of degree 1.

16. Find all graphs G such that both G and G^c are trees.

17. Let G be a connected graph with edge weights any real numbers. For vertices u and v of G, prove that there is a shortest path between u and v if and only if no path from u to v contains a cycle, the sum of whose edge weights is negative.

18. Here is the idea of Borůvka's original minimum-weight spanning tree algorithm.

STEP 1. Input G, a connected weighted graph with n vertices
STEP 2. Set T equal to the n vertices of G
STEP 3. Repeat
    STEP 4. For each component C of T do
        STEP 5. Select a minimum-weight edge joining a vertex of C with a vertex of $G - C$ and add it to T
    Until T is a spanning tree of G
STEP 6. Output T and stop.

Prove that this algorithm produces a minimum-weight spanning tree of G. Compare this algorithm with KRUSKAL. Find examples where it produces the same and where it produces different minimum-weight spanning trees. What is its complexity?

19. Here is another version of a minimum-weight spanning tree algorithm due to Prim.

STEP 1. Input G, a weighted connected graph with n vertices
STEP 2. Set $T = \{v\}$, where v is a vertex of G
STEP 3. For $i = 1$ to $n - 1$ do
    STEP 4. Select a minimum-weight edge e joining a vertex x not in T with a vertex in T; set $T = T + e + x$
STEP 4. Output T and stop.

Prove that this algorithm produces a minimum-weight spanning tree of G. Compare the algorithm with KRUSKAL and with Borůvka's algorithm of Exercise 18. What is the complexity of this algorithm?

20. A subset I of the vertex set of a graph G is said to be **independent** if no two vertices in I are joined by an edge in G. The **independence number** of a graph G, denoted by $\boldsymbol{\alpha(G)}$, is defined to be the maximum number of vertices in an independent set in G. Find $\alpha(G)$ for each of the following graphs:

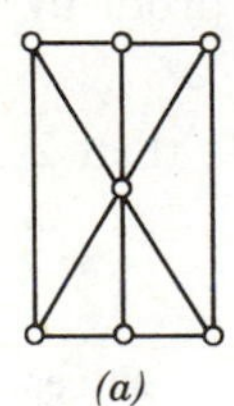
(a)

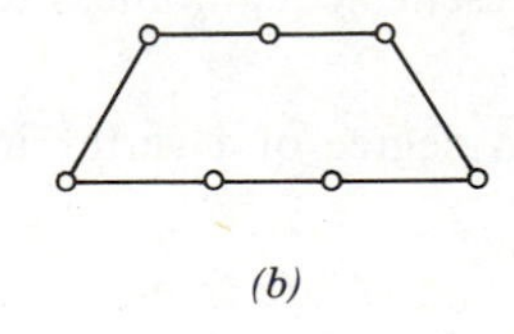
(b)

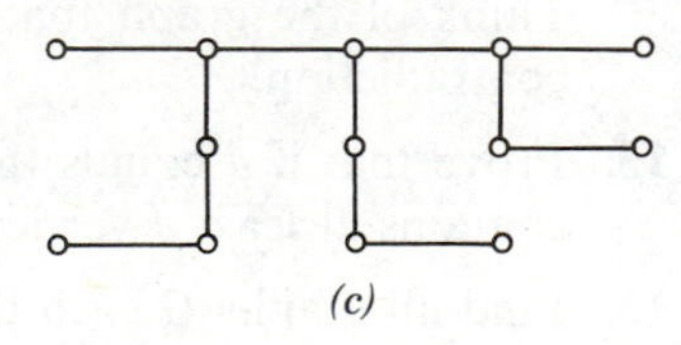
(c)

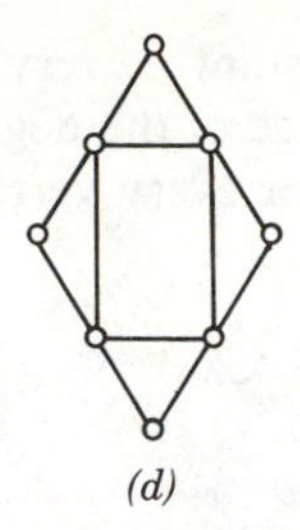
(d)

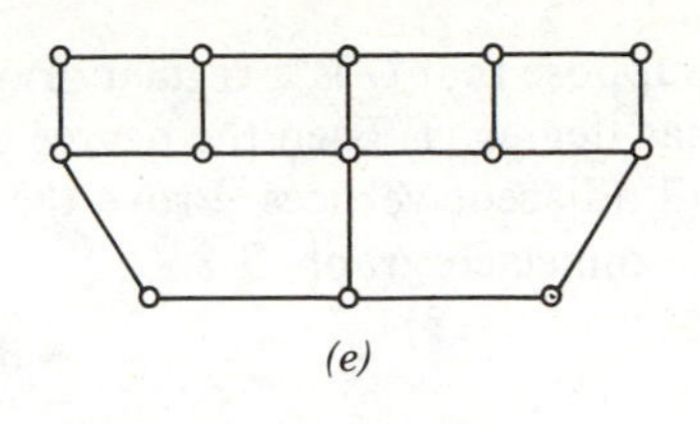
(e)

21. Show that if F is a forest, then $\alpha(F) \geq V/2$. Find an example of a forest with $V = 10$ and $\alpha = 5$.

22. Here is an algorithm to find an independent set.

Algorithm IND

STEP 1. Input G; set I to be empty
STEP 2. While G is nonempty do
Begin
STEP 3. Find a vertex x with $\deg(x, G)$ minimum
STEP 4. Set $I := I + \{x\}$
STEP 5. Set $G := G - \{x\} - \{y: (x, y) \in E(G)\}$
End {step 2}
STEP 6. Output I and stop.

Run IND on each of the following graphs.

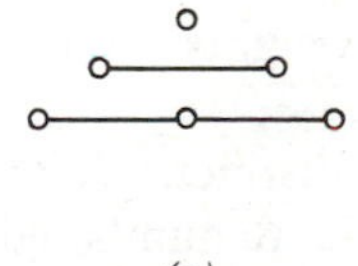
(a)

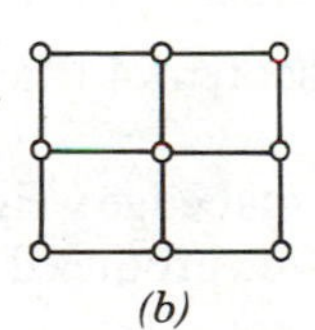
(b)

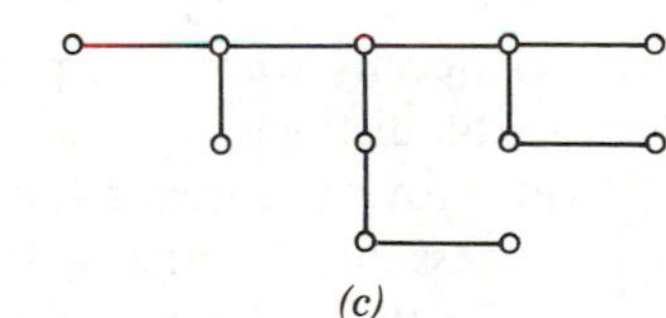
(c)

23. Prove that Algorithm IND works, that is, it finds an independent set in a graph G.

24. Find an example where IND does not find a maximum independent set.

25. Show that if the input to Algorithm IND is a forest, then the output I will be a maximum independent set.

26. Let A be the $V \times V$ adjacency matrix of a graph. What information does the matrix A^2, the product of A with itself, contain about the graph?

27. Let A be the adjacency matrix of a graph G and let i and j be in $V(G)$. Then prove that the least integer k such that A^k contains a positive entry in the (i,j)th position equals $d(i,j)$, the distance between i and j.

28. Suppose that G is a **regular graph** (i.e., for some fixed constant r, every vertex has degree r). Then the degree of each vertex is the average of the degrees of all adjacent vertices. Prove the converse: Suppose that for every vertex v of a connected graph G

$$\deg(v) = \frac{\deg(x_1) + \cdots + \deg(x_r)}{\deg(v)},$$

where the $x_1, \ldots, x_r$ are all of the vertices adjacent to v. Then prove that G is regular.

29. Suppose that f is a function with domain $V(G)$ and target the real numbers for some connected graph G, and that f satisfies the following property: For every vertex v of $V(G)$

$$f(v) = \frac{f(x_1) + \cdots + f(x_j)}{\deg(v)},$$

where the sum is taken over all vertices x adjacent to v. Then prove that f is a constant function, that is, $f(v) = c$ for some constant c for all v in $V(G)$. Is the result true if G is not connected?

30. Suppose that the number of comparisons made in a graph algorithm A is given by $g(V)$ and that $g(V) > r^V$ for some positive constant r. If $B = V^2$, then show that $r^{(B^{1/2})}$ is a lower bound on the number of comparisons made in the algorithm A. Prove that the function $h(B) = r^{(B^{1/2})}$ is not $O(p(B))$ for any polynomial p and that $h(B) = O(s^B)$ for $s > 1$. What can you conclude about whether or not A is a good algorithm in terms of the input size B?

31. **(*a*)** Specify where the proof of Theorem 4.1 fails if the edge weights are not all distinct.

(*b*) Prove Theorem 4.1 in the case that edge weights are not all distinct. (*Hints:* Assume that the edges of any tree produced by KRUSKAL are numbered in the order in which they were selected. Further suppose that F is a minimum-weight spanning tree that has the greatest initial agreement with T. Then complete the proof along the lines of the proof of Theorem 4.1.)

32. Suppose that G is a graph and $\sim$ the corresponding relation on $V(G)$ (as defined in Section 2). For what graphs G is $\sim$ symmetric? Transitive?

33. The **transitive closure** of a graph G is defined to be the graph G' with $V(G') = V(G)$, $E(G) \subseteq E(G')$, and additional edges of $E(G')$ given by: Whenever (a, b) and $(b, c) \in E(G)$, then $(a, c) \in E(G')$. Explain why the corresponding relation $\sim$ defined on $V(G)$ is a transitive relation.

34. Characterize all graphs G such that there is an equivalence relation $\sim$ on a set S whose corresponding graph is G.

6

SEARCHING AND SORTING

6:1 INTRODUCTION: RECORD KEEPING

A college's financial aid office has just created the job of Director of Student Employment. The responsibilities of this position include the organization of student employment information. Until now this information has been kept in the following fashion. Each student employee has been assigned a record card on which is written payroll information, including the student's social security number. These record cards are organized in a file drawer arranged in alphabetical order of the students' last names. Each time the treasurer's office issues a payroll check the director receives a memo containing the payee's social security number, the total amount of the check, and the amount withheld for various taxes. Of course, she wishes that these memos also contained the payee's name; however, the particular computer program that the treasurer's office uses to cut checks doesn't have that capability. When a payroll memo arrives, the director examines each record card in turn to determine if the social security number on the card is identical to the number on the memo.

Question 1.1. Suppose that there were 20 cards in the director's file drawer. (a) When a payroll memo arrives, what is the minimum number of cards that the director might have to check? (b) What is the maximum number of cards that the director might have to check? (c) About how many cards (on average) would you expect the director to have to check?

Question 1.2. Suppose that each of the 20 students whose cards are in the file drawer receives exactly one payroll check each week. (a) What is the total number of social security number comparisons that the director will have to make to record all of the payroll transactions? (b) If it takes 2 seconds to make a comparison and 1 minute to record all the information on a file card, will the director spend more time making comparisons or recording information?

Now let's answer the previous questions if there are n record cards in the director's drawer. The minimum possible number of comparisons occurs when the payee happens to be the individual whose card is first in the file, the one whose name is alphabetically first. In this instance there is just one comparison to make. The largest number of comparisons occurs when the payee is the individual whose card is last in the file. In this case there would be n comparisons to make. It is plausible to think that the average number of comparisons should be the average of the smallest number and the largest number. Here that number would be $(n + 1)/2$. In fact, this is correct as we see by the following explicit computation.

If every individual in the file is paid exactly once, we can count the total number of comparisons in the following manner. First, note that the payee who is listed first alphabetically will require just one comparison to locate. We don't know which memo corresponds to this first payee, but whichever one it is, it will still take just one comparison. Similarly, the payee who is listed second alphabetically will take exactly two comparisons to locate. In general, the payee who is listed kth alphabetically will take exactly k comparisons to locate (regardless of when this memo is processed). Thus the total number of comparisons will be

$$1 + 2 + 3 + \cdots + k + \cdots + n = \frac{n(n + 1)}{2}.$$

Since the total number of comparisons needed is $n(n + 1)/2$, the average number of comparisons needed will be this total divided by the number of payees. This yields $(n + 1)/2$ comparisons on average. The total time needed for comparisons will be $n(n + 1)$ seconds while the time required to record the payroll information will be $60n$ seconds. Thus if $n \geq 60$, more time will be spent finding the correct file than writing information to it.

Let's formalize the director's task.

Problem. Given an array $A = \langle a_1, a_2, \ldots, a_n \rangle$ and an object S, determine S's position in A, that is, find an index i such that $a_i = S$ (if such an i exists).

Algorithm SEQSEARCH

STEP 1. Input A and S.
STEP 2. For $i = 1$ to n do
    STEP 3. If $a_i = S$, then output i and stop.
STEP 4. Output "S not in A" and stop.

If we count the comparisons in step 3, then the worst case will occur either if S is not in A or if $S = a_n$. In this instance SEQSEARCH requires n comparisons. Thus the complexity of this algorithm is $O(n)$. Note that in our particular example with social security numbers, S and the elements in the array A are numbers; however, all that is required for this algorithm to work is that we can determine whether $a_i = S$. Thus SEQSEARCH would work equally well when the entries of A are words.

The director decides that record keeping would be more efficient if the record cards were kept in order of their social security numbers. The director begins the sorting process by finding the card with the smallest social security number. She does this by comparing the number on the first card with the number on the second. She keeps the smaller of the two and then compares it with the number on the third card. She picks the smaller and now has the smallest number from the first three cards.

Question 1.3. In a drawer of 20 record cards, how many comparisons would be required to be certain of finding the card with the smallest social security number?

We formalize the problem and the response.

Problem. Given an array of numbers $A = \langle a_1, a_2, \ldots, a_n \rangle$, sort these numbers into increasing order, that is, arrange the numbers within the array so that $a_1 \leq a_2 \leq a_3 \leq \cdots \leq a_n$.

Algorithm SELECTSORT

STEP 1. Input A, an array of n numbers
STEP 2. For $i = 1$ to $n - 1$ do {Find the correct ith number.}
  Begin
  STEP 3. Set TN $:= a_i$ {TN = temporary number}
  STEP 4. For $j = i + 1$ to n do
    STEP 5. If $a_j <$ TN, switch a_j and TN
  STEP 6. Set $a_i :=$ TN
  End {Step 2}
STEP 7. Output A and stop.

Example 1.1. Table 6.1 gives a trace of SELECTSORT applied to the array $A = \langle 4, 7, 3 \rangle$. Notice that all the action occurs at step 5.

Since the smallest remaining element is repeatedly selected, this method is called Selection sort. See Exercises 11 to 13 for a comparison with the sorting algorithm known as Bubblesort.

Table 6.1

Step No.	i	j	a_1	a_2	a_3	TN
3	1	?	4	7	3	4
5	1	2	4	7	3	4
5	1	3	4	7	4	3
6	1	3	3	7	4	3
3	2	3	3	7	4	7
5	2	3	3	7	7	4
6	2	3	3	4	7	4

Question 1.4. Apply SELECTSORT to the array $A = \langle 6, 4, 2, 3 \rangle$. Exhibit the values assigned to i, j, TN, and each location in A after every execution of step 5.

Theorem 1.1. SELECTSORT is a $O(n^2)$ algorithm.

Proof. We count the comparisons, which only occur in step 5. When i is assigned the value 1, j varies from 2 to n. Thus there are $n - 1$ different values assigned to j and $n - 1$ comparisons when $i = 1$. When i is assigned the value 2, j varies from 3 to n. Thus there are $n - 2$ comparisons. For general i, j varies from $i + 1$ to n. In this case there are $(n - (i + 1) + 1) = n - i$ comparisons. Hence the total number of comparisons equals

$$(n - 1) + (n - 2) + \cdots + (n - i) + \cdots + 1 = \frac{n(n-1)}{2} = O(n^2). \qquad \square$$

A bit analysis of SELECTSORT would begin by noting that each of the n numbers in the input array could be represented by M bits. Thus the total input size would be nM. Every comparison of two M bit numbers would require, in the worst case, M bit comparisons. Thus the total number of bit comparisons would be $Mn(n - 1)/2$. If M is constant, then SELECTSORT is quadratic in the bit analysis also.

Notice that SELECTSORT could work equally well on arrays of words using alphabetical ordering. In a subsequent section we shall see that SELECTSORT can operate on sets with more general orderings. We'll also find that there are more efficient algorithms to perform sorting as well as searching; however, for small arrays SEQSEARCH and SELECTSORT are worth using, in part because they are so simple.

Here is the terminology we shall use throughout the rest of the chapter. Each unit of information to be sorted is called a **record**. The set of records is called a **file**. The element in the record with which the sorting is done is called the **key**.

Thus in the employment director's office, her drawer contains the file. Each card in the file is a record and the key is the social security number on the card. To keep numerical examples simple, we shall often consider a record that consists only of the key, but in applications the record will contain more information. Consequently, interchanging two records in a file will be a more time-consuming process than that of switching two numbers. If the records are stored in computer memory and accessed by a language that admits the use of pointers, then the pointers will be changed rather than the records.

EXERCISES FOR SECTION 1

1. Apply SEQSEARCH to the following arrays and objects S; record the output of the algorithm.
   (*a*) $A = \langle 1, 2, 3, \ldots, 17 \rangle$, $S = 15$.
   (*b*) $A = \langle 1, 2, 3, \ldots, 17 \rangle$, $S = 12.5$.
   (*c*) $A = \langle$apple, banana, cantaloupe, kiwi, mango, papaya$\rangle$, $S =$ strawberry.
   (*d*) $A = \langle a, b, c, \ldots, z \rangle$, $S = h$.
   (*e*) $A = \langle a, b, c, \ldots, z \rangle$, $S = \&$.
   (*f*) $A = \langle 1, 2, 3, a, b, c, \#, \$, \%, \hat{} \rangle$, $S = \$$.

(*Note:* In Exercises 2 to 7 we assume that, as in Questions 1.1 to 1.3, the record cards are listed alphabetically and the payroll memos come identified by social security number.)

2. Suppose that there are 40 student employees who each receive 2 checks per month. How many comparisons does the director make in a month using SEQSEARCH? If it takes 2 seconds to make a comparison and 1 minute to record the payroll information, which requires more time, comparing or recording information?
3. Suppose that there are 20 student employees and exactly 10 receive a check in any given week. What is the minimum and maximum number of comparisons that the director might make in a week?
4. Suppose that there are n student employees who each receive k checks per month. How many comparisons will the director make in one month?
5. Suppose that there are $2n$ student employees and that exactly n of these students receive a check in a given week. What is the minimum and maximum number of comparisons that might be performed? What can be said about the average number of comparisons that will be made?
6. Suppose that there are n student employees who each receive one check per week. If it takes 3 seconds to make a comparison and 30 seconds to record the

salary information, for what values of n is more time spent on comparisons than on recording?

7. Suppose that there are n student employees who each receive one check per week. If it takes x seconds to make a comparison and y seconds to record information, then for what values of x and y do comparing and recording take the same amount of time? For what values of x and y does comparing take more time than recording?

8. Apply SELECTSORT to the arrays $\langle 1,2,3\rangle$, $\langle 3,2,1\rangle$, and $\langle 3,1,2,1\rangle$. Trace out the values assigned to i, j, TN, and every location in A after each execution of step 5.

9. Write an algorithm that, given an array of numbers, (a) selects the largest number and places it in the last position, (b) selects the next largest number and places it in the next to last position, and (c), in general, finds the largest remaining number and places it in the last unfilled position. Analyze the complexity of your algorithm.

10. Write an algorithm that finds the largest and the smallest entry in $A = \langle a_1, a_2, \ldots, a_n\rangle$, an array of real numbers. Count the number of comparisons made in the worst case.

11. Look back at the algorithm BUBBLES, Exercise 2.4.13. Recall that this algorithm found the largest entry in an array of n elements and placed it in the last location. BUBBLES can be readily transformed into a procedure that can be repeatedly called to sort the entire array. Here is an algorithm that does just this.

Algorithm BUBBLESORT

STEP 1. Input m, a positive integer, and the array $X = \langle x_1, \ldots, x_m\rangle$
STEP 2. For $n = m$ down to 2 do
    STEP 3. Call BUBBLES $(n, x_1, \ldots, x_n)$
STEP 4. Output $\langle x_1, x_2, \ldots, x_m\rangle$ and stop.

Apply BUBBLESORT to the following arrays, exhibiting the values of the array, n and j (the index in BUBBLES) throughout.

(*a*) $A = \langle 4,7,3\rangle$.
(*b*) $B = \langle 2,1,4,3,6,5\rangle$.
(*c*) $C = \langle 4,3,2,1\rangle$.

12. Count the number of comparisons made by BUBBLESORT. Compare the number of comparisons made in BUBBLESORT and SELECTSORT. Is one algorithm more efficient than the other?

13. How might you modify BUBBLESORT to recognize when the array X was already in order?

14. One way a record can contain more than the key, is using a 2-dimensional array $A = \langle a_{i,j}: i = 1, \ldots, m, j = 1, \ldots, n \rangle$. This can be pictured as a matrix with m rows and n columns:

$$\begin{array}{cccccc} a_{1,1} & a_{1,2} & \cdots & a_{1,j} & \cdots & a_{1,n} \\ a_{2,1} & a_{2,2} & \cdots & a_{2,j} & \cdots & a_{2,n} \\ \cdots & & & & & \\ a_{i,1} & a_{i,2} & \cdots & a_{i,j} & \cdots & a_{i,n} \\ \cdots & & & & & \\ a_{m,1} & a_{m,2} & \cdots & a_{m,j} & \cdots & a_{m,n} \end{array}$$

Each row might represent the record of one student, and different columns contain different types of information. Suppose that the key for each record is stored in the first column so that the key for ith record is the entry $a_{i,1}$ for $i = 1, 2, \ldots, m$. Use the idea of SELECTSORT to design an algorithm to sort the array A so that the rows of A are rearranged to have their first entries (the keys) listed in increasing order $a_{1,1} \leq a_{2,1} \leq \cdots \leq a_{m,1}$. How many comparisons does the algorithm use? In the worst case how many assignment statements are there? Your answers will depend on n and m.

15. Write an algorithm that finds the second smallest entry in an array $A = \langle a_1, a_2, \ldots, a_n \rangle$ of real numbers. Count the number of comparisons made.

16. Here is the idea for an algorithm to find the kth smallest entry in an array A of n numbers: Find the smallest entry of A, then find the second smallest entry, and so on, until the kth smallest entry is found. Write an algorithm that implements this idea and count the number of comparisons; your answer will be in terms of n and k.

17. Here is another algorithmic solution to the problem of finding the kth smallest entry in an array A (see Exercise 16): First order the array using SELECTSORT and then find the kth entry of the sorted array. Compare the number of comparisons made by this algorithm with that of Exercise 16; which is more efficient?

18. Suppose that you have a balance scale with which you can determine which (if either) of a pair of given coins is lighter in weight. Given n supposedly identical coins, but such that one weighs less than the others, give a technique suggested by SELECTSORT to find the light coin. How many comparisons will your technique require in the worst case?

19. Suppose that you have 16 supposedly identical coins, exactly one of which weighs less. Using a balance scale, each pan of which can hold as many coins as you like, how can you find the light coin with only 4 weighings?

6:2 SEARCHING A SORTED FILE

We return to the employment director's problem of transcribing payroll information. We assume that there are n employees whose record cards are filed now in the order of increasing social security number. When a memo arrives from the payroll office, the director searches for the record whose social security number is the same as the one on the memo. Suppose she selects the mth record from the file, or drawer, and compares the two social security numbers. If the two numbers are equal, then the director writes the information on the selected record. If the number on the memo is less than the number on the mth record, then the correct record must be located in the front portion of the file. Otherwise, the correct record must be located behind the mth record.

Of course, the director hopes to pick the correct record on the first try. However, she does not believe in her own good luck. Furthermore (with a touch of pessimism), the director believes that when she picks a record to compare with the payroll memo, the record she really wants will be in the larger part of the remaining records. Thus the director wants to choose a record in the mth position so that there are about as many records in front of the mth record as there are behind the mth record. If the drawer has n records, the director picks the record roughly in the middle, the record in the mth position, where $m = \lfloor (n+1)/2 \rfloor$. The director has, of course, assumed a worst-case scenario.

Question 2.1. Find $m = \lfloor (n+1)/2 \rfloor$ if $n = 136, 68, 34, 17, 9, 5$, and 3.

Question 2.2. If the drawer contains n records and the mth record is selected, where $m = \lfloor (n+1)/2 \rfloor$, when is it the case that there are exactly the same number of records before and after the mth record? When these two numbers differ, by how much do they differ? After examining this mth record, what is the largest number of records that still must be searched?

Let's assume for the moment that the director has selected m and the number on the memo is less than the number on the mth record. Then she begins the search all over again, confining her attention to that portion of the file that is in front of the mth record. In pseudocode she sets $n := m - 1$ and then chooses m to be (as before) $\lfloor (n+1)/2 \rfloor$. On the other hand, if the number on the memo is greater than the number on the mth record, then the correct record must be in position j, where $(m+1) \leq j \leq n$. As above she begins the search all over again, concentrating on the records in positions $m+1, \ldots, n$. The next record to select is the one that, as nearly as possible, divides the remaining records into equal piles.

Question 2.3. For the following pairs (i, j) find the number that will be the index of the entry that, as nearly as possible, divides $\langle a_i, \ldots, a_j \rangle$ into two equal pieces: $(6, 8)$, $(10, 17)$, $(18, 33)$, $(35, 67)$, and $(69, 136)$.

In general, as the director progresses, she narrows down the possible records that might correspond with the memo to a subarray $\langle a_i, a_{i+1}, \ldots, a_j \rangle$ of the original array A. She wants to select the "middle" record of this subarray. The index of the "middle" record is essentially the average of the indices of the end records. We say essentially because the average might not be an integer. However, a record that, as nearly as possible, divides the subarray into two equal pieces has index $m = \lfloor (i + j)/2 \rfloor$. With this insight we can now formulate the director's algorithm.

Problem. Given an array $A = \langle a_1, a_2, \ldots, a_n \rangle$ whose elements are numbers listed in increasing order and a number S, determine S's position in A, that is, find an index i (if it exists) such that $a_i = S$.

Algorithm BINARYSEARCH

STEP 1. Input A, an array of n numbers in increasing order, and a number S
STEP 2. Set first := 1, last := n
STEP 3. While first $\leq$ last do
  Begin
  STEP 4. Set mid := $\lfloor$(first + last)/2$\rfloor$
  STEP 5. If $S = a_{\text{mid}}$, then output "found S at location mid" and stop.
  STEP 6. If $S < a_{\text{mid}}$, then set last := mid − 1,
   Else set first := mid + 1
  End {Step 3}
STEP 7. Output "S is not in A" and stop.

Note that in step 6 exactly one of two assignment statements is executed, depending on the result of the comparison in that step.

Example 2.1. Table 6.2 is a trace of BINARYSEARCH, where $A = \langle 3, 4, 6, 7, 9, 11 \rangle$ and $S = 9$. We begin after the first encounter with step 4.

Table 6.2

Step No.	*first*	*last*	*mid*	a_{mid}
4	1	6	3	6
5	1	6	3	6
6	4	6	3	6
4	4	6	5	9
5	4	6	5	9

Question 2.4. Trace BINARYSEARCH if A consists of the first eight primes in increasing order and (a) $S = 5$, (b) $S = 10$, and (c) $S = 17$. In each case how many elements in the array do you examine?

BINARYSEARCH can find S without examining all the entries in A because the elements of A are numbers listed in increasing order. Actually, this algorithm will work on any set that is totally ordered. See Exercises 4.6.12 to 4.6.14. Since A is totally ordered, either $S < a_{\text{mid}}$ or $a_{\text{mid}} \leq S$. Consequently, the value of (last–first) decreases with each loop and so BINARYSEARCH must terminate. Furthermore, the transitive property allows the algorithm to check S against a_{mid} and discard about half of the ordered list at each pass through the loop. Exercise 6 asks you to modify BINARYSEARCH so that it works on the set of all English words in alphabetical order.

Theorem 2.1. BINARYSEARCH requires at most $3\lfloor \log(n) \rfloor + 4$ comparisons to search an ordered array of n numbers.

Proof. First note that each of steps 3, 5, and 6 requires exactly one comparison. Thus each time we execute the loop beginning at step 3, we use no more than three comparisons. The proof will be by induction on the number of elements in the array. We begin with the base case $n = 1$. Given the array $A = \langle a_1 \rangle$, the algorithm uses two comparisons if $S = a_1$. If $S \neq a_1$, then the algorithm cycles through the loop once and executes step 3 one additional time. Thus a total of four comparisons is needed in this case.

The inductive hypothesis will be that BINARYSEARCH can search any ordered array of t elements with at most $3\lfloor \log(t) \rfloor + 4$ comparisons for any $t < n$. We suppose that A is an ordered array with n elements. If we find equality the first time at step 5, we are done, using 2 comparisons. Otherwise, we return to step 3 with a smaller array, having performed three comparisons. The new array contains no more than half of the elements of the original array. (See Question 2.2.) By the inductive hypothesis it takes at most $3\lfloor \log(n/2) \rfloor + 4$ comparisons to search the new array. Thus the total number of comparisons needed to search the original array is at most

$$\begin{aligned} 3 + (3\lfloor \log(n/2) \rfloor + 4) &= 7 + 3\lfloor \log(n) - 1 \rfloor \\ &= 4 + 3\lfloor \log(n) \rfloor. \end{aligned}$$ □

Question 2.5. For each of $n = 2$, 3, and 4 find two examples of arrays and a number S, one that requires a full $3\lfloor \log(n) \rfloor + 4$ comparisons and one that requires fewer.

Question 2.6. Suppose that the director's file has 1000 records in it. In the worst case, how many comparisons will it take to find a record with a particular social security number on it if (a) SEQSEARCH is used and (b) BINARYSEARCH is used?

How did we originally find the bound $3\lfloor \log(n) \rfloor + 4$ of Theorem 2.1? This expression works in the inductive proof, but why? Suppose that $B(n)$ denotes the maximum number of comparisons made by BINARYSEARCH on an array of n elements. Then in the worst case we perform three comparisons (in steps 3, 5, and 6) and then face a smaller array with $\lfloor n/2 \rfloor$ elements in which to search for S. $B(\lfloor n/2 \rfloor)$ denotes the maximum number of comparisons needed to search this smaller array and so

$$B(n) = 3 + B(\lfloor n/2 \rfloor) \qquad \text{and} \qquad B(1) = 4. \tag{*}$$

This fact doesn't solve the problem immediately but can lead to a solution as outlined in Exercises 13 to 15. In Chapter 7 we pursue a systematic study of how, given an equation like that of line (*), we can find an expression for the number of comparisons (or other significant operations) performed in the worst case of an algorithm.

From Theorem 2.1 we can get an estimate of the amount of work the director must do each week. If each week one memo arrives for each of the n student employees, then the next result gives an upper bound on the number of comparisons necessary.

Corollary 2.2. BINARYSEARCH requires at most

$$3n\lfloor \log(n) \rfloor + 4n = O(n \log(n))$$

comparisons to search an ordered array for each of the n files located in it.

This result is an immediate consequence of Theorem 2.1; however, $3n\lfloor \log(n) \rfloor + 4n$ is really an overestimate. A tighter upper bound on the number of comparisons, but one that is still $O(n \log(n))$, is derived in Exercises 9 to 11. In any case the worst-case behavior of BINARYSEARCH is significantly better than that of SEQSEARCH. Indeed the worst-case performance of BINARYSEARCH is better than the average-case performance of SEQSEARCH. The average-case performance of BINARYSEARCH is analyzed in Exercise 12. In its defense it should be emphasized that SEQSEARCH will work on any set in an array A regardless of whether or not the elements of A form a totally ordered set.

In the next section we use the ideas of binary search to construct a more efficient sorting algorithm.

EXERCISES FOR SECTION 2

1. Let $A = \langle 1, 2, \ldots, 7 \rangle$, $B = \langle 2, 4, 6, \ldots, 16 \rangle$, and $C = \langle 1, 3, 7, 15, 31, 63 \rangle$. Trace BINARYSEARCH to find ***(a)*** $S = 3$ in A, ***(b)*** $S = 8$ in A, ***(c)*** $S = 6$ in B, ***(d)*** $S = 7$ in B, ***(e)*** $S = 31$ in C, and ***(f)*** $S = 14$ in C.
2. Suppose that the number on the director's memo is less than that on the $\lfloor (n+1)/2 \rfloor$nd record. What is the index of the next record she consults? Express

this as a function of n. If the number is greater than that on the $\lfloor (n+1)/2 \rfloor$nd record, what is the index of the next record she consults?

3. Find all values of n for which SEQSEARCH uses fewer comparisons in the worst case than BINARYSEARCH.
4. Find a value of N such that SEQSEARCH uses at least twice as many comparisons in the worst case as BINARYSEARCH. Show that for every $n > N$ SEQSEARCH will always use at least twice as many comparisons in the worst case as BINARYSEARCH.
5. In the worst case, how many subintervals of the form $\langle a_{\text{first}}, \ldots, a_{\text{last}} \rangle$ does BINARYSEARCH examine in an array with n entries?
6. Suppose that A is an array containing n words, where each word is a (finite) sequence of letters taken from the English alphabet. Suppose further that your computer can answer the following questions:

   given words w and w', does $w = w'$?
   does w precede w' alphabetically?

   Write a version of BINARYSEARCH that upon input of A, an array of words listed in alphabetical order, and a word w, searches for w in A.
7. Suppose that we are searching an ordered array of n elements for an element that is in position k (but we don't know that). For what values of k will SEQSEARCH use fewer comparisons than BINARYSEARCH?
8. Modify BINARYSEARCH so that, given an array A with entries in increasing order ($a_1 \leq \cdots \leq a_n$) and a number S, it finds all indices i such that $a_i = S$.
9. Let $n = 2^k - 1$. Suppose that payroll memos for n students come into the financial aid office in random order and that records for these n students are arranged by increasing social security number. For each memo BINARYSEARCH is used to locate the appropriate record. At some point, the memo for the $\lfloor (n+1)/2 \rfloor$nd student arrives and requires only two comparisons to find the correct record. Memos for two other students will require exactly five comparisons.
   (*a*) Which numbered students are these?
   (*b*) How many memos require exactly eight comparisons to locate their records?
   (*c*) What is the next smallest number of comparisons needed and how many students need this many?
   (*d*) For each possible value of i, determine the number of memos that require exactly i comparisons.
10. Prove that

$$1 \cdot 2 + 2 \cdot 5 + 4 \cdot 8 + \cdots + 2^{i-1}(3i - 1) + \cdots + 2^{k-1}(3k - 1) = (3k - 4)2^k + 4.$$

11. Suppose that $n = 2^k - 1$. Then explain why using BINARYSEARCH to search an ordered array for each of n records requires

$$\begin{aligned}(3k - 4)2^k + 4 &= 3n\lfloor \log(n) \rfloor - n + 3\lfloor \log(n) \rfloor + 3 \\ &= O(n \log(n))\end{aligned}$$

comparisons. Is this bound on the number of comparisons better than that given in Corollary 2.2?

12. Use the results of the preceding exercises to obtain the average number of comparisons used per record in BINARYSEARCH in the case $n = 2^k - 1$. Compare this average with that of SEQSEARCH.

13. Suppose that

$$B(n) = B(\lfloor n/2 \rfloor) + 3 \qquad \text{for } n > 1, \tag{*}$$

and

$$B(1) = 4.$$

Determine the value of $B(n)$ for $n = 2, 3, 4, 5, 8$, and 16.

14. Suppose that $n = 2^k$. Use (*) repeatedly to determine a formula for $B(n)$. Prove your formula correct by using induction and the equation in (*).

15. Verify that $f(n) = 3\lfloor \log(n) \rfloor + 4$ gives the same values as those obtained for $B(n)$ in Exercise 14. Then prove by induction that $B(n) = f(n)$ satisfies the equation in (*).

6:3 SORTING A FILE

We have seen that searching for one record in an unsorted file with n records in it requires $O(n)$ comparisons in the worst case. This contrasts with a worst case of $O(\log(n))$ comparisons in searching a sorted file. A natural question to ask is whether or not it's better to sort before searching or not. For the moment let's return to the problem of searching the file for each of the n records during every payroll period. If there are t payroll periods and the file remains unsorted, the total number of comparisons required will be $O(tn^2)$. On the other hand, if the director uses SELECTSORT to place the file in order, then the total number of comparisons will be

$$O(n^2) + O(tn \log(n)). \tag{A}$$

If, for example, there were n payroll periods (so $t = n$), then the number of comparisons would be $O(n^3)$ without sorting and $O(n^2 \log(n))$ with sorting. Thus, if the number of payroll periods is large, sorting before searching pays off. Suppose, for contrast, that the number of payroll periods is a small constant. Is it better to sort before searching or not? If the only sorting algorithm available were SELECTSORT, then both solutions be $O(n^2)$. However, if there were a better sorting algorithm, then one could expect sorting before searching to be faster.

SEQSEARCH requires, on average $(n + 1)/2$ comparisons to position a record correctly within a file containing n records. To sort more economically, we need a way to position a record correctly using fewer comparisons. BINARYSEARCH provides just such a mechanism.

Problem. Given an ordered array of numbers $A = \langle a_1, a_2, \ldots, a_r \rangle$ with $a_1 \leq a_2 \leq \cdots \leq a_r$ and a number D, insert D in the ordered list.

We develop the procedure BININSERT that will insert a number D into its correct position in an ordered array. The parameters of the procedure are $(r, a_1, \ldots, a_r, a_{r+1})$. We assume that upon calling the procedure the r numbers $a_1, \ldots, a_r$ are in order and that a_{r+1} equals D. Upon return $a_1, \ldots, a_{r+1}$ should be in order. Within the procedure we repeatedly compare D with the midpoint of a subarray in order to find its correct location. Once D's correct location is determined, the elements that should follow it are shifted over one space in order to make room for D. We make this algorithm a procedure, since we shall use it within BINARYSORT, which will be our first efficient sorting routine.

Procedure BININSERT $(r, a_1, \ldots, a_r, a_{r+1})$

{The initial segment of the procedure finds the correct location for a_{r+1}.}

STEP 1. Set first := 1, last := r
STEP 2. While first $\leq$ last do
  Begin
  STEP 3. Set mid := $\lfloor(\text{first} + \text{last})/2\rfloor$
  STEP 4. If $a_{r+1} < a_{\text{mid}}$, then set last := mid − 1,
    Else set first := mid + 1
  End {Step 2}

{At this point first equals last + 1, and first gives the correct position for a_{r+1}. The next segment creates a space for and inserts a_{r+1}.}

STEP 5. If first = r + 1, then Return. {a_{r+1}'s place is correct.}
STEP 6. Set temp := a_{r+1} {save a_{r+1}}
STEP 7. For $j = r + 1$ down to (first + 1) do
  STEP 8. $a_j := a_{j-1}$
STEP 9. Set a_{first} := temp
STEP 10. Return.

Example 3.1. Table 6.3 is a trace of the procedure BININSERT given the array $A = \langle 3, 5, 8, 10, 14 \rangle$, $r = 5$, and $D = 11$.

Table 6.3

Step No.	*first*	*last*	*mid*	a_{mid}	*j*	*A*
3	1	5	3	8		$\langle 3, 5, 8, 10, 14, 11\rangle$
4	4	5	3	8		
3	4	5	4	10		
4	5	5	4	10		
3	5	5	5	14		
4	5	4				
8	5				6	$\langle 3, 5, 8, 10, 14, 14\rangle$
9	5				6	$\langle 3, 5, 8, 10, 11, 14\rangle$

Question 3.1. Trace BININSERT if $A = \langle 2, 5, 7, 9, 13, 15, 19\rangle$ and $D =$ (a) 1, (b) 4, (c) 14, and (d) 23.

Notice the similarity between BINARYSEARCH and BININSERT. The test for equality has been eliminated because if $a_{r+1} = a_{\text{mid}}$, this procedure correctly inserts a_{r+1} in position mid + 1 or higher. Exercise 12 outlines a proof that BININSERT works correctly.

Question 3.2. If $A = \langle 2, 5, 7, 9, 13, 15, 19\rangle$, trace BINARYSEARCH and BININSERT with $S = D = 16$. Compare the two algorithms.

Before discussing the complexity of BININSERT, we use this procedure to develop an algorithm to totally order an array.

Problem. Given an array of n numbers $\langle a_1, a_2, \ldots, a_n\rangle$, place them in increasing order.

Algorithm BINARYSORT

STEP 1. Input n and an array $\langle a_1, \ldots, a_n\rangle$
STEP 2. For $m = 2$ to n do {insert mth item}
    STEP 3. Call BININSERT $((m-1), a_1, \ldots, a_m)$
STEP 4. Stop.

Question 3.3. Given the array $\langle 13, 23, 17, 19, 18, 28\rangle$ trace out the algorithm BINARYSORT.

Once we determine the complexity of the procedure BININSERT, the complexity of algorithm BINARYSORT will be easy to analyze, since BININSERT is used $n - 1$ times in BINARYSORT. The steps in BININSERT are either assignments or comparisons. We count the latter.

Theorem 3.1. BININSERT requires at most $2\lfloor \log(r) \rfloor + 4$ comparisons to insert the $(r+1)$st term into an already sorted list of r items.

Proof. The only steps containing comparisons are steps 2, 4, and 5, and each of these executes exactly one comparison. We proceed by induction. If $r = 1$, then after the first execution of step 4, either first $= 1$ and last $= 0$ or first $= 2$ and last $= 1$, depending on whether a_2 is less than a_1 or not. Step 2 is repeated to check this. Step 5 is required to rearrange the array. Thus four comparisons are used in total.

The induction hypothesis will be that for $t < r$ BININSERT requires at most $2\lfloor \log(t) \rfloor + 4$ comparisons to insert the $(t+1)$st item into any already sorted list with t items.

We suppose that A is an ordered array with r elements and $a_{r+1} = D$ is to be inserted. It takes two comparisons to execute through step 4 the first time. After the first execution of step 4, if $a_{r+1} < a_{\text{mid}}$ then last is assigned the value mid $- 1$. Thus we restrict our attention to $\langle a_1, \ldots, a_{\text{mid}-1}, a_{r+1} \rangle$. There are mid $- 1$ ordered values in this array. Now

$$\text{mid} - 1 = \left\lfloor \frac{1+r}{2} \right\rfloor - 1 = \left\lfloor \frac{r-1}{2} \right\rfloor < \frac{r}{2}.$$

After the first execution of step 4 if $a_{r+1} \geq a_{\text{mid}}$, then first is assigned the value mid $+ 1$. Thus we restrict our attention to $\langle a_{\text{mid}+1}, \ldots, a_r, a_{r+1} \rangle$. The number of elements in this smaller ordered array is $(r - (\text{mid} + 1) + 1) = r - \text{mid}$. Now if $r = 2j$,

$$\begin{aligned} r - \text{mid} &= r - \left\lfloor \frac{1+r}{2} \right\rfloor \\ &= 2j - \left\lfloor \frac{1+2j}{2} \right\rfloor = 2j - j = j = \frac{r}{2}. \end{aligned}$$

On the other hand, if $r = 2j + 1$,

$$\begin{aligned} r - \text{mid} &= 2j + 1 - \left\lfloor \frac{1+2j+1}{2} \right\rfloor \\ &= 2j + 1 - (j+1) = j < \frac{r}{2}. \end{aligned}$$

Thus in either case the smaller ordered array has no more than $r/2$ entries. By the inductive hypothesis we can insert D into the new array using at most $2\lfloor \log(r/2) \rfloor + 4$ comparisons. Thus the total number of comparisons required will be at most

$$2 + (2\lfloor \log(r/2) \rfloor + 4) = 2\lfloor \log(r) - 1 \rfloor + 6$$
$$= 2\lfloor \log(r) \rfloor + 4. \qquad \square$$

Question 3.4. For each of the examples in Question 3.1 count the number of comparisons and verify that these are each no more than $2\lfloor \log(7) \rfloor + 4$.

The formula $2\lfloor \log(n) \rfloor + 4$ in the preceding complexity analysis appears out of the blue. That BININSERT and BINARYSEARCH have similar complexity analyses is not surprising. To motivate the particular formula we obtain, we examine the proof of Theorem 3.1. If $C(n)$ denotes the maximum number of comparisons made when BININSERT inserts a number into a sorted array of length n, then

$$C(n) = C(\lfloor n/2 \rfloor) + 2,$$

since two comparisons are performed, and then the algorithm proceeds to work on an array containing at most $\lfloor n/2 \rfloor$ entries. This equation for $C(n)$ is like that of line (*) of Section 6.2 and can be used to derive the formula $C(n) = 2\lfloor \log(n) \rfloor + 4$. This derivation will be discussed in depth in Chapter 7.

Theorem 3.2. The number of comparisons required by BINARYSORT to order an array of n numbers is $O(n \log(n))$.

Proof. The only comparisons in BINARYSORT are performed within the BININSERT procedure. BININSERT is called $n - 1$ times. The number of comparisons in each call is at most $2\lfloor \log(n - 1) \rfloor + 4$. Thus the total number of comparisons will be no more than

$$(n - 1)(2\lfloor \log(n - 1) \rfloor + 4) < n\{2(\log(n)) + 4\}$$
$$\leq 6n \log(n). \qquad \square$$

BINARYSORT is thus considerably more efficient than SELECTSORT, a $O(n^2)$ algorithm.

Question 3.5. Count the number of comparisons made in Question 3.3 and compare this number with $(n - 1)(2\lfloor \log(n - 1) \rfloor + 4)$ for $n = 6$.

It is instructive to contrast the analyses presented in Theorems 3.1 and 3.2. We showed that BININSERT required at most $2\lfloor \log(n) \rfloor + 4$ comparisons to insert the $(n + 1)$st item into an already sorted array. In Exercises 9 and 10 you will see that this bound is sharp. We mean that there are problem instances where $2\lfloor \log(n) \rfloor + 4$ comparisons are, in fact, required. Thus there can be no upper bound for the number of comparisons that is always better than the one given in Theorem 3.1.

Notice that our analysis of BINARYSORT was not so sharp. In particular, we assumed that each call to BININSERT needed the full $2\lfloor \log(n-1) \rfloor + 4$ comparisons whereas we really need only $2\lfloor \log(1) \rfloor + 4$ comparisons for the first insertion, $2\lfloor \log(2) \rfloor + 4$ comparisons for the second, and, in general, $2\lfloor \log(i) \rfloor + 4$ comparisons for the ith insertion. Thus the total number of comparisons we perform is at most

$$\begin{aligned}
&(2\lfloor \log(1) \rfloor + 4) + \cdots + (2\lfloor \log(n-1) \rfloor + 4) \\
&\quad \le (2\log(1) + 4) + \cdots + (2\log(n-1) + 4) \\
&\quad = 4(n-1) + 2\{\log(1) + \cdots + \log(n-1)\} \\
&\quad = 4(n-1) + 2\log((n-1)!). \qquad \text{(B)}
\end{aligned}$$

Exercise 13 asks you to use equation (B) to provide a smaller upper bound than the one obtained so far for BINARYSORT. However, no analysis of the complexity of BINARYSORT can demonstrate that it is more efficient than $O(n\log(n))$. The goal of Section 5 is to show that SELECTSORT, BINARYSORT, and every sorting method that uses comparisons must perform at least a constant times $n\log(n)$ comparisons in the worst case. Before we get to that, we shall see in the next section that trees provide an illustrative model of these searching and sorting algorithms.

What effect does BINARYSORT have on the employment director's work load, as presented in the first paragraph of this section? If she first sorts the employment file using BINARYSORT, using $O(n\log(n))$ comparisons, and then during t time periods processes information using BINARYSEARCH with $O(tn\log(n))$ comparisons, then the total number of comparisons is

$$O(n\log(n)) + O(tn\log(n)) = O(tn\log(n)). \qquad \text{(C)}$$

Comparing the results of (C) with those of (A), we see that the latter process is at least as efficient as the former and for some values of t is more efficient.

EXERCISES FOR SECTION 3

1. Trace BININSERT on the following data:
   (*a*) $A = \langle 1,2,3 \rangle$, $D = 2.5$.
   (*b*) $A = \langle 1,2,3 \rangle$, $D = 0$.
   (*c*) $A = \langle 1,2,4 \rangle$, $D = 2$.
   (*d*) $A = \langle 2,3,5,7,11,13,17,19 \rangle$, $D = 12$.
   (*e*) $A = \langle 2,4,6,8,10 \rangle$, $D = 5$.

2. Count the number of comparisons made in each part of Exercise 1. Compare this number with $2\lfloor\log(r)\rfloor + 4$ for the appropriate values of r.

3. Here is another algorithm to search for D in an array A:

STEP 0. Input $A = \langle a_1, a_2, \ldots, a_r\rangle$, set $a_{r+1} := D$
STEP 1. Set first := 1, last := r
STEP 2. While first < last do
  Begin
  STEP 3. Set mid := $\lfloor(\text{first} + \text{last})/2\rfloor$
  STEP 4. If $a_{r+1} \leq a_{\text{mid}}$, then set last := mid
    Else set first := mid + 1
  End
STEP 5. If $a_{\text{first}} = a_{r+1}$, then output "found D at location first" and stop.
  Else output "D is not in A" and stop.

Run this algorithm and BINARYSEARCH on $A = \langle 1, 2, 3, 4, 5, 6, 7\rangle$ with $D = 1.5$, 2, and 2.5.

4. Compare the algorithm BINARYSEARCH and that given in the preceding exercise. Determine which one performs fewer comparisons.

5. Suppose that the employment director uses SEQSEARCH on an unsorted file of n records to register n students' payroll data during t time periods, making $f(n, t)$ comparisons as described in the first paragraph of this section. Let $g(n, t)$ denote the number of comparisons made if the file is first sorted using SELECTSORT and then the same recordings are made using BINARYSEARCH [see line (A) in text]. Find the smallest value of t such that $g(n, t) < f(n, t)$.

6. Let $g(n, t)$ be as defined in the preceding problem and let $h(n, t)$ be the number of comparisons made if the file is first sorted using BINARYSORT and then the memos are recorded for n students on n records in t time periods using BINARYSEARCH. Compare $g(n, t)$ and $h(n, t)$.

7. Run BINARYSORT on each of the following: **(*a*)** $A = \langle 1, 2, 3\rangle$, **(*b*)** $A = \langle 2, 1, 3\rangle$, **(*c*)** $A = \langle 3, 1, 2\rangle$, and **(*d*)** $A = \langle 3, 2, 1\rangle$.

8. Count the number of comparisons made in the preceding exercise and compare this number with $(n-1)(2\lfloor\log(n-1)\rfloor + 4)$ for the appropriate value of n.

9. For an n of your choice find an example of an array A of n numbers and a number D on which BININSERT performs exactly $2\lfloor\log(n)\rfloor + 4$ comparisons.

10. Let $n = 2^k$ with k an arbitrary positive number. Describe an array A of n numbers and another number D on which BININSERT performs exactly $2\lfloor\log(n)\rfloor + 4 = 2k + 4$ comparisons.

11. Explain why BINARYSORT will always perform fewer than $6n\log(n)$ comparisons when sorting an array of length n. Will BINARYSORT perform fewer than $(n-1)(2\lfloor\log(n-1)\rfloor+4)$ comparisons on any or all arrays of length n?

12. Prove that BININSERT works correctly by proving each of the following statements.

(*a*) While first $\leq$ last, a_{r+1} should be stored in one of the entries a_{first}, $a_{\text{first}+1}, \ldots,$ or $a_{\text{last}+1}$. In particular, check that this is so when $a_{r+1} = a_{\text{mid}}$.

(*b*) Eventually either last equals first or first + 1.

(*c*) If last equals first or first + 1, then BININSERT places a_{r+1} in the correct position.

13. Stirling's formula, discussed in Chapter 3, implies that

$$n! = O\left(\sqrt{n}\left(\frac{n}{e}\right)^n\right).$$

Use this result together with equation (B) to derive an upper bound on the number of comparisons made in BINARYSORT. How does this upper bound compare with the upper bound derived in the text?

14. Suppose that A is an array of n elements that is already sorted (but we may not know that in advance). Which algorithm works faster on A, SELECTSORT or BINARYSORT? Explain.

6:4 SEARCH TREES

Suppose that we search an ordered array $A = \langle a_1, a_2, \ldots, a_7\rangle$ for a particular object S. BINARYSEARCH would have us first compare S with a_4. There are three possible outcomes of such a comparison. If $S = a_4$, we're done. If $S < a_4$ and S is in A, then it must be one of a_1, a_2, or a_3. Finally, if $S > a_4$ and S is in A, then it must be one of a_5, a_6, or a_7. In this section we show how to use a tree structure to illuminate these logical alternatives.

Recall from Chapter 5 that we can think of a graph as a set of points in the plane and a set of line segments or arcs joining pairs of these points. A graph that is both connected and acyclic is called a tree.

Here is how BINARYSEARCH as applied to the seven-element set $A = \langle a_1, a_2, \ldots, a_7\rangle$ can be modeled by a path within a tree of seven vertices. Begin with a single vertex labeled a_4. Think of two edges coming out of a_4, one labeled by "$<$" and the other labeled by "$>$." (See Figure 6.1.) The $<$ edge joins a_4 with a_2 and the $>$ edge joins a_4 with a_6. In terms of BINARYSEARCH if S equals a_4, we stay at the vertex labeled a_4 and we are done. If $S < a_4$ we proceed along the

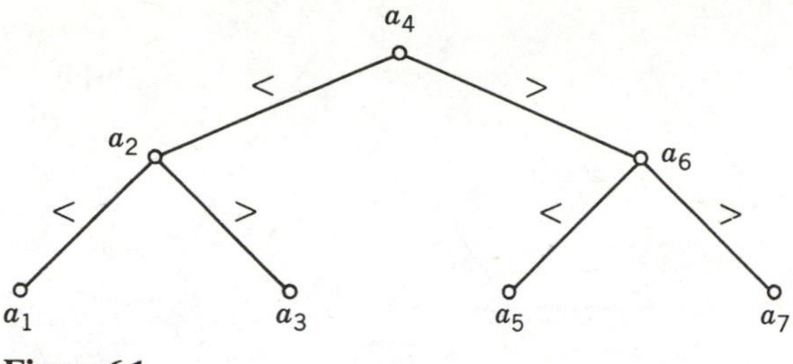

Figure 6.1

edge labeled $<$ to the vertex labeled a_2. If $S > a_4$ we proceed along the edge labeled $>$ to the vertex labeled a_6. The vertices a_2 and a_6 each have two additional edges coming out of them labeled with $<$ and $>$. The new edge from a_2 labeled $<$ terminates at a_1 while the new edge from a_2 labeled $>$ terminates at a_3. Similarly, the new edges from a_6 terminate at a_5 and a_7. For example, if $S = a_5$, BINARYSEARCH would examine a_4, followed by a_6 and then a_5. If S is not present in A, we should also perform comparisons, for example, with a_4, a_6, and a_5 and then deduce that S was not in A. In all cases these comparisons correspond to a path within the so-called **search tree** shown in Figure 6.1.

Question 4.1. Draw a search tree to illustrate a binary search of an array of 15 elements.

Recall that within a graph the degree of a vertex is the number of edges incident with that vertex. In a tree or forest a vertex of degree 1 is called a **leaf**.

Definition. A tree is called **binary** if

(1) it possesses a distinguished vertex called the **root** whose degree is either 2 or 0, and

(2) every vertex of the tree other than the root has degree either 3 or 1.

Note that the tree in Figure 6.1 is binary. It is customary to draw a binary tree "upside down" with the root at the top, as in Figure 6.1. From the root (if its degree is not 0) there is a left edge down and a right edge down. Similarly, every other vertex that is not a leaf has a left and a right edge down. One of the nice properties that binary trees with three or more vertices have is that if the root and its incident edges of a binary tree are erased, then two smaller binary trees are formed. These are called the **left** and **right subtrees** of the original tree.

Example 4.1. Figure 6.2 exhibits a binary tree with five vertices.

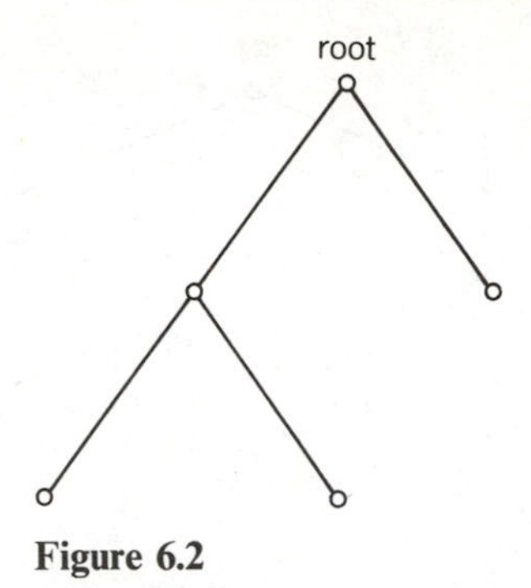

Figure 6.2

Question 4.2. (a) Draw all binary trees with fewer than eight vertices; (b) draw all binary trees with two, three or four leaves.

Question 4.3. Draw the left and right subtrees that are formed when the root and its two incident edges are deleted from (a) the tree in Figure 6.1; (b) the tree in Figure 6.2; and (c) every tree with two or three leaves (see Question 4.2).

Definition. In any tree with a designated root the **depth** or **level of a vertex** x is the number of edges in the unique path from x to the root. The **depth of the tree** is the maximum depth of any vertex in the tree. Alternatively, it is the length of a longest path from the root.

In the tree shown in Figure 6.1, a_4, the root, is at level 0; a_2 and a_6 are at level 1; and a_1, a_3, a_5, and a_7 are all at level 2. Thus the tree has depth 2.

Question 4.4. If T is a binary tree of depth $d > 0$ and T' is the left subtree of T, what can you say about the depth of T'?

Theorem 4.1. A binary tree has at most 2^d vertices at depth d.

Proof. The proof is by induction on d. The root is the only vertex at level 0. By definition there are either two or zero vertices adjacent to the root, and these vertices are at level 1. We assume that there are no more than 2^k vertices at level k in a binary tree. Consider level $k + 1$. Every vertex at this level must be adjacent to exactly one vertex at level k by the definition of tree (see Exercise 1). Since each vertex at level k has degree 1 or 3, it is adjacent to either zero or two vertices at level $k + 1$. Thus the number of vertices at level $k + 1$ can be no more than twice the number of vertices at level k. If N_k denotes the number of vertices at level k, we have

$$N_{k+1} \leq 2N_k \leq 2(2^k) = 2^{k+1}. \qquad \square$$

Corollary 4.2. A binary tree of depth d contains at most $2^{d+1} - 1$ vertices.

Proof. A binary tree of depth d has vertices at levels $0, 1, \ldots, d$. By the preceding theorem there are at most 2^k vertices at level k. Thus the total number of vertices in the tree is at most

$$
\begin{aligned}
1 + 2 + 4 + \cdots + 2^k + \cdots + 2^d &= \frac{1 - 2^{d+1}}{1 - 2} \qquad \text{by Question 2.3.3} \\
&= 2^{d+1} - 1. \qquad \square
\end{aligned}
$$

A binary tree of depth d with $2^{d+1} - 1$ vertices is called a **full binary tree**.

Question 4.5. Determine the depth and the number of vertices in the smallest full binary tree that has n or more leaves.

Now we specify the connection between binary trees and our problem of searching a sorted array. Suppose that the array A contains $n = 2^k - 1$ elements in order. Then the corresponding binary tree will be a full binary tree of depth $k - 1$ with n vertices. In BINARYSEARCH the first element of the array A with which we compare S is the element a_{mid}. Note that mid $= 2^{k-1}$. The element a_{mid} will label the root of the binary tree that we are about to search. If we find that $S = a_{\text{mid}}$ (in step 5), then the algorithm stops. We can similarly stop searching the tree. If $S < a_{\text{mid}}$, we set

$$\text{last} := \text{mid} - 1 = 2^{k-1} - 1$$

and

$$\text{mid} := \left\lfloor \frac{(\text{first} + \text{last})}{2} \right\rfloor = 2^{k-2}.$$

This corresponds with traversing the edge from the root of the binary tree down to the root of the left subtree; this vertex is labeled with the new $a_{\text{mid}} = a_{2^{k-2}}$. Similarly, if $S > a_{\text{mid}}$, we traverse an edge to the right subtree. If we have not found S, we repeat this process. Each time we examine a new a_{mid} it will be the root of a subtree of the original binary tree. Each such subtree will contain $2^j - 1$ vertices for some j. In the end either we find S and terminate our path down from the root of the tree or we reach a leaf without finding S and stop. The number of comparisons made in BINARYSEARCH is the same as the number of vertices visited on the corresponding path in the tree.

More generally, if A contains n elements where

$$2^{k-1} \leq n < 2^k,$$

then

$$k - 1 \leq \log(n) < k,$$

and

$$k - 1 = \lfloor \log(n) \rfloor.$$

Set

$$n' = 2^k - 1.$$

We model the search of A by a search of the full binary tree of depth $k - 1$ containing n' vertices, labeled as before. If $n' > n$, some of the labels of vertices, namely $a_{n+1}, \ldots, a_{n'}$, do not correspond with array elements.

Question 4.6. Compute n' if $n = 15$, 26, and 31. Show that, in general, $n' \geq n$.

Question 4.7. Draw and label the tree that corresponds with a binary search of a 23-element ordered array.

One advantage of the binary tree model of BINARYSEARCH is that it supports a simple complexity analysis. Suppose that we are searching an ordered array of n elements. If $2^{d-1} \leq n < 2^d$, then the elements of the array correspond with some of the vertices of a full binary tree of depth $d - 1$. Comparing S with array elements corresponds with visiting vertices in the tree. Since each time we traverse an edge to the root of a new subtree, the depth of the visited vertex increases, we shall in the worst case visit vertices at depth $0, 1, \ldots,$ and $(d - 1)$. Correspondingly, we need to compare S with no more than d elements of the array. Since a search tree with n vertices has depth $\lfloor \log(n) \rfloor$, we can determine whether S is present in an ordered array of n elements by examining no more than $\lfloor \log(n) \rfloor$ elements of the array. If it takes just a constant number of steps for each such examination, then it is immediate that BINARYSEARCH is $O(\log(n))$. In contrast in Exercise 15 we explore how SEQSEARCH can be modeled by searching a graph that is just a path.

It is possible to think of a binary tree model of the first four steps of BININSERT in much the same way. Recall that these steps determine the location of the next element to be inserted. Continuing the model is awkward because the insertion of a single element can cause a radical change in the binary tree. Exercise 13 illustrates this.

Although it is difficult to use trees to model BINARYSORT, there is an elegant sorting method called TREESORT that is based on binary trees. Suppose that we want to sort $A = \langle a_1, \ldots, a_n \rangle$, where the entries of A are distinct. (See Exercise 20 for the case of repeated elements.)

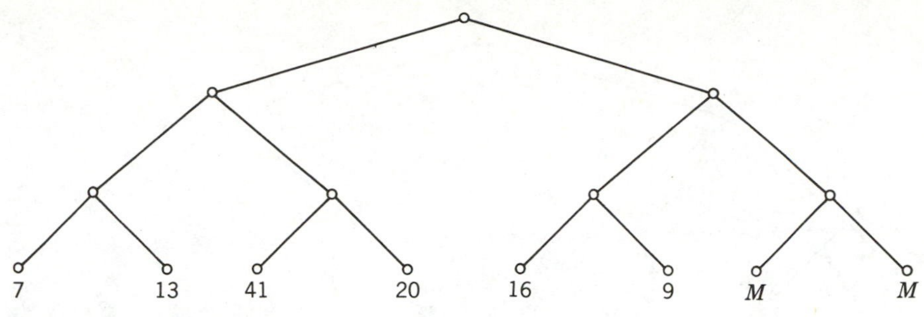

Figure 6.3

Algorithm TREESORT

STEP 1. Set $k = \lceil \log(n) \rceil$ and construct the full binary tree of depth k. {As you saw in Question 4.5, this tree has at least n leaves.} Assign each element in the array to a leaf. Pick a number M that is greater than any element in the array and assign M to every blank leaf.

Example 4.2. Given $A = \langle 7, 13, 41, 20, 16, 9 \rangle$, $k = \lceil \log(6) \rceil = 3$. Figure 6.3 exhibits the full binary tree of depth 3 with its leaves labeled.

STEP 2. For $j = k - 1$ down to 0 do
Assign to each vertex at level j the minimum of the two values assigned to its neighbors at level $j + 1$.

Example 4.2 (continued). We show in Figure 6.4 the full binary tree after step 2. (At this stage the minimum value in the array is assigned to the root of the binary tree.)

STEP 3. Set $b_1 :=$ value assigned to the root
STEP 4. For $i = 2$ to n do
Begin
STEP 5. Erase every occurrence of b_{i-1} from nodes of the tree.
STEP 6. Assign M to the leaf that originally was labeled with b_{i-1}
STEP 7. For $j = k - 1$ down to 0 do
Assign to the vertex at level j that used to be labeled b_{i-1} the minimum of the two values assigned to its neighbors at level $j + 1$
STEP 8. Set $b_i :=$ value assigned to the root
End
STEP 9. Stop.

Example 4.2 (continued again). We exhibit in Figure 6.5 the labeled tree after the first execution of step 7; $b_1 = 7$.

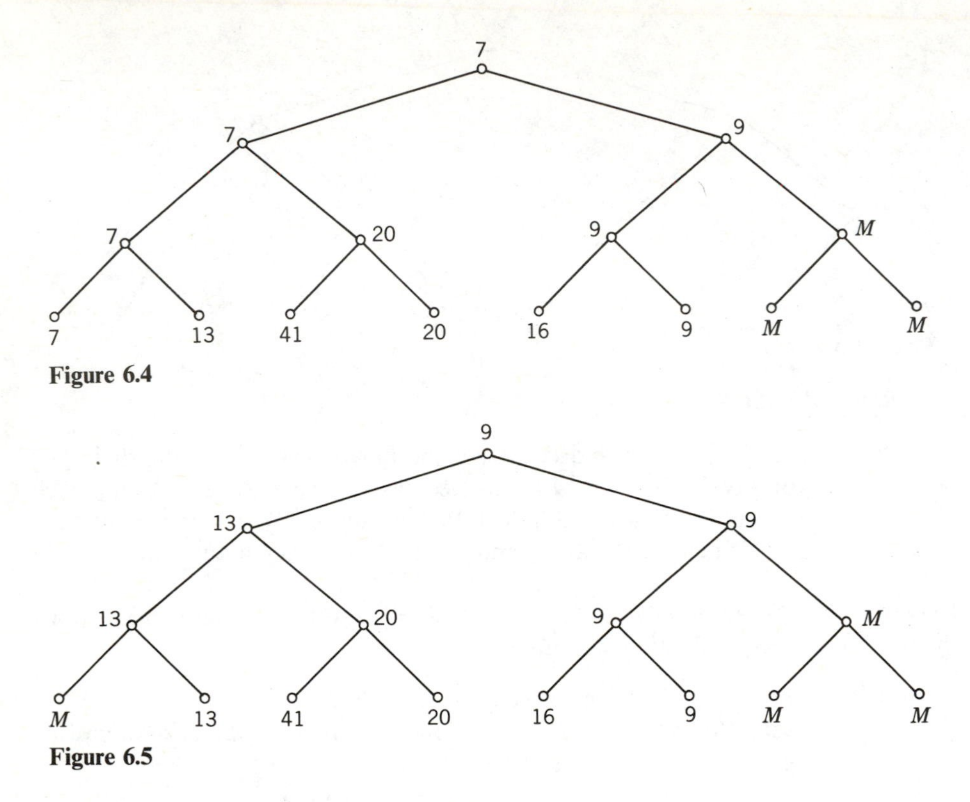

Figure 6.4

Figure 6.5

After the execution of step 7 the ith smallest value in the array is assigned to the root and thus in step 8 is assigned to b_i.

Question 4.8. Complete this execution of TREESORT.

Exercises 17 to 19 ask you to show that TREESORT is a $O(n \log(n))$ algorithm.
Binary trees are useful models for many topics in combinatorics and computer science, for example, see Exercises 5 and 6. We shall use binary trees again in the following sections and in Chapter 8.

EXERCISES FOR SECTION 4

1. In the proof of Theorem 4.1 we claimed that every vertex at level $k + 1$ is adjacent to exactly one vertex at level k. Why is this so?
2. If x is a vertex in a rooted tree, let $l(x)$ denote the level of x. Show that if v is adjacent to v', then $l(v) = l(v') + 1$ or $l(v) = l(v') - 1$. Give a proof or a counterexample to the converse of this statement.

3. If T is a binary tree of depth d, what is the smallest number of vertices that T might have at level k (for $k = 1, \ldots, d$)?

4. What is the smallest number of vertices that a binary tree of depth d might have?

5. For what integers n is there no binary tree with exactly n vertices? For what integers n is there no binary tree with exactly n leaves? Prove your answers by induction.

6. We can also represent the subsets of a set $\langle a_1, a_2, \ldots, a_n \rangle$ with a full binary tree. Suppose that we label the root of the tree with $\varnothing$, the empty set. Then the left subtree will correspond with subsets not containing a_1 and the right subtree subsets containing a_1. Similarly, the left subtree within the left subtree will correspond with subsets containing neither a_1 nor a_2, whereas its right subtree will correspond with subsets containing a_2 but not a_1. Each node can be labeled with a subset, representing the choices of elements made along the path from the root to that node. Using this idea, construct the binary tree associated with all subsets of a 3-set and of a 4-set.

7. In a full binary tree there are 2^k vertices at level k. Find a correspondence between the subsets of a k-set and the vertices at the kth level of a full binary tree.

8. Prove Theorem 4.1 by "erasing the root."

9. Prove Corollary 4.2 by "erasing the root."

10. Suppose that $n = 2^k - 1$ and consider a full binary tree with vertices labeled with the elements of an array $A = \langle a_1, \ldots, a_n \rangle$ corresponding with BINARYSEARCH. Which elements label vertices at depth 1? At depth 2? What are the labels of the leaves?

11. Repeat Exercise 10 in the case of arbitrary n. How can you tell from i and j if a_i and a_j label vertices at the same level?

12. Trace out the path corresponding to BINARYSEARCH when this algorithm is applied to the following arrays, modeled by binary trees, and elements S:
(*a*) A as in Figure 6.1, $S = a_1$.
(*b*) A as in Figure 6.1, $a_3 < S < a_4$.
(*c*) A as in Question 4.7, $S = a_{16}$.
(*d*) A as in Question 4.7, $S = a_{23}$.
(*e*) A as in Question 4.7, $a_{16} < S < a_{17}$.

13. Suppose that you want to insert $D = 31$ into the sorted array $A = \langle 3, 6, 9, 12, 15, 18, 21, 24, 27, 30, 33, 36 \rangle$. Construct a binary tree to model the sorted array A both before and after the insertion of D. In how many locations do these two trees differ?

14. If T is a binary tree with q leaves, how many vertices of degree 3 does T have?

15. Explain how the algorithm SEQSEARCH can be modeled by traversing a graph that is a path.
16. Run TREESORT on the following arrays, showing the binary tree and its node values at the end of each execution of steps 3 and 8: **(*a*)** $A = \langle 1, 2, 3 \rangle$; **(*b*)** $A = \langle 2, 1, 3 \rangle$; and **(*c*)** $A = \langle 1, 5, 2, 6, 3, 4 \rangle$.
17. How many comparisons are performed in step 2 of TREESORT?
18. In TREESORT how may b_{i-1}'s get erased the ith time through the loop? How many comparisons will you need to relabel the tree?
19. Show that TREESORT is a $O(n \log(n))$ algorithm.
20. Rewrite TREESORT so that it sorts arrays with repeated entries.

6:5 LOWER BOUNDS ON SORTING

A new employment director wants to improve the efficiency of the student employment office. After conversations with the previous director, the new director is convinced that she should sort her payroll drawer at the beginning of the year. She learns that the previous director originally used the quadratic algorithm SELECTSORT and then switched to the $n \log(n)$ algorithm BINARYSORT, but the new director does not want to keep switching algorithms each year when the local algorithm experts come up with new and faster (and possibly more complex) sorting algorithms. She decides she would like once and for all to find a fastest sorting algorithm and guesses that there must be a linear algorithm, one that runs in $O(n)$-time on an array of n elements. So she calls her friends taking the computer science course on algorithms and asks for such a linear-time sorting algorithm.

The algorithm students report that they haven't learned about such an algorithm yet, but maybe they will later in the semester. In the meantime they suggest that the director might like to try Treesort, Shakersort, or Mergesort. These are all $n \log(n)$ algorithms. The director rejects these offers. She is trying to run an efficient office and is not interested in becoming an algorithmics specialist herself. However, she decides that she will, on her own, search for a linear-time sorting algorithm or else prove that there is no such algorithm.

The new director has studied some discrete mathematics and begins with small examples of arrays. If she has an array like $\langle a_1, a_2, a_3 \rangle$, then in how many different orders might the array appear? For example, the array might be "in order" so that $a_1 \leq a_2 \leq a_3$ or "out of order" with $a_1 \leq a_3 \leq a_2$ or $a_2 \leq a_3 \leq a_1$, and so on. How many comparisons must be made to sort these elements into increasing order? The element a_1 can be compared with a_2 and with a_3, and the elements a_2 and a_3 can be compared with each other. Are all these comparisons necessary?

Question 5.1. For $n = 3$, 4, and 5, given an array of n distinct items $A = \langle a_1, a_2, \ldots, a_n \rangle$ decide in how many different possible orderings these elements might appear. Then determine the total number of pairwise comparisons that can be made among the members of the set.

Proposition 5.1. There are at most $n!$ different possible orderings of an array of n elements. In such an array there are $n(n-1)/2$ distinct pairs of items that might be compared.

Proof. If the array entries are distinct, there are the same number of orderings as permutations of an n-set. If the entries are not distinct, the number of different orderings is less than $n!$, since some permutations produce the same ordering. (See also Exercises 2 and 3.) In either case the number of possible comparisons between pairs is the same as the number of 2-subsets of an n-set or (equivalently) the number of edges in an n-clique. □

Proposition 5.1 tells us (and the employment director) that if we make all possible comparisons, we have a $O(n^2)$ algorithm, an algorithm as slow as SELECTSORT. We know we can use $cn \log(n)$ comparisons for some constant c, but can we use even fewer than this?

Any sorting algorithm that uses comparisons contains a sequence of comparisons, say $C_1, C_2, \ldots, C_k$. Regardless of the particular sorting algorithm used, what can we say about the value of k, the number of comparisons, in the worst case? We model the problem with a **binary search tree**.

Example 5.1. Given three distinct objects, say a_1, a_2, and a_3, we make a binary search tree labeled with possible comparisons and possible outcomes. Suppose that we begin by comparing a_1 and a_2. There are two possible outcomes, either $a_1 \leq a_2$ or $a_1 > a_2$. We label the root of the binary tree with $(a_1 : a_2)$ for this comparison; the edge from the root to the left subtree is labeled "$\leq$" to denote the first possible outcome and the edge to the right subtree is labeled "$>$" for the second outcome. See Figure 6.6. Suppose that we next compare a_1 with a_3 and label the two nodes on level 1 with $(a_1 : a_3)$ and their left edges with "$\leq$" and their right edges with "$>$." Notice that in two of the four possibilities we know the correct ordering and have written that in as a leaf of the tree, but in the remaining two cases we need to make the additional comparison of a_2 with a_3.

The results of Example 5.1 show that at least with this order of comparisons three comparisons are needed in the worst case.

Question 5.2. As in Example 5.1 construct and label a binary search tree when comparisons are made in the following order:
(a) a_1 with a_2, a_2 with a_3, a_1 with a_3.
(b) a_1 with a_3, a_1 with a_2, a_2 with a_3.

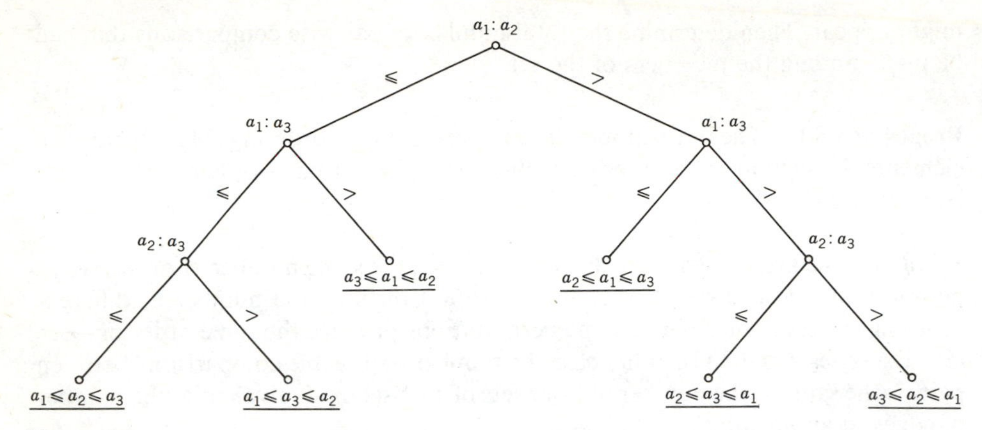

Figure 6.6

The results of Example 5.1 and Question 5.2 convince the employment director that three comparisons are needed to sort a three-element array. This finding is inconclusive from the complexity point of view, since all possible comparisons are needed in the worst case. On the other hand, three comparisons for a three-element array might indicate the possibility of a linear-time algorithm.

Example 5.2. Suppose that we sort $A = \langle a_1, a_2, a_3, a_4 \rangle$. Here is part of the search tree (Figure 6.7). Notice that once it is known that $a_3 < a_1 \leq a_2$, then it is impos-

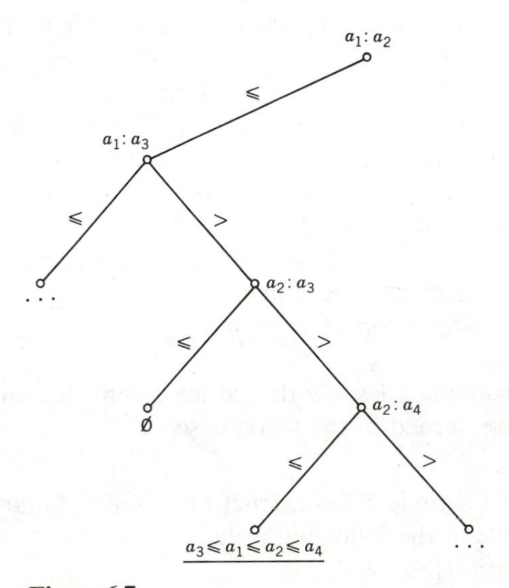

Figure 6.7

sible to have $a_2 \leq a_3$. The path corresponding to this impossibility terminates with a leaf labeled with the empty set.

The director also realizes that she can analyze any sorting algorithm using the binary tree structure. If any algorithm makes comparisons $C_1, C_2, \ldots, C_k$ in that order, then we represent this algorithm as a binary search tree of depth k. The root is labeled C_1, the two nodes at depth 1 are labeled C_2, and, in general, the nonleaf nodes at depth i are labeled C_{i+1} for $i = 0, \ldots, k-1$. The leaves are each labeled with either the empty set or with one of the $n!$ possible orderings of the array. The label on a leaf at the end of a path from the root is the ordering specified by the series of $\leq$ and $>$s on that path; as in Example 5.2 there may be no such ordering. If k comparisons are made in the worst case, then the binary tree representing these comparisons has depth k. By Theorem 4.1 there are at most 2^k leaves in such a tree. Thus we must have

$$2^k \geq n!, \qquad \text{or} \qquad k \geq \log(n!). \qquad \square$$

We have proved the following theorem.

Theorem 5.2. It requires at least $\log(n!)$ comparisons to sort an n-element set in the worst case.

Another approach to the proof of this theorem, using a game and binary numbers, is given in Exercises 16 and 17.

The director realizes now that an algorithm that performs sorting by comparisons must in the worst case do at least $\log(n!)$ comparisons. Her next goal is to determine the size of $\log(n!)$ or at least a lower bound on $\log(n!)$. The next arithmetic lemma will lead to a lower bound on $\log(n!)$.

Lemma 5.3. If $n \geq i \geq 1$, then $i(n + 1 - i) \geq n$.

Proof. Since $n \geq i$ and $(i - 1) \geq 0$, we have

$$(i - 1)n \geq (i - 1)i.$$

Adding n to both sides and subtracting $(i - 1)i$ yields

$$in - i(i - 1) \geq n,$$

which simplifies to

$$i(n - i + 1) \geq n. \qquad \square$$

We now use the lemma to estimate $\log(n!)$.

$$\begin{aligned}\log(n!) &= \log(n(n-1)\cdot\cdots\cdot 3\cdot 2\cdot 1)\\ &= \log\{[(n)1][(n-1)2][(n-2)3]\cdots[(n-i+1)i]\cdots\}\end{aligned}$$

by regrouping factors

$$= \log[(n)1] + \log[(n-1)2] + \cdots + \log[(n-i+1)i] + \cdots \qquad \text{(A)}$$

by the additive property of logs

$$\geq \log(n) + \log(n) + \cdots + \log(n) + \cdots \qquad \text{(B)}$$

by Lemma 5.3.

Note that if n is even, the sum in (A) ends with $\log[((n/2)+1)(n/2)]$. Hence there are $n/2$ terms and so (B) equals $(n/2)\log(n)$. Thus

$$\log(n!) \geq \left(\frac{n}{2}\right)\log(n).$$

(For the case of n odd, see Exercise 9.)

Corollary 5.4. If $S(n)$ is defined to be the number of comparisons needed in the worst case to sort a file with n items using comparisons, then

$$S(n) \geq \log(n!) \qquad \text{and} \qquad n\log(n) = O(S(n)).$$

Corollary 5.4 is often referred to as the information theoretic lower bound on sorting. What it means is that there cannot be a sorting method based on pairwise comparisons whose complexity is of order less than $n\log(n)$. In other words, every such sorting algorithm will be big oh of some function in the functional hierarchy (as developed in Chapter 2) that is at least as big as $n\log(n)$. So far, we have seen three sorting procedures, one that is $O(n^2)$ and two that are $O(n\log(n))$.

Example 5.3. Suppose that we want to sort $A = \langle a_1, a_2, a_3, a_4\rangle$. As we have already seen, three elements in an array can be sorted with three or fewer comparisons. Suppose that we find that $a_1 \leq a_2 \leq a_3$ and we want to find the position of a_4. We could compare a_4 with each of the first three to find its position or we could use the idea of BINARYSORT. Then we would compare a_4 with a_2. If $a_4 \leq a_2$, we compare a_4 with a_1. If $a_4 > a_2$, then we compare a_4 with a_3 and so determine the final order. In total, five comparisons have been made in the worst case. Since $\lceil\log(4!)\rceil = 5$, Corollary 5.4 tells us that a file with four items cannot be sorted using fewer than five comparisons.

Question 5.3. Draw the binary search tree for a four-element array when the comparisons are made as suggested in Example 5.3. Count the number of comparisons in the worst case. (Note that to parallel Example 5.3 we allow different nodes on the same level of the binary tree to receive different comparison labels.)

Notice that we talk repeatedly about "sorting by comparisons." How else might a sorting algorithm proceed? In fact, in some special cases comparisons are not necessary; these ideas are explored in Section 8 and in Supplementary Exercises 29 to 32. For example, the algorithm BUCKETSORT is a linear-time sorting algorithm but is of limited applicability because of its excessive storage requirements.

The director is now convinced that there is no algorithm that uses comparisons and is essentially better than BINARYSORT, and so she decides to stick with this $n\log(n)$ algorithm. However, due to increased tuition and decreased financial aid the number of student employees doubles and then grows to triple the original number of students.

Question 5.4. Suppose that the number of student employees increases from 600 to 1200. If the employment director first sorts her file with $n\log(n)$ comparisons and then performs $n\log(n)$ more comparisons for each of the four payroll periods in one semester, how many comparisons are performed for 600 and for 1200 students? And for 1800 students?

The employment director finds she can't keep up with the explosion of work in her office as the number of employees doubles and then triples. She petitions the president's office for an assistant. The petition is granted, and the president even offers, if need be, to provide a second assistant. The next sections will present ways in which the director can use her assistants effectively.

EXERCISES FOR SECTION 5

1. Given an array of n distinct elements, we have said that there are $n!$ possible initial orderings of the n elements. For $n = 3$ and $n = 4$, give examples of $n!$ arrays, one corresponding to each of the possible orderings.
2. Suppose that in the array $A = \langle a_1, a_2, a_3, a_4 \rangle$ it is known that $a_3 = a_4$. Then how many different possible (unsorted) orderings are there?
3. Suppose that it is known that in the array $A = \langle a_1, a_2, \ldots, a_n \rangle$ two elements are the same but otherwise the elements are distinct. How many different (unsorted) orderings are there?
4. Suppose that we have an array $\langle a_1, a_2, a_3 \rangle$ and ask the questions

"Is $a_3 \le a_2$? and Is $a_2 \le a_1$?"

Give an example of values for a_1, a_2, and a_3 for which the answers to these two questions do determine the order correctly. Then give an example of values for which the answers are not sufficient to determine the order correctly.

5. Suppose that we have an array $\langle a_1, a_2, a_3, a_4 \rangle$. Find four questions that sometimes do and sometimes do not determine the correct order.

6. If an algorithm makes k comparisons, explain why the corresponding binary tree has depth k.

7. Given an array containing n elements, is it ever possible to ask $n - 2$ or fewer questions of the form "Is $a_i \le a_j$?" and from the answers to learn the correct order? Explain.

8. Suppose you have an n-element array and ask the $n - 1$ questions "Is $a_1 \le a_2$?," "Is $a_2 \le a_3$?," ..., "Is $a_{n-1} \le a_n$?." If it is possible to determine the order of the array from the answers to these questions, what can you say about the entries in the array?

9. Find a lower bound for $\log(n!)$ in the case that n is odd, similar to the one found in the text when n is even.

10. Stirling's formula from Chapter 3 tells us that

$$\sqrt{n}\left(\frac{n}{e}\right)^n = O(n!).$$

Use this to obtain a lower bound on $\log(n!)$. Does this lead to a better (larger) lower bound than that derived in the text?

11. Draw a binary tree for sorting $A = \langle a_1, a_2, a_3, a_4 \rangle$ that begins by comparing a_1 with a_2, then a_2 with a_3, and a_3 with a_4, and that has depth as small as possible. (As in Question 5.3 you may make different comparisons at different nodes on the same level.)

12. Suppose that a binary search tree for $A = \langle a_1, \ldots, a_5 \rangle$ begins with the comparisons a_1 with a_2, then a_1 with a_3, then a_1 with a_4 and a_1 with a_5. How many leaves are there at depth 4? Give an example of a leaf at depth 5. Give an example of a leaf at depth 6 or more.

13. Suppose that $tn\log(n)$ comparisons are made on records for n student employees during t time periods. If n is doubled to $2n$, does the number of comparisons double? Does it triple? Suppose that the number of students triples from n to $3n$? How many times the original number of comparisons is the new number of comparisons?

14. If $f(n) = n\log(n)$, find k such that for n sufficiently large

$$kn\log(n) \le 2n\log(2n) \le (k+1)n\log(n).$$

15. Corollary 5.4 says that a file with five items in it cannot be sorted with fewer than $\lceil \log(5!) \rceil = 7$ comparisons. If one sorts a file with five records by first sorting four records and then inserting the fifth record, it takes a total of eight comparisons. First five comparisons are needed to correctly sort four records and then three more to insert the fifth item into the ordered list of four items. Decide whether the minimum number of comparisons that will necessarily sort a file with five items is seven or eight.

16. Here is a sorting game, played by two players on an array $A = \langle a_1, \ldots, a_n \rangle$. Player 1 picks two elements a_i and a_j and asks player 2 to compare these values and to say which is smaller. Player 2 then assigns values to a_i and a_j in any way and answers, for example, that $a_i \leq a_j$. Player 1 next picks another pair to compare, and player 2 again assigns values and reports the answer. (Once player 2 has picked and used a value for some a_i, the value cannot be changed, but values do not need to be selected until player 1 brings them up.) Player 1's goal is to determine the order as quickly as possible; player 2's goal is to keep the order obscure as long as possible. Play the sorting game for both $n = 4$ and $n = 5$. How many comparisons can player 2 force player 1 to make?

17. In the sorting game if player 1 asks for the results of k comparisons, then player 2 must give k different pieces of information, in this case either "$\leq$" or "$>$." There are 2^k different patterns of answers that player 1 may receive from player 2. Use this to explain why to sort a set of n objects with k comparisons, it must be the case that $2^k \geq n!$.

6:6 RECURSION

The director of the student employment office hopes that with an assistant she can delegate more of the routine work. For example, at the beginning of the year she might sort half of the student records, give the other half to an assistant and then merge the two sorted files into one.

To train an assistant with an easy comparison task, the director begins with the job of finding the minimum entry of $A = \langle a_1, a_2, \ldots, a_n \rangle$, an array of real numbers. She asks the assistant to find the minimum entry of $A' = \langle a_1, a_2, \ldots, a_{n-1} \rangle$ because she knows she can compare the assistant's minimum with a_n to find the overall minimum of A. Now the assistant catches on quickly and realizes that he can use another assistant to find the minimum of $A'' = \langle a_1, a_2, \ldots, a_{n-2} \rangle$ and then compare that minimum with a_{n-1} to find the minimum of A'. If each of the assistants has an assistant (or a friend to help with the work), each assistant can pawn off the work of finding the minimum of a smaller array. Eventually, the array under consideration will have just one entry, which will be the minimum value of that array. This minimum value will be passed up and probably changed until the director receives the correct minimum of A' and then with one comparison finds the minimum of A.

This fanciful idea is an example of what is known as a **recursive algorithm** or **recursive procedure**. The word *recur* means to show up again, and that's exactly what happens in a recursive procedure: The procedure shows up again, or is used, within itself.

We now formalize a recursive procedure that carries out the idea described above. The procedure MIN will find the minimum entry in an array A. The input to MIN is A and n, the length of A, and the output from A is k, the value of the index of the minimum entry of A.

Procedure $MIN(A, n, k)$

STEP 1. If $n = 1$, then set $k := 1$
Else
Begin
STEP 2. $n := n - 1$
STEP 3. Procedure MIN(A, n, k)
STEP 4. If $a_{n+1} < a_k$, then $k := n + 1$
End {step 1}
STEP 5. Return.

In step 3 we call the procedure MIN, but on an array of smaller size. This is the essence of a recursive procedure. We repeatedly call MIN until $n = 1$. When $n = 1$, the array has one element and we actually find the minimum, successfully completing step 3. Every time step 3 is completed, we proceed to step 4 with the value of k just received and with the value of n equal to what it was when that instance of the procedure MIN was called.

Example 6.1. Table 6.4 is a trace of MIN with $A = \langle 4.2, 2.1, 3.5, 0.9 \rangle$.

When $n = 1$, we set $k := 1$ and return to (C) to complete steps 4 and 5. Notice that when we then return to (B), n is reset to 2, its value at the time of this execution of step 3.

Question 6.1. Trace the procedure MIN on the array $\langle 4, 3, 2, 1, 5 \rangle$.

There are two properties essential for a recursive procedure to be correct. These are dictated by the requirement that an algorithm must terminate after a finite number of steps. At each call of the procedure within the procedure, the value assigned to some variable, say P ($P = n =$ number of elements in the array in Example 6.1) must decrease. When the value assigned to this variable is sufficiently small, there must be a "termination condition" that instructs the procedure what to do in this final case. It is common to think of a recursive procedure as operating on different levels. If the procedure begins with the parameter P initially assigned the value q, one might think of beginning at the qth story of a building. With

Table 6.4

Step No.	n	k	a_k	a_{n+1}	
1.	4				
2.	3				
3.	{Call MIN($\langle 4.2, 2.1, 3.5\rangle, 3, k$)}				(A)
1.	3				
2.	2				
3.	{Call MIN($\langle 4.2, 2.1\rangle, 2, k$)}				(B)
1.	2				
2.	1				
3.	{Call MIN($\langle 4.2\rangle, 1, k$)}				(C)
1.	1	1			
5.	{Return to (C)}				
	1	1	4.2	2.1	
4.	1	2			
5.	{Return to (B)}				
	2	2	2.1	3.5	
4.	2	2			
5.	{Return to (A)}				
	3	2	2.1	0.9	
4.	3	4			
5.	{Return with $k = 4$}.				

each call the value assigned to P is decreased, and one descends to a lower story until, say, P is assigned the value 1. On the first floor some real calculation or comparison is performed and the message is sent back up through the floors to the qth story, where the final answer is assembled.

A recursive program is also analogous to an induction proof. The "termination condition" corresponds to checking the base case. The call of the procedure within itself corresponds to using the inductive hypothesis.

Example 6.2. Here is an example of a recursive procedure that calculates the nth Fibonacci number (see Section 4.4).

Procedure FIB(n, F) {This procedure has n as input and F as output.}

STEP 1. If $n \leq 1$, then $F := n$
Else
Begin
STEP 2. Procedure FIB($n - 1, F'$)
STEP 3. Procedure FIB($n - 2, F''$)
STEP 4. $F := F' + F''$
End {step 1}
STEP 7. Return.

In step 1 we use the fact that $F_0 = 0$ and $F_1 = 1$. Notice that we can call FIB with input $n - 1$ or $n - 2$; the parameter n does not have to be decreased before the call, as was done in MIN. And the answers will be stored in F' and F'' as directed.

Another classic example of the use of recursive procedures is in calculating the greatest common divisor of two integers (see Algorithm EUCLID from Chapter 4). The next procedure is based on the fact that $\gcd(b, c) = \gcd(r, b)$, where $r = c - \lfloor c/b \rfloor b$. The procedure takes b and c as input and produces $g = \gcd(b, c)$ as output.

Procedure GCD(b, c, g)

STEP 1. If $b = 0$, then $g := c$
Else
Begin
STEP 2. $r := c - \lfloor c/b \rfloor * b$
STEP 3. Procedure GCD(r, b, g)
End {step 1}
STEP 4. Return.

COMMENT. The values of b and c used in computing r in step 2 come from the input parameters of the procedure. They equal the original b and c only in the first execution of step 2.

Question 6.2. Trace GCD with $b = 13$ and $c = 21$. How many recursive calls does it make?

In the exercises you will see examples and problems on recursive procedures for the algorithms SUBSET, JSET, PERM, BtoD, among others. Some of these will be more efficient than before, others no more so.

We conclude with a recursive version of SELECTSORT based on an extension of MIN. The plan is to use basically the same ideas as in SELECTSORT, only to allow a director-sorter to delegate work to assistants. First we rewrite MIN so that upon input of an array A and two integers start $\leq$ finish, it proceeds recursively to find the index k of the minimum entry in the subarray $\langle a_{\text{start}}, a_{\text{start}+1}, \ldots, a_{\text{finish}} \rangle$.

Procedure MIN(A, start, finish, k)

STEP 1. If start = finish, then $k :=$ start
Else
Begin
STEP 2. Procedure MIN(A, start, finish $- 1, k$)
STEP 3. If $a_{\text{finish}} < a_k$, then $k :=$ finish
End {step 1}
STEP 4. Return.

Question 6.3. Trace MIN on $A = \langle -1, 0.333, 5.2, -10, 6.001, 17 \rangle$ for: (a) start = 2, finish = 3; (b) start = 3, finish = 6; and (c) start = 1, finish = 6.

Algorithm R-SELECTSORT

STEP 1. Input an array A and its length n
STEP 2. For start := 1 to $n - 1$ do
  Begin
  STEP 3. Procedure MIN(A, start, n, k)
  STEP 4. If $k \neq$ start, then switch the values of a_{start} and a_k
  End {step 2}
STEP 5. Stop.

We don't claim that R-SELECTSORT is an improvement over SELECTSORT, but it is good training for the recursive sorting algorithm in the next section. In fact, R-SELECTSORT performs about twice as many comparisons as SELECTSORT as we shall see in the following discussion.

First we count the number of comparisons performed by MIN on an array of n elements, that is, when n = finish − start + 1; denote this number by $M(n)$. Then $M(1) = 1$. For $n > 1$, first one comparison is performed in step 1, then MIN is applied to an array with one fewer entry, and finally one additional comparison is made in step 3. Thus

$$M(n) = M(n-1) + 2. \tag{D}$$

In other words, each additional array entry requires two more comparisons. Thus

$$M(2) = M(1) + 2 = 3, \qquad M(3) = M(2) + 2 = 5,$$

and apparently $M(n) = 2n - 1$. To be sure, we prove that this formula is correct by induction. Since $M(1) = 1$, the base case is correct. Then

$$\begin{aligned} M(n) &= M(n-1) + 2 && \text{by (D)} \\ &= 2(n-1) - 1 + 2 && \text{by the inductive hypothesis} \\ &= 2n - 1. \end{aligned}$$

Complexity results for recursive procedures are often similarly established using induction.

Now in R-SELECTSORT we call the procedure MIN(A, start, n, k) for start $= 1, \ldots, n-1$. Thus the total number of comparisons performed is

$$(2n-1) + (2n-3) + \cdots + 3 = n^2 - 1 = O(n^2),$$

(see Exercise 14), giving the same big oh complexity as for SELECTSORT. Comparing the more precise count of comparisons (see Theorem 1.1) shows the recursive version to be less efficient.

EXERCISES FOR SECTION 6

1. Trace (the second version) of MIN$(A, 1, n, k)$ if
 (*a*) $A = \langle -3, -2, -1 \rangle$, $n = 3$.
 (*b*) $A = \langle -10, 10, -3, 3 \rangle$, $n = 4$.
 (*c*) $A = \langle 1, 2, 3, 5, 7 \rangle$, $n = 5$.

2. Trace GCD if
 (*a*) $b = 3$, $c = 5$.
 (*b*) $b = 1$, $c = 10$.
 (*c*) $b = 0$, $c = 5$.
 (*d*) $b = 3$, $c = 14$.

3. (*a*) Write a recursive procedure MINMAX$(A, n, \text{min}, \text{max})$ to find the minimum and maximum entry of an array of n numbers.
 (*b*) Determine the number of comparisons made in the worst case.

4. Using the fact that $n! = n(n-1)!$ write a recursive procedure FACT(n, F) that upon input of a nonnegative integer n, calculates $F = n!$. Trace the procedure for $n = 4$.

5. What do the following recursive procedures compute?
 (*a*) *Procedure NUM1(n, ans)*

```
STEP 1. If n = 0, then ans := 0
        Else
        Begin
        STEP 2. Procedure NUM1(n + 1, ans)
        STEP 3. ans := ans − (n + 1)
        End
STEP 4. Return.
```

 (*b*) *Procedure NUM2(n, ans)*

```
STEP 1. If n = 0, then ans := 0
        Else
        Begin
        STEP 2. Procedure NUM2(n − 1, ans)
        STEP 3. ans := ans + n
        End
STEP 4. Return.
```

 (*c*) *Procedure NUM3(n, ans)*

```
STEP 1. If n = 0, then ans := 0
        Else Procedure NUM3(n − 2, ans)
STEP 2. Return.
```

6. Find an equation relating the two binomial coefficients $\binom{n}{k}$ and $\binom{n}{k-1}$. Use this to write a recursive procedure that calculates $\binom{n}{k}$. What is the termination condition?

7. Here is the classic relation between binomial coefficients:

$$\binom{n}{k} = \binom{n-1}{k} + \binom{n-1}{k-1}.$$

Use this to write a recursive procedure to calculate $\binom{n}{k}$; you may need to use more than one termination condition.

8. Compare the algorithms in Exercises 6 and 7 by counting multiplications, divisions, additions, and the number of calls to the procedure. Which is more efficient?

9. Trace FIB with $n = 5$. Count the number of recursive calls. Comment on the efficiency of this method of calculating Fibonacci numbers as compared with methods learned in Chapter 4.

10. Suppose that $A(n)$ equals the number of additions performed in FIB(n, F). Then $A(0) = A(1) = 0$ and $A(2) = 1$.

(*a*) Show that $A(n) = A(n-1) + A(n-2) + 1$.
(*b*) Compute $A(n)$ for $n = 3, 4, 5, 6, 7, 8$.
(*c*) Compare these values with the nth Fibonacci numbers for $n = 2, 3, \ldots, 8$.
(*d*) Determine a formula for $A(n)$ and then prove that FIB(n, F) performs this many additions.
(*e*) Is FIB a polynomial algorithm?

11. Suppose that we define $F_0^{\#} = 0$, $F_1^{\#} = 1$ and $F_2^{\#} = 1$, and for $n > 2$, $F_n^{\#} = F_{n-1} + F_{n-2} + F_{n-3}$. Write a recursive procedure FIB$\#(n, F^{\#})$ that calculates $F_n^{\#}$ and stores it in $F^{\#}$.

12. Here is an attempt to improve the efficiency of the procedure FIB; to determine the nth Fibonacci number the numbers n, $s = 0$, and $t = 1$ should be input and F will be the output, containing the nth Fibonacci number.

Procedure FIB2(n, s, t, F)

STEP 1. If $n = 0$, then $F := s$
Else Procedure FIB2$(n-1, t, s+t, F)$
STEP 2. Return.

Run this procedure with $n = 1, 2, 3, 4, 5$. Then explain why FIB2$(n, 0, 1, F)$ correctly returns F as the nth Fibonacci number.

13. Compare the number of recursive procedure calls made by FIB2 and by FIB.

14. Prove that $(2n - 1) + (2n - 3) + \cdots + 3 = n^2 - 1$.

15. Prove that the number of multiplications and divisions performed by the recursive version of GCD on b and c is at most $4\lfloor \log(b) \rfloor$. (*Hint:* Reread the complexity analysis of the algorithm EUCLID given in Chapter 4.)

16. Here is a recursive procedure to form a list L of all subsets of an n-set A.

Procedure R-SUBSET(A, n, L)

STEP 1. If $n = 0$, then $L := \{\emptyset\}$
    Else
    Begin
    Procedure R-SUBSET$(A, n - 1, L)$
    STEP 2. For each set S in L, add $S \cup \{a(n)\}$ to L
    End
STEP 3. Return.

(*a*) Trace this algorithm on $A = \langle 1, 2, \ldots, n \rangle$ for $n = 2$, 3, and 4.
(*b*) Explain why the algorithm works correctly.
(*c*) If a step is considered to be the formation of a set, prove by induction that R-SUBSET(A, n, L) performs 2^n steps.

17. Here is a recursive version of the algorithm JSET, presented in Chapter 3. This procedure receives n and j and then stores all j-subsets of the n-set $\{1, 2, \ldots, n\}$ in the list L.

Procedure R-JSET(n, j, L)

STEP 1. If $j = 0$, then $L := \{\emptyset\}$
    Else if $j = n$, then $L := \{\{1, 2, \ldots, n\}\}$
    Else
    Begin
    STEP 2. Procedure R-JSET$(n - 1, j, L_1)$
    STEP 3. Procedure R-JSET$(n - 1, j - 1, L_2)$
    STEP 4. For each set S in L_2,
        set $S := S \cup \{n\}$
    STEP 5. $L := L_1 \cup L_2$
    End
STEP 6. Return.

Run R-JSET on the following data: (*a*) $n = 2, j = 1$; (*b*) $n = 3, j = 1$; (*c*) $n = 3, j = 2$; (*d*) $n = 4, j = 2$; and (*e*) $n = 5, j = 3$.

18. What is the mathematical idea behind R-JSET that makes it work correctly? Count the number of assignment statements made in R-JSET. Is it a good algorithm?

19. Write a recursive version of TREESORT.

6:7 MERGESORT

In this section we present an efficient, recursive sorting algorithm, known as MERGESORT. This algorithm is particularly well suited to the situation when a set of records must be added to a large already sorted set of records. It has the disadvantage that to sort an array of n elements an additional array of size n is used to keep the programming simple and the element shuffling to a minimum.

Here is the idea of MERGESORT from the point of view of the employment director. Suppose that there is now enough work to employ two assistants. At the beginning of the semester the director receives a large file from the payroll office, containing one card for each student employee, listed in alphabetical order. She wants to sort these by social security number. To divide up the work, she splits the file, giving half to each assistant to sort. The director will then merge the two smaller sorted files into one large sorted file.

Algorithms that proceed by dividing the problem in half, working on one or both halves, and then constructing the final solution from the solutions to the smaller problems are known as **divide-and-conquer algorithms**. The algorithms BINARYSEARCH and BININSERT also follow this approach.

Back in the employment office, the assistants remember the principle of recursion. If they each have two assistants or friends, they will give half of their file to each for sorting and then merge the resulting sorted files. The halves or pieces to be sorted will get smaller until an array of one element is reached, say $\langle a_i \rangle$, and this array is sorted as it is. Notice that this process can be modeled by a binary tree with the root labeled with the director, the roots of the left and right subtrees labeled with the assistants, and so on.

Question 7.1. Let A be an array containing eight numbers. Using the ideas of the preceding paragraphs, draw the corresponding binary tree for this case. What is the depth of the tree? How many vertices does it contain? In total, how many assistants are employed in the task?

This approach will be good, provided that we can efficiently merge two sorted files into one sorted file. We shall see that such a merger can be performed in time linear in the total number of elements to be merged.

Here specifically is how to merge two sorted files. Assume that we have an array C such that $c_1, \ldots, c_{\text{mid}}$ is in sorted order as is $c_{\text{mid}+1}, \ldots, c_n$. (If we had two separate, sorted arrays A and B, we could place them in C with A listed before B.) The goal is to rearrange C so that it becomes a sorted array. We use an auxiliary array D into which we sort the elements of C; in the end we transfer the sorted D back into C.

First we compare the first entries in the sorted subarrays, c_1 and $c_{\text{mid}+1}$, and place the smaller in d_1. Next, depending on the outcome of the first comparison, we compare c_2 with $c_{\text{mid}+1}$ or c_1 with $c_{\text{mid}+2}$, placing the smaller in d_2. We continue until either the first subarray or the second has been entirely placed in D.

Then we fill up D with the remaining elements of the other subarray and finally copy D into C.

Procedure MERGE (*C, start, mid, finish*) {C is an array with entries $c_{\text{start}}, c_{\text{start}+1}, \ldots, c_{\text{mid}}$ in increasing order and entries $c_{\text{mid}+1}, \ldots, c_{\text{finish}}$ also in increasing order.}

STEP 1. Set $i := \text{start}$ and $j := \text{mid} + 1$ {i and j index the entries of C being compared}
Set $k := \text{start}$ {k indexes the entry of D being filled}
STEP 2. While ($i \leq \text{mid}$) and ($j \leq \text{finish}$) do
STEP 3. If $c_i \leq c_j$, then do
Begin
STEP 4. $d_k := c_i$
STEP 5. $i := i + 1$
STEP 6. $k := k + 1$
End
Else
Begin
STEP 7. $d_k := c_j$
STEP 8. $j := j + 1$
STEP 9. $k := k + 1$
End
{Right now one of the subarrays is in D}
STEP 10. If $i > \text{mid}$, then do {Transfer remaining entries into D}
For index $:= j$ to finish do
Begin
STEP 11. $d_k := c_{\text{index}}$
STEP 12. $k := k + 1$
End
Else
For index $:= i$ to mid do
Begin
STEP 13. $d_k := c_{\text{index}}$
STEP 14. $k := k + 1$
End
STEP 15. For index := start to finish do {transfer D to C}
$c_{\text{index}} := d_{\text{index}}$
STEP 16. Return.

Example 7.1. Table 6.5 is a trace of MERGE run on the array $C = \langle 1, 2, 3, 4, -2, 0, 2, 4, 6 \rangle$ with start = 1, mid = 4, and finish = 9. We show the array D after the completion of each execution of step 3.

Table 6.5

Step No.	*i*	*j*	*k*	*D*
1	1	5	1	
3	1	6	2	$\langle -2, \ldots$
	1	7	3	$\langle -2, 0, \ldots$
	2	7	4	$\langle -2, 0, 1, \ldots$
	3	7	5	$\langle -2, 0, 1, 2, \ldots$
	3	8	6	$\langle -2, 0, 1, 2, 2, \ldots$
	4	8	7	$\langle -2, 0, 1, 2, 2, 3, \ldots$
	5	8	8	$\langle -2, 0, 1, 2, 2, 3, 4, \ldots$
10				$\langle -2, 0, 1, 2, 2, 3, 4, 4, 6\rangle$
15	$C = \langle -2, 0, 1, 2, 2, 3, 4, 4, 6\rangle$.			

Notice that when equal entries occur, the entry of the first half is inserted in D first.

Question 7.2. Trace MERGE on $C = \langle 0.1, 0.2, 0.3, 0, 0.09, 0.19, 0.29, 0.39, 0.49\rangle$.

How efficient is MERGE? Three comparisons occur at every execution of steps 2 and 3, except for the final time when only the two comparisons in step 2 occur. These steps happen at most n times, where n is the length of the array. Then counting the additional comparison of step 10, at most $3n = O(n)$ comparisons are performed in total. MERGE is a linear algorithm.

With MERGE and the assurance of its efficiency, we plan MERGESORT. We begin with an unsorted array C of length n. We divide C at roughly the midpoint, setting mid equal to $\lfloor n/2 \rfloor$. We sort the first half of C recursively and the second half of C recursively and then use MERGE to combine them in sorted order. This will be accomplished by calling the recursive procedure below with start = 1 and finish = n.

Procedure MERGESORT$(C, start, finish)$

STEP 1. If start = finish, then Return.
Else
Begin
STEP 2. Set mid := $\lfloor (\text{start} + \text{finish})/2 \rfloor$
STEP 3. Procedure MERGESORT (C, start, mid)
STEP 4. Procedure MERGESORT (C, mid + 1, finish)
STEP 5. Procedure MERGE (C, start, mid, finish)
End {step 1}
STEP 6. Return.

The main trick in tracing a procedure like this is to remember where to return upon the completion of a procedure and what the values assigned to the variables are at the return. For example, if we call MERGESORT(C, i, j) in step 3, the procedure receives as input whatever subarray is currently stored in entries i through j of C, sorts it and returns it at the end of step 3 to the same subarray of C. All this bookkeeping is done for us in a programming language like Pascal.

Example 7.2. Table 6.6 is a trace of MERGESORT on $C = \langle 0.3, 0.1, 0.2 \rangle$. The results of the procedure MERGE are just written under the call statement, since we have seen how this works before.

Procedure MERGESORT $(\langle 0.3, 0.1, 0.2 \rangle, 1, 3)$

Table 6.6

Step No.	*C*	*start*	*mid*	*finish*	
1, 2	$\langle 0.3, 0.1, 0.2 \rangle$	1	2	3	
3	{Call MERGESORT(C, 1, 2)}				(A)
1, 2	$\langle 0.3, 0.1 \rangle$	1	1	2	
3	{Call MERGESORT(C, 1, 1)}				(B)
1	$\langle 0.3 \rangle$	1		1	
	Return to (B)	1	1	2	
4	{Call MERGESORT(C, 2, 2)}				(C)
1	$\langle 0.1 \rangle$	2		2	
	Return to (C)	1	1	2	
5	{Call MERGE(C, 1, 1, 2)}				
	$\langle 0.1, 0.3 \rangle$				
	Return to (A)				
	$\langle 0.1, 0.3, 0.2 \rangle$	1	2	3	
4	{Call MERGESORT(C, 3, 3)}				(D)
1	$\langle 0.2 \rangle$	3		3	
	Return to (D)	1	2	3	
5	{Call MERGE(C, 1, 2, 3)}				
	$\langle 0.1, 0.2, 0.3 \rangle$				
6	Return.				

Question 7.3. Trace MERGESORT on
(a) $C = \langle 1, 0 \rangle$ with start = 1 and finish = 2.
(b) $C = \langle 22, 24, 23 \rangle$ with start = 1 and finish = 3.
(c) $C = \langle 1.1, 3.3, 2.2, 4.4 \rangle$ with start = 1 and finish = 4.

We now verify the efficiency of MERGESORT. The origins of this complexity bound are explored in Exercise 7.

Theorem 7.1. MERGESORT is a $O(n \log(n))$ algorithm.

Proof. We begin by proving that if $n = 2^k$, then the number of comparisons executed by MERGESORT is $3n\log(n) + 2n - 1$. The proof is by induction on k. If $k = 0$, then C contains one entry and with one comparison in step 1 the procedure is finished. Since $1 = 3 \cdot 1\log(1) + 2 \cdot 1 - 1$, the base case is established.

We assume that the result holds for all exponents less than k and consider an array C with 2^k entries. Then initially mid equals 2^{k-1}, and in steps 3 and 4 MERGESORT is called on arrays of $n' = 2^{k-1}$ entries each. By the inductive hypothesis MERGESORT performs

$$\begin{aligned} 3n'\log(n') + 2n' - 1 &= 3 \cdot 2^{k-1}\log(2^{k-1}) + 2 \cdot 2^{k-1} - 1 \\ &= 3 \cdot 2^{k-1}(k-1) + 2 \cdot 2^{k-1} - 1 \\ &= 2^{k-1}(3k-1) - 1 \end{aligned}$$

comparisons on each smaller array. The total number of comparisons is 1 (from step 1) plus the number performed on the first half of C plus the number performed on the second half of C plus $3n$, the number of comparisons used by MERGE, or

$$\begin{aligned} 2(2^{k-1}(3k-1) - 1) + 3n + 1 &= 2^k(3k-1) + 3 \cdot 2^k - 1 \\ &= 2^k 3k + 2 \cdot 2^k - 1 \\ &= 3n\log(n) + 2n - 1. \end{aligned}$$

Now suppose that C is an array of n elements, where n is not necessarily a power of 2. Set $r = \lceil \log(n) \rceil$ and $m = 2^r$. We know

$$n \le m = 2^r < 2^{\log(n)+1} = 2n.$$

Create C', an array of m elements by appending $m - n$ new elements to the end of C. Suppose that all these elements are assigned a very large value, a number larger than all entries in C. When MERGESORT is applied to C' we know that the number of comparisons is at most

$$\begin{aligned} 3m\log(m) + 2m - 1 < 3 \cdot (2n)\log(2n) + 2 \cdot (2n) - 1 &= 6n(\log(n) + 1) + 4n - 1 \\ &= 6n\log(n) + 10n - 1 \\ &< 16n\log(n) = O(n\log(n)). \end{aligned}$$

Now C' and C have been sorted with $O(n\log(n))$ comparisons; had we applied MERGESORT to C alone, perhaps fewer comparisons would have been performed. □

An alternative, tighter upper bound on the number of comparisons in MERGESORT is outlined in Supplementary Exercises 27 and 28.

Question 7.4. Look at Question 7.3(a) and (c) and verify that exactly $3n\log(n) + 2n - 1$ comparisons were performed.

Question 7.5. Verify that the number of comparisons MERGESORT performs on an array of size 3 is 20. Is this number less than $3n\log(n) + 2n - 1$ with $n = 3$? Show that this number is less than $6n\log(n) + 10n - 1$ when $n = 3$.

EXERCISES FOR SECTION 7

1. Trace MERGE on the following data. In each case count the number of comparisons made and compare with $3n$, where n is the length of the array.
(*a*) $C = \langle 1, 1, 3, 5 \rangle$, start $= 1$, mid $= 1$ and finish $= 4$.
(*b*) $C = \langle 0.1, 0.2, 0.3, 0, 0.2, 0.4, 0.6 \rangle$, start $= 1$, mid $= 3$ and finish $= 7$.
(*c*) $C = \langle 1, 2, 3, 4, 1, 2, 3, 4 \rangle$, start $= 1$, mid $= 4$ and finish $= 8$.
(*d*) $C = \langle 5, 1, 2, 3, 4 \rangle$, start $= 1$, mid $= 1$ and finish $= 5$.
(*e*) $C = \langle 1, 2, 3, 4, 5 \rangle$, start $= 1$, mid $= 4$ and finish $= 5$.
(*f*) $C = \langle 1, 2, 3, 4, 0 \rangle$, start $= 1$, mid $= 4$ and finish $= 5$.

2. What happens if you run MERGE with the subarray $c_1, \ldots, c_{\text{mid}}$ not sorted?

3. Trace MERGESORT on each of the following arrays:
(*a*) $A = \langle 2, 4, 6, 8, 10 \rangle$.
(*b*) $A = \langle 10, 8, 6, 4, 2 \rangle$.
(*c*) $A = \langle 2, 6, 4, 10, 8 \rangle$.
(*d*) $A = \langle 1, 3, 1, 5, 4, 5 \rangle$.
(*e*) $A = \langle 2, 2, 2, 2, 2 \rangle$.

4. Draw a binary tree that corresponds to the divisions into subarrays in Example 7.2 and Question 7.3. In general, for what arrays of length n is the corresponding tree a full binary tree?

5. Count the number of comparisons made in each case of Exercise 3. Compare these numbers with $3n\log(n) + 2n - 1$ and with $6n\log(n) + 10n - 1$ for appropriate values of n.

6. From the numerical evidence of Questions 7.4 and 7.5 and Exercise 3, conjecture whether the following is true or false: MERGESORT performs at most $3n\lceil\log(n)\rceil + 2n - 1$ comparisons to sort an array of n elements. (See also Supplementary Exercises 27 and 28.)

7. Let $M(n)$ denote the maximum number of comparisons made in MERGESORT, applied to an array of n elements, and suppose that $n = 2^k$. Then $M(1) = 1$ and for $n > 1$ MERGESORT proceeds by calling itself on two arrays of size $n/2 = 2^{k-1}$ and then MERGE-ing the two sorted arrays with $3n$ additional comparisons. Thus

$$M(n) = 2M(n/2) + 3n + 1.$$

Use this relation to determine $M(n)$ for $n = 2, 4, 8$, and 16.

Then find an expression for $M(n)$ in terms of $M(n/4)$ and in terms of $M(n/8)$. Explain why this leads to the formula

$$\begin{aligned} M(n) &= 2^k M(1) + k3n + n - 1 \\ &= 3n \log(n) + 2n - 1 \qquad \text{(still assuming that } n = 2^k\text{).} \end{aligned}$$

8. Suppose that A is an array containing 2^n numbers. If the array is divided in half for each of two assistants to sort and if they each divide their half in half for two additional assistants to sort, and so on, until finally the assistants receive arrays of length one, then how many assistants in all are used? How many levels of assistants are used?

9. Answer the same question as in Exercise 8 when the array contains m numbers, where $2^n \leq m < 2^{n+1}$ for some integer n.

10. Write a procedure that inputs an array $A = \langle a_1, a_2, \ldots, a_n \rangle$ (not necessarily sorted) and rearranges A so that if mid $= \lfloor (1 + n)/2 \rfloor$ and $S = a_{\text{mid}}$, then all entries preceding S are less than or equal to S and all entries following S are greater than or equal to it. (Note that S may need to be moved to a different position.)

11. Write a sorting algorithm that splits the input array A using the preceding exercise and then recursively sorts the parts preceding and following S.

12. Write an algorithm 4-MERGE that takes four sorted arrays and merges them into one sorted array. Compare the complexity of your algorithm with that of using MERGE three times to combine these four arrays into one.

6:8 SORTING IT ALL OUT

The art of searching and sorting is an extremely important and highly developed one in computer science and applications. These are processes used in almost all record-keeping tasks. Not only do telephone companies, banks, the IRS, and so forth, perform these tasks repeatedly, but now even writers find these tasks indispensible in their word processing programs. For example, searching was done repeatedly in the preparation of this text. Every time a theorem, a question, an example, or an exercise was renumbered, a search was run to find all occurrences of the changed number. A spelling checker program also searches for spelling errors and is equally useful because it picks up most typing mistakes. Sorting is also important, for example, in alphabetizing the index of a book.

One important theme of this chapter is the difference between $O(n^2)$ and $O(n \log(n))$ algorithms. Both kinds are good algorithms, but the latter are noticeably more efficient. Except for cases with small data sets (small like most that we've considered in examples and exercises), the faster algorithms make a significant difference in general real-life applications. Of course, there are exceptions to every rule, and two such exceptions are BUBBLESORT (Exercises 1.11 to 1.13) and INSERTIONSORT (Supplementary Exercise 7). In the worst case these are $O(n^2)$ algorithms, but when given a nearly sorted array, they can run in linear time. For example, when an array is nearly sorted except that some adjacent pairs of elements are transposed, both BUBBLESORT and INSERTIONSORT are able to benefit from the nearly sorted arrangement. In contrast an algorithm like BINARYSORT will perform the same number of comparisons on a nearly sorted array as on a randomly ordered array.

Another theme of this chapter is the difference between algorithms with the same big oh complexity. When sorting files with large individual records, algorithms should be used that minimize record transfers; for example, SELECTSORT would be perferred to BUBBLESORT if a $O(n^2)$ algorithm were being used. MERGESORT should be avoided if the file length is so long or the records so large that there is not room for a duplicate array. However, MERGESORT is a good choice when two smaller sorted files are to be sorted into one. No algorithm using comparisons can be faster than $O(n \log(n))$, and the number of different $n \log(n)$ algorithms confirms that these must have been designed for varying needs. TREESORT uses the most sophisticated data structure among the algorithms we've seen. This algorithm and algorithms based on storing data in tree structures have wide applicability in these and other combinatorial settings.

We have alluded to the existence of a linear-time sorting algorithm, known as BUCKETSORT or "distribution counting." If we want to sort an array of n elements whose entries are integers from 0 to M for some small number M, like $M = O(n)$, then we can make one pass through the array and can store the record with key a_i in the a_ith entry of a new array. (More picturesquely, we think of tossing the record in the a_ith "bucket.") Then in one additional sweep through the new array, we can pick up the elements in order. We have performed $n + M$ assignments and no comparisons. This algorithm has limited applicability; for example, using this algorithm the employment director would store the record cards of, say, 600 students in a new array of length 999,999,999, since there are this many possible social security numbers. This approach would necessitate an inappropriately large array. More sophisticated versions of such an algorithm are known as hashing.

Finally, the concept of recursive procedures is an important one. This is the computer scientists' analogue of induction. In Chapter 7 we shall study solutions of recurrence relations and counting problems that arise from recursive procedures.

This chapter has been only an introduction to a deep and well-understood theory, which merits further study.[1]

SUPPLEMENTARY EXERCISES FOR CHAPTER 6

1. Devise an algorithm TRISECTSEARCH that upon input of an array A of n numbers in increasing order and a number S, searches by thirds of A for S. Specifically, first the algorithm should see whether S equals the $\lfloor (n+1)/3 \rfloor$rd entry in A. If not and S is smaller, then it begins again with the first third of A. If S is larger than this entry, it compares S with the $\lfloor 2(n+1)/3 \rfloor$rd entry. If S is smaller, it proceeds with the middle third of A; if S is larger, it proceeds with the last third of A. Determine the worst-case complexity of your algorithm.

2. Write an algorithm that upon input of an ordered array A of n numbers and a number S, searches for S and if found, deletes it. Determine the worst-case complexity of your algorithm.

3. Write an algorithm that upon input of an ordered array A of n numbers and a number S, searches for S and if it is not found, inserts it in the correct order. Determine the worst-case complexity of your algorithm.

4. Rewrite a version of BININSERT, called BININSERT2, that tests whether $a_{r+1} = a_{\text{mid}}$ after step 3, and if so, immediately inserts a_{r+1} at the (mid)th entry of the array. Are there arrays on which BININSERT2 will run faster than BININSERT? Are there arrays on which BININSERT2 will run slower than on BININSERT? Determine the worst-case complexity of BININSERT2.

5. Use BININSERT2 to form a new version of BINARYSORT, called BINARYSORT2. Run both BINARYSORT and BINARYSORT2 on $\langle 1, 5, 1, 1, 1 \rangle$ and compare the efficiency of these algorithms on this array.

6. In this exercise you are asked to compare the number of assignment statements in SELECTSORT and in BINARYSORT. The significant assignment statements are those involving array elements, not just index counters in loops.

(a) Rewrite SELECTSORT so that step 5 is expanded and actually carries out the details of switching a_j and TN. Call this X-SELECTSORT.

(b) Count the number of assignments of elements a_i and TN in the worst case in X-SELECTSORT.

(c) In the procedure BININSERT with $r = 1, 2, 3$, and 4 find examples in which $r + 2$ assignments of elements a_i and temp are made.

(d) Explain why the maximum number of assignments of elements a_i and temp in BININSERT is $r + 2$.

[1] A good next source is a large book on the subject: D. E. Knuth, *Sorting and Searching*, Volume 3 of *The Art of Computer Programming*, Addison-Wesley, Reading, Mass., 1973.

(*e*) Use the result of part (d) to determine the maximum number of assignment statements performed in BINARYSORT.

(*f*) Which of X-SELECTSORT and BINARYSORT performs more assignments?

7. INSERTIONSORT is another sorting algorithm; it is based on the idea of how one often sorts a hand of playing cards: with the left end of the hand sorted, the remaining cards are inserted in order, one at a time.

(*a*) Write a procedure INSERT($r, a_1, \ldots, a_r, a_{r+1}$) that has a sorted array of length r, $\langle a_1, a_2, \ldots, a_r \rangle$, and an element a_{r+1} as input and that outputs the array $\langle a_1, \ldots, a_{r+1} \rangle$ in sorted order. The procedure should search through the input array sequentially until the position for inserting a_{r+1} is found; then a_{r+1} should be inserted there.

(*b*) Here is the algorithm INSERTIONSORT:

Algorithm INSERTIONSORT

STEP 1. Input n and an array $\langle a_1, a_2, \ldots, a_n \rangle$
STEP 2. For $m = 2$ to n do {insert mth entry}
  STEP 3. Procedure INSERT $((m - 1), a_1, \ldots, a_m)$
STEP 4. Stop.

Trace this on $\langle 1, 3, 2, 5, 4, 6 \rangle$.

(*c*) Compare this sorting algorithm with SELECTSORT and BINARYSORT. Describe arrays on which INSERTIONSORT works more efficiently than the others and arrays on which it is less efficient.

(*d*) Determine the complexity of INSERTIONSORT.

8. A sorting algorithm is said to be **stable** if whenever $a_i = a_j$ for some indices $i < j$, then in the sorted array a_i precedes a_j. Is either SELECTSORT or BINARYSORT stable? Explain. If not, can they be rewritten (easily) so that they are stable?

9. Is INSERTIONSORT a stable sorting algorithm?

10. Suppose that A is a sorted array of n elements. How does the speed of INSERTIONSORT on A compare with the speed of SELECTSORT and BINARYSORT?

11. Suppose that you have an (unsorted) array with n items and another item D. What is the minimum number of comparisons necessary to determine whether D is contained in the array or not?

12. Looking up a telephone number in a directory is an example of a typical search through a large ordered list. If the name you are looking for is, say, Smith, you wouldn't turn to the exact middle of the directory despite the high quality of BINARYSEARCH. The reason is that you have some knowledge concerning how the names in the directory are distributed. If you are looking for the name Smith, you will look more toward the back of the book because you expect that more names come before Smith than after it. You might use a strategy like the following: Since S is the 19th letter of the alphabet, you might look at the page numbered m, where $m = \lfloor 19n/26 \rfloor$ and n is the total number of pages in the directory. Develop an algorithm, called **weighted binary search**, that exploits this idea. When should you use weighted binary search and when should you definitely avoid it? (This kind of approach is also known as interpolated search.)

13. Here is a recursive version of the algorithm DtoB from Chapter 1 that upon input of a nonnegative integer m determines its binary expansion s.

Procedure R-DtoB(m, s)

STEP 1. If $m \leq 1$, then set $s := m$
Else
Begin
STEP 2. Procedure R-DtoB($\lfloor m/2 \rfloor, s$)
STEP 3. If m is even, then set s equal to s with a 0 added at the end,
Else set s equal to s with a 1 added at the end
STEP 4. Return.

(*a*) Trace this algorithm for $m = 1, 3, 6, 8$.
(*b*) Show that this algorithm is correct.
(*c*) Prove by induction that the number of divisions in R-DtoB is at most $\log(m)$.

14. Reread the algorithm EXPONENT in Chapter 2. Then use the fact that $x^n = x \cdot x^{n-1}$ to write a recursive version of EXPONENT. Compare the number of multiplications in EXPONENT and the recursive version.

15. Write a recursive version of FASTEXP, called R-FASTEXP(x, n, ans) that upon input of x and n will calculate x^n and store it in ans. Is this version faster or slower than FASTEXP?

16. Look back in Chapter 3 at the algorithm PERM. Write a recursive version of this algorithm.

17. Does the following correctly compute the greatest common divisor of b and c? Explain.

Procedure GCD3(b, c, g)

STEP 1. If $b = c$, then $g := b$
Else if $b \leq c - b$, then Procedure GCD3($b, c - b, g$)
Else Procedure GCD3($c - b, b, g$)
STEP 2. Return.

18. Here is an idea for a recursive version of BINARYSEARCH: Given an array $\langle a_i, \ldots, a_j \rangle$ of numbers and a number S, determine whether S is less than the middle entry of the array and, if so, search the first half. If not, search the second half. Write a recursive version of BINARYSORT.

19. Suppose that we have a sorted array A of length n and an unsorted array B of length m that we wish to merge into A to form a final sorted array A of length $n + m$. Here are some different approaches:

(*a*) Add B to the end of the array A and then use BINARYSORT on this array.

(*b*) Add B to the end of the array A and then use MERGESORT on this array.

(*c*) Add B to the end of the array A and then use INSERTIONSORT (see Exercise 7), replacing step 2 with "For $m = n + 1$ to $n + m$ do."

(*d*) Use MERGESORT on B and then use MERGE on A and B.

Comment on the pros and cons of these approaches. In particular, decide which one you would pick for best efficiency.

20. Compare the efficiency (i.e., number of comparisons performed) of SELECT-SORT, BINARYSORT, and MERGESORT on the following types of arrays:

(*a*) A sorted array.

(*b*) An array listed in reverse order.

(*c*) An array that is nearly sorted except for the interchange of some adjacent pairs of numbers (like $\langle 1, 3, 2, 5, 4, 6 \rangle$).

(*d*) An array with many repeated numbers.

(*e*) An array with its first half sorted and its second half sorted.

21. Is MERGESORT a stable sorting algorithm? (See Exercise 8.)

22. Develop the following idea into an algorithm to sort $A = \langle a_1, a_2, \ldots, a_n \rangle$.

(1) Find the least integer i such that $\langle a_1, \ldots, a_i \rangle$ is sorted, but $\langle a_1, \ldots, a_i, a_{i+1} \rangle$ is not.

(2) Find the next least integer j such that $\langle a_{i+1}, \ldots, a_j \rangle$ is sorted, but $\langle a_{i+1}, \ldots, a_j, a_{j+1} \rangle$ is not.

(3) Merge $\langle a_1, \ldots, a_i \rangle$ and $\langle a_{i+1}, \ldots, a_j \rangle$.

(4) Set $i := j$ and if $j < n$, go to line 2.

Implement this as an algorithm and run it on the following data:

(*a*) $A = \langle 1, 3, 2, 5, 4, 6 \rangle$.

(*b*) $A = \langle 1, 2, 3, 5, 4, 6 \rangle$.

(*c*) $A = \langle 2, 4, 6, 3, 5, 7 \rangle$.

(*d*) $A = \langle 1, 2, 3, 4, 5, 6 \rangle$.

(*e*) $A = \langle 6, 5, 4, 3, 2, 1 \rangle$.

Determine the worst-case complexity of your algorithm.

23. Here is an alleged sorting algorithm that is supposed to take an array of length n with start $= 1$ and finish $= n$ and to rearrange A in increasing order:

Procedure MYSTERY(A, start, finish)

STEP 1. If start < finish, then
Begin
STEP 2. test := a_{start}
STEP 3. i := start + 1
STEP 4. j := finish
STEP 5. Repeat
Begin
STEP 6. While test < a_j, set $j := j - 1$
STEP 7. Whilc test > a_i and i < finish, set $i := i + 1$
STEP 8. Switch a_i and a_j
Until $j \leq i$ {end of step 5}
STEP 9. Switch a_i and a_j {undoing the last switch}
STEP 10. Switch a_{start} and a_j
STEP 11. Procedure MYSTERY(A, start, $j - 1$)
STEP 12. Procedure MYSTERY(A, $j + 1$, finish)
End {step 1}
STEP 13. Return.

Run this algorithm on a variety of arrays and then answer the following equations:

(*a*) MYSTERY finds an index j, places some entry in it, and then recursively goes to work on the array in front of j and behind j. What value of j does it determine and what entry is placed in a_j?

(*b*) Describe in words how MYSTERY works.

(*c*) Determine the worst-case complexity of MYSTERY.

24. Why does BINARYSEARCH require more comparisons than BININSERT in the worst case?

25. Rewrite BINARYSEARCH so that the maximum number of comparisons it performs in searching an array of n items is $2\log(n) + c$, where c is a constant.

26. In the complexity analysis of BINARYSORT we proved that the maximum number of comparisons performed on an array of n elements is $4(n - 1) + 2\log((n - 1)!)$. First prove that for $i = 1, \ldots, n - 1$,

$$i(n + 1 - i) \leq \frac{(n + 1)^2}{4}.$$

Then use this to derive an upper bound on $\log((n - 1)!)$ that is $O(n\log(n))$.

27. Let $T(n)$ denote the maximum number of comparisons performed by MERGESORT on an array of n entries. Then explain why

$$T(n) = T(\lfloor n/2 \rfloor) + T(\lceil n/2 \rceil) + 3n + 1 \qquad \text{for } n > 1$$

and

$$T(1) = 1.$$

Calculate $T(i)$ for $i \leq 8$ and compare these results with those of Questions 7.4 and 7.5 and Exercise 7.5.

28. Use the results of the previous exercise to prove that

$$T(n) \leq 3n\lceil \log(n) \rceil + 2n - 1.$$

29. Suppose that a file of n records is to be sorted and the keys of these records are known to be precisely the numbers $1, 2, \ldots, n$. Here is an algorithm to accomplish a sort on the keys a_i:

Algorithm BUCKETSORT

STEP 1. Input $A = \langle a_1, a_2, \ldots, a_n \rangle$ containing distinct entries from $1, 2, \ldots, n$
STEP 2. For $i := 1$ to n do
  STEP 3. $B(a_i) := a_i$
STEP 4. For $i := 1$ to n do
  STEP 5. $a_i := B(i)$
STEP 6. Output $\langle a_1, a_2, \ldots, a_n \rangle$.

Run a trace on this algorithm with input $A = \langle 2, 1, 5, 4, 3, 6, 7 \rangle$.

30. Count the number of assignment statements made in BUCKETSORT when run on an array of size n; in this algorithm these are the most time-consuming statements.

31. Write an algorithm BUCKETSORT2 that has as input an array A of n distinct numbers whose entries lie between 0 and some constant M. The algorithm should first do a "bucketsort" of A into an array B of length M and then transfer the sorted elements back into A. Count the number of assignment statements made in BUCKETSORT2.

32. Suppose that a comparison takes twice as long as an assignment statement. Compare the time needed to run BUCKETSORT2 when $M = 2n$, kn for some constant k, $n\log(n)$, and n^2 with the time needed for BINARYSORT.

33. The Pancake Problem asks the following: Given a stack of pancakes of varying diameters, rearrange them into a stack with decreasing diameter (as you move up the stack) using only "spatula flips." With a spatula flip you insert the spatula and invert the (sub)stack of pancakes above the spatula. Design an algorithm that correctly solves the pancake problem for a stack of n pancakes with at most $2n$ flips. Count exactly how many flips your algorithm uses in the worst case.

7

RECURRENCE RELATIONS

7:1 BEGINNINGS OF SEQUENCES

A few years ago an advertisement on the London subway (or tube) system read as follows:

"If you can determine the next number in each of the following lists before you arrive at your stop, come in and we'll offer you a job!

1	2	4	8	16	32	_
0	1	3	6	10	15	_
1	1	2	3	5	8	_
2	9	28	65	126	_	
2	3	5	7	11	13	_"

Question 1.1. Find a plausible next entry in as many of the above lists as you can.

This chapter explores intrinsic properties of lists or sequences of numbers. Given a partial list, we would like to determine its next entry. More generally, we would like to find a formula for the nth number on the list. From this we can determine the growth rate of the sequence, that is, we can discover whether the nth number in the sequence grows like a polynomial or an exponential function of n. Many of the sequences we shall study have appeared before in the course; others are important in combinatorics and algorithms.

Definition. A **sequence** is a function whose domain is the positive integers and whose target is the real numbers. A sequence whose target is the integers is called an **integer sequence**.

In the London subway puzzle we see initial segments of five integer sequences.
We represent a sequence

$$S: a_1, a_2, a_3, \ldots, a_n, \ldots$$

by listing the values of the sequence at the integers $1, 2, 3, \ldots$. Even though a sequence is a function, it is common and convenient to use the subscript notation. We call a_1 the first term of S, a_2 the second term, and in general a_n the ***n*th term of the sequence**. Occasionally, we shall extend the domain of a sequence to include zero (e.g., Fibonacci numbers). In this case we talk about the 0th term of a sequence.

Example 1.1. Here are two common sequences:

$$S_1: 1, 2, 3, 4, 5, 6, \ldots,$$
$$S_2: 2, 4, 8, 16, 32, 64, \ldots.$$

For these sequences a formula for the nth term is not hard to guess: for S_1, $a_n = n$ and for S_2, $a_n = 2^n$. In many situations a formula for a sequence may not be immediately apparent.

Example 1.2. In the London subway problem, you probably recognized the Fibonacci numbers; let S_3 be this sequence:

$$S_3: 1, 1, 2, 3, 5, 8, \ldots.$$

We know that each Fibonacci number (after the first two) is the sum of its two immediate predecessors. Knowing this pattern, we can in principle calculate any Fibonacci number. However, the formula for the nth term is not obvious. Recall that, in Chapter 4, we verified this formula by induction but deferred until this chapter how such a formula could be discovered.

Question 1.2. Here are the initial segments of some (possibly familiar) integer sequences:

$$S_4: 1, 2, 6, 24, 120, \ldots$$
$$S_5: 0, 1, 3, 6, 10, 15, \ldots$$
$$S_6: 1, 3, 7, 15, 31, \ldots$$
$$S_7: 2, 3, 5, 7, 11, 13, 17, \ldots$$
$$S_8: 1, 4, 9, 16, 25, 36, \ldots$$
$$S_9: 1, -3, 9, -27, 81, \ldots$$

For at least two of the sequences S_4 to S_9, find a formula that generates the initial segment of the sequence as listed above.

Notice that there might be ambiguity when we see only the initial segment of a sequence. We can't be sure about the numbers that appear in the ... until we have a precise description of the sequence.

Example 1.3. Here are three different functions that each generate the initial segment of S_4:

$$f_n = n!$$

$$\begin{aligned} g_n = (\tfrac{1}{24})[&(n-2)(n-3)(n-4)(n-5) - 8(n-1)(n-3)(n-4)(n-5) \\ &+ 36(n-1)(n-2)(n-4)(n-5) - 96(n-1)(n-2)(n-3)(n-5) \\ &+ 120(n-1)(n-2)(n-3)(n-4)] \qquad \{\text{try it!}\} \end{aligned}$$

$$h_n = \begin{cases} n! & \text{if } n \leq 5 \\ 0 & \text{if } n > 5. \end{cases}$$

Question 1.3. Find two functions that each produce the first three values of the sequence: $5, 9, 17, \ldots$.

Formulas are not the only way to describe sequences. Look at S_7: A moment's reflection leads one to conjecture that this is the sequence of all prime numbers. Although this description is exact, we cannot write down a formula for a function that generates the primes. On the other hand, we could write down an algorithm using the Sieve of Eratosthenes (as described in Supplementary Exercise 4 of Chapter 4) to determine the next prime.

Another way that words can describe a sequence is by identifying patterns within the sequence. We know that each Fibonacci number is the sum of the two preceding ones:

$$F_n = F_{n-1} + F_{n-2},$$

and this pattern completely specifies the Fibonacci numbers once we know the first two values. Such a pattern is known as a recurrence relation.

Example 1.4. Notice that in $S_5: 0, 1, 3, 6, 10, 15, \ldots$, the difference of successive terms is 1, 2, 3, and so on. In symbols,

$$a_n = a_{n-1} + (n-1) \qquad \text{for } n > 1. \tag{A}$$

Thus, like the Fibonacci numbers, the sequence S_5 is completely determined once we specify this pattern and the fact that $a_1 = 0$.

Question 1.4. For the sequence S_1 find an equation that relates a_n and a_{n-1}. Then do the same for S_2.

Example 1.5. Suppose that an algorithm SECRET performs M_n steps upon input of a positive integer n. Trial runs show that $M_1 = 1$, $M_5 = 3$, and $M_8 = 4$. Suppose that by analyzing SECRET we find the following pattern or recurrence relation:

$$M_n = M_{[n/2]} + 1 \quad \text{if } n > 1; \qquad M_1 = 1. \tag{B}$$

Is SECRET good or exponential? The known values of M_n together with others we could calculate specify the start of a sequence; however, we don't know a formula for M_n. In Section 5 we shall discover such a formula and deduce that SECRET is a good algorithm.

Question 1.5. Starting with $a_1 = 0$, use equation (A) to calculate $a_2, \ldots, a_6$. Then use equation (B) to calculate $M_1, \ldots, M_8$.

Question 1.6. Here are two different recurrence relations for a sequence known as the **harmonic numbers**.

Let $H'_1 = 1$, and for $n > 1$ let $H'_n = H'_{n-1} + \dfrac{1}{n}$.

Let $H''_1 = 1$, and for $n > 1$ let

$$H''_n = \frac{1}{n}[H''_{n-1} + H''_{n-2} + \cdots + H''_1] + 1.$$

Determine the first five values of H'_n and H''_n.

Question 1.7. Let $C_1 = 1$, and for $n > 1$ let

$$C_n = C_1 C_{n-1} + C_2 C_{n-2} + \cdots + C_{n-1} C_1.$$

Determine the first five values of C_n. These are known as the **Catalan numbers**.

This chapter will present techniques to discover and verify formulas for the nth term of a sequence given an initial segment and a recurrence relation.

EXERCISES FOR SECTION 1

1. Give an example of a function that is not a sequence. Give an example of a sequence that is not an integer sequence.
2. Which of the following prescribes a sequence?
   (*a*) $2, 4, 6, 8, \ldots, 2n, \ldots$.
   (*b*) $f_n = 3n - 1$.

(*c*) $g_n = 1/n$.
(*d*) $h_n = \pm\sqrt{n}$.
(*e*) $f_n = \dfrac{1}{n^3 - 14n^2 + 64n - 90}$.
(*f*) $G_1 = 1$, $G_2 = 2$ and $G_n = G_{n-1} + 2G_{n-2}$ for $n \geq 3$.
(*g*) $H_n = H_{n-1} + H_{n-2}$.
(*h*) A list of all positive integers that are perfect cubes.
(*i*) The set of all real numbers.

3. Which of the sequences in the preceding problem are integer sequences?

4. Match the following initial segments of sequences and formulas:

(*a*) 1 3 12 60 360 ⋯	(i) $f_n = n^2 - 6n + 8$
(*b*) 3 3 6 9 15 24 ⋯	(ii) $f_n = n^2 - 6n + 10$
(*c*) 2 4 8 14 22 32 ⋯	(iii) $f_n = n^2 + n - 2$
(*d*) 4 1 0 1 4 ⋯	(iv) $f_n = n^2 - n + 2$
(*e*) 1 4 10 20 35 ⋯	(v) $f_n = 2F_n$, where F_n is the nth Fibonacci number
(*f*) 0 2 6 14 30 ⋯	(vi) $f_n = 3F_n$
(*g*) 1 8 27 64 125 ⋯	(vii) $f_n = 2\binom{n}{2}$
(*h*) 5 2 1 2 5 ⋯	(viii) $f_n = 2^n - 1$
	(ix) $f_n = \binom{n+2}{3}$
	(x) $f_n = n^2$
	(xi) $f_n = (n+1)!/2$
	(xii) $f_n = 2n!$
	(xiii) $f_n = n^2$
	(xiv) $f_n = (n-3)^2$
	(xv) $f_n = 3n^2$
	(xvi) $f_n = n!$
	(xvii) $f_n = n^3$
	(xviii) $f_n = 2^{n-2}$
	(xix) $f_n = 2^n - 2$
	(xx) $f_n = (-1)^{n-1} \cdot 3^{n-1}$
	(xxi) $f_n = (n^2 - n)/2$

5. For each of S_4, S_5, S_6, S_8, and S_9 in Question 1.2 find a formula among those listed in Exercise 4 that generates the sequence.

6. For how many of their first entries do $f_n = n^3 - 3n^2 + 2n$ and $g_n = n^4 - 13n^3 + 56n^2 - 92n + 48$ agree?

7. The recurrence relation $F_n = F_{n-1} + F_{n-2}$ produces the Fibonacci numbers with the initial values $F_1 = F_2 = 1$. Give an example of initial values that produce a different sequence. Then find initial values that produce some Fibonacci numbers, but not all of them. Characterize the initial values that produce a sequence whose entries are each a Fibonacci number.

8. Define the extended Fibonacci numbers by $G_1 = 1$, $G_2 = 1$, $G_3 = 2$, and for $n > 3$

$$G_n = G_{n-1} + G_{n-2} + G_{n-3}.$$

Are the following statements true or false about G_n?

(i) For $n \geq 2$, $G_n = \binom{n}{2} + 1$.

(ii) For $n > 3$, $G_n \neq F_n$.

(iii) G_n is a Fibonacci number, but not necessarily F_n.

(iv) $G_n = F_n + G_{n-3}$ for $n > 3$.

(v) $G_n = O(2^n)$.

9. Here is a famous sequence of letters; identify the pattern:

$$o \quad t \quad t \quad f \quad f \quad s \quad s \quad e \quad n \quad t \quad e \quad \cdots.$$

10. Find two functions f and g such that $f_1 = g_1 = 3$ and $f_3 = g_3 = 5$, but $f_2 \neq g_2$.

11. For each of the following, find a formula that expresses the nth term f_n as a function of n:

(i) $f_1 = 1$, $f_n = f_{n-1} + 2$ for $n > 1$.

(ii) $f_1 = 2$, $f_n = f_{n-1} + 2n - 1$ for $n > 1$.

(iii) $f_1 = 1$, $f_n = f_{n-1} + 2n$ for $n > 1$.

12. Verify that for $n = 1, 2, \ldots, 5$, $H_n = 1 + (1/2) + \cdots + (1/n)$ satisfies both recurrence relations in Question 1.6. Use induction to show that this formula works in general.

13. If H_n is the nth harmonic number, then show that

$$H_{2^m} \geq 1 + \frac{m}{2}.$$

Explain why for every positive integer M there is an integer N such that $H_n \geq M$ for all $n \geq N$. Show also that

$$H_{2^m - 1} \leq m.$$

14. Let H_n be the nth harmonic number. Show that

$$H_n \leq \frac{(n+1)}{2}$$

in two different ways: (i) by induction using the recurrence relation for H_n'' and (ii) using the formula for H_n given in Exercise 12.

15. Show that $1 + \frac{\lfloor \log(n) \rfloor}{2} \leq H_n \leq 1 + \lfloor \log(n) \rfloor$ for $n > 0$.

16. The Catalan numbers of Question 1.7 satisfy the formula

$$C_n = \frac{1}{n}\binom{2n-2}{n-1}.$$

Verify this for $n \leq 5$.

17. Show that $\frac{1}{n}\binom{2n-2}{n-1}$ is always an integer.

18. The **Bernoulli numbers** are defined by $B_0 = 1$ and for $n > 0$,

$$B_n = -\frac{1}{n+1}\left[\binom{n+1}{n-1}B_{n-1} + \binom{n+1}{n-2}B_{n-2} + \cdots + \binom{n+1}{0}B_0\right].$$

Determine the next five values of B_n.

19. Check that $B_7 = B_9 = 0$.

20. Find the first Bernoulli number that is greater than 1.

21. Although the Bernoulli numbers start out small, the even ones grow very quickly. It can be shown that

$$|B_{2n}| = \frac{O((2n)!)}{(2\pi)^{2n}},$$

and that

$$\frac{(2n)!}{(2\pi)^{2n}} = O(|B_{2n}|).$$

Use the bounds on $n!$ derived from Stirling's formula (in Section 3.4) to obtain bounds on the growth rate of $|B_{2n}|$.

7:2 ITERATION AND INDUCTION

In this section we begin to explore ways to deduce a formula for a sequence given a recurrence relation that the terms of the sequence satisfy.

When we look at the London subway puzzles or the sequences S_1 to S_9, listed in the previous section, we can, without too much difficulty, find patterns in the sequence entries. We found that the sequence S_5 satisfies the recurrence relation,

$$a_n = a_{n-1} + (n - 1), \qquad n > 1. \tag{A}$$

In Question 1.4 you determined that for S_1

$$a_n = a_{n-1} + 1,$$

and for S_2

$$a_n = 2a_{n-1}.$$

Here is more precisely what we are looking for.

Definition. Suppose that S is the sequence

$$S: a_1, a_2, a_3, \ldots, a_n, \ldots.$$

If the nth term of S can be expressed as a function of previous terms in the sequence:

$$a_n = f(a_1, a_2, \ldots, a_{n-1}), \tag{B}$$

then equation (B) is called a **recurrence relation**, and we say that the sequence S **satisfies** that recurrence relation.

The function in (B) may depend on only some of the previous entries or it may depend upon all of them. The former happens frequently, but Questions 1.6 and 1.7 illustrate the latter possibility.

In a sense, once we have found a recurrence relation underlying a sequence we are done. We can use this relation to find the next (or any subsequent) term. However, it might get tedious to calculate a_{100}. It is important to look for a formula that would give us a_n directly.

Example 2.1. Let S_{10} be the sequence

$$S_{10}: 1, 3, 4, 7, 11, 18, \ldots,$$

where, like the Fibonacci numbers, each term is the sum of the preceding two terms. Thus a sequence is not completely specified by its recurrence relation.

Question 2.1. Find recurrence relations for the sequences $S_6: 1, 3, 7, 15, 31, \ldots,$ and $S_9: 1, -3, 9, -27, 81, \ldots$. In each case, find a different sequence satisfying the same recurrence relation.

Once we have enough initial values of a sequence together with the recurrence relation, the sequence is determined. For example, if a sequence begins with $a_1 = 1$ and $a_2 = 1$ and then obeys

$$a_n = a_{n-1} + a_{n-2}, \tag{C}$$

we get the Fibonacci numbers; however, if $a_1 = 1$ and $a_2 = 3$ then the sequence S_{10} results.

Definition. Let k be the least integer such that once values are assigned to $a_1, a_2, \ldots, a_k$, then (B) prescribes a unique value for each a_n with $n > k$. Then the values of $a_1, a_2, \ldots, a_k$ are called the **initial conditions** of the recurrence relation. We say that the recurrence relation together with its initial conditions **generates** the sequence

$$S: a_1, a_2, \ldots, a_n, \ldots.$$

Typically, a recurrence relation will be given in the form

$$a_n = f(a_1, \ldots, a_{n-1}) \qquad \text{for } n > k,$$

where k is some fixed integer. The bound "$n > k$" specifies the range over which the recurrence relation holds, and the initial conditions that must be assigned are the values of $a_1, a_2, \ldots,$ and a_k.

Example 2.2. The recurrence relation (A) requires only one initial condition, the value of a_1. If $a_1 = 0$, the sequence generated is that given in S_5. If $a_1 = 2$, the sequence generated is

$$2, 3, 5, 8, 12, \ldots$$

(which is not a subsequence of Fibonacci numbers!). The recurrence relation

$$a_n = a_{n-4} + a_{n-2} \qquad \text{for } n > 4$$

expresses a_n in terms of two previous values but requires four initial conditions before all values of the sequence are uniquely defined.

Question 2.2. In the recurrence relation $a_n = a_{n-4} + a_{n-2}$, why are fewer than four initial conditions not enough to define a_n for all larger n? If $a_1 = a_2 = a_3 = a_4 = 1$, find the first 10 terms of the sequence determined by this recurrence relation. Describe the resulting sequence. Then determine an explicit formula for a_n.

Question 2.3. For each of the following, determine the number of initial conditions that must be assigned so that a unique sequence is generated:
(i) $a_n = na_{n-2}$
(ii) $a_n = a_{n-1} + a_{n-3}$
(iii) $a_n = 2a_{\lfloor n/2 \rfloor}$.

Example 2.3. Consider the recurrence relation (A) with initial condition $a_1 = 0$. Repeated application of (A) will lead to a formula. Since $a_{n-1} = a_{n-2} + (n-2)$, substitution in (A) yields

$$a_n = a_{n-2} + (n-2) + (n-1). \tag{D}$$

Since $a_{n-2} = a_{n-3} + (n-3)$, substitution in (D) yields

$$a_n = a_{n-3} + (n-3) + (n-2) + (n-1).$$

Continuing until we reach $a_1 = 0$, we get

$$\begin{aligned} a_n &= a_1 + 1 + 2 + \cdots + (n-2) + (n-1) \\ &= 0 + 1 + 2 + \cdots + (n-2) + (n-1) \\ &= \frac{n(n-1)}{2}, \qquad \text{(see Example 2.3.2)} \end{aligned}$$

a formula for the nth term of the sequence S_5.

The process used in Example 2.3 is known as **iteration** and in straightforward cases will lead to a formula for the sequence.

Question 2.4. Use iteration on each of the following recurrence relations and initial conditions to obtain a formula for the sequence they generate:
(i) $a_n = na_{n-1}$ for $n > 1$, $a_1 = 1$.
(ii) $b_n = b_{n-1} + 2$ for $n > 1$, $b_1 = 1$.

It seems clear that the formulas we come up with using iteration are correct, but to be certain we need to use induction. In Example 2.3 we decided that $a_n = n(n-1)/2$; now we prove that this formula is correct.

Example 2.3 (continued). Here is an inductive proof.

Theorem. $a_n = n(n-1)/2$ satisfies (A) with initial condition $a_1 = 0$.

First the base case: $a_1 = 1 \cdot 0/2 = 0$. We want to use the assumption that $a_k = k(k-1)/2$ to prove that

$$a_{k+1} = \frac{(k+1)(k+1-1)}{2} = \frac{(k+1)k}{2}.$$

To accomplish this, we begin with the recurrence relation:

$$\begin{aligned} a_{k+1} &= a_k + k \\ &= \frac{k(k-1)}{2} + k && \text{by the inductive hypothesis} \\ &= k\left[\frac{(k-1)}{2} + 1\right] && \text{by arithmetic} \\ &= \frac{k(k+1)}{2} \end{aligned}$$

just as we wanted. □

Question 2.5. Prove by induction that the formulas you obtained in Question 2.4 are correct.

Definition. A recurrence relation is called **homogeneous** if it is satisfied by the sequence that is identically zero (i.e., $a_n = 0$ for all n). Otherwise, it is called **inhomogeneous**.

Example 2.4. To test whether a recurrence relation is homogeneous, replace every a_j with zero and see if, for all n, a valid identity remains. For example, $a_n = a_{n-1} + a_{n-2}$ becomes $0 = 0 + 0$ and so is homogeneous, but $a_n = a_{n-1} + (n-1)$ becomes $0 = 0 + n - 1$ and is consequently inhomogeneous.

Question 2.6. Which of the recurrence relations of Questions 2.3 and 2.4 are homogeneous and which inhomogeneous?

Our only suggestion for solving inhomogeneous recurrence relations is the method of iteration and induction. If that fails, then it is time to consult a book specializing in recurrence relations.

How well does iteration and induction work on homogeneous recurrence relations? Let's try to use it to obtain the formula for the Fibonacci numbers knowing that they satisfy (C) with initial conditions $a_1 = a_2 = 1$. Since $a_{n-1} = a_{n-2} + a_{n-3}$, and $a_{n-2} = a_{n-3} + a_{n-4}$, we substitute these into (C) to obtain

$$a_n = a_{n-2} + 2a_{n-3} + a_{n-4}. \tag{E}$$

No formula is yet apparent, so let's keep substituting in the right-hand side of (E) using

$$a_{n-2} = a_{n-3} + a_{n-4},$$
$$a_{n-3} = a_{n-4} + a_{n-5},$$

and

$$a_{n-4} = a_{n-5} + a_{n-6}.$$

We get

$$\begin{aligned} a_n &= a_{n-3} + a_{n-4} + 2(a_{n-4} + a_{n-5}) + a_{n-5} + a_{n-6} \\ &= a_{n-3} + 3a_{n-4} + 3a_{n-5} + a_{n-6}. \end{aligned}$$

Still no formula has emerged, although the coefficients seem familiar. In fact, continuing in this vein never leads to the correct formula.

In summary, we have learned one technique that sometimes obtains a formula for a sequence. Given a sequence we first find a recurrence relation that it satisfies. Then we try iteration and induction to derive a formula for the sequence and to prove it correct. This technique is most likely to work for inhomogeneous recurrence relations. In Section 5 we shall use this technique to solve recurrence relations related to algorithms from Chapter 6. If iteration does not work (easily!) on a homogeneous recurrence relation, then we can use the techniques of the next sections.

EXERCISES FOR SECTION 2

1. Write down three new recurrence relations and specify the number of initial conditions. Which are homogeneous?

2. Here are two recurrence relations and initial conditions:
(i) $c_n = 2c_{n-1}$ for $n > 1$, $c_1 = 1$.
(ii) $d_n = d_{n-1} - d_{n-2}$ for $n > 2$, $d_1 = 1$, $d_2 = 2$.

For each, find different initial conditions that produce a sequence that is a subset of the original sequence. Then find initial conditions that produce a sequence that has no number in common with the original sequence.

3. For each of the following, determine the number of initial conditions:
   (i) $a_n = 2a_{n-1}a_{n-2}$
   (ii) $a_n = a_{n-2} - a_{n-3}$
   (iii) $a_n = a_1 + 2^n$
   (iv) $a_n = \begin{cases} 3a_{n/3} & \text{if 3 divides } n \\ 2a_{(n-1)/3} & \text{if 3 divides } (n-1) \\ a_{(n-2)/3} & \text{if 3 divides } (n-2). \end{cases}$

4. Use iteration on each of the following recurrence relations to obtain a formula for the sequence they generate:
   (i) $b_n = 2b_{n-1}$ for $n > 1$, $b_1 = 1$.
   (ii) $c_n = c_{n-1} + (2n - 2)$ for $n > 1$, $c_1 = 0$.
   (iii) $d_n = 2d_{n-1} + 2$ for $n > 1$, $d_1 = 1$.
   (iv) $e_n = e_{n-1} + (2n - 1)$ for $n > 1$, $e_1 = 1$.
   (v) $f_n = f_{n-1} + 3^n$ for $n > 1$, $f_1 = 3$.
   (vi) $g_n = g_{n-1} + k$ for $n > 0$ and k constant, $g_0 = 1$.
   (vii) $h_n = h_{n-1} + (-1)^{n+1}n$ for $n > 1$, $h_1 = 1$.
   (viii) $j_n = (n - 2)j_{n-1}$ for $j > 2$, $j_1 = 5$, $j_2 = 10$.
   (ix) $k_n = (4n^2 - 2n)k_{n-1}$ for $n > 0$, $k_0 = 1$.
   (x) $l_n = l_{n-1}l_{n-2}$ for $n > 2$, $l_1 = l_2 = 2$.
   (xi) $m_n = m_{n-1} + (n - 1)^2$ for $n > 1$, $m_1 = 0$.

5. Prove by induction that your formulas in the preceding exercise are correct.

6. Notice that if the formula for the nth term of a sequence is known, then it is easy to detect recurrence relations for the sequence. For example if $a_n = n!$, then $a_n = na_{n-1}$. Explain why the following equations give a means of finding recurrence relations from formulas:

$$a_n = \frac{a_n}{a_{n-1}} a_{n-1},$$

$$a_n = a_{n-1} + (a_n - a_{n-1}).$$

7. Each of the following formulas generates an integer sequence. For each find a recurrence relation that is satisfied by the sequence.
   (i) $a_n = n(n - 1)$.
   (ii) $a_n = 2n - 1$.
   (iii) $a_n = 2^n + 3^n$.
   (iv) $a_n = 2^n - 1$.

8. Find recurrence relations that are satisfied by the sequences formed from the following functions:
(i) $a_n = n^2 - 6n + 8$.
(ii) $a_n = n!/15!$.
(iii) $a_n = n!/[15!(n-15)!]$ for $n > 14$.
(iv) $a_n = \binom{n}{j}$, where j is a fixed integer between 0 and n.
(v) $a_n = n^3 + 3n^2 + 3n + 1$.

9. Which of the recurrence relations in Exercises 2, 3, and 4 are homogeneous and which inhomogeneous?

10. Sometimes iteration works on homogeneous recurrence relations. Use this technique to find formulas satisfying the following:
(i) $a_n = a_{n-1}$ for $n > 1$, $a_1 = 1$.
(ii) $b_n = 2b_{n-1}$ for $n > 1$, $b_1 = 2$.
Then prove that your formulas are correct.

11. At the end of this section we saw that the Fibonacci numbers satisfy all the following equations:

$$F_n = F_{n-1} + F_{n-2},$$
$$F_n = F_{n-2} + 2F_{n-3} + F_{n-4},$$

and

$$F_n = F_{n-3} + 3F_{n-4} + 3F_{n-5} + F_{n-6}.$$

Find a similar expression for F_n in terms of $F_{n-4}, F_{n-5}, \ldots, F_{n-8}$. Then for k, an arbitrary positive integer less than n, find and prove a formula that expresses F_n in terms of F_{n-k} and smaller Fibonacci numbers.

12. **(i)** Suppose that $T_n = T_{\lfloor n/2 \rfloor} + 2$ for $n > 1$, $T_1 = 1$. If n is a power of 2, use iteration to deduce a formula for T_n. Is this formula also valid for values of n that are not powers of 2? If so, prove your result; if not, find and prove a formula that is valid for these values of n.
(ii) Repeat for $S_n = 2S_{\lfloor n/2 \rfloor}$ for $n > 1$, $S_1 = 1$.
(iii) Repeat for $U_n = 2U_{\lfloor n/2 \rfloor} + 2$ for $n > 1$, $U_1 = 2$.

13. Given n lines in the plane no two of which are parallel and no three of which intersect in a point, how many regions do these lines create?

14. Let $H_n = H_{n-1} + H_{n-2} + 1$ for $n > 2$, $H_1 = H_2 = 1$. Find $H_3, \ldots, H_8$. Guess a relationship between H_n and F_n, then prove it by induction.

15. Let $Q_n = Q_{n-1} + Q_{n-2} + 2$ for $n > 2$, $Q_1 = Q_2 = 1$. Find $Q_3, \ldots, Q_8$. Guess a relationship between Q_n and F_n, then prove it by induction.

7:3 LINEAR HOMOGENEOUS RECURRENCE RELATIONS WITH CONSTANT COEFFICIENTS

The title of this section is a mouthful that describes the kind of recurrence relation that the Fibonacci numbers satisfy.

Definition. A recurrence relation of the form

$$a_n = k_1 a_{n-1} + k_2 a_{n-2} + \cdots + k_r a_{n-r} \qquad \text{for } n > r, \tag{A}$$

where $k_1, k_2, \ldots, k_r$ are constants is called a **linear homogeneous recurrence relation with constant coefficients**. We denote these by **LHRRWCC**. We assume that $k_r \neq 0$ and call r the **order** of the recurrence relation.

Here's what all these words mean. First *linear* refers to the fact that every term containing an a_i has exactly one such factor and it occurs to the first power. We introduced *homogeneous* in the previous section. The words *constant coefficients* mean that each of the k_is is a constant. In contrast, the recurrence relation $a_n = a_{n-1}^2$ is not linear although it is homogeneous, and the recurrence relation $b_n = nb_{n-1}$ does not have constant coefficients, but it is a first-order linear homogeneous recurrence relation.

The sequences S_2 and S_6 satisfy first- and second-order LHRRWCCs, respectively:

$$S_2: 2, 4, 8, 16, \ldots \qquad a_n = 2a_{n-1} \qquad \text{for } n > 1$$

$$S_6: 1, 3, 7, 15, 31, \ldots \qquad a_n = 3a_{n-1} - 2a_{n-2} \qquad \text{for } n > 2.$$

(By the way, can you now guess a formula for the nth term of S_6? If not, try comparing S_6 with S_2.)

Question 3.1. Which of the following are LHRRWCCs? For those that are not, explain why they fail to satisfy the definition.

(i) $a_n = a_{n-1} + 1$ for $n > 1$.
(ii) $a_n = a_{n-4} a_{n-2}$ for $n > 4$.
(iii) $a_n = a_{n-1} + n^2$ for $n > 1$.

Given a sequence that satisfies a LHRRWCC, we can find an explicit formula for the nth term of the sequence. The derivation seems magical, so we begin by working out the details for the sequences S_2 and S_6.

Example 3.1. The sequence $S_2: 2, 4, 8, 16, \ldots$ satisfies the recurrence relation

$$a_n = 2a_{n-1} \qquad \text{for } n > 1, \tag{B}$$

a first-order LHRRWCC, and also satisfies the formula $a_n = 2^n$. Note that 2 is the root of the equation

$$x - 2 = 0.$$

It is also a root of the equation obtained from the previous one by multiplying every term by x^{n-1}:

$$x^n - 2x^{n-1} = 0$$

or

$$x^n = 2x^{n-1}. \tag{C}$$

Notice the similarities between (B) and (C). One involves subscripts and as while the other involves superscripts and xs. Is this coincidence?

Example 3.2. The sequence $S_6: 1, 3, 7, 15, 31, \ldots$ satisfies the second-order LHRRWCC

$$a_n = 3a_{n-1} - 2a_{n-2}, \qquad n > 2, \tag{D}$$

and has nth term formula

$$a_n = 2^n - 1 = 2^n - 1^n,$$

a difference of two exponentials. Now 2 and 1 are roots of the equation

$$(x - 2)(x - 1) = 0$$

or

$$x^2 - 3x + 2 = 0.$$

Multiplying by x^{n-2}, we get

$$x^n - 3x^{n-1} + 2x^{n-2} = 0$$

or

$$x^n = 3x^{n-1} - 2x^{n-2}. \tag{E}$$

Again, notice the similarities between equations (D) and (E). These are not by chance. We turn now to the theory that connects LHRRWCCs, polynomial equations, and their roots.

Given a LHRRWCC (A) we create the corresponding equation

$$x^n - k_1x^{n-1} - k_2x^{n-2} - \cdots - k_rx^{n-r} = 0.$$

Next we divide through by the common factor x^{n-r} to get

$$x^r - k_1x^{r-1} - \cdots - k_{r-1}x - k_r = 0. \qquad \text{(F)}$$

Why do we do this? Because (F) will be helpful in solving (A).

Definition. Given a LHRRWCC (A), the equation (F) is called the **characteristic equation** of the recurrence relation. The left-hand side of the characteristic equation is a polynomial (often called the **characteristic polynomial**) whose degree equals the order of the recurrence relation. This polynomial has r roots $q_1, q_2, \ldots, q_r$ (either real or complex numbers) called the **characteristic roots** of the recurrence relation.

Notice that no characteristic root q_i is zero. This is because we assume in (A) that k_r is not zero. Thus $x = 0$ is not a root of (F).

Example 3.2 (continued). Using the recurrence relation for S_6:

$$a_n = 3a_{n-1} - 2a_{n-2},$$

we form the characteristic equation as follows:

$$x^n = 3x^{n-1} - 2x^{n-2}$$
$$x^n - 3x^{n-1} + 2x^{n-2} = 0$$
$$x^2 - 3x + 2 = 0,$$

which is the desired equation. In this case we easily find the characteristic roots, since (as we saw before)

$$x^2 - 3x + 2 = (x - 1)(x - 2).$$

Thus the characteristics roots are $q_1 = 1$ and $q_2 = 2$.

Question 3.2. For each of the following LHRRWCCs find the characteristic equation and the characteristic root or roots:

(i) $a_n = 2a_{n-1}$
(ii) $a_n = a_{n-1} + 6a_{n-2}$
(iii) $a_n = 2a_{n-1} - a_{n-2}$.

The next theorem demonstrates that the results of Examples 3.1 and 3.2 were not just coincidence.

Theorem 3.1. Let q be a nonzero real or complex number. Then

$$a_n = q^n$$

is a solution (also called a **basic solution**) to the recurrence relation (A) if and only if q is a characteristic root of the recurrence relation (i.e., a root of the characteristic polynomial).

Proof. The sequence $a_n = q^n$ is a solution to (A)

$$a_n = k_1 a_{n-1} + \cdots + k_r a_{n-r}$$

if and only if

$$q^n = k_1 q^{n-1} + \cdots + k_r q^{n-r}$$

if and only if

$$q^{n-r}[q^r - k_1 q^{r-1} - \cdots - k_r] = 0.$$

Since q is not zero, the last equation holds if and only if

$$q^r - k_1 q^{r-1} - \cdots - k_r = 0.$$

This last equation is true if and only if q is a characteristic root of the recurrence relation. □

Example 3.1 (continued). S_2:2, 4, 8, 16, . . . satisfies the LHRRWCC

$$a_n = 2a_{n-1}$$

which has (from Question 3.2) the characteristic equation

$$x - 2 = 0$$

and characteristic root $q_1 = 2$. By Theorem 3.1, $a_n = 2^n$ is a basic solution to this recurrence relation, and that's just what we've known all along!

Example 3.2 (continued). We can now find formulas that satisfy the recurrence relation

$$a_n = 3a_{n-1} - 2a_{n-2}. \tag{D}$$

By the previous version of Example 3.2 we know that the characteristic roots of this recurrence relation are $q_1 = 1$ and $q_2 = 2$. By Theorem 3.1 both $a_n = 1^n = 1$ and $a_n = 2^n$ are basic solutions to the recurrence relation. But neither of these solutions gives a formula that generates the sequence given by S_6. Notice, however, that if we were to use (D) with initial condition $a_1 = a_2 = 1$, then we would get the sequence $1, 1, 1, \ldots$, and the formula for this is clearly $a_n = 1^n$. If we were to use the initial conditions $a_1 = 2$ and $a_2 = 4$, then we would get $2, 4, 8, \ldots$ and the formula for this is $a_n = 2^n$, the other formula uncovered by Theorem 3.1. The point is that had the initial conditions been different than they are in S_6, we might have found the generating formula. We have more work to do in this example.

Our goal is to find a formula for a sequence generated by a given recurrence relation with any set of initial conditions. The technique will be to combine basic solutions.

Theorem 3.2. If f_n and g_n both satisfy the recurrence relation (A), then for any constants c and d so does

$$a_n = c f_n + d g_n.$$

Proof. Since f and g are each solutions to (A), we have that

$$f_n = k_1 f_{n-1} + \cdots + k_r f_{n-r}, \text{ and}$$
$$g_n = k_1 g_{n-1} + \cdots + k_r g_{n-r}.$$

If we multiply the first equation by c and the second by d and then add them, we get

$$\begin{aligned} c f_n + d g_n &= c[k_1 f_{n-1} + \cdots + k_r f_{n-r}] \\ &\quad + d[k_1 g_{n-1} + \cdots + k_r g_{n-r}] \\ &= k_1[c f_{n-1} + d g_{n-1}] + \cdots + k_r[c f_{n-r} + d g_{n-r}], \end{aligned}$$

and this shows that $a_n = c f_n + d g_n$ is a solution to (A). □

More generally, it can be proved by induction that if $f_n^1, f_n^2, \ldots, f_n^s$ are all solutions of (A), then so is

$$a_n = c_1 f_n^1 + c_2 f_n^2 + \cdots + c_s f_n^s$$

for any constants $c_1, \ldots, c_s$. (See Exercises 14 and 15.) In this event a_n is said to be a **linear combination** of the fs. Usually, the fs will be basic solutions.

Example 3.2 (continued again). Theorem 3.2 says that the basic solutions $a_n = 1$ and $a_n = 2^n$ to the recurrence (D) can be combined so that $a_n = c1 + d2^n$ is also a solution for any constants c and d. To produce a formula that yields the specific sequence

$$S_6: 1, 3, 7, 15, 31, \ldots,$$

we need to find the correct values of c and d. The sequence S_6 comes with the initial conditions $a_1 = 1$ and $a_2 = 3$. We use this information to find c and d. If the correct formula for this sequence is given by $a_n = c1 + d2^n$, then we must have

$$1 = a_1 = c + d2^1 = c + 2d,$$

and

$$3 = a_2 = c + d2^2 = c + 4d.$$

If we subtract the first equation from the second, we get $2 = 2d$. From this $d = 1$ and then $c = -1$. Thus $a_n = 2^n - 1$ is a formula that meets the initial conditions and gives a solution to the recurrence relation.

Question 3.3. Prove by induction that $a_n = 2^n - 1$ satisfies the recurrence relation $a_n = 3a_{n-1} - 2a_{n-2}$ with initial conditions $a_1 = 1$ and $a_2 = 3$.

Here is a summary of when the procedure followed in Example 3.2 works. It does not succeed in all cases. Given a recurrence relation as in (A), we find the characteristic equation

$$x^r - k_1 x^{r-1} - \cdots - k_r = 0.$$

The general theory of equations tells us that this equation has r real or complex roots, and so the equation can be factored into

$$(x - q_1)(x - q_2) \cdots (x - q_r) = 0,$$

where $q_1, q_2, \ldots, q_r$ are the roots. For example, we might get,

$$(x - 1)(x - 2) = 0, \qquad (x - 3)(x - 3)(x - 4) = 0, \qquad \text{or} \qquad (x - i)(x + i) = 0,$$

where i is the **imaginary number**, $\sqrt{-1}$. In the first and third examples the roots are distinct, but in the second example the root 3 appears twice. It is then called a **multiple root** and the root 3 is said to have **multiplicity** 2. It turns out that we must treat the two cases differently.

Theorem 3.3. Suppose that the LHRRWCC as shown in (A) has r distinct characteristic roots $q_1, q_2, \ldots, q_r$. Then every solution to (A) is a linear combination of the basic solutions:

$$a_n = c_1 q_1^n + c_2 q_2^n + \cdots + c_r q_r^n \tag{G}$$

where $c_1, c_2, \ldots, c_r$ are constants.

The proof of this theorem essentially requires knowing that if all the roots of (F) are distinct, then it is possible to solve r equations in r unknowns to find the constants $c_1, c_2, \ldots, c_r$. The equations are determined by substituting $n = 1, 2, \ldots, r$ into (G). We see

$$\begin{aligned} a_1 &= c_1 q_1 + c_2 q_2 + \cdots + c_r q_r \\ a_2 &= c_1 q_1^2 + c_2 q_2^2 + \cdots + c_r q_r^2 \\ &\cdots \\ a_r &= c_1 q_1^r + c_2 q_2^r + \cdots + c_r q_r^r. \end{aligned}$$

In these equations the unknowns are $c_1, \ldots, c_r$. A complete proof of Theorem 3.3 requires knowledge of linear algebra and so is omitted.

Sometimes (like now) the arithmetic involved in solving for the constants c_1, $c_2, \ldots, c_r$ will be simplified if we consider sequences that begin with a zeroth term,

$$S: a_0, a_1, \ldots, a_n, \ldots.$$

Any sequence can be transformed into this type by working backward with the recurrence relation to find a value for a_0 that is consistent. For example, look at Example 3.2. Since

$$a_n = 3a_{n-1} - 2a_{n-2},$$

we want the value of a_0 to be such that

$$a_2 = 3a_1 - 2a_0 \qquad \text{or} \qquad 3 = 3 \cdot 1 - 2a_0.$$

If we give a_0 the value of 0, then the sequence

$$S_6': 0, 1, 3, 7, 15, 31, \ldots$$

satisfies the same recurrence relation but with initial conditions $a_0 = 0$ and $a_1 = 1$. It also has the same formula $a_n = 2^n - 1$. By beginning at 0, the arithmetic in solving r equations in r unknowns might be easier. This can be especially convenient when doing small examples by hand.

Question 3.4. Suppose the recurrence relations in Question 3.1 have the following initial conditions:
(i) $a_1 = 1$.
(ii) $a_i = 2$ for all $i \leq 4$.
(iii) $a_1 = 1$.
In each case determine a value for a_0 that satisfies the same recurrence relation.

Example 3.3. The Fibonacci formula (at last!). Let's use the machine we've just built to discover the formula for the Fibonacci numbers that appeared out of the blue in Section 4.4. These numbers satisfy the recurrence relation

$$a_n = a_{n-1} + a_{n-2} \qquad \text{for } n > 1$$

with initial conditions $a_0 = 0$ and $a_1 = 1$. The characteristic equation of the LHRRWCC is

$$x^2 - x - 1 = 0,$$

and this has distinct characteristic roots $q_1 = (1 + \sqrt{5})/2$, which we called ϕ, and $q_2 = (1 - \sqrt{5})/2$, called ϕ'. Thus the general formula that solves this recurrence relation is

$$\begin{aligned} a_n &= c\left(\frac{1+\sqrt{5}}{2}\right)^n + d\left(\frac{1-\sqrt{5}}{2}\right)^n \\ &= c\phi^n + d(\phi')^n \end{aligned}$$

for some constants c and d. Notice that this looks like the Fibonacci formula we had earlier, but we need to determine the constants c and d from the initial conditions. (Here's where beginning at 0 makes life easier.)

$$\begin{aligned} 0 = a_0 &= c\phi^0 + d(\phi')^0 = c + d \\ 1 = a_1 &= c\phi^1 + d(\phi')^1 \\ &= c\,\frac{1+\sqrt{5}}{2} + d\,\frac{1-\sqrt{5}}{2}. \end{aligned}$$

From the first equation we get that $c = -d$. Substituting into the second equation, we get that

$$c = \frac{1}{\sqrt{5}} \qquad \text{and} \qquad d = \frac{-1}{\sqrt{5}}.$$

In conclusion we have the formula for the Fibonacci numbers as

$$a_n = \frac{\phi^n - (\phi')^n}{\sqrt{5}}.$$

Question 3.5. For each of the following recurrence relations find the formula for the sequence of numbers generated if the characteristic equation has distinct roots. (These are recurrence relations from Question 3.2.)

(i) $a_n = a_{n-1} + 6a_{n-2}$ for $n > 1$, $a_0 = 2$, $a_1 = 1$.
(ii) $a_n = a_{n-1} + 6a_{n-2}$ for $n > 1$, $a_0 = 1$, $a_1 = 3$.
(iii) $a_n = 2a_{n-1} - a_{n-2}$ for $n > 1$, $a_0 = 2$, $a_1 = -1$.

In the next section we consider LHRRWCCs whose characteristic equations have multiple roots. The techniques will be similar.

EXERCISES FOR SECTION 3

1. **(i)** Give an example of a LHRRWCC.
(ii) Give an example of a linear homogeneous recurrence relation with coefficients that are not constant.
(iii) Give an example of a linear recurrence relation with constant coefficients that is inhomogeneous.
(iv) Give an example of a homogeneous recurrence relation with constant coefficients that is not linear.

2. Which of the following are LHRRWCCs? For those that are not, explain why they fail to satisfy the definition.
(i) $a_n = 2a_{n-1}$ for $n > 1$.
(ii) $a_n = 2a_{n-1} + 1$ for $n > 1$.
(iii) $a_n = a_{n-4} - a_{n-2}$ for $n > 4$.
(iv) $a_n = a_{n-1}^3 + 3a_{n-2}$ for $n > 2$.
(v) $a_n = a_{n-1} + a_1$ for $n > 1$ and a_1 a constant.
(vi) $a_n = a_{n-4} - a_{n-3} + a_{n-2} - a_{n-1}$ for $n > 4$.
(vii) $a_n = 6a_{n-1} - 11a_{n-2} + 6a_{n-3}$ for $n > 3$.
For each LHRRWCC determine its order and find its characteristic equation.

3. Let $f(x)$ be the fourth degree polynomial

$$f(x) = (x - 1)(x + 2)(x - 3)(x + 4).$$

What are the roots of f? Are they distinct?

4. Find LHRRWCCs with each of the following characteristic equations:
 (i) $f(x) = (x - 1)(x + 2)(x - 3)$.
 (ii) $g(x) = (x + 3)(x - 3)$.
 (iii) $h(x) = (x - \frac{1}{2})(x + \frac{3}{2})$.

5. Find a cubic polynomial whose roots are 5, -1, and 3. Then find a LHRRWCC with this characteristic polynomial.

6. **(i)** For the LHRRWCC of the preceding exercise find initial conditions such that $a_n = 5^n$ is the formula for the sequence produced by the recurrence relation.
 (ii) Repeat part (i) finding initial conditions for the formula $a_n = 5^n + (-1)^n$.
 (iii) Repeat part (i) finding initial conditions for the formula $a_n = 5^n + 2(-1)^n - 3^n$.

7. Prove that the quadratic equation

$$x^2 + bx + c = 0$$

has roots $\dfrac{-b + \sqrt{b^2 - 4c}}{2}$ and $\dfrac{-b - \sqrt{b^2 - 4c}}{2}$.

8. Find a formula for the roots of the equation

$$x^4 + bx^2 + c = 0.$$

9. Suppose that the recurrence relations in Exercise 2 have the following initial conditions. Find values of a_0 that satisfy the same recurrence relation.
 (i) $a_1 = 0$.
 (ii) $a_1 = 1$.
 (iii) $a_1 = a_2 = 1$, $a_3 = a_4 = 2$.
 (iv) $a_1 = -1$, $a_2 = -2$.
 (v) $a_1 = 3$.
 (vi) $a_1 = \frac{1}{2}$, $a_2 = -1$, $a_3 = -\frac{1}{2}$, $a_4 = 1$.
 (vii) $a_1 = 1$, $a_2 = 2$, $a_3 = 4$.

10. Find the characteristic equation and characteristic roots of the following relations:
 (i) $a_n = 4a_{n-1} - 4a_{n-2}$.
 (ii) $a_n = -a_{n-1}$.
 (iii) $a_n = 5a_{n-2}$.
 (iv) $a_n = -8a_{n-3}$.
 (v) $a_n = 2a_{n-2} - a_{n-4}$.
 Which of the above relations have distinct roots and which have multiple roots?

11. Is the following statement true or false?

If both $a_n = f_n$ and $a_n = g_n$ are formulas that satisfy a given linear inhomogeneous recurrence relation with constant coefficients, then so is $a_n = c f_n + d g_n$ for every choice of constants c and d.

Either explain why this is true or find a counterexample.

12. If possible, for each situation listed give an example of a fourth-degree polynomial with

(i) Four distinct roots.
(ii) Two distinct roots, one of which has multiplicity one and the other multiplicity three.
(iii) Exactly two distinct roots, one of multiplicity one and one of multiplicity two.
(iv) Two distinct roots, each of multiplicity two.
(v) One root of multiplicity four.
(vi) One root of multiplicity two and one root of multiplicity three.

13. Find a recurrence relation that is satisfied by both

(i) $a_n = 1$ and $a_n = 3^n$,
(ii) $a_n = (-1)^n$ and $a_n = 2^n$,
(iii) $a_n = 2^n$ and $a_n = 3 \cdot 2^n + 2 \cdot 3^n$.

14. Prove that if f_n, g_n and h_n are three functions that each satisfy

$$a_n = k_1 a_{n-1} + k_2 a_{n-2} + \cdots + k_r a_{n-r},$$

then for any constants c, d, and e, the function

$$s_n = c f_n + d g_n + e h_n$$

also satisfies this recurrence relation.

15. Prove by induction (on j) that if $f^1_n, f^2_n, \ldots, f^j_n$ are j functions that each satisfy

$$a_n = k_1 a_{n-1} + k_2 a_{n-2} + \cdots + k_r a_{n-r},$$

then for any constants $c_1, c_2, \ldots, c_j$, the function

$$g_n = c_1 f^1_n + c_2 f^2_n + \cdots + c_j f^j_n$$

also satisfies this recurrence relation.

16. Find a formula for a function a_n that satisfies the following recurrence relation with given initial conditions:

(i) $a_n = -a_{n-1}$ for $n > 0$, $a_0 = 1$.
(ii) $a_n = 4a_{n-2}$ for $n > 1$, $a_0 = 0$, $a_1 = 1$.
(iii) $a_n = -8a_{n-3}$ for $n > 2$, $a_0 = a_1 = a_2 = 1$.

17. The recurrence relation $a_n = a_{n-1} + a_{n-2} - a_{n-3}$ with $a_0 = 2$, $a_1 = 1$, and $a_2 = 4$ has characteristic roots 1 and -1. Show that the generating formula is not of the form

$$f_n = c1^n + d(-1)^n.$$

18. The **Lucas numbers** are defined by $L_1 = 1$, and

$$L_n = F_{n+1} + F_{n-1} \qquad \text{for } n > 1,$$

where F_n is the nth Fibonacci number. Find the first eight Lucas numbers. Find a recurrence relation for the Lucas numbers and then find a formula for L_n.

19. Is the following proof correct? Explain your answer.

Theorem. For all positive n, $L_n = F_n$.

Proof (by induction): $L_1 = F_1 = 1$. Assuming that the result is true for $n \leq k$, we examine L_{k+1}:

$$\begin{aligned} L_{k+1} &= L_k + L_{k-1} && \text{from the recurrence relation found in Exercise 18} \\ &= F_k + F_{k-1} && \text{by induction} \\ &= F_{k+1} && \text{by definition.} \end{aligned}$$

20. Suppose that $T_n = 12T_{n-1} - 35T_{n-2}$ for $n > 1$ while $T_0 = 0$ and $T_1 = 2$. Find a formula for T_n.

21. How many n-bit binary sequences have no two consecutive zeros?

22. Suppose that the second floor of the firehouse has two poles to the first floor. Suppose that every higher floor of the firehouse has five poles. Two of these poles go down one floor while the remaining three poles go down two floors. If you slide down a pole that goes down two floors, you cannot get off at the intermediate floor. How many different ways are there to get from the nth floor to the first floor?

7:4 LHRRWCCS WITH MULTIPLE ROOTS: MORE ABOUT RABBITS

We reconsider the rabbit breeding model from Chapter 4 that led to the Fibonacci numbers. Suppose that each pair of newborn rabbits produces exactly one pair of bunnies after one month and this is all of their offspring. If rabbits are still assumed to be immortal, how many pairs of rabbits are there at the end of each month? Let the number of pairs of rabbits at the end of n months be denoted by b_n. Thus $b_1 = 1$, $b_2 = 2$, $b_3 = 3$ (since only the younger pair produces a new pair of bunnies), and $b_4 = 3 + (3 - 2) = 4$. In general, the number of rabbit pairs at time n equals the number of pairs at time $(n - 1)$ plus the number of new bunny pairs produced in the year $(n - 1)$. Thus

$$\begin{aligned} b_n &= b_{n-1} + (b_{n-1} - b_{n-2}) \\ &= 2b_{n-1} - b_{n-2} \qquad \text{for } n > 2, \end{aligned} \tag{A}$$

a second-order LHRRWCC with initial conditions $b_1 = 1$ and $b_2 = 2$.

Question 4.1. Give an inductive proof that $b_n = n$.

Although we have the solution to this rabbit problem, we continue the example, since it illustrates the case of LHRRWCCs with multiple roots. The recurrence relation (A) is a LHRRWCC whose characteristic equation is

$$x^2 - 2x + 1 = 0 \qquad \text{or} \qquad (x - 1)^2 = 0.$$

Thus 1 is a root of multiplicity 2, and from Theorem 3.1 we know that $b_n = 1^n = 1$ is a basic solution to (A). This does not satisfy the initial conditions of the current problem. Since there are no other roots to the characteristic equation, it must be the case that the solution to the recurrence takes a form different from that of the previous section. Here is the pertinent result for characteristic equations with multiple roots.

Theorem 4.1. Suppose that the following LHRRWCC has a characteristic root q of multiplicity $m > 1$:

$$a_n = k_1 a_{n-1} + k_2 a_{n-2} + \cdots + k_r a_{n-r} \tag{B}$$

for $n > r$. Then the following are (basic) solutions to (B):

$$\begin{aligned} a_n &= q^n \\ a_n &= nq^n \\ a_n &= n^2 q^n \\ &\cdots \\ a_n &= n^{m-1} q^n. \end{aligned}$$

Note that each root of (B) supplies as many basic solutions as its multiplicity. Thus there will be a total of r basic solutions to (B). Once again the number of basic solutions equals the order of the recurrence relation.

Proof. We prove the result for $m = 2$. Theorem 3.1 gives that $a_n = q^n$ is a solution to (B). Thus we must show that

$$a_n = nq^n$$

is also a solution. If $t(x)$ denotes the characteristic polynomial of the recurrence relation (B), set $p(x) = t(x)x^{n-r}$. Thus

$$p(x) = x^n - [k_1x^{n-1} + k_2x^{n-2} + \cdots + k_rx^{n-r}].$$

Since q is a root of multiplicity 2 for $t(x)$, the characteristic polynomial, q is also a root of multiplicity 2 for $p(x)$. Set

$$D(x) = \frac{p(x) - p(q)}{x - q}.$$

(For those of you with a calculus background, $D(x)$ is the **difference quotient** that leads to the derivative.) Since q is a root of $p(x)$, the quantity $p(q)$ is just a fancy way to write zero. Furthermore, since q is a multiple root of $p(x)$, when we divide $(x - q)$ into $p(x)$ we are left with a polynomial that still has q as a root. Thus $D(q) = 0$. This is the heart of the proof. What is left is an algebraic rearrangement of $D(x)$ after which we substitute $x = q$ and find that $a_n = n \cdot q^n$ is a solution of (B).

First we collect the terms from $D(x)$ that have the same exponent to get:

$$D(x) = \frac{x^n - q^n}{x - q} - k_1\frac{x^{n-1} - q^{n-1}}{x - q} - \cdots - k_j\frac{x^{n-j} - q^{n-j}}{x - q} - \cdots. \tag{C}$$

Question 4.2. Construct $D(x)$ if $p(x) = x^2 - 2x + 1$. Be sure to leave the characteristic root as q (rather than substitute its value). Simplify $D(x)$ by dividing $x - q$ into each term. (Here $n = r = 2$.)

We simplify (C) term by term. From Exercise 5 in the Supplementary Exercises for Chapter 2 (or by multiplying out the right-hand side), we note that

$$x^n - q^n = (x - q)[x^{n-1} + \cdots + x^{n-1-i}q^i + \cdots + q^{n-1}].$$

Notice that there is a convenient factor of $x - q$ in the above expression and that the exponents of each term sum to $n - 1$. Thus

$$\frac{x^n - q^n}{x - q} = x^{n-1} + \cdots + x^{n-1-i}q^i + \cdots + q^{n-1}. \tag{E}$$

The right-hand side of (E) is just one term in the expansion of $D(x)$. We eventually want to compute $D(q)$, so we substitute $x = q$ into the right-hand side of (E) just to see what happens. Every term becomes q^{n-1}. Since the exponent on x decreases from $n - 1$ to 0 in steps of 1, there are n terms and so the total contribution from (E) will be nq^{n-1}.

Next we simplify the general term in (C):

$$-k_j \frac{x^{n-j} - q^{n-j}}{x - q} = \frac{-k_j}{x - q}[x^{n-j} - q^{n-j}]$$

$$= \frac{-k_j}{x - q}(x - q)[x^{n-j-1} + \cdots + x^{n-j-1-i}q^i + \cdots]. \quad \text{(F)}$$

Notice that there is a convenient factor of $x - q$ in this expression and that the exponents of each term sum to $n - j - 1$. As above we substitute $x = q$ into the right-hand side of (F). Every term becomes q^{n-j-1}. Since the exponent on x decreases from $n - j - 1$ to 0 in steps of 1, there are $n - j$ terms and so the total contribution from (F) will be

$$-k_j(n - j)q^{n-j-1}. \quad \text{(G)}$$

Thus $D(q)$ contains a term (G) for each j with $0 \leq j \leq r$. Since $D(q) = 0$, we get

$$0 = nq^{n-1} - [k_1(n - 1)q^{n-2} + \cdots + k_j(n - j)q^{n-j-1} + \cdots]. \quad \text{(H)}$$

We multiply both sides of (H) by q and rearrange the terms to obtain

$$nq^n = k_1(n - 1)q^{n-1} + \cdots + k_j(n - j)q^{n-j} + \cdots. \quad \text{(I)}$$

Finally, (I) is the same as (B) after substituting

$$a_{n-j} = (n - j)q^{n-j}$$

for $j = 0, \ldots, r$. Thus $a_n = nq^n$ is a solution to the original recurrence relation. □

We stop the proof of Theorem 4.1 with the completion of the case $m = 2$. For $m > 2$ a proof using calculus is outlined in Supplementary Exercises 19 and 20.

Example 4.1. We analyze the rabbit recurrence (A). The characteristic equation is

$$(x - 1)^2 = 0,$$

so by Theorem 4.1 both $b_n = 1$ and $b_n = n1^n = n$ are basic solutions to this recurrence relation. The initial conditions $b_1 = 1$ and $b_2 = 2$ show that the second solution, $b_n = n$, is exactly the one we want.

Question 4.3. Find the characteristic equation and root or roots of the second-order recurrence relation

$$b_n = 4b_{n-1} - 4b_{n-2} \qquad \text{for } n \geq 2.$$

Check that both $b_n = 2^n$ and $b_n = n2^n$ are solutions to this.

Question 4.4. Show that the characteristic equation of

$$c_n = -3c_{n-1} - 3c_{n-2} - c_{n-3} \qquad \text{for } n \geq 3$$

has -1 as a characteristic root of multiplicity 3. Check that each of $c_n = (-1)^n$, $c_n = n(-1)^n$ and $c_n = n^2(-1)^n$ is a solution to this recurrence relation.

Example 4.1 (altered). Suppose that we want to solve the recurrence relation given in (A) with initial conditions $b_1 = 1$ and $b_2 = 3$. As before, both $b_n = 1$ and $b_n = n$ are basic solutions. Theorem 3.2 (which isn't restricted to recurrence relations whose characteristic roots are distinct) tells us that

$$b_n = c1 + dn$$

is also a solution for all constants c and d. Using the initial conditions, we find that

$$1 = b_1 = c1 + d1$$
$$3 = b_2 = c1 + d2.$$

Subtracting the first equation from the second, we deduce that

$$c = -1, \qquad d = 2, \qquad \text{and} \qquad b_n = 2n - 1.$$

Question 4.5. Find a solution to the recurrence relation of Question 4.4 with initial conditions $a_0 = 1$, $a_1 = -2$ and $a_2 = 1$.

Finally, we reach the generalization of Theorem 3.3 (which we also do not prove).

Theorem 4.2. Given a sequence $S: a_1, \ldots, a_n, \ldots$ whose terms satisfy a LHRRWCC of order r, then a_n is a linear combination of the r basic solutions given by Theorems 3.1 and 4.1.

Example 4.2. Suppose that we have a recurrence relation of order 6 (so $r = 6$ and the characteristic equation has degree 6) and suppose that we (luckily) find that the characteristic equation factors as

$$f(x) = (x - 1)^3(x + 1)^2(x - 17).$$

Then by Theorem 4.1 we have the following basic solutions to the recurrence relation:

$$a_n = 1^n,\ a_n = n1^n,\ a_n = n^2 1^n,$$
$$a_n = (-1)^n,\ a_n = n(-1)^n,$$

and

$$a_n = 17^n.$$

Using Theorem 4.2, we see that every solution is of the form

$$a_n = c_1 1 + c_2 n + c_3 n^2 + c_4(-1)^n + c_5 n(-1)^n + c_6 17^n.$$

In a concrete situation we would use the six initial conditions and solve six equations to find $c_1, \ldots, c_6$.

Exercises 9 through 12 present other variations of the rabbit-breeding model. However, the study of LHRRWCCs has not been developed for the interest of rabbit breeders. There are important uses of recurrence relations in combinatorics and in computer science. Often the complexity analyses of recursive algorithms lead to recurrence relations that must be solved. In the next section we'll meet a type of recurrence relation that occurs frequently in the "divide-and-conquer" algorithms.

EXERCISES FOR SECTION 4

1. Write down equations of degree 2, 3, and 4, each with a multiple root. Specify the root and its multiplicity. Then write down an equation that has one root of multiplicity 1, one root of multiplicity 2, one root of multiplicity 3, and no other roots.
2. Which of the following equations have multiple roots?
   (i) $f(x) = x^2 - 1$.
   (ii) $f(x) = x^2 + 2x + 1$.
   (iii) $f(x) = x^2 + x - 12$.
   (iv) $f(x) = x^2 - 6x + 9$.
   (v) $f(x) = x^4 - 2x^2 + 1$.
   (vi) $f(x) = x^4 + 2x^2 + 1$.
   (vii) $f(x) = x^4 + 2x^3 - 3x^2 - 4x + 4$.

3. For each of the functions in Exercise 2 write down a recurrence relation with characteristic equation $f(x) = 0$. Then find a formula that satisfies the recurrence relation.

4. Show that the recurrence relation

$$a_n = 10a_{n-1} - 40a_{n-2} + 80a_{n-3} - 80a_{n-4} + 32a_{n-5}$$

has the characteristic equation $(x-2)^5 = 0$. Then check that $a_n = 2^n$, $a_n = n2^n$, $a_n = n^2 2^n$, $a_n = n^3 2^n$ and $a_n = n^4 2^n$ all satisfy the recurrence relation.

5. For each of the following, find a recurrence relation with initial conditions that has this as a solution:

(i) $a_n = 3^n + n3^n$.
(ii) $a_n = 3^n + 2n3^n$.
(iii) $a_n = 2n - 1 + 2^n$.
(iv) $a_n = 1 + (-1)^n + 2^n$.
(v) $a_n = 4(\frac{1}{2})^n + 8n(\frac{1}{2})^n$.

6. For each recurrence relation in list A find its characteristic equation in list B:

List A

(i) $a_n = 5a_{n-2} + 4a_{n-4}$.
(ii) $a_n = 7a_{n-1} - 17a_{n-2} + 17a_{n-3} - 6a_{n-4}$.
(iii) $a_n = 8a_{n-1} - 23a_{n-2} + 28a_{n-3} - 12a_{n-4}$.
(iv) $a_n = 9a_{n-1} - 29a_{n-2} + 39a_{n-3} - 18a_{n-4}$.
(v) $a_n = 6a_{n-1} - 13a_{n-2} + 12a_{n-3} - 4a_{n-4}$.
(vi) $a_n = 5a_{n-1} - 9a_{n-2} + 7a_{n-3} - 2a_{n-4}$.
(vii) $a_n = 7a_{n-1} - 18a_{n-2} + 20a_{n-3} - 8a_{n-4}$.

List B

(a) $f(x) = x^4 - x^3 + x^2 - x + 1$.
(b) $f(x) = (x-1)(x-2)(x-3)(x-4)$.
(c) $f(x) = x^4 + 10x^3 + 25x^2 + 20x + 4$.
(d) $f(x) = x^4 - 8x^3 - 23x^2 - 28x - 12$.
(e) $f(x) = x^4 - 8x^3 + 23x^2 - 28x + 12$.
(f) $f(x) = (x-1)^2(x-2)^2(x-3)$.
(g) $f(x) = (x+1)^2(x+2)^2$.
(h) $f(x) = (x+1)^2(x-2)^2$.
(i) $f(x) = (x-1)^2(x-2)^2$.
(j) $f(x) = x^4 + 7x^3 + 18x^2 + 20x + 8$.
(k) $f(x) = (x-1)(x-2)^3$.
(l) $f(x) = (x-1)(x-3)^2$.
(m) $f(x) = (x-1)^3(x-2)$.
(n) $f(x) = x^4 + 7x^3 - 17x^2 + 17x - 6$.

(*o*) $f(x) = (x - 1)(x + 1)(x - 2)(x + 2)$.
(*p*) $f(x) = (x - 1)^2(x - 2)(x - 3)$.
(*q*) $f(x) = (x - 1)(x - 2)(x - 4)^2$.
(*r*) $f(x) = (x - 2)(x - 3)(x - 4)^2$.
(*s*) $f(x) = (x - 2)(x - 3)^2(x - 4)$.
(*t*) $f(x) = (x - 2)^2(x - 3)(x - 4)$.
(*u*) $f(x) = (x - 1)(x - 2)(x - 3)^2$.

Then for each recurrence relation in list A find the most general form of a solution to it.

7. Use Theorems 4.1 and 4.2 to find solutions as general as possible to the following recurrence relations:
(i) $a_n = 2a_{n-1} - a_{n-2}$.
(ii) $a_n = 3a_{n-1} - 3a_{n-2} + a_{n-3}$.
(iii) $a_n = 4a_{n-1} - 5a_{n-2} + 2a_{n-3}$.

8. Find a formula for the solution of the following recurrence relations:
(i) $a_n = 3a_{n-1} - 3a_{n-2} + a_{n-3}$ for $n > 4$ with $a_0 = a_1 = a_2 = 1$.
(ii) $a_n = 4a_{n-1} - 5a_{n-2} + 2a_{n-3}$ for $n > 4$ with $a_0 = 3$, $a_1 = 4$ and $a_2 = 7$.

9. Suppose that at the end of each month a rabbit pair produces a pair of bunnies, but that after two sets of offspring they produce no more. Write down the recurrence relation with initial conditions that describes this model, beginning with one pair.

10. Suppose that at the end of one month a rabbit pair produces one pair of bunnies, but that during the next month the (older) rabbit pair dies. Beginning with one pair, write down the number of rabbit pairs at the end of each month for the first five months. Then write down a recurrence relation for the number of rabbit pairs at the end of each month.

11. Suppose that at the end of each month a pair of rabbits produces one new bunny pair, but that rabbits die during their third month after having produced bunnies twice. Write down the recurrence relation that decribes this model and the initial conditions assuming that we begin with one pair.

12. We return to the original Fibonacci model of rabbit breeding: A pair of rabbits requires a month to mature to the age of reproduction and then they mate and produce two bunnies. We now do not assume that these are one male and one female, and furthermore we assume that whatever sex they are, a mate is found for each from another warren of rabbits. Thus at the beginning of the first and second months we have one pair, but at the beginning of the third month we have three pairs of rabbits, one old and two new young pairs. How many pairs do we have at the beginning of the fourth and fifth months? Write down a recurrence relation with initial conditions that describes this model.

13. For each of the following determine whether $a_n = O(2^n)$ and whether $2^n = O(a_n)$:
  (i) $a_n = 2a_{n-1} - a_{n-2}, a_1 = 1, a_2 = 2.$
  (ii) $a_n = 2a_{n-1}, a_1 = 1.$
  (iii) $a_n = 4a_{n-1} - 4a_{n-2}, a_0 = 0, a_1 = 2.$
  (iv) $a_n = a_{n-1} + a_{n-2}, a_1 = a_2 = 1.$

14. Is the following true or false? Explain.
A formula a_n that satisfies a LHRRWCC will always be exponential in n; that is, there will always be constants $1 < r < s$ such that $r^n = O(a_n)$ and $a_n = O(s^n)$.

15. Find solutions to the following recurrence relations:

$$a_n = 4a_{n-1} - a_{n-2} \text{ for } n > 1 \text{ with } a_0 = 0 \text{ and } a_1 = 1.$$
$$b_n = 4b_{n-1} - b_{n-2} \text{ for } n > 1 \text{ with } b_0 = 2 \text{ and } b_1 = 4.$$

16. Prove by induction that the sequences a_n and b_n of Exercise 15 satisfy

$$b_n = a_{n+1} - a_{n-1},$$

and

$$a_{m+n} = a_m a_{n+1} - a_{m-1} a_n.$$ [1]

7:5 DIVIDE-AND-CONQUER RECURRENCE RELATIONS

The goal of this section is to formulate and solve recurrence relations that generate the complexity functions of divide-and-conquer algorithms like the searching and sorting procedures from Chapter 6.

Example 5.1. In the algorithm BINARYSEARCH we are given an ordered array of length n and an element S to search for. We begin by comparing S with the middle entry of the array. If these are not equal, we search half of the original array. This leads to the recurrence relation

$$B_n = 3 + B_{\lfloor n/2 \rfloor} \text{ for } n > 1, \qquad B_1 = 4, \tag{A}$$

where B_n denotes the maximum number of comparisons needed in BINARYSEARCH with input an array of n elements. (Reread Section 6.2.)

[1] For an application of the results of Exercises 15 and 16 to the Lucas–Lehmer test for Mersenne primes, see D. E. Knuth, *Seminumerical Algorithms*, Volume 2 of *The Art of Computer Programming*, Addison-Wesley, Reading, Mass., 1973, pp. 356–359.

Question 5.1. Explain why (A) is a recurrence relation for B_n. Use (A) to obtain B_n for $n = 2, 3, 4$, and 5. Compare these numbers with the derived complexity result, $3\lfloor \log(n) \rfloor + 4$.

Example 5.2. The idea behind BININSERT is similar to that of BINARY-SEARCH. The input to the procedure is an array of $n + 1$ entries with the first n in increasing order. The goal is to insert the $(n + 1)$st entry into the correct position of the array. Again we compare with the middle entry and then search half of the array. We repeat this process until we find the correct position. After shifting elements, the $(n + 1)$st entry is inserted in the correct position.

This leads to the recurrence relation

$$C_n = C_{\lfloor n/2 \rfloor} + 2 \text{ for } n > 1, \qquad C_1 = 4, \tag{B}$$

where C_n is the number of comparisons performed by BININSERT on an array of length n. [Reread Section 6.3 to remind yourself why (B) is the recurrence relation for C_n.]

Example 5.3. In the algorithm MERGESORT we begin with an unsorted list of n elements, divide the list in half, sort each half, and then merge the two parts. Thus if M_n denotes the number of comparisons performed in the worst case of MERGESORT, then in the case that $n = 2^k$

$$M_n = 2M_{n/2} + (3n + 1) \text{ for } n > 1, \qquad M_1 = 1, \tag{C}$$

gives the recurrence relation for M_n.

Question 5.2. Explain why (C) is a recurrence relation for M_n. Use (C) to obtain M_n for $n = 2, 4$, and 8. Compare these numbers with the complexity bound derived for the case $n = 2^k$, namely $3n \log(n) + 2n - 1$.

Each of the above algorithmic problems is solved by dividing it into smaller problems, solving the smaller problems and then combining these solutions; we have called these divide-and-conquer algorithms. Suppose that a_n is the number of steps in the worst case of a divide-and-conquer algorithm. Then $a_{\lfloor n/2 \rfloor}$ or $a_{\lfloor n/d \rfloor}$ gives the maximum number of steps needed to solve a problem of half or one dth the size. The number of steps needed to solve some or all of the smaller problems plus the number needed to combine these solutions into a final one is given by a so-called **divide-and-conquer recurrence relation** like

$$a_n = k\,a_{\lfloor n/d \rfloor} + cn + e, \tag{D}$$

where c, d, e, and k are constants. We shall not solve the most general version of (D); however, the text and the exercises contain the most important cases.

Question 5.3. Find constants to show that (A), (B), and (C) are special cases of (D).

Question 5.4. Suppose that we have the recurrence relation

$$a_n = a_{\lfloor n/3 \rfloor} + 1.$$

How many initial conditions must be specified before this relation gives a value for all positive values of n? Then using the recurrence for a_n, specify a set of initial conditions and determine the resulting values of a_n for all $n \le 7$.

Notice that the presence of the floor function in these recurrence relations could lead to some computational awkwardness. For instance, if we want to show that $B_n = 3\lfloor \log(n) \rfloor + 4$, then working with $B_{\lfloor n/2 \rfloor}$ would require consideration of two cases depending on the parity of n. One way to avoid this problem is to consider the special case of $n = 2^i$ (or $n = d^i$) and then to try to generalize the solution to the arbitrary case.

Example 5.4. Consider a recurrence relation of the form

$$a_n = a_{\lfloor n/2 \rfloor} + c,$$

where c is a constant. We'll try iteration and induction, since this is an inhomogeneous recurrence relation, and we'll experiment with the special case when $n = 2^i$.

$$\begin{aligned}
a_n &= a_{\lfloor n/2 \rfloor} + c \\
&= a_{n/2} + c && \text{since } n/2 \text{ is an integer} \\
&= a_{n/(2^2)} + 2c && \text{since } a_{n/2} = a_{\lfloor n/(2^2) \rfloor} + c = a_{n/(2^2)} + c \\
&\quad \cdots \\
&= a_1 + ic \\
&= a_1 + \log(n)c.
\end{aligned}$$

Question 5.5. Prove by induction that $a_n = a_1 + \log(n)c$ is a solution of the recurrence relation $a_n = a_{\lfloor n/2 \rfloor} + c$ if $n = 2^i$.

Example 5.4 worked out nicely using iteration because we assumed that $n = 2^i$, and it seems reasonable to conjecture that this bound is correct for all values of n. Thus we attempt to prove the same result for arbitrary n by induction. To avoid problems with the floor function, we'll shift now to inequalities. That is, we use the fact that $\lfloor x \rfloor \leq x$. This will lead to upper bounds on the solution function, like $B_n \leq 3\lfloor \log(n) \rfloor + 4$; however, such an upper bound is often satisfactory, since it leads to big oh results, like $B_n = O(\log(n))$.

Theorem 5.1. If a_n is the nth term of an integer sequence that satisfies

$$a_n = a_{\lfloor n/2 \rfloor} + c,$$

where c is a constant, then

$$a_n \leq c\lfloor \log(n) \rfloor + a_1.$$

Proof. (We do not assume that n is a power of 2.) The base case holds with $n = 1$:

$$a_1 \leq c\lfloor \log(1) \rfloor + a_1 = a_1.$$

The inductive hypothesis is that

$$a_n \leq c\lfloor \log(n) \rfloor + a_1$$

for all $n < k$, and we try to obtain the same bound for a_k. We know that

$$\begin{aligned}
a_k &= a_{\lfloor k/2 \rfloor} + c \\
&\leq c\lfloor \log(\lfloor k/2 \rfloor) \rfloor + a_1 + c && \text{by induction} \\
&\leq c\lfloor \log(k/2) \rfloor + a_1 + c && \text{since } \lfloor k/2 \rfloor \leq k/2 \\
&= c\lfloor \log(k) - 1 \rfloor + a_1 + c \\
&= c\lfloor \log(k) \rfloor - c + a_1 + c \\
&= c\lfloor \log(k) \rfloor + a_1.
\end{aligned}$$

□

Notice that we obtain a slightly smaller bound by using $\lfloor \log(k) \rfloor$ in place of $\log(k)$.

Example 5.1 (concluded). When the results of Theorem 5.1 are applied to the recurrence relation for B_n with $c = 3$, and $B_1 = 4$, we have that $B_n \leq 3\lfloor \log(n) \rfloor + 4 = O(\log(n))$, just as in Theorem 2.1 of Chapter 6.

Question 5.6. Apply Theorem 5.1 to (B) and compare the result with that of Theorem 3.1 in Chapter 6.

The exercises ask you to solve a number of special cases of the generic divide-and-conquer recurrence relation (D). Here is one more case that will yield an alternative analysis of MERGESORT.

Example 5.3 (varied). Suppose that $n = 2^k$ for some integer k. Then the recurrence relation in (C) holds for MERGESORT:

$$M_n = 2M_{n/2} + (3n + 1), \qquad \text{with } M_1 = 1.$$

Instead of solving the above recurrence, we consider an inequality version:

$$M_n \leq 2M_{n/2} + 4n, \tag{C'}$$

since $(3n + 1) \leq 4n$ for $n \geq 1$. This will be easier to solve and will lead to an upper bound on M_n. In the next theorem we solve a more general form of recurrence relation of which this is a special case.

Question 5.7. Use iteration and induction to verify that for the case $n = 2^k$, $M_n \leq 4n\log(n) + M_1 n$ satisfies (C′).

More generally, if we use a divide-and-conquer algorithm that solves d smaller problems each of which is $(1/d\text{th})$ of the original, then the complexity analysis might involve a recurrence relation of the form

$$a_n = d\, a_{\lfloor n/d \rfloor} + cn,$$

or

$$a_n \leq d\, a_{\lfloor n/d \rfloor} + cn, \tag{E}$$

where c and d are constants, $d > 1$. Manipulation of (E) is simplified if we use the logarithm to the base d, denoted by $\log_d$.

Theorem 5.2. If c and d are constants with $d > 1$ and

$$a_n \leq d\, a_{\lfloor n/d \rfloor} + cn,$$

then

$$a_n \leq cn\log_d(n) + a_1 n.$$

Proof. We prove this by induction on n. For the base case we have that $a_1 \leq 0 + a_1 1$. We assume that the theorem is true for all $n < k$ and we examine a_k.

$$
\begin{aligned}
a_k &\le d\, a_{\lfloor k/d \rfloor} + ck \\
&\le d\left[c\left\lfloor \frac{k}{d}\right\rfloor \log_d\left(\left\lfloor \frac{k}{d}\right\rfloor\right) + a_1\left\lfloor \frac{k}{d}\right\rfloor\right] + ck && \text{by induction} \\
&\le d\left[c\left(\frac{k}{d}\right)\log_d\left(\frac{k}{d}\right) + a_1\left(\frac{k}{d}\right)\right] + ck && \text{since } \left\lfloor \frac{k}{d}\right\rfloor \le \frac{k}{d} \\
&= ck\log_d\left(\frac{k}{d}\right) + a_1 k + ck && \text{by algebra} \\
&= ck\left[\log_d(k) - 1\right] + a_1 k + ck && \text{by properties of } \log_d \\
&= ck\log_d(k) + a_1 k. && \text{by algebra}
\end{aligned}
$$

□

Example 5.3 (last thoughts). If $d = 2$ and $c = 4$, Theorem 5.2 gives the following bound on the complexity of MERGESORT.

$$M_n \le 4n\log(n) + n.$$

Note that the recurrence relations (C) and (C′) and hence this bound hold only when $n = 2^i$. You should check that this is a larger upper bound than that of Theorem 7.1 from Chapter 6.

In the most general divide-and-conquer recurrence relation

$$a_n = k\, a_{\lfloor n/d \rfloor} + cn + e,$$

we have just seen that if $k = d$ and $e = 0$, then $a_n = O(n\log(\mathrm{n}))$. Exercise 11 demonstrates that if $k < d$, then $a_n = O(n)$. In contrast Exercise 12 shows that if $k > d$, then $a_n = O(n^q)$, where $q = \log_d(k)$. Thus the complexity of a recursive procedure is quite sensitive to small changes in the constants.

EXERCISES FOR SECTION 5

1. Suppose that $a_0 = 1$ is the initial condition for each of the following recurrence relations. Then list the first five terms of the sequence generated.

- **(i)** $a_n = a_{\lfloor n/2 \rfloor} + 1$.
- **(ii)** $a_n = a_{\lfloor n/3 \rfloor} - 1$.
- **(iii)** $a_n = a_{\lfloor n/5 \rfloor}$.
- **(iv)** $a_n = 2a_{\lfloor n/2 \rfloor}$.
- **(v)** $a_n = 3a_{\lfloor n/3 \rfloor} + 1$.
- **(vi)** $a_n = a_{\lfloor n/4 \rfloor} + 1$.
- **(vii)** $a_n = 5a_{\lfloor n/5 \rfloor} - n$.
- **(viii)** $a_n = a_{\lfloor n/3 \rfloor} + 3n$.

2. How many initial conditions are needed for each of the following recurrence relations?

(i) $a_n = a_{\lfloor n/2 \rfloor} + a_{\lfloor n/4 \rfloor}$.

(ii) $a_n = a_{\lfloor n/4 \rfloor} + 1$.

(iii) $a_n = a_{n-1} + a_{\lfloor n/2 \rfloor}$.

(iv) $a_n = a_{n-2} + a_{\lfloor n/2 \rfloor}$.

3. **(i)** Suppose that $h_n = 2h_{\lfloor n/2 \rfloor} + s$, where s is a constant. Use iteration and induction to solve this recurrence relation in the case that $n = 2^i$.

(ii) For arbitrary n find an upper bound on h_n.

4. Consider the recurrence relation $a_n = a_{\lfloor n/d \rfloor} + c$, where c and d are constants, $d > 1$. Show that if $n = d^i$, then $a_n = a_1 + c \log_d(n)$. What happens for arbitrary n?

5. Suppose that $n = d^i$, $d > 1$. Use iteration and induction to deduce that the recurrence relation

$$a_n = d\, a_{\lfloor n/d \rfloor} + c$$

is satisfied by

$$a_n = d^i a_1 + d^{i-1} c + \cdots + dc + c.$$

Then explain why

$$a_n = \left(a_1 + \frac{c}{d-1}\right) n - \frac{c}{d-1}.$$

6. Given the recurrence $z_n = k\, z_{\lfloor n/d \rfloor}$ with $d > 1$, solve for z_n.

7. Explain why Theorem 5.2 and the preceding exercises contain the condition that $d > 1$.

8. Why is the following not a valid proof?

Theorem. If a_n is the nth term of a sequence that satisfies

$$a_n = d\, a_{\lfloor n/d \rfloor} + c$$

for some constants c and d with $d > 1$, then

$$a_n = O(n).$$

Proof. We must show that $a_n \leq sn$ for some constant s. Let $s = a_1$ so that the base case is met: $a_1 \leq s1 = a_1$. Then assume that for all $n < k$, $a_n \leq sn$.

From the recurrence relation we have

$$\begin{aligned} a_n &= d\,a_{\lfloor n/d \rfloor} + c \\ &\le d\left(s\left\lfloor \frac{n}{d} \right\rfloor\right) + c \qquad \text{by induction} \\ &\le \frac{dsn}{d} + c \\ &= sn + c \\ &= O(n). \end{aligned}$$

9. Prove the Theorem of the preceding exercise (correctly). (*Hint:* Use the result from Exercise 5.)

10. Use iteration and induction to find a function f_n such that $a_n = O(f_n)$ for each of the following:

(i) $a_n = k\,a_{\lfloor n/d \rfloor} + c$, where c, d, and k are constants such that $k \neq d$ and $1 < d$.

(ii) $a_n = a_{\lfloor n/d \rfloor} + \log(n)$, where d is a constant greater than 1.

(iii) $a_n = d\,a_{\lfloor n/d \rfloor} + n^2$, where $1 < d$.

11. Show that if c, d, and k are constants such that $k \neq d$, and

$$a_n = k\,a_{\lfloor n/d \rfloor} + cn,$$

then

$$a_n \le sn^{\log_d(k)} + \left(\frac{dc}{d-k}\right)n \qquad \text{where } s \text{ is a constant.}$$

12. Show that if a_n is as given in the previous problem and $k < d$, then

$$a_n = O(n).$$

13. For each of the recurrence relations in Exercise 1 find a function f_n such that $a_n = O(f_n)$.

14. For each of the following recurrence relations decide whether $a_n = O(1)$, $O(\log^n(n))$, $O(n)$, $O(n\log(n))$, $O(n^2)$ or $O(2^n)$:

(i) $a_n = a_{\lfloor n/3 \rfloor}$, $a_0 = 1$.

(ii) $a_n = a_{\lfloor n/4 \rfloor} + 1$, $a_0 = 1$.

(iii) $a_n = a_{\lfloor n/d \rfloor} - 1$ for some constant $d > 1$, $a_0 = 1$.

(iv) $a_n = 3a_{\lfloor n/3 \rfloor}$, $a_0 = 3$.

(v) $a_n = 3a_{\lfloor n/3 \rfloor} + 3$, $a_0 = 3$.

(vi) $a_n = a_{\lfloor n/3 \rfloor} + 3n$, $a_0 = 1$.
(vii) $a_n = 3a_{\lfloor n/3 \rfloor}$, $a_0 = 2$.
(viii) $a_n = 3a_{\lfloor n/3 \rfloor} + 1$, $a_0 = 0$.
(ix) $a_n = 3a_{\lfloor n/3 \rfloor} + n$, $a_0 = 1$.
(x) $a_n = 2a_{\lfloor n/3 \rfloor}$, $a_0 = 1$.
(xi) $a_n = 2a_{\lfloor n/3 \rfloor} + 1$, $a_0 = 1$.
(xii) $a_n = 2a_{\lfloor n/3 \rfloor} - 1$, $a_0 = 1$.
(xiii) $a_n = 4a_{\lfloor n/3 \rfloor}$, $a_0 = 1$.
(xiv) $a_n = 4a_{\lfloor n/3 \rfloor} + 1$, $a_0 = 1$.

15. Reread Example 1.5 and explain why SECRET is a good algorithm.

16. Let the recurrence relation for P_n be defined by

$$P_n = P_{n-1} + P_{n-2} + \cdots + P_1$$

for $n > 1$ with $P_1 = c$, some constant. Is this a LHRRWCC? If so, write down its characteristic equation. In any case, determine the first eight values of P_n, in terms of c. Then guess and prove a formula for P_n as a function of n and c.

17. Here is the algorithm MAX from Exercise 4.12 of Chapter 2. Given an array of n real numbers, it finds the maximum number and stores it in the variable max.

Algorithm MAX

STEP 1. Input n, a positive integer, and $x_1, \ldots, x_j, \ldots, x_n$, real numbers
STEP 2. Set max := x_1
STEP 3. For $j = 2$ to n do
  STEP 4. If $x_j >$ max then max := x_j
STEP 5. Output max and stop.

Explain why MAX always make $(n - 1) = O(n)$ comparisons.

18. In comparison with MAX, here is the idea for a recursive divide-and-conquer algorithm to find the maximum entry in an array of n numbers. If the list has one element, then max equals this entry. Otherwise, we divide the list in half:

$$L_1 = x_1, \ldots, x_{\lfloor n/2 \rfloor}$$
$$L_2 = x_{\lfloor n/2 \rfloor + 1}, \ldots, x_n.$$

Let m_1 be the maximum entry in L_1 and m_2 the maximum in L_2. Then we compare m_1 and m_2, and the larger is the overall maximum. If M_n is the number of comparisons performed using this idea on a list of length n, then find a divide-and-conquer recurrence relation that M_n satisfies. What are the values of $M_1, \ldots, M_8$?

19. Solve the recurrence relation of the preceding exercise. (*Hint:* first let $n = 2^k$. Otherwise, use the trick of MERGESORT to extend the array to one with 2^k entries.) Is this algorithm more efficient than MAX?

7:6 RECURRING THOUGHTS

In this chapter recurrence relations have come up in the definitions of integer sequences, in mathematical models, and in the complexity analysis of algorithms. With naturally specified initial conditions, the goal is to find a formula (or at least an upper bound) for the nth term of the sequence. Once we have found such a formula, then it is not difficult to prove this result by induction. In fact, recurrence relations are ideally suited to inductive proofs, using either ordinary or complete induction, because they give the nth term as a function of preceding terms. Thus the hard question is generally to find the solution to the recurrence relation.

The first commonsense approach is to use iteration. This technique works well on inhomogeneous recurrence relations, especially on the divide-and-conquer recurrence relations. In general, it does not work so well on homogeneous recurrence relations.

For the special case of linear homogeneous recurrence relations with constant coefficients, we have presented a complete solution using characteristic equations and their roots. In theory, we can find the solution of any LHRRWCC.

The general technique used to solve LHRRWCCs is an important one with wider application in mathematics. We look for "basic" or "linearly independent" solutions and combine them in "linear combinations" to derive all possible solutions. This technique is used whenever the underlying mathematical structure is a "linear space." For example, the field of linear algebra deals with the solution of homogeneous systems of linear equations, and the field of differential equations studies the solution of linear homogeneous differential equations. It is not by chance (or bad planning) that the same words appear repeatedly in different fields; the underlying ideas and solution techniques are really the same.

Iteration and induction is the technique of choice for the divide-and-conquer recurrence relations. Typically, we use iteration on a simplified case, as when $n = 2^k$ or $n = d^k$, and then find that the resulting formula gives a bound for a solution of the general recurrence relation. We can solve or get tight upper bounds on essentially all recurrence relations of the form

$$a_n = k\, a_{\lfloor n/d \rfloor} + cn + e,$$

where c, d, e, and k are constants. These techniques will also work on other, more irregular recurrence relations.

SUPPLEMENTARY EXERCISES FOR CHAPTER 7

1. Let L be the list of all positive integers that begin with a 7, listed in increasing order. Write down the first 12 entries of L. Can you (within, say, 5 minutes) find a formula L_n that give the nth entry of L as a function of n?

2. Suppose that n points are placed around a circle and that every pair of points is joined by a line, either straight or curved, but drawn so that at most two lines cross each other at the same point. Into how many regions is the interior of the circle divided? Call this number R_n. Does $R_n = 2^{n-1}$ for all positive n?

3. Let S_n be defined by $S_n = S_{n-1} + 1/n^2$ for $n > 1$, with $S_1 = 1$. Use iteration and induction to find a formula for S_n.

4. Let SR_n be defined by $SR_n = SR_{n-1} + 1/\sqrt{n}$ for $n > 1$, with $SR_1 = 1$. Find and justify a formula for SR_n.

5. Refer to the definitions of S_n and SR_n in the preceding exercises. Which of the following are true and which false? Justify.

(i) $S_n = O(1)$
(ii) $SR_n = O(1)$
(iii) $1 = O(S_n)$
(iv) $1 = O(SR_n)$
(v) $S_n = O(\log(n))$
(vi) $SR_n = O(\log(n))$
(vii) $\log(n) = O(S_n)$
(viii) $\log(n) = O(SR_n)$
(ix) $S_n = O(n)$
(x) $SR_n = O(n)$
(xi) $n = O(S_n)$
(xii) $n = O(SR_n)$

6. The **Towers of Hanoi** puzzle consists of a board with three pegs rising from the base. On one peg there are six circular disks of differing size. The largest disk is on the bottom and the others are stacked above it in order of decreasing size. These disks are to be transferred, one at a time, onto another peg so that at no time is a larger disk placed above a smaller one. What is the minimum number of moves needed to move the six disks?

7. We consider the abstract n-fold Tower of Hanoi puzzle in which we suppose that n disks are stacked on one peg and must be moved to another peg, as described in Exercise 6. Let H_n denote the minimum number of moves required to transfer the n disks. Then $H_1 = 1$ and $H_2 = 3$. Find a recurrence relation that expresses H_n in terms of H_{n-1}, the number of moves needed to move the top $(n - 1)$ disks. Then find a formula for H_n and prove that it is correct.

8. Suppose that we consider a variant on the Tower of Hanoi puzzle in which there are four pegs with n disks stacked on one peg. Let M_n denote the minimum number of moves needed to move the stack of n disks to another peg. Calculate M_n for $n = 2$, 3, 4, and 5. Do these values agree with those of H_n? Find a recurrence relation for M_n and find as small a function f_n as possible such that $M_n = O(f_n)$.

9. Explain why the polynomial

$$p(x) = x^r + c_{r-1}x^{r-1} + \cdots + c_1x + c_0$$

has a root s if and only if $p(x)$ can be factored as

$$p(x) = (x - s)q(x),$$

where $q(x)$ is a polynomial of degree $(r - 1)$. [*Hint:* Suppose that when $p(x)$ is divided by $(x - s)$, $q(x)$ is the quotient and $r(x)$ the remainder. In other words, $p(x) = (x - s)q(x) + r(x)$.]

10. The complex (or imaginary) number i has the property that

$$i^2 = -1.$$

Explain why $(-i)^2 = -1$ and $(-i)i = +1$. What is the value of i^3. $(-i)^3$, i^4 and $(-i)^4$? In general, what is the value of i^{2n-1}, $(-i)^{2n-1}$, i^{2n} and $(-i)^{2n}$?

11. **(i)** The sequence $2, 0, -2, 0, 2, 0, -2, 0, \ldots$ satisfies

$$a_n = -a_{n-2} \text{ for } n \geq 2, \qquad a_0 = 2,\ a_1 = 0.$$

Use the methods of Section 3 to find a formula for the nth term of this sequence. By inspection we can see that the following is also a formula for the nth term of the sequence:

$$f_n = \begin{cases} 2 & \text{if 4 divides } n \\ 0 & \text{if 4 divides } (n-1) \text{ or } (n-3) \\ -2 & \text{if 4 divides } (n-2). \end{cases}$$

Do these agree?

(ii) Repeat the problem in part (i) with the sequence

$$1, 3i, -1, -3i, 1, 3i, -1, -3i, \ldots$$

What formula can you derive for this by inspection? Does it agree with your formula obtained through a recurrence relation?

12. A generalized Fibonacci number is defined as follows: For k a fixed integer greater than 2,

$$F_n^k = F_{n-1}^k + F_{n-2}^k + \cdots + F_{n-k}^k \qquad \text{for } n \geq k$$

with initial conditions $F_0^k = F_1^k = \cdots = F_{k-2}^k = 0$ and $F_{k-1}^k = 1$.

(i) For $k = 3$, 4, and 5 write out the first 10 generalized Fibonacci numbers.

(ii) For $k = 3$, 4, and 5 find the characteristic equation of F_n^k.

(iii) For $k = 3$ find the approximate values of the characteristic roots.

13. **(i)** Let Y_n denote the number of strings of length n, containing 0s, 1s and (-1)s with no two consecutive 1s and no two consecutive (-1)s. Determine Y_1, Y_2, and Y_3 by listing all such strings.

(ii) Find a recurrence relation for Y_n. Then, if possible, solve it using the initial conditions found in part (i).

14. Let a_n be the recurrence relation defined by

$$a_n = a_{n-1} + a_{n-2} - n + 3 \qquad \text{for } n \geq 2$$

with initial conditions $a_0 = 0$ and $a_1 = 2$. Find a formula for a_n expressed in terms of F_n, the nth Fibonacci number.

15. For m a fixed positive integer, consider

$$a_n = a_{n-1} + a_{n-2} + \binom{n}{m} \qquad \text{for } n \geq 2$$

with initial conditions $a_0 = 0$ and $a_1 = 1$.

(i) For $m = 5$, find the first 10 entries of the sequence a_n. Express each entry in terms of Fibonacci numbers. (Recall that $\binom{n}{m}$ is equal to 0 when $m > n$.)

(ii) For arbitrary m, find a formula for a_n, expressed in terms of F_n, the nth Fibonacci number.

16. Suppose that $q = r/s$ is a rational number that is a root of

$$x^3 + bx^2 + cx + d = 0,$$

where $\gcd(r, s) = 1$ and where b, c, and d are all integers. Explain why $s = 1$ and r is a divisor of d. Then explain why when searching for a rational root of a cubic equation of the form above, one needs to check only the divisors of d.

17. Find all rational roots of the following equations.

(i) $x^3 - 2x^2 + x - 2 = 0$.

(ii) $x^3 + x^2 + x - 3 = 0$.

(iii) $x^3 - 3x^2 + 2x = 0$.
(iv) $x^3 + x^2 + x + 1 = 0$.
(v) $x^3 - 4x^2 + x + 6 = 0$.
(vi) $x^3 + 2x^2 - 3x + 7 = 0$.

18. Explain why every cubic polynomial has some real number as a root.

CAVEAT. The following problems, 19 and 20, require some knowledge of calculus, specifically knowing how to find the derivative of a polynomial and the product rule.

19. By definition we know that if s is a root of a polynomial $p(x)$ of multiplicity $m > 1$, then $p(x)$ can be factored as

$$p(x) = (x - s)^m q(x),$$

where $q(x)$ is a polynomial. Prove that if s is a root of $p(x)$ of multiplicity $m > 1$, then s is also a root of $p'(x)$ of multiplicity $(m - 1)$, where $p'(x)$ is the derivative of $p(x)$.

20. This exercise is a general proof of Theorem 4.1. If the LHRRWCC

$$a_n = k_1 a_{n-1} + k_2 a_{n-2} + \cdots + k_r a_{n-r} \qquad \text{for } n > r \tag{A}$$

has a characteristic root q of multiplicity $m > 1$, then the following are all solutions to the recurrence relation:

$$a_n = q^n$$
$$a_n = nq^n$$
$$a_n = n^2 q^n$$
$$\cdots$$
$$a_n = n^{m-1} q^n.$$

From the text we know that $a_n = q^n$ and $a_n = nq^n$ are solutions.

CASE **1.** We repeat the case where $m = 2$, since the technique here generalizes more readily than the one given in the text. Let $p(x)$ be the characteristic polynomial of (A). Calculate the function $xp'(x)$, where $p'(x)$ is the derivative of $p(x)$. Show that q is a root of $xp'(x)$ and determine its multiplicity. From this deduce that $a_n = nq^n$ is a solution to (A).

CASE **2** ($m > 2$). Calculate the function $x(xp'(x))'$. Using the results of Case 1, show that q is a root of this function and find its multiplicity. Then deduce that $a_n = n^2 q^n$ is a solution of (A).

CASE 3 ($m > 3$). Let $P_i(x)$ be the polynomial obtained from $p(x)$ by i times repeating the process of taking the derivative of $p(x)$ and multiplying by x, then taking the derivative of this new function and multiplying by x:

$$P_1(x) = xp'(x),$$
$$P_2(x) = x(xp'(x))',$$

and so on. Calculate $P_3(x)$ and $P_4(x)$ starting with $p(x)$ the characteristic equation of (A). Then write out the general form of $P_i(x)$. Prove by induction on i that if q is a root of the characteristic equation of (A) of multiplicity $m \geq i \geq 1$, then q is a root of multiplicity $(m - i)$ of $P_i(x)$. From this deduce that for $i = 1, \ldots, m - 1$, $a_n = n^i q^n$ is a solution of (A).

21. Reread Exercises 6 to 8 about the Tower of Hanoi puzzle. Suppose as in Exercise 8 that the puzzle has four pegs with the disks on the first peg, and suppose that we move the disks as follows:

(*a*) Move the top $\lfloor n/2 \rfloor$ disks to the second peg, one by one following the rules of the puzzle.

(*b*) Move all but the last of the remaining disks to the third peg by a legal series of moves.

(*c*) Move the largest disk to the fourth peg.

(*d*) Move the bottom half from the third peg to the fourth peg.

(*e*) Move the top half from the second peg to the fourth peg.

If H'_n denotes the minimum number of moves needed to transfer n disks in this version of the Tower of Hanoi puzzle, then find a recurrence relation for H'_n and solve for H'_n.

NOTE. From Section 1 it seems that the odd-indexed Bernoulli numbers are zero, starting with B_3. Here is a sequence of exercises that shows why $B_{2k+1} = 0$ for $k \geq 1$.

22. A function is called **even** if $f(x) = f(-x)$ for all values of x. Which of the following functions are even?

(i) $f(x) = c$, c a constant.
(ii) $f(x) = x$.
(iii) $f(x) = x^2$.
(iv) $f(x) = x^3$.
(v) $f(x) = x^4$.
(vi) $f(x) = x^5$.
(vii) $f(x) = x^{2n}$.
(viii) $f(x) = x^{2n+1}$.
(ix) $f(x) = 2^x$.
(x) $f(x) = \sqrt{x}$.

23. Prove that the function

$$f(x) = \frac{x}{(e^x - 1)} + \frac{x}{2}$$

is even.

24. An infinite polynomial of the form

$$p(x) = c_0 + c_1 x + \cdots + c_n x^n + \cdots,$$

where $c_0, c_1, \ldots, c_n, \ldots$ are constants, is said to be even if $p(x) = p(-x)$. Explain why $p(x)$ is even if and only if all the odd-indexed terms $c_1, c_3, \ldots, c_{2k+1}$ are zero.

25. Use the results of the preceding two exercises to explain why every other Bernoulli number starting with B_3 is zero.

26. One form of the so-called "**Ballot problem**" asks what the probability is that in an election between candidates A and B the number of votes for A always exceeds that for B until the last ballot is cast when the votes are tied. Suppose that a vote for A is denoted by $+1$ and a vote for B by -1. Then there must be an even number of voters, say $2m$. We want to determine the number of strings of m $+1$s and m -1s such that every partial sum, from 1 to $i < 2m$ is positive. Write down all such strings for $m = 1$, 2, and 3. Then check that the number with a final tie is given by the mth Catalan number as defined in Question 1.7.

NOTE. The Catalan numbers arise in a number of fundamental problems of computer science including the problem of having a computer evaluate an arithmetic expression; Exercises 27–29 explore these connections. To a computer each of the operations +, −, *, /, and ^ is a "binary operation." Each operation requires two numbers upon which to act. Parentheses tell us exactly which two numbers are to be combined into one by each operation.

27. Explain why the addition of parentheses makes a difference in the following expressions.
(i) $a - b - c$.
(ii) $x/y/z$.
(iii) r^{s^t}.

28. Calculate the number of ways to parenthesize expressions with three, four, and five variables by listing all possibilities. Show that these numbers are given by the corresponding Catalan numbers.

29. Suppose that we have an expression combining n variables, like

$$x_1 \mathbin{?} x_2 \mathbin{?} x_3 \mathbin{?} \cdots \mathbin{?} x_n$$

where ? stands for one of the usual arithmetic operations. Show that the number of different parenthesizations of an expression with n variables satisfies the Catalan recurrence relation.

30. Use the fact that

$$C_n = \frac{1}{n}\binom{2n-2}{n-1}$$

to derive a recurrence relation for C_n in terms of C_{n-1}, using no other Catalan number.

31. Show that

$$C_n \geq \frac{1}{n}2^{n-1}$$

for all positive n. Is there a positive integer k such that $C_n = O(n^k)$?

32. A tree is called a **planted planar** tree if one vertex of degree 1 is designated as the root r and then the tree is drawn in the plane. For example, the following diagram shows all different planted planar trees with 1, 2, and 3 edges. Let PT_n denote the number of different planted planar trees with n edges. Determine PT_4 and PT_5, and for $n = 1, 2, \ldots, 5$ show that $PT_n = C_n$.

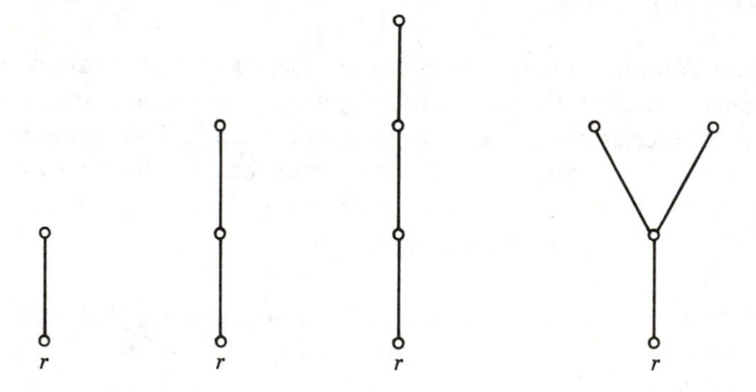

8

MORE GRAPH THEORY

8:1 MINIMUM-DISTANCE TREES

This chapter discusses five real-world problems that can be solved using graphs. Reflecting the current state of knowledge concerning graph algorithms, some of our problems have good solutions, while others have no known good algorithmic solution. In the latter case we present exponential algorithms. In Section 4 we present an **approximation algorithm**, that is, an algorithm that runs in polynomial time but doesn't necessarily give a best possible answer.

Our first application is plowing snow off the streets of a city. We envision two problems. The first consists of clearing roads connecting important city services along shortest routes. The second consists of finding a route that traverses every remaining street at least once but that is overall as short as possible. In a variation we seek a shortest route that visits a designated set of points in a city. As a fourth problem we design a program to position a laser bit to drill thousands of holes in a sheet of material. Finally, we consider storage allocation in computer memory.

To begin the first snowplowing problem, imagine that we are in charge of plowing the snow off the streets of a city all of whose essential services (e.g., police, fire protection, ambulance, and snowplow) are located in one building called City Hall. Within the city there are special facilities (e.g., hospitals and schools) that we would like to be able to reach with the essential services. How should we plow the streets to enable our vehicles to reach the special facilities as quickly as possible?

At first glance it seems reasonable to think of trees. In a tree every pair of distinct vertices is joined by a unique path. Here too it seems that we don't need more than one clear route joining different locations; however, a minimum-weight spanning tree (defined in Section 5.3) is not what we want. (See also Exercise 12.)

In particular, the snowplowing problem has a distinguished location, City Hall. Hence our graph model needs a distinguished vertex called the **root**. The remaining vertices of the graph correspond with principal intersections within the city. For simplicity assume that each of the special facilities is at one of these intersections. Two vertices of the graph will be joined by an edge if there is a direct road connection between the corresponding intersections. The weight attached to an edge will represent the length of that connection. Thus the resulting weighted graph models the streets of the city, and as a plan for emergency snowplowing we want a spanning tree of this graph with the property that the distance from the root to each of the special facility vertices along edges of the spanning tree is minimized. In fact, we find a spanning tree that contains a minimum-distance path from the root to every vertex. Such a subgraph is called a **minimum-distance spanning tree**.

Here are precise formulations of ideas from the preceding paragraph. Recall from Section 5.3 that the weight of a path P, denoted by $w(P)$, is the sum of the weights of all edges in P; we call this the **length** of P. Recall also that each edge of a weighted graph has positive weight.

Definition. In a weighted connected graph G, the **distance between two vertices x and y**, denoted by $d(x, y)$, is the minimum value of $w(P)$, taken over all paths P from x to y. (Informally, the distance from x to y is the length of the shortest path between them.) A **minimum-distance spanning tree** in a weighted connected graph G with root r is a tree T such that for each vertex v of G, the length of the unique path in T from r to v equals $d(r, v)$.

Example 1.1. In Figure 8.1 we show weighted graphs G and H with root r, their minimum-distance spanning tree, and their minimum-weight spanning tree. Note that the two types of trees may differ.

Problem. Given a weighted connected graph G and a root vertex r, find a minimum-distance spanning tree of G.

Question 1.1. For each vertex $v \neq r$ in the graph shown in Figure 8.2, find the shortest path from v to r. Does the union of these paths form a minimum-distance spanning tree? Pick a different vertex for the root and find all shortest paths to this new vertex. Does the union of these paths form the same tree?

There is a good algorithm to solve the minimum-distance problem, due to E. W. Dijkstra. It is not obvious that every connected weighted graph contains a minimum-distance spanning tree; it is conceivable that the union of shortest paths from r to different vertices contains a cycle. However, one consequence of Dijkstra's algorithm and the proof that it works is that minimum-distance spanning trees always exist. The fundamental idea is simple (and in a sense greedy). We shall

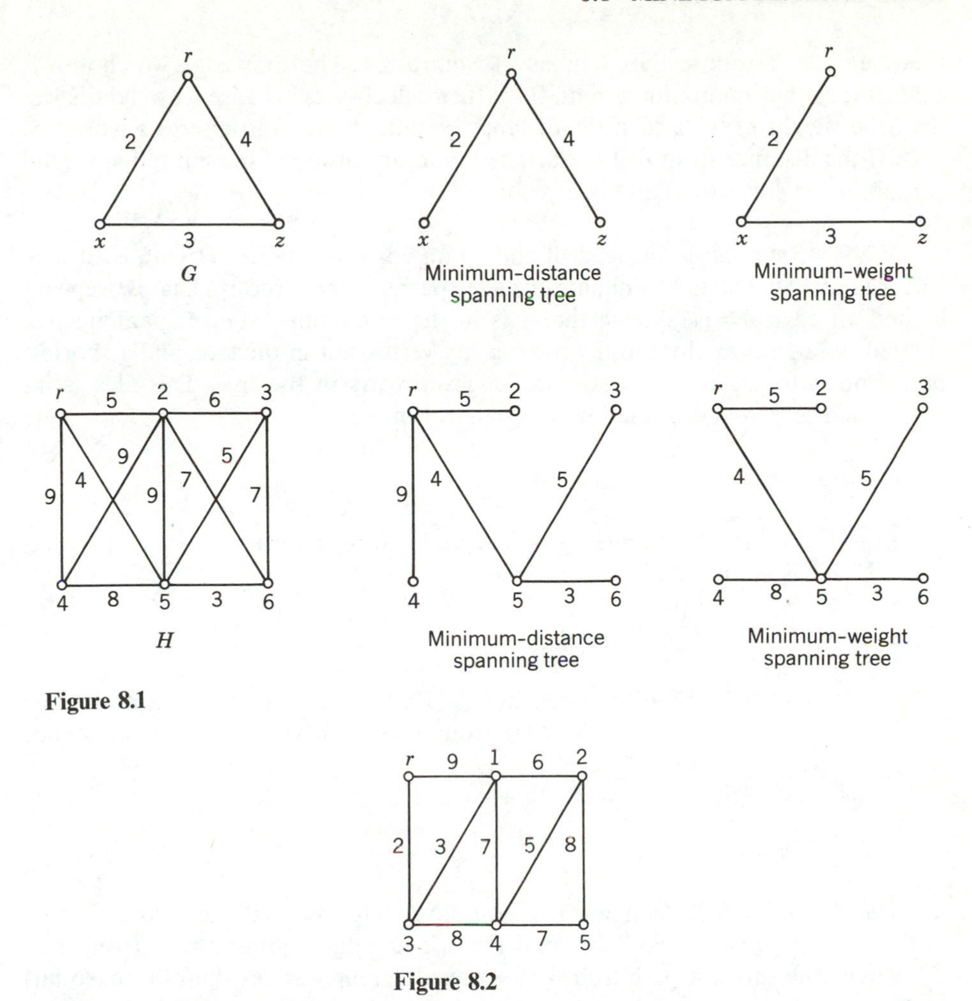

Figure 8.1

Figure 8.2

describe the algorithm informally and leave details of implementation to the exercises.

Begin with the root r. If we examine the edges incident with r and select one of smallest weight, say $e = (r, x)$, then the shortest path from r to x is the edge e, since we assume that all edge weights are positive. So far the minimum-distance tree consists of e and the vertices r and x. Next we want to extend this tree so that it remains a minimum-distance spanning tree for a subgraph of G. There are two kinds of edges that we might pick to add to the tree. We might select an edge of the form $f = (r, y)$ or $g = (x, z)$. Among the edges incident with r assume that f has minimum weight (other than e). Among the edges incident with x assume that g has minimum weight. A naive choice would be to select whichever of f and g has minimum weight. Unfortunately, this will not work in all cases.

Example 1.2. Suppose that G is as in Figure 8.1. The first edge we choose is (r, x), since it has minimum weight. If we then select (x, z) because its weight is less than the weight of (r, z), then the distance (within the tree) from r to z would be 5. In G the distance from r to z is 4. The minimum-distance spanning tree should consist of the edges (r, x) and (r, z).

What we should do, instead of adding an edge of smallest weight, is to pick a new edge that creates a minimum-distance path to a new vertex. That is, we want to find an edge $e = (y, z)$ such that y is in the minimum-distance spanning tree created so far, z is as close to the root as any vertex not in the tree, and a shortest path from z to the root r uses e and edges already in the tree. This idea is incorporated into step 4 of the following algorithm.

Algorithm DIJKSTRA

STEP 1. Input the weighted graph G and the root vertex r {Assume that G is connected.}
STEP 2. Set $T := \{r\}$
STEP 3. For $j = 1$ to $V - 1$ do
Begin
STEP 4. Find z, a vertex in $G - T$ whose distance from r is minimum; let e be the edge from z to T in some minimum-distance path from z to r
STEP 5. Set $T := T + z + e$
End
STEP 6. Output T and stop.

Step 4 in DIJKSTRA might raise a question. Suppose that z is a closest vertex of $G - T$ to r. How do we know that there is an edge e joining z to a vertex of T? Maybe the shortest path from z to r uses different vertices than those (so far) in T? That this problem will not arise is a consequence of the proof of Theorem 1.1.

Example 1.3. We trace Dijkstra on the graph shown in Figure 8.3. See Table 8.1.

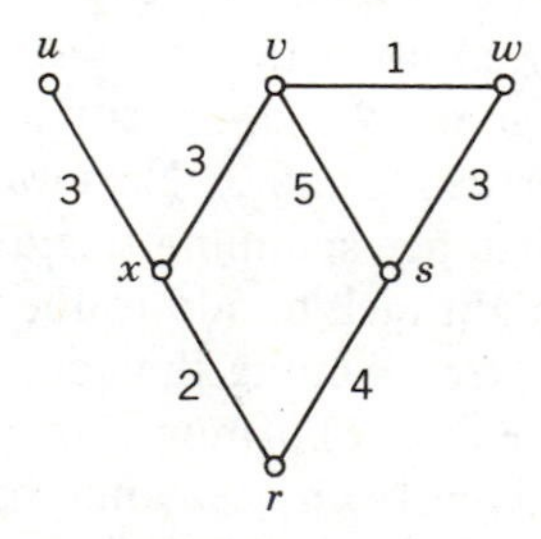

Figure 8.3

Table 8.1

Step No.	j	z	$V(T)$	$E(T)$
2	?	?	$\{r\}$	$\varnothing$
4	1	x		
5	1	x	$\{r, x\}$	$\{(r, x)\}$
4	2	s		
5	2	s	$\{r, x, s\}$	$\{(r, x), (r, s)\}$
4	3	u		
5	3	u	$\{r, x, s, u\}$	$\{(r, x), (r, s), (x, u)\}$
4	4	v		
5	4	v	$\{r, x, s, u, v\}$	$\{(r, x), (r, s), (x, u), (x, v)\}$
4	5	w		
5	5	w	$\{r, x, s, u, v, w\}$	$\{(r, x), (r, s), (x, u), (x, v), (v, w)\}$

Note that when we have a choice, as between u and v in the third application of step 4, we may choose either vertex.

Question 1.2. Given the weighted graph in Figure 8.4 with root r as shown, use DIJKSTRA to find the minimum-distance tree.

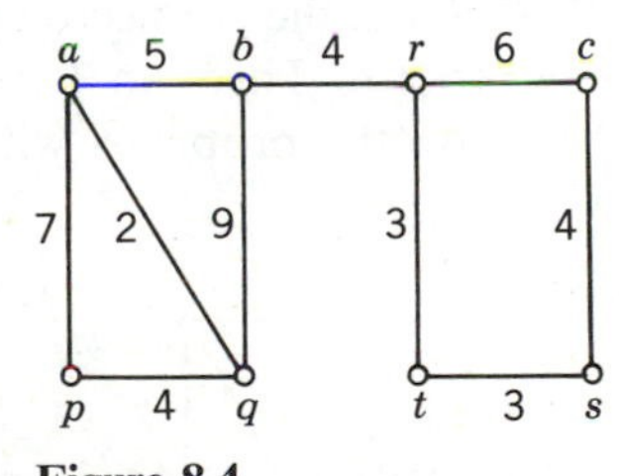

Figure 8.4

Theorem 1.1. DIJKSTRA produces a minimum-distance spanning tree T of a connected weighted graph G with root r.

Proof. We must prove that T is a spanning tree and that for each vertex v of G the distance from v to the root r along the edges of T equals $d(r, v)$, the length of a shortest path in G joining r and v. We prove by induction on $|V(T)|$ that at each stage T is a tree containing a minimum-distance path from each vertex of T to the root r. Thus when DIJKSTRA stops and $|V(T)| = V = |V(G)|$, T is a minimum-distance spanning tree.

Example 1.3 (reexamined). Look at T after the third completion of step 5. T contains four vertices and three edges and is a minimum-distance spanning tree of the subgraph of G that contains the vertices $\{r, x, s, u\}$ and the edges (r, x), (r, s) and (x, u).

Initially, $|V(T)| = 1$ since $V(T) = \{r\}$. In step 5 we add the edge $e = (r, x)$ of least weight and the vertex x to T. Then T is a tree with two vertices and one edge, and this edge provides the shortest path from x to r. Thus the base case is safely accounted for.

Assume that T is a tree containing minimum-distance paths whenever $|V(T)| < k$, and suppose that $|V(T)| = k$. T received a kth vertex, say v, and a $(k-1)$st edge, say e, in step 5. Then $T' = T - v - e$ was stored as T in the previous execution of step 5. Since T' contains $(k-1)$ vertices, by the inductive hypothesis it is a tree that contains minimum-distance paths from each of its vertices to r. At the next occurrence of step 4, the vertex v in $G - T$ was selected as a vertex of minimum distance to r. Since v is not in T, the addition of v and e does not create a cycle and T remains acyclic.

Suppose that $P = \langle v, x, \ldots, r \rangle$ is a minimum-distance path from v to r in G beginning with edge e. Since x is closer to r than v, x is in T'. Otherwise, DIJKSTRA would have selected x before v. Then we add v and $e = (v, x)$ to T', and $T = T' + v + e$ is acyclic and connected, hence a tree. Furthermore, the shortest path from x to r in T' plus e will be a shortest path from v to r in T. (See also Exercise 11.) This proves the inductive step. Thus the tree output by DIJKSTRA is a minimum-distance spanning tree. □

Question 1.3. Where in DIJKSTRA is the connectivity of the graph G essential? Find at least two places in the proof of Theorem 1.1 where we use the fact that edge weights are positive. What are the problems with running DIJKSTRA on the graph shown in Figure 8.5?

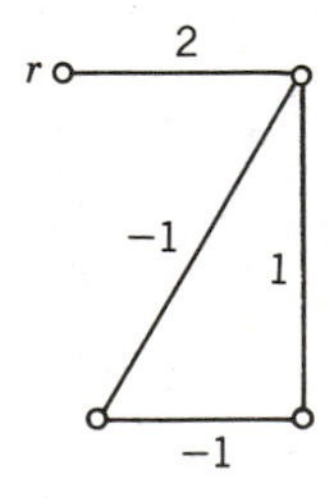

Figure 8.5

We now show that the complexity of DIJKSTRA is $O(V \cdot E)$. We need to perform comparisons and additions to find the minimum-distance paths; however, as in other graph complexity results we count only comparisons. The loop at step 3 occurs $V - 1$ times. In step 4 we need to check at most E edges that join a vertex of T with one in $G - T$ to find the next shortest path and the vertex z. Thus there are no more than $(V - 1)E$ comparisons needed in total. In Exercises 19 to 21 a more detailed version of DIJKSTRA is presented, and in that version we can see that only $O(V)$ comparisons are needed within the equivalent of step 4 so that DIJKSTRA has an overall complexity of $O(V^2)$. This was the original complexity bound obtained by Dijkstra. There has been considerable interest in this minimum-distance spanning tree problem, and variations of this algorithm using more sophisticated data structures have been developed, including one that has complexity $O(E\log(V))$.

With a slight change DIJKSTRA can be applied to unweighted graphs. Recall that distance in an unweighted graph has been defined to be the fewest number of edges in a path joining two vertices. (See Section 5.3.) Thus if we assign a weight of 1 to each edge of an unweighted connected graph and choose a root r, DIJKSTRA will find a minimum-distance spanning tree. In this context the resulting tree is known as a **breadth-first-search** (or **BFS**) **spanning tree**. Notice that when DIJKSTRA is applied to an unweighted graph, first it "visits" and adds in to the tree T all vertices adjacent to the root r. Next it "visits" and adds in all vertices adjacent to vertices adjacent to r, that is, it "visits" all vertices at distance 2 from r, and then successively "visits" all vertices at distance j from r for $j = 3, 4, \ldots$. "Visiting" vertices in a graph in this order is known as breadth-first search.

Example 1.4. Figure 8.6 shows a graph G and two BFS spanning trees of G. Working on the tree G with each edge weight 1, DIJKSTRA first adds edges $(r, 2)$ and $(r, 6)$ to the tree, since vertices 2 and 6 are at distance 1 from the root r. Next the vertices 3 and 5 at distance 2 from r are added to the tree; there is a choice of edges here. Finally, vertex 4 at distance 3 from r is added. Two possible breadth-first-search spanning trees are shown in Figure 8.6.

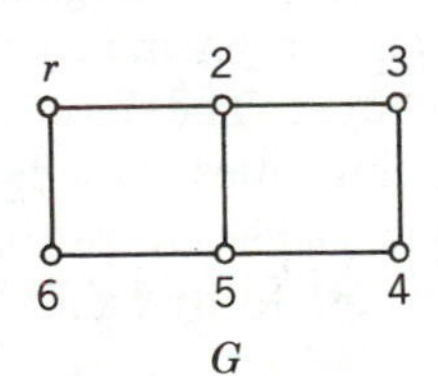

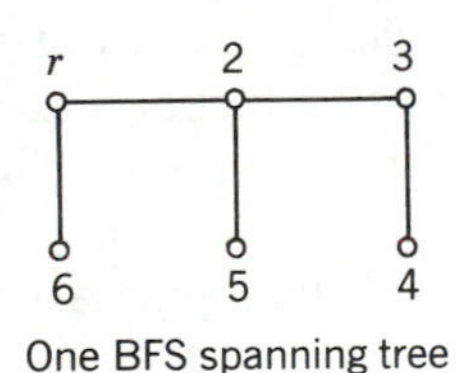

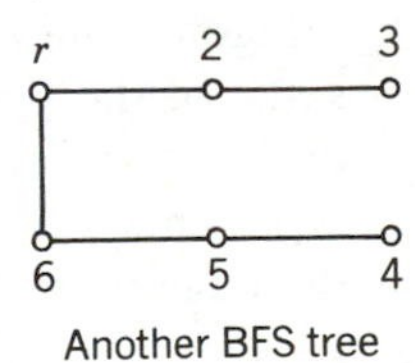

Figure 8.6

Here is a breadth-first-search algorithm, modeled upon DIJKSTRA. The set T contains the vertices and edges of the BFS tree.

Algorithm BREADTHFIRSTSEARCH (BFS)

STEP 1. Input the unweighted graph G and root r
STEP 2. $T := \{r\}$
STEP 3. For $j = 1$ to $V - 1$ do
  STEP 4. For each vertex v in $G - T$ adjacent to a vertex at distance $(j - 1)$ from r do
    STEP 5. Select w, one neighbor of v in T;
    set $T := T + (v, w) + v$
STEP 6. Output T and stop.

Notice that when applied to a disconnected graph, BFS visits and constructs a spanning tree on precisely the vertices in the same component as the root r. In applications it is common for BFS to perform some calculation when it visits a vertex and to output more than just the spanning tree. Breadth-first search is an important algorithmic technique that will be used again in Section 5.

EXERCISES FOR SECTION 1

1. Find a minimum-distance tree for each of the following graphs.

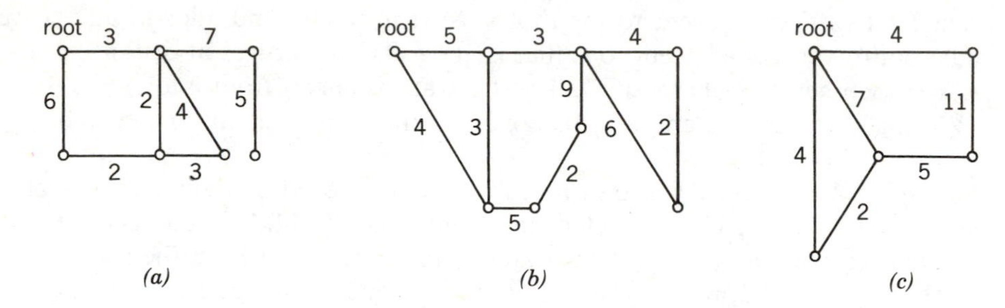

2. There do not exist direct flights from Bradley Field to all other airports in the United States. If you wish to fly from Bradley to, for example, Eugene, Oregon, you will have to change planes at least once. The graph indicates some of the possible connecting flights that you might choose. The vertices of the graph are labeled with the names of the corresponding cities. An edge represents a direct flight between the two corresponding cities. The weight on the edge indicates the cost of the flight. Assume that the cost to fly from A

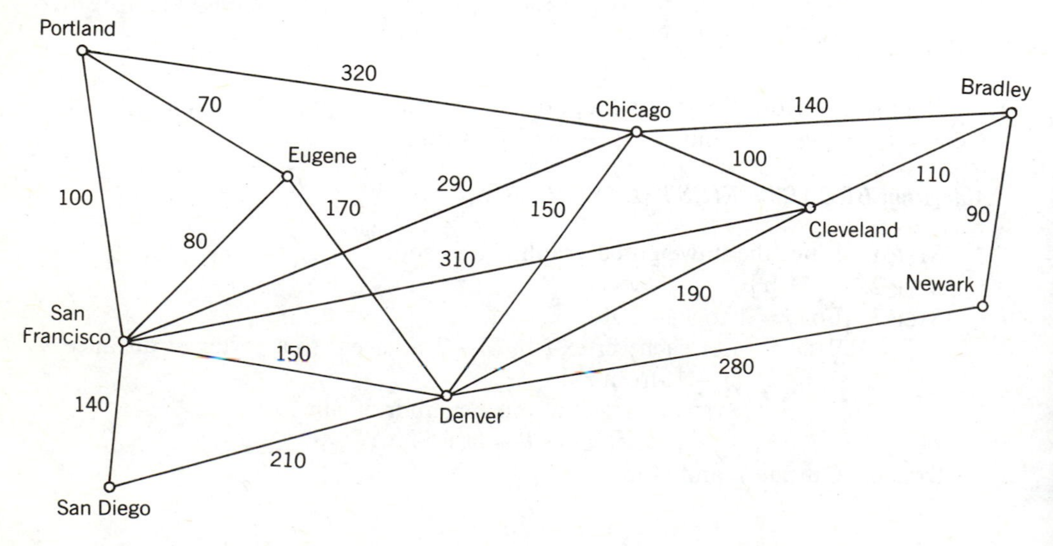

to C changing planes at B equals the cost to fly from A to B plus the cost to fly from B to C. Find a minimum-cost trip from Bradley to Eugene.

3. Rewrite DIJKSTRA so that upon input of a weighted connected graph G and two vertices x and y, it finds a shortest path from x to y. Is your algorithm necessarily more efficient than (the original version of) DIJKSTRA?

4. Find an unweighted connected graph G so that every spanning tree of G is a minimum-distance spanning tree for some choice of the root.

5. Find a weighted graph G so that no matter what vertex is chosen for the root of G, the minimum-distance spanning tree is heavier than the minimum-weight spanning tree (i.e., for every vertex v the sum of the edge weights of a minimum-distance spanning tree with root v is larger than the weight of a minimum-weight spanning tree).

6. Rewrite DIJKSTRA so that it finds a minimum-distance spanning tree if the graph is connected, or else reports that the graph is not connected.

7. Here is a table of costs of some intercity flights; a zero indicates no direct flight. Find the cost of a cheapest trip between every pair of cities.

	C_1	C_2	C_3	C_4	C_5
C_1	—	100	0	150	210
C_2	100	—	0	90	0
C_3	0	0	—	50	280
C_4	150	90	50	—	0
C_5	210	0	280	0	—

8. Run the unweighted version of DIJKSTRA on each of the graphs in Exercise 1 with the edge weights discarded (i.e., run it on the underlying unweighted graphs.)

9. A new commuter airline called Capital Cities offers flights between the capital cities of every pair of states that share a border. So, for example, there is a flight from Pierre, South Dakota, to Bismark, North Dakota, since these two states share a common boundary. Each such flight costs $25. Using Capital Cities, what is the cost of a cheapest trip from Boston to Sacramento? Bismark to Trenton?

10. Suppose that G is a weighted connected graph and that one path $P = \langle x_1, x_2, \ldots, x_j, r\rangle$ to the root is designated as top priority. Rewrite DIJKSTRA so that it finds a spanning tree that includes P. For every vertex v not in P, the algorithm should find as short a path from v to r as possible.

11. Suppose that $P = \langle x, y, \ldots, r\rangle$ is a shortest path from the vertex x to the root r. Explain why the same path, minus x and starting at y, $\langle y, \ldots, r\rangle$, is a shortest path from y to r. Suppose that P_1 is a shortest path from a vertex x to a vertex z and P_2 is a shortest path from z to r. Is P_1 followed by P_2 a shortest path from x to r?

12. Suppose that a city wants a snowplowing plan to connect City Hall with each of the designated special facilities; however, the plowing budget is greatly overspent. If the sole criterion for choosing plowing routes is that the total plowing cost should be a minimum, then explain why a minimum-weight spanning tree rather than a minimum-distance spanning tree provides the best plan.

13. Prove that at the end of the algorithm BFS precisely those vertices in the same component as the root r are contained in T.

14. Rewrite BFS so that the vertices are assigned a number giving the order in which they become visited. (*Note:* This ordering is not unique but depends on arbitrary choices made within the algorithm.)

15. Verify that BFS performs at most $O(V^2)$ comparisons given a graph with V vertices.

16. Rewrite BFS so that it performs a breadth-first search on each connected component of G.

17. Construct a BFS algorithm that given a graph computes the eccentricity of every vertex. The eccentricity is defined in Exercise 5.4.14.

18. The **radius** $r(G)$ and **diameter** $d(G)$ of a graph G are respectively the minimum and maximum value of the eccentricity (see the preceding exercise). Construct an algorithm that, given a graph G, outputs the radius and diameter of G. Show that $d(G)/2 \leq r(G) \leq d(G)$ for any graph G. Find graphs to show that there are no better bounds than those given by the inequality above.

19. Here is a detailed version of DIJKSTRA that specifies the equivalent of step 4.

Algorithm DIJKSTRA2

STEP 1. Input the weighted graph G and the root vertex r {Assume that G is connected.}
STEP 2. Set $d(r) = 0$ {$d(x)$ denotes the distance of x from the root in the partial tree.}
STEP 3. Set $V(T) := \{r\}$; $E(T) = \varnothing$ {These will contain, respectively, the vertices and edges of T.}
STEP 4. For $j = 1$ to $V - 1$ do
Begin
STEP 5. For each (t, x) in $E(G)$ with t in $V(T)$ and x in $V(G) - V(T)$ do
STEP 6. Set $c(t, x) := d(t) + w(t, x)$

{At this point $c(t, x)$ indicates the length of the path from r to x using a path in T together with the edge (t, x).}

STEP 7. Set $c_{\min} :=$ minimum value of $c(t, x)$

STEP 8. Set $x_{\min}$ and $t_{\min}$ equal to the vertices that achieve the minimum of step 7

STEP 9. Set $V(T) := V(T) + x_{\min}$

STEP 10. Set $E(T) := E(T) + (t_{\min}, x_{\min})$

STEP 11. Set $d(x_{\min}) := c_{\min}$

End {step 4}

STEP 12. Output $E(T)$ and stop.

Run DIJKSTRA2 on each of the graphs in Exercise 1.

20. Explain why in DIJKSTRA2 when a vertex $x_{\min}$ and the edge $(t_{\min}, x_{\min})$ are added to the tree T, T is a tree containing a minimum-distance path from $x_{\min}$ to r.

21. Count the maximum number of additions and comparisons performed in DIJKSTRA2 and show that each is $O(V^2)$.

22. In the remarks following DIJKSTRA's algorithm we asserted that there is an algorithm for the minimum-distance spanning tree problem that runs in time $O(E \log(V))$. For what graphs is this a better bound than the $O(V^2)$ complexity bound that is obtained in the preceding exercise?

23. Modify DIJKSTRA2 so that the shortest path from each vertex to r is maintained as well as the distance of that path.

24. Using BFS construct an algorithm to check whether a connected graph is bipartite or not. (See Section 5.2 for the definition of bipartite. See also Supplementary Exercise 10 of Chapter 5.)

25. Modify DIJKSTRA so that for every pair of distinct vertices v and w in a weighted connected graph, the shortest path between v and w and its length is found. This is known as the **All Pairs Problem**. Show that the complexity of the All Pairs Problem is at most $O(V)$ times the complexity of the minimum-distance spanning tree problem.

8:2 EULERIAN CYCLES

We continue with snowplowing. To repeat the setting, suppose that a weighted graph is drawn to model city streets. Each vertex represents an intersection, and two vertices are joined by an edge if the corresponding intersections are joined by a direct road connection. The weight of an edge is the length of the road. The problem is to plow the streets efficiently (or if, as in the preceding section, certain streets are already clear, to plow the remaining streets efficiently). More precisely, the problem is to devise a plan to travel along each unplowed street at least once in as short a trip as possible, beginning and ending at City Hall. Of course, it would be most efficient to plow the streets with no repetitions. Is this possible, and if so, how can such a plan be found?

Example 2.1. Consider Figure 8.7. The graph G contains a cycle (for example $\langle r, x, c, y, x, b, a, r \rangle$) that traverses every edge exactly once. Although the graph H contains no such cycle, it does contain the path $\langle r, u, s, v, r, s \rangle$ that traverses every edge exactly once. The graph I is a 3×2 grid graph. In Chapter 3 we saw that it was impossible to traverse each edge of this graph exactly once.

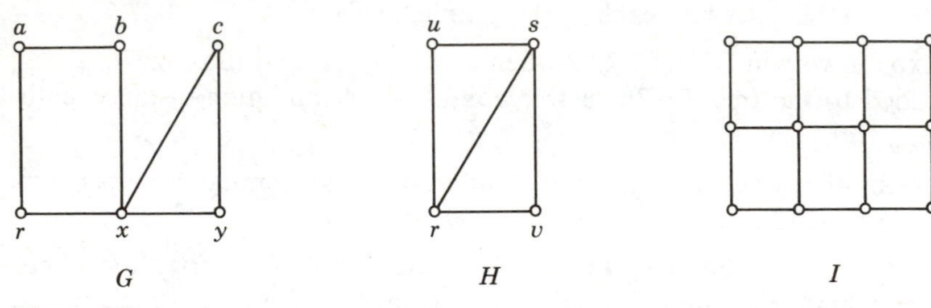

Figure 8.7

Definition. A path or cycle that includes every edge of a graph exactly once is called **Eulerian**. A graph that contains an Eulerian cycle is called an **Eulerian graph**.

Question 2.1. (a) Find an Eulerian graph with four vertices. (b) Find a graph with eight vertices that is not Eulerian but contains an Eulerian path. (c) Find a connected graph with six vertices that does not contain an Eulerian path.

The problem of characterizing the graphs that contain Eulerian paths led to the first graph theory paper, written in 1736 by Leonhard Euler. Euler was visiting Königsberg, a town with seven bridges and demonstrated that it was impossible to take a walk crossing every bridge exactly once.

Question 2.2. Which of the graphs in Figure 8.8 are Eulerian? Which contain Eulerian paths but not Eulerian cycles?

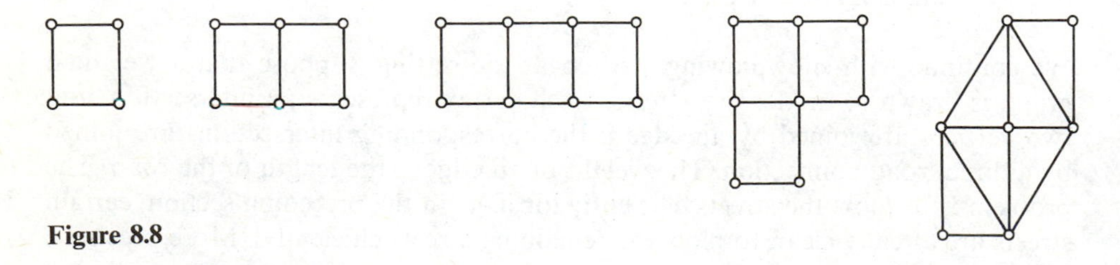

Figure 8.8

Theorem 2.1 (Euler's Theorem). A connected graph is Eulerian if and only if every vertex has even degree.

Euler's theorem tells the snowplow planners that they can plow each street exactly once if and only if an even number of streets comprise every intersection. Note that one can think of an Eulerian cycle as starting and ending at any vertex, in particular City Hall. Our proof of Euler's theorem will lead to an efficient algorithm for finding Eulerian paths and cycles. Then we shall consider ways to modify this algorithm to be useful in diverse settings.

Proof of Euler's theorem. First notice that an Eulerian graph must be connected. Let C be an Eulerian cycle in a graph G. Pick an arbitrary vertex of G, say x. We can assume that C begins at x, leaving on an edge, say e_1, and at some point returns to x on, say e_2. If that is the end of C, then $\deg(x) = 2$, an even number. Otherwise, C leaves x again on, say e_3, and later returns on, say e_4. Each time that C leaves x on an edge e_i, it returns on a different edge e_{i+1}. Since the edges at x can be paired, e_1 with e_2, e_3 with e_4, and so on, there must be an even number of them. Thus $\deg(x)$ is even. Since x was chosen arbitrarily, every vertex of G must have even degree.

Next we prove the converse, that a connected graph with all vertices of even degree contains an Eulerian cycle. We prove this by induction on V. If $V = 1$, then the graph contains no edges and vacuously satisfies the conclusion. A connected graph with $V = 2$ consists of one edge and so does not have vertices of even degree. If $V = 3$, then the only connected graph with all vertices of even degree is the 3-clique, K_3, and this graph contains an Eulerian cycle.

Question 2.3. Find every connected graph with four or five vertices all of whose degrees are even. Show that each such graph is Eulerian.

Assume that every connected graph with fewer than k vertices all of whose degrees are even contains an Eulerian cycle. Let G be a connected graph with k vertices all of even degree. Pick a vertex, say x, and create a path P beginning at x. Extend P, appending incident unused edges at its end, until this is no longer possible. We claim that P is a cycle, ending at x. Since every vertex of G has even degree, when P arrives at a vertex $v \neq x$ there is always an unused edge on which to leave v. Thus P must end at x; hence we rename P as C, since it is a cycle. If C traverses every edge of G, then it is the sought-after Eulerian cycle.

Example 2.2. Consider the graph shown in Figure 8.9. The cycle $C' = \langle r, 2, 5, 6, r \rangle$ can be extended further at r. The cycle $C = \langle r, 2, 5, 6, r, 7, 8, r \rangle$ cannot be extended further at r; however, C is not an Eulerian cycle.

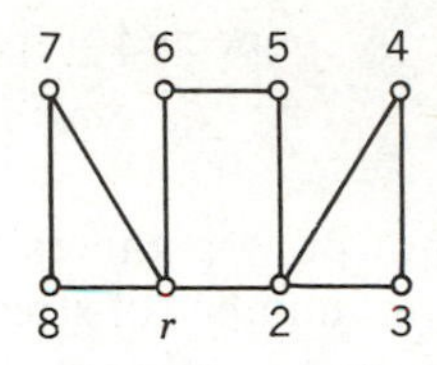

Figure 8.9

If C is not an Eulerian cycle, then we construct G' from G by erasing all edges of C. Since C is a cycle, we erase an even number of edges at each vertex. Since all vertices had even degree originally, their degree remains even in G'. In G' the vertex x has degree 0. Thus each connected component of G' contains fewer than k vertices and is consequently an Eulerian graph.

Question 2.4. Let C be as in Example 2.2. Identify the graph G' obtained by deleting the edges of C.

If H is a component of G', there is a vertex h of H that is also in C, since G was connected. By the inductive hypothesis an Eulerian cycle D can be found on H, beginning and ending at h. Then the original cycle C can be extended by inserting D at the vertex h. This extension can be done for each component of G'. The resulting cycle will traverse every edge of G and so is an Eulerian cycle. □

Example 2.3. In the graph shown in Figure 8.10 let $C = \langle r, 6, 7, r \rangle$ and let H be the component consisting of the four vertices $\{3, 4, 5, 6\}$ and their incident edges. Then with $h = 6$ let $D = \langle 6, 3, 4, 5, 6 \rangle$, an Eulerian cycle on H. This can be merged with C to form the larger cycle $C' = \langle r, 6, 3, 4, 5, 6, 7, r \rangle$.

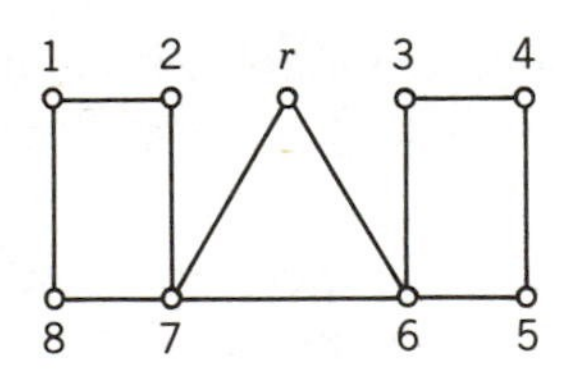

Figure 8.10

Theorem 2.1 also leads to conditions under which a graph contains an Eulerian path but not an Eulerian cycle. One such example was the graph H in Figure 8.7.

Corollary 2.2. A connected graph contains an Eulerian path, but not an Eulerian cycle, if and only if exactly two vertices have odd degree.

Proof. Suppose that x and y are the end vertices of the Eulerian path. Let G' be the graph obtained from G by creating a vertex r adjacent to x and y but to no other vertex. If G contains an Eulerian path, then this path together with the two edges incident with r form an Eulerian cycle. Since by Theorem 2.1 G' has all vertices of even degree, G has exactly two vertices of odd degree. Conversely, if G is a connected graph with exactly two vertices of odd degree, then G', formed by adding a vertex r adjacent to the two vertices of odd degree, is connected and has every vertex of even degree. By Theorem 2.1 G' has an Eulerian cycle. We can imagine that this cycle begins at x and proceeds to r and then y. In G this cycle becomes an Eulerian path from y to x. □

Question 2.5. The graph in Figure 8.11 contains exactly two vertices of odd degree. Create an Eulerian graph as in the proof of Corollary 2.2. Find an Eulerian cycle on the larger graph and from that an Eulerian path on the original graph.

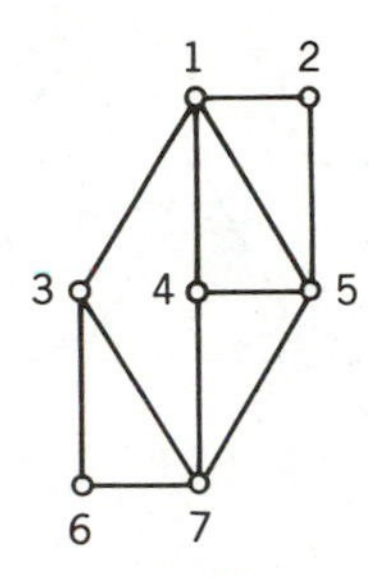

Figure 8.11

The proof of Theorem 2.1 is constructive. That is, it gives us the idea for an efficient algorithm for finding an Eulerian cycle.

Algorithm EULER

STEP 1. Input G, a connected graph with all vertices of even degree; set $C := \langle x \rangle$ $\{x$ arbitrary in $V(G)\}$
STEP 2. While $|E(G)| > 0$ do
Begin
STEP 3. Pick x in $V(C)$ with $\deg(x, G) > 0$
STEP 4. Create a maximal cycle D beginning at x $\{D$ cannot be made longer by appending edges at its end.$\}$
STEP 5. Set $E(G) := E(G) - E(D)$
STEP 6. Set $C :=$ cycle obtained from C by inserting D at x
End
STEP 7. Output C and stop.

Example 2.3 (again). Here is a trace (Table 8.2) of EULER, run on the graph shown in Figure 8.12.

Table 8.2

Step No.	x	C	D
1	r	$\langle r \rangle$	
3	r		
4			$\langle r, 6, 7, r \rangle$
6		$\langle r, 6, 7, r \rangle$	
3	6		
4			$\langle 6, 3, 4, 5, 6 \rangle$
6		$\langle r, 6, 3, 4, 5, 6, 7, r \rangle$	
3	7		
4			$\langle 7, 2, 1, 8, 7 \rangle$
6		$\langle r, 6, 3, 4, 5, 6, 7, 2, 1, 8, 7, r \rangle$	

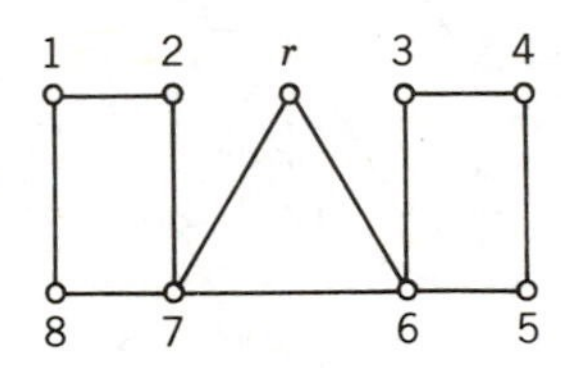

Figure 8.12

Notice that the cycles C and D can be created in any way consistent with the algorithm. For example, the initial cycle C might be $\langle r, 7, 8, 1, 2, 7, 6, r \rangle$ and the first cycle D might be $\langle 6, 5, 4, 3, 6 \rangle$. If this algorithm were implemented on a computer, which particular cycles C and D are created depends on how the graph G is stored. When tracing these cycles by hand, we can choose edges however we wish as long as a cycle is formed.

Question 2.6. Trace the algorithm EULER on the graph in Figure 8.13.

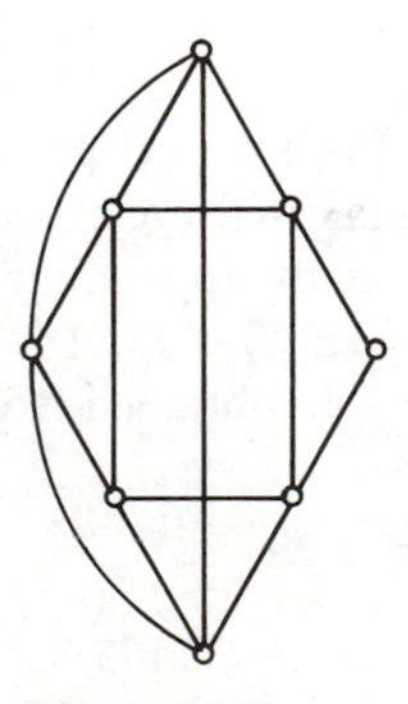
Figure 8.13

We have already proved that EULER works correctly, since it follows the proof of EULER's theorem.

Of course, to use EULER on an arbitrary graph G we would need to verify that G is connected. Since G is connected if and only if it contains a spanning tree by Exercise 5.3.13, connectivity could be checked using the algorithm SPTREE of Exercise 5.3.19 or KRUSKAL or BFS (of Section 8.1.) If G is connected, we could then determine if its vertex degrees are all even. If so, we could run EULER.

Before returning to the snowplowing problem, we consider the complexity of the algorithm EULER. We choose to count comparisons and note that step 4 is the critical step to examine. Suppose that the graph is input as an adjacency matrix. Creating cycle D in step 4 can be accomplished one edge at a time by finding a 1 in the row of the matrix that corresponds with the current vertex. It will take no more than V such comparisons to find the next edge. Thus step 4 may require $O(V \cdot E)$ comparisons. Actually, with appropriate data structures EULER can be made linear in E. (Details appear in Exercise 12.)

The algorithm EULER is designed to run on unweighted graphs. In the original snowplowing problem the related graph was a weighted one. The theory and algorithm so far deal only with a special case, when all vertices of the graph derived from the street system have even degree. In this case we can plow each city street exactly once, and regardless of the street lengths (or edge weights) we have found a minimum-weight cycle without any repetition.

Otherwise, what are the possibilities? Question 2.3 of Chapter 5 states that every graph contains an even number of vertices of odd degree. Thus there cannot be just one vertex of odd degree, but there might be exactly two vertices of odd degree in the graph of the streets. Suppose that City Hall is located at one vertex of odd degree. Then we could use an Eulerian path algorithm (see Exercise 13) to plow each street exactly once, ending at the other vertex of odd degree; call it z. To conclude, the plows would travel home from z to City Hall on a shortest path. This shortest path from z to City Hall could be found using DIJKSTRA with z as the root. In Exercises 8 to 10 you are asked to verify that this scheme does produce the overall most-efficient plowing plan when there are only two vertices of odd degree.

When there are four or more vertices of odd degree, the most-efficient snowplowing plan involves paths between pairs of these vertices of odd degree. (See Exercises 5 to 7). In fact, there is an efficient algorithm to solve this problem in full generality. It consists of pairing up the vertices of odd degree so that the sum of the distances of the minimum-distance paths between pairs is minimized. This is known as the **Minimum Weight Matching Problem** or the Chinese Postman Problem (named after the Chinese mathematician, M-K. Kwan, who first considered this problem.) The solution is complex, beyond the scope of this chapter.

But wait—there's something unsatisfactory about the snowplowing model. Most snowplows don't plow streets just once, but rather twice, once in each direction to clear both sides of the street. Can we devise an algorithm to traverse each edge of a graph in this more realistic way? The answer is yes, quite easily, given our experience with Eulerian graphs.

We model this new situation with a directed graph, that is, a graph in which each edge is given an orientation or direction from one incident vertex to the other. In directed graphs paths and cycles must traverse each edge in its given orientation.

Definition. A **directed graph** G consists of a finite set $V(G)$ of vertices and a finite set $E(G)$ of edges (also called arcs) such that each edge consists of an ordered pair of distinct vertices of $V(G)$. We think of the edge $e = (x, y)$ as being directed from x to y.

Example 2.5. Figure 8.14 shows some directed graphs.

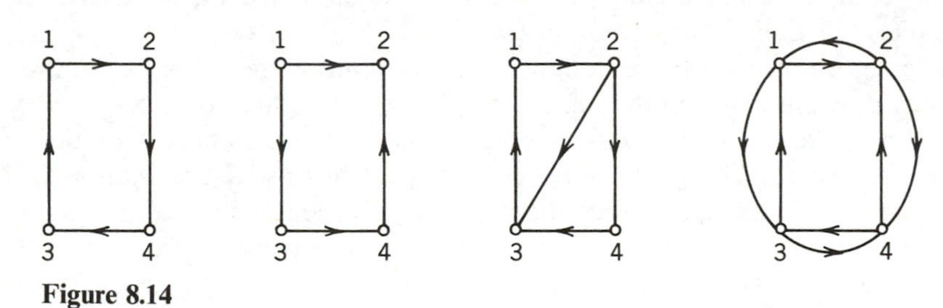

Figure 8.14

Definition. A **path in a directed graph** G from x to y is a sequence of distinct edges, $e_1, e_2, \ldots, e_k$, such that $e_1 = (x, x_1)$, $e_2 = (x_1, x_2), \ldots, e_k = (x_{k-1}, y)$ for some vertices $x_1, x_2, \ldots, x_{k-1}$. *A* **cycle in a directed graph** is a path from a vertex to itself. An **Eulerian path** (respectively, **Eulerian cycle**) is a path (respectively, cycle) that includes every edge exactly once.

Question 2.7. For each graph in Example 2.5 find an Eulerian path or cycle if there is one.

Theorem 2.3. A directed graph whose underlying undirected graph is connected contains an Eulerian cycle if and only if at every vertex v the number of edges directed in to v equals the number of edges directed out of v.

Proof. Follow the same proof as given for Theorem 2.1. □

Theorem 2.3 can be applied to solve the snowplow problem completely. If G is the (undirected) graph derived from the city street plan, let D be the directed graph created by replacing every edge of G by two directed edges, one in each direction. Then an efficient snowplowing plan would be an Eulerian cycle on D.

Since each edge in G is replaced by two directed edges in D, a vertex of degree k in G becomes a vertex in D with k edges directed in to it and k edges directed out of it. Hence the conditions of Theorem 2.3 are met and D contains an Eulerian cycle. How do we find an Eulerian cycle? Exercise 18 asks you to modify EULER so that it works on directed graphs.

EXERCISES FOR SECTION 2

1. Which of the following graphs are Eulerian?

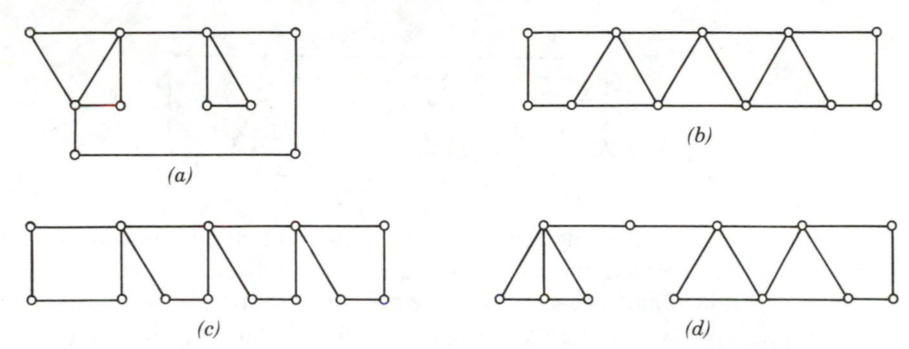

2. Can an Eulerian graph contain an odd number of edges? Either find an example or prove that there is none.
3. Suppose that $C = \langle x_1, x_2, \ldots, x_k, x_1 \rangle$ is a cycle in an Eulerian graph G with no repeated vertices (except for the start and finish at x_1). Show that there is an Eulerian cycle on G that visits the vertices $x_1, x_2, \ldots, x_k$ in the same order as in C, that is, for each $i = 1, 2, \ldots, k - 1$, vertex x_i is reached before vertex x_{i+1}.
4. An Eulerian graph is called **arbitrarily traceable at a vertex** x if every cycle beginning and ending at x can be extended to an Eulerian cycle by continuing the cycle at x. Prove that G is arbitrarily traceable if and only if every cycle of G passes through x.
5. Show that the following graphs can have their edges divided into two paths, each joining two vertices of odd degree.

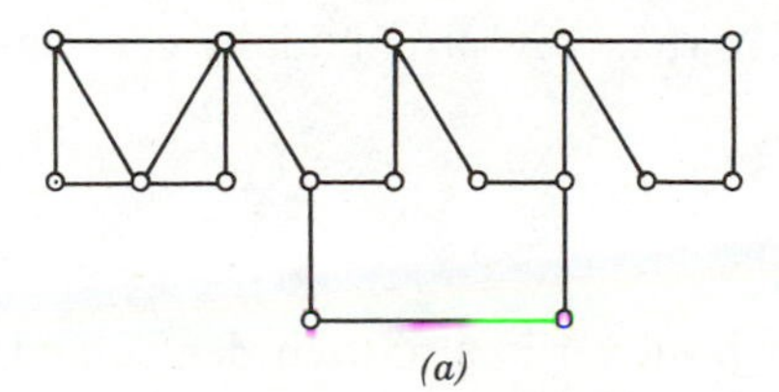

(a)

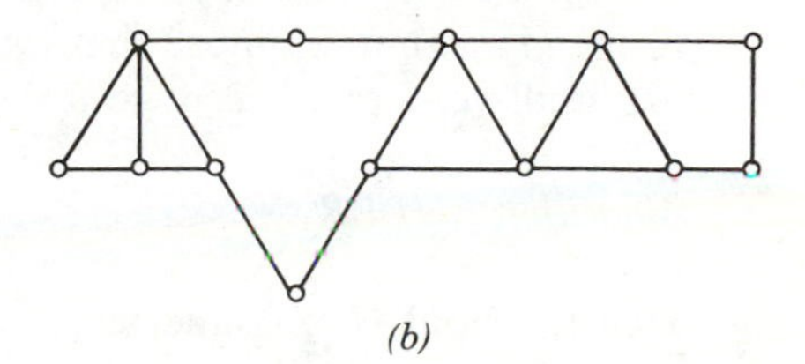

(b)

6. Show that any graph with exactly four vertices of odd degree can have its edges divided into two paths, one path joining a pair of the vertices of odd degree and the other path joining the other pair.
7. Show that a graph with exactly $2k$ vertices of odd degree can have its edges divided into k paths, each path joining a different pair of vertices of odd degree.
8. In the following graph, show that the shortest "snowplowing plan" that traverses each edge at least once consists of an Eulerian path from r to z and then a minimum-distance path from z to r.

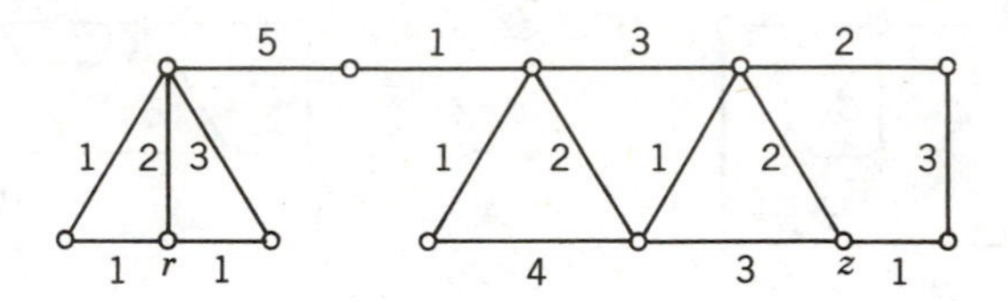

9. Suppose that G is a connected (unweighted) graph that contains an Eulerian path but not an Eulerian cycle. Let r and s be the vertices of odd degree. Show that the "trail" that traverses every edge of G at least once, starting and ending at r, and uses the fewest edges, consists of an Eulerian path from r to s plus a shortest path back from s to r. (*Hint:* Let C be a minimum trail that traverses every edge at least once, and let $G^{\#}$ be the graph formed by adding an edge for each repeated use of an edge by C. Thus C is an Eulerian cycle on the graph $G^{\#}$. Study the graph formed from the new edges in $G^{\#}$. Actually, $G^{\#}$ will be a **multigraph**, i.e., some pairs of vertices can be joined by more than one edge.)
10. Answer the same question as in Exercise 9, only assume that the graph G is weighted and path length, as usual, is the sum of the edge weights of edges in the path.
11. Suppose that G is a connected weighted graph with exactly two vertices of odd degree and let v be an arbitrary vertex of G. Describe a shortest trail on G that begins and ends at v and covers each edge at least once. Prove that your trail is the shortest.
12. Here is a closer look at the complexity of EULER with an expanded version of the algorithm. Suppose that for each vertex v, $Nbor(v)$ initially contains a list of all vertices adjacent to v.

Algorithm EULER

STEP 1. Input G, a connected graph with all vertices of even degree and E edges; set $C = \langle x \rangle$ $\{x$ arbitrary in $V(G)\}$

STEP 2. While $|E(G)| > 0$ do
Begin
STEP 3. Pick x in $V(C)$ with $\deg(x, G) > 0$
STEP 4. Call Trace (D, x) {The procedure Trace finds a maximal cycle D beginning at x}
STEP 5. Set $C :=$ cycle obtained from C by inserting D at x
End
STEP 6. Output C and stop.

Procedure Trace (D, x)

STEP 1. Set $z := x$
STEP 2. While $Nbor(z) \neq \varnothing$ do
Begin
STEP 3. Pick w in $Nbor(z)$
STEP 4. Add w to D
STEP 5. $Nbor(z) := Nbor(z) - \{w\}$; $Nbor(w) := Nbor(w) - \{z\}$
STEP 6. $E(G) := E(G) - (z, w)$
STEP 7. Set $z := w$
End
STEP 8. Return

In this version comparisons are made in steps 2 and 3 in the main program and in step 2 of Trace. Suppose that the algorithm cycles through (the main) step 2 s times, creating the cycles $D_1, D_2, \ldots, D_s$, which together form the Eulerian cycle. Find an upper bound on s in terms of E and then show that the total number of comparisons is $O(E)$.

13. Modify EULER so that upon input of a connected graph with exactly two vertices of odd degree, x and y, it traces an Eulerian path, beginning at x and ending at y.

14. Find conditions under which a directed graph contains an Eulerian path but not an Eulerian cycle. Prove your result.

15. Prove that an undirected graph G is Eulerian if and only if there is a way to direct its edges so that the resulting directed graph contains an Eulerian cycle.

16. Give an example of an undirected Eulerian graph G and then a direction on each of its edges so that the resulting directed graph does not contain an Eulerian cycle.

17. Explain how a directed graph can be stored in a $V \times V$ adjacency matrix.

18. Rewrite the algorithm EULER so that upon input of a directed graph, whose underlying undirected graph is connected, it finds an Eulerian cycle if there is one.

19. A **spanning in-tree** of a directed graph G with root r is defined to be a spanning tree of the underlying undirected graph such that each path P from a vertex y to r within the spanning tree is a directed path from y to r in G. Find examples of directed graphs that do and do not contain a spanning in-tree. Find an example of a graph G with root r that has a spanning in-tree but such that with some other vertex s as root there is no spanning in-tree.

20. Explain why in a spanning in-tree with root r, each vertex, except for r, has exactly one edge of the tree directed out of the vertex.

21. Let G be a directed graph that contains an Eulerian cycle and r an arbitrary vertex. Prove that G has a spanning in-tree with root r.

22. Devise an algorithm that upon input of a directed graph that contains an Eulerian cycle finds a spanning in-tree.

23. Suppose that G is a directed graph that contains an Eulerian cycle. Explain why the following is a valid algorithm for finding an Eulerian cycle in such a graph. First find a spanning in-tree with root r, as in the previous exercise. Then construct a cycle by appending incident, unused edges in any way except that at each vertex $v \neq r$ the unique out-directed edge in the spanning in-tree should be saved for the last exit from v.

24. A directed graph is called **strongly connected** if for every pair of vertices x and y there is a path from x to y and a path from y to x. A directed graph is called **connected** if for every pair of vertices x and y there is a path from one to the other. A directed graph is called **weakly connected** if the underlying undirected graph is connected. Find examples of the following types of directed graphs.

(*a*) The graph is strongly connected.
(*b*) The graph is connected but not strongly connected.
(*c*) The graph is weakly connected but not connected.
(*d*) The graph satisfies none of the connectivity definitions.

25. Explain why a graph that is strongly connected is also connected and why a graph that is connected is also weakly connected.

26. Prove that a directed graph that contains an Eulerian cycle is strongly connected. Is the same true for a directed graph that contains an Eulerian path? If so, explain why; if not, find additional conditions which when met by the graph ensure that it is strongly connected.

8:3 HAMILTONIAN CYCLES

Suppose that a mail carrier wants to pick up mail from every mailbox in town, or an inspector wants to check the traffic signals at every intersection. Can these jobs be accomplished efficiently by visiting each location exactly once? These problems are modeled by constructing an appropriate graph with a vertex for each location and two vertices joined by an edge if there is a street connection that passes through no intermediate location. The problem is to find a cycle that passes through each vertex of the graph exactly once.

This graph theory problem is known as the **Hamiltonian cycle problem**, named for Sir W. R. Hamilton, the inventor of a related game (see Exercise 3). It is one of graph theory's most demanding unsolved problems. In this section we search for conditions that guarantee the existence of a Hamiltonian cycle and an efficient algorithm to find such a cycle. We also learn the important algorithmic technique of depth-first search. This can be used to solve the Hamiltonian cycle problem for an arbitrary graph, although not efficiently. In the following section we turn to the equally challenging problem of finding a minimum-weight Hamiltonian cycle in a weighted complete graph.

Question 3.1. Which of the graphs in Figure 8.15 contains a cycle that visits each vertex exactly once?

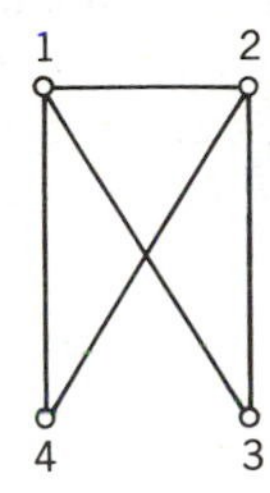

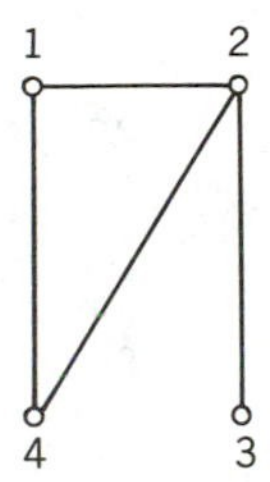

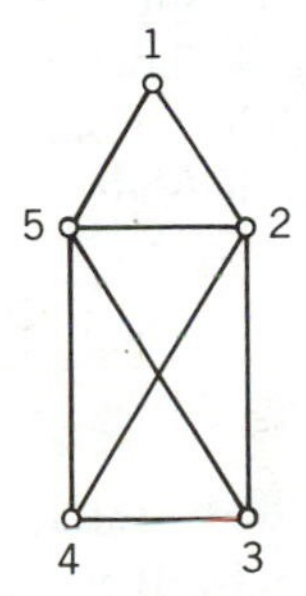

Figure 8.15

Definitions. In a graph G with V vertices, a path (or cycle) that contains exactly $V - 1$ (respectively, V) edges and spans G is called **Hamiltonian**. A graph is called **Hamiltonian** if it contains a Hamiltonian cycle.

Problem. Given a graph G, is it Hamiltonian? If so, find a Hamiltonian cycle.

Notice how similar this problem seems to be to that of Eulerian cycles (of Section 8.2.) In that case we found necessary and sufficient conditions for a connected graph to be Eulerian. In the case of Hamiltonian graphs life is not so simple and no such nice characterization is known (or likely to be discovered).

First we remark that since a Hamiltonian graph must contain a spanning cycle, every vertex must have degree at least 2. Exercises 4 and 5 develop more involved conditions that a graph must satisfy in order to be Hamiltonian.

Question 3.2. Find an example of a connected graph that is not Hamiltonian but does not contain any vertices of degree 1.

The r-clique, with $r \geq 3$, contains a Hamiltonian cycle, namely $\langle 1, 2, \ldots, r, 1 \rangle$. More generally, graphs with all vertices of relatively high degree contain Hamiltonian cycles as seen in the first theorem.

Theorem 3.1. If G has $V \geq 3$ vertices and every vertex has degree at least $V/2$, then G is Hamiltonian.

Proof. The proof is constructive and will lead to an algorithm HAMCYCLE. There are two principal steps. In the first we take a maximal path (i.e., a path that cannot be extended at either end) and find a cycle on the same set of vertices. In the second step we take any vertex not on the cycle and construct a maximal path using it and all of the vertices of the cycle. This new path will be longer than the original one. We continue alternating these two steps until all of the vertices of the graph are in the maximal path whence the next cycle we create will be Hamiltonian.

We first note that any maximal path must contain more than half the vertices of the graph. Suppose that $\langle x = x_1, x_2, \ldots, x_k = y \rangle$ forms a maximal path within the graph G. Since the path is maximal, x cannot be adjacent to any vertex off the path. Thus x has at most $k - 1$ neighbors. Since the degree of x is at least $V/2$, we know that

$$k - 1 \geq \frac{V}{2} \qquad \text{or} \qquad k \geq 1 + \frac{V}{2}.$$

Now we want to find a cycle whose vertex set is the same as that of the maximal path. If x is adjacent to y, then the vertices of the path form a cycle in their natural order. If there exist vertices x_i and x_{i+1} such that x is adjacent to x_{i+1} and y is adjacent to x_i, then $\langle x, x_{i+1}, x_{i+2}, \ldots, y, x_i, x_{i-1}, \ldots, x_2, x \rangle$ is a cycle containing all the vertices of the original path. See Figure 8.16. On the other hand, if there were no such pair x_i and x_{i+1}, then whenever x is adjacent to x_{i+1}, y is not adjacent to x_i. Since the path is maximal, neither x nor y can be adjacent to any vertex off the path. Thus if $\deg(x) = s$, then $\deg(y) \leq k - 1 - s$. Since $\deg(x) \geq V/2$ and $\deg(y) \geq V/2$,

$$V \leq \deg(x) + \deg(y) \leq s + (k - 1 - s) = k - 1 < V,$$

a contradiction.

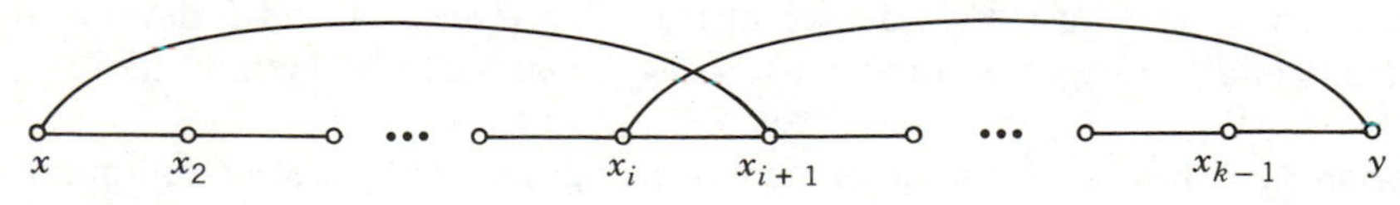

Figure 8.16

Thus we can create a cycle on any vertex set from a maximal path; we relabel vertices so that the resulting cycle is $\langle x_1, \ldots, x_k, x_1 \rangle$.

To justify the second step, let z be any vertex not contained in the cycle $\langle x_1, \ldots, x_k, x_1 \rangle$. Since, as we saw above,

$$k \geq 1 + \frac{V}{2} \qquad \text{and} \qquad \deg(z) \geq \frac{V}{2},$$

z must be adjacent to at least one vertex on the cycle. If z is adjacent to x_i, then

$$\langle z, x_i, x_{i+1}, \ldots, x_k, x_1, \ldots, x_{i-1} \rangle$$

forms a path in G with $k + 1$ vertices. This can be extended to a maximal path. □

Here is an algorithm suggested by the proof of Theorem 3.1.

Algorithm HAMCYCLE

STEP 1. Input G, a graph with V vertices all of degree $\geq V/2$
STEP 2. Set $P := \varnothing$
STEP 3. Repeat
Begin
STEP 4. Pick z in $V(G) - V(P)$; set $P :=$ a maximal path containing $V(P)$ and z
STEP 5. Find C a cycle on $V(P)$
End
until $|V(C)| = V$
STEP 6. Output C and stop.

Example 3.1. Table 8.3 is a trace of HAMCYCLE as applied to the graph in Figure 8.17.

Table 8.3

Step No.	z	P	C
2		$\varnothing$	
4	x	$\langle x, v, t, w, y \rangle$	
5			$\langle x, t, w, y, v, x \rangle$
4	u	$\langle u, t, w, y, v, x \rangle$	
5			$\langle u, w, y, v, x, t, u \rangle$

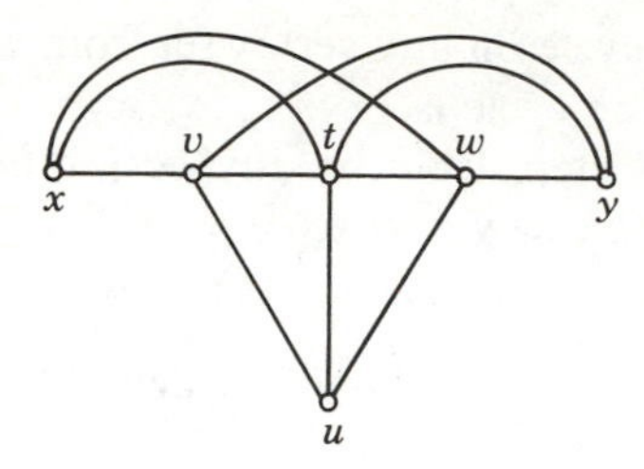

Figure 8.17

Question 3.3. Trace HAMCYCLE on the graph in Figure 8.18.

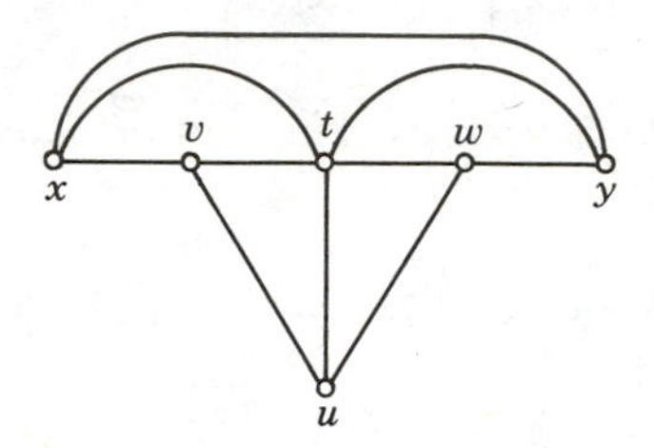

Figure 8.18

How efficient is HAMCYCLE? It is a $O(V^2)$ algorithm; here's why. Constructing the first maximal path in step 4 will require fewer than V comparisons to add each vertex and thus fewer than V^2 comparisons in total. The path has length at least $V/2$ and so the Repeat . . . Until loop repeats at most $V/2$ times. In subsequent executions of step 4 the addition of the vertex z will require no more than one comparison for each vertex in P and so fewer than V in total. Constructing a maximal path in step 4 requires fewer than V comparisons for each additional vertex. Thus the total number of comparisons in step 4 is $V^2 + (V/2)(V + V) = 2V^2$. Finding C in step 5 can be accomplished with no more than two comparisons for each edge in P and thus fewer than $2V$ for the step. Thus the total number of comparisons in step 5 is V^2 and HAMCYCLE is $O(V^2)$.

Suppose that we want to find a Hamiltonian cycle in an arbitrary graph, one with all sorts of different degrees. We might try to list all cycles in the graph, but that certainly sounds like the basis of an exponential algorithm. (Recall that algorithms that list all possible subsets, like J-SET and BADMINTREE, are exponential.) Why don't we just try to build as long a path as possible, and then check that it can be completed to a cycle? (The phrase, "Why don't we just . . . ," is a famous one in algorithms. Often there seems to be a simple way to proceed, but the heart of the matter is then proving that the resulting algorithm always works and is efficient.)

We pursue the idea of hunting for a longest path. This technique, known as **depth-first search** (or **DFS**), is a method for systematically visiting all vertices of a graph by traversing paths that are as long as possible.

Example 3.2. Suppose that we want to visit all vertices in the graph shown in Figure 8.19.

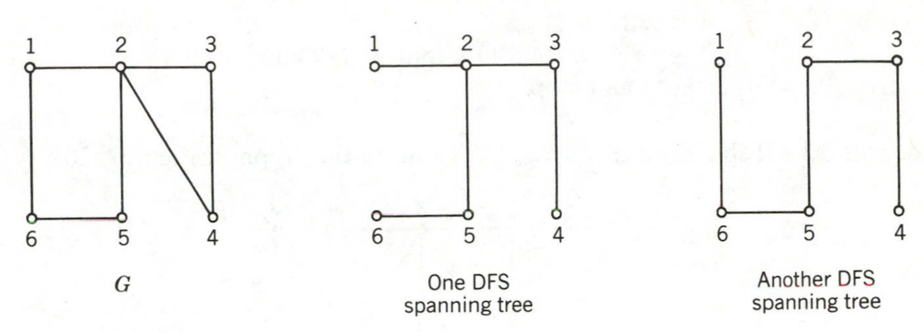

Figure 8.19

Beginning at vertex 1 we might create a path $P = \langle 1, 2, 3, 4 \rangle$. From vertex 4 we can visit no additional new vertices. We backup to vertex 3 from which we also cannot visit new vertices. Then we backup to vertex 2 and visit vertices 5 and then 6. If we keep track of the edges traversed in this process, we find a spanning tree. This tree is known as a **depth-first-search** (or **DFS**) **spanning tree**. As with breadth-first search, a DFS spanning tree is not uniquely determined.

First we present this technique as an algorithm, designed to visit vertices and to construct a spanning tree if possible; in later applications we shall embellish upon this fundamental depth-first-search procedure. As vertices are visited, they and their adjoining edges are placed in T and $E(T)$, the vertices and edges of a DFS tree. We may use the edges in $E(T)$ to backup if need be.

Algorithm DEPTHFIRSTSEARCH (DFS)

STEP 1. Initialize
  Input G, a graph with vertices $1, \ldots, V$ and edge set $E(G)$
  Set $J := 1$ {J will index the vertex currently visited.)
  Set $T := \{1\}$ {T will contain the visited vertices.}
  Set $E(T) := \varnothing$ {$E(T)$ will contain the edges of the DFS tree.}
STEP 2. While $|T| < V$ do
  STEP 3. If there is a K in $G - T$ such that (J, K) is in $E(G)$, then do
    Begin
    STEP 4. $T := T + K$
    STEP 5. $E(T) := E(T) + (J, K)$
    STEP 6. $J := K$
    End

Else {no such K}
STEP 7. If $J \neq 1$, then do {backup}
Find (I, J) in $E(T)$ and set $J := I$
Else $\{J = 1\}$
STEP 8. Output T, $E(T)$ and stop.
STEP 9. Output $E(T)$ and stop.

Example 3.3. Table 8.4 is a trace of DFS run on the graph in Figure 8.20.

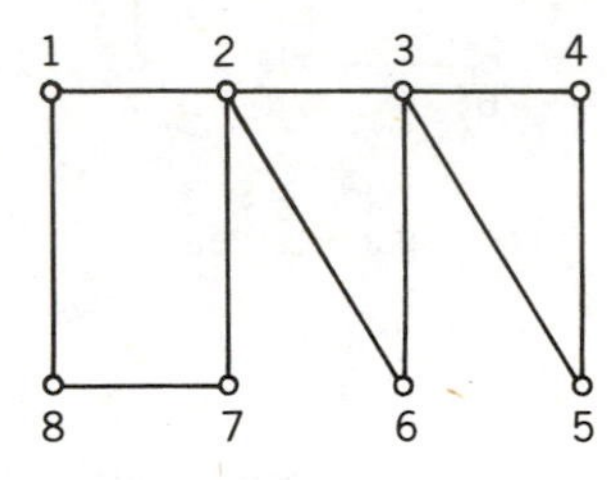

Figure 8.20

Table 8.4

Step No.	*J*	*K*	*T*	*E(T)*
1	1		$\{1\}$	$\varnothing$
3	1	2		
4–6	2	2	$\{1,2\}$	$\{(1,2)\}$
3	2	3		
4–6	3	3	$\{1,2,3\}$	$\{(1,2),(2,3)\}$
3	3	4		
4–6	4	4	$\{1,2,3,4\}$	$\{(1,2),(2,3),(3,4)\}$
3	4	5		
4–6	5	5	$\{1,2,3,4,5\}$	$\{(1,2),(2,3),(3,4),(4,5)\}$
3	5	{no K}		
7	4			
3	4	{no K}		
7	3			
3	3	6		
4–6	6	6	$\{1,2,3,4,5,6\}$	$\{(1,2),(2,3),(3,4),(4,5),(3,6)\}$
3	6	{no K}		
7	3			
3	3	{no K}		
7	2			
3	2	7		
4–6	7	7	$\{1,2,3,4,5,6,7\}$	$\{(1,2),(2,3),(3,4),(4,5),(3,6),(2,7)\}$
3	7	8		
4–6	8	8	$\{1,2,3,4,5,6,7,8\}$	$\{(1,2),(2,3),(3,4),(4,5),(3,6),(2,7),(7,8)\}$
9	Stop.			

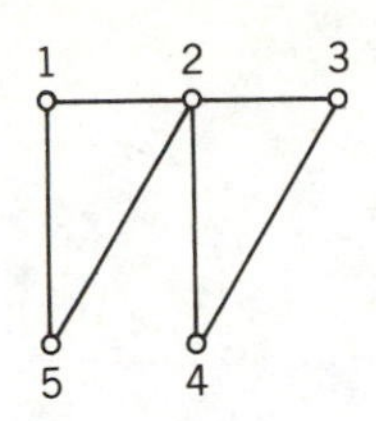

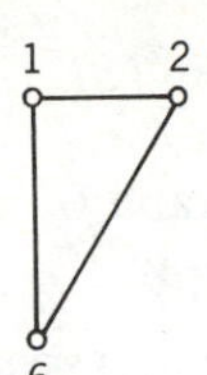

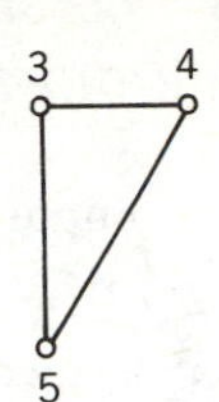

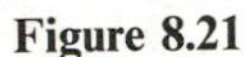
Figure 8.21

Question 3.4. Trace DFS on the graphs in Figure 8.21.

Theorem 3.2. DEPTHFIRSTSEARCH terminates with T containing precisely those vertices that are in the same component as vertex 1. The algorithm performs $O(V^2)$ comparisons.

Proof. Note that if vertex v is not in the same component as vertex 1, then there is no path connecting 1 and v and so v cannot be added to the tree T.

Conversely, suppose that v is a vertex in G that is in the same component as the vertex 1, yet v is not in T after DFS is executed. There is a path from 1 to v. Assume that the edge (u, v) is on that path and that u is in T. (Otherwise, replace v by u in this argument.) Since v is never added to T, $|T| < V$ and the algorithm terminates in step 8. Yet whenever DFS examines the vertex u in step 3, there is a vertex K available, namely $K = v$, and so DFS must add v to T before J is decreased back to 1 in step 8.

To count comparisons in DFS, we first note that the loop at step 2 is executed fewer than V times. Each pass through the loop requires one comparison at steps 2 and 7 and fewer than V comparisons at step 3. Thus DFS requires fewer than $V(V + 2) = O(V^2)$ comparisons in total. □

In contrast we also offer a recursive version of DFS; here the internal backing up (in step 7 of DFS) is managed by the recursive calls.

Procedure R-DFS(J) {The procedure begins a depth-first search at vertex J.}

STEP 1. $T := T + J$
STEP 2. For each edge (J, K) do
  STEP 3. If K is in $G - T$, then do
    Begin
    STEP 4. $E(T) := E(T) + (J, K)$
    STEP 5. Call Procedure DFS(K)
    End
STEP 6. Return.

Then a depth-first search is performed by the following.

Algorithm R-DEPTHFIRSTSEARCH

STEP 1. Input the graph G

STEP 2. $T := \varnothing$; $E(T) := \varnothing$

STEP 3. Call Procedure DFS(1)

STEP 4. Output T, $E(T)$, and stop.

More analysis of this recursive approach is contained in Exercise 21.

Our immediate aim is to use DEPTHFIRSTSEARCH (or R-DEPTHFIRST-SEARCH) to try essentially all possible ways to construct a Hamiltonian cycle. First we create as long a path as possible. If this can be completed to a Hamiltonian cycle, we are done. If not, we backup, throwing away vertices on the path and trying to find another way to extend.

Example 3.4. Here is the idea of a depth-first search for a Hamiltonian cycle on the graph shown in Figure 8.22. We begin creating as long a path as possible: $\langle 1, 2, 3, 4 \rangle$. This cannot be completed to a cycle. We backup to 3, but there is no way to make a new path. When we backup to 2, we can start a new path $\langle 1, 2, 4 \rangle$, which gets completed to the Hamiltonian cycle $\langle 1, 2, 4, 3, 1 \rangle$.

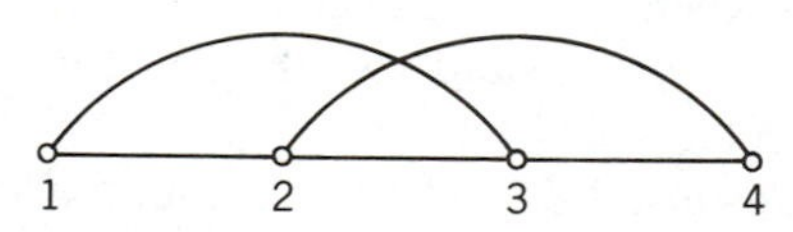

Figure 8.22

In this variation on depth-first search, the variable PATH will contain the vertices of a path that we hope can be completed to a Hamiltonian cycle; entries are initialized as 0. When we backup, we must delete vertices from PATH, resetting entries to 0. The procedure BUILD is used to find the next entry to be added to PATH. As long as vertices are being added to PATH the variable FORWARD equals TRUE, but when backing up FORWARD is FALSE.

Algorithm DFS-HAMCYCLE

STEP 1. Initialize
Input G, a graph with vertices $1, \ldots, V$
Set PATH(1) := 1, PATH(I) := 0 for $I = 2, \ldots, V + 1$
Set $J := 2$ {J indexes the entry of PATH that is currently sought.}
Set FORWARD := TRUE

STEP 2. While $J < V + 1$ do
Begin
STEP 3. If FORWARD = TRUE do
Call Procedure BUILD

```
        Else {FORWARD = FALSE}
                Begin
                STEP 4.  PATH(J) := 0
                STEP 5.  J := J − 1
                STEP 6.  If J ≠ 1, then
                             Call Procedure BUILD
                Else {J = 1}
                             Output "No Ham cycle" and stop.
                End {step 3}
        STEP 7.  If J = V + 1, then do
                STEP 8.  If (1, PATH(V)) is in E(G), then
                             PATH(J) := 1
                Else
                             FORWARD := FALSE
        End {step 2}
STEP 9. Output PATH and stop.
```

Procedure BUILD

```
STEP 1. Find smallest K > PATH(J) such that K ≠ PATH(I) for I < J and
        (PATH(J − 1), K) is in E(G)
STEP 2. If there is no such K, then
        Set FORWARD := FALSE
Else
        Begin
        STEP 3.  PATH(J) := K
        STEP 4.  J := J + 1
        STEP 5.  FORWARD := TRUE
        End {step 2}
End.
```

Note that this algorithm parallels DFS; however, the K chosen in step 1 of the procedure is specified.

Example 3.5. Table 8.5 is a trace of DFS-HAMCYCLE on the graph from Figure 8.22.

Question 3.5. Trace DFS-HAMCYCLE on one graph of Figure 8.15.

In Exercises 13 and 14 you are asked to show that DFS-HAMCYCLE is correct. It would be nice if it were $O(V^2)$. Unfortunately, the seemingly small changes from DEPTHFIRSTSEARCH make the algorithm no longer a polynomial algorithm. (See Exercise 24.)

The Hamiltonian cycle problem of this section is not just an exercise in frustration for the reader—it is equally frustrating for researchers in graph theory and

Table 8.5

Step No.		*J*	*K*	*FORWARD*	*PATH*
Main	1	2		TRUE	⟨1, 0, 0, 0, 0⟩
BUILD	1	2	2		
	3	2			⟨1, 2, 0, 0, 0⟩
	4	3			
	5			TRUE	
BUILD	1	3	3		
	3–5	4	3	TRUE	⟨1, 2, 3, 0, 0⟩
BUILD	1	4	4		
	3–5	5	4	TRUE	⟨1, 2, 3, 4, 0⟩
Main	7, 8	5		FALSE	
Main	4	5			⟨1, 2, 3, 4, 0⟩
	5	4			
BUILD	1	4	{no K}		
	2			FALSE	
Main	4	4			⟨1, 2, 3, 0, 0⟩
	5	3			
BUILD	1	3	4		
	3–5	4	4	TRUE	⟨1, 2, 4, 0, 0⟩
BUILD	1	4	3		
	3–5	5	3	TRUE	⟨1, 2, 4, 3, 0⟩
Main	7, 8	5			⟨1, 2, 4, 3, 1⟩
Main	9	STOP			

graph algorithms! This is one of a collection of problems for which there is no known polynomial algorithm and for which it has not been proved that there cannot be a polynomial algorithm, as yet undiscovered. The question of the complexity of the Hamiltonian cycle problem has been shown to be equivalent to a large collection of similarly unsolved problems; these are known as the **NP-Complete problems**. Two other NP-Complete problems are the Traveling Salesrepresentative Problem, mentioned in Section 5.5, and the Satisfiability Problem of Section 1.10. The NP-Complete problems are equivalent in the sense that if a good algorithm is found for one of these problems, then there are good algorithms for all NP-Complete problems. Conversely, if one of these problems is intractable, then the same is true for all the NP-Complete problems. The NP-Complete problems are intensively studied, and rumors of proofs (and of false proofs!) circulate through the research community. There is an expectation that the issue should be resolved soon. Most algorithmics experts would be surprised if there were good algorithms for the NP-complete problems. (We place no bets on how soon or which way these problems will be resolved . . . .)

In the next two sections we shall discuss instances of two other NP-Complete problems and shall demonstrate different ways of coping with the unsettled state of affairs.

EXERCISES FOR SECTION 3

1. Which of the following graphs contain a Hamiltonian cycle? Which contain a Hamiltonian path but not a Hamiltonian cycle? Give reasons for your answers.

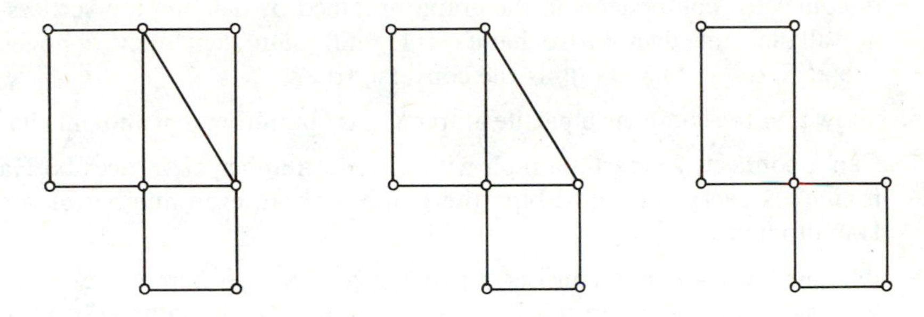

2. Find examples of the following types of graphs:
(a) Eulerian and Hamiltonian.
(b) Eulerian but not Hamiltonian.
(c) Hamiltonian but not Eulerian.
(d) connected but neither Hamiltonian nor Eulerian.

3. Here is the original graph on which Hamilton based his game; find a Hamiltonian cycle in this graph. One version of the original game consisted of one player selecting a path with five vertices and the second player attempting to

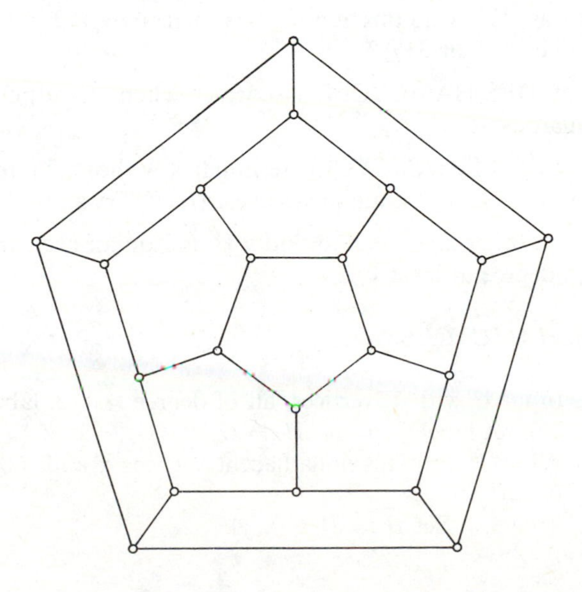

extend the path to a Hamiltonian cycle. Is there a path with five vertices that cannot be extended to a Hamiltonian cycle?

4. Show that if G, a connected graph, contains a vertex v whose removal leaves a disconnected graph, then G is not Hamiltonian. Is the converse true?

5. Let S be a set of vertices of a graph G and let $c(G - S)$ denote the number of connected components in the graph obtained by deleting the vertices of S and all incident edges. Prove that if G is Hamiltonian, then for every nonempty subset S, $c(G - S) \leq |V(S)|$. Is the converse true?

6. Show that the complete bipartite graph $K_{r,s}$ is Hamiltonian if and only if $r = s$.

7. Can a connected bipartite graph with an odd number of vertices be Hamiltonian? Is every connected bipartite graph with an even number of vertices Hamiltonian?

8. Imagine the graph that consists of two copies of $K_{V/2}$ (where V is even) joined by exactly one edge. Use this graph to argue that the hypothesis of Theorem 3.1 (that the degrees are at least $V/2$) cannot be weakened and still have the conclusion hold.

9. Explain why a graph with V vertices and all vertices of degree at least $V/2$ is connected.

10. Suppose that G is a graph with V vertices such that for every pair of nonadjacent vertices, say x and y, the degree of x plus the degree of y is at least V. Show that G is Hamiltonian. Compare this result with Theorem 3.1.

11. Where in the proof of Theorem 3.1 is the assumption that $V \geq 3$ needed?

12. Where does the algorithm HAMCYCLE use the fact that all vertices have degree at least $V/2$? Can this number be replaced by $V/2 - 1$ or by any smaller number, like $2V/5$ or $3V/8$?

13. Prove that DFS-HAMCYCLE is correct when it outputs the edges of a Hamiltonian cycle.

14. Prove that if DFS-HAMCYCLE terminates without finding a Hamiltonian cycle, then G does not contain such a cycle. (*Hint:* Try a proof by contradiction.)

15. Here is an another algorithm to find a Hamiltonian cycle in a graph with all vertices of degree at least $V/2$.

Algorithm HAMCYCLE2

STEP 1. Input G with V vertices, all of degree $\geq V/2$, label each edge with a 0, set label $:= 1$ and $H := G$

STEP 2. While H contains nonadjacent vertices x and y do
Begin
STEP 3. Set $H := H + (x, y)$

STEP 4. Label (x, y) with the value of label
STEP 5. Set label := label + 1
End
STEP 6. Find C a Hamiltonian cycle in H and let k be the largest label on an edge of C
STEP 7. Repeat
Begin
STEP 8. Find the edge (x, y) of label k
STEP 9. Renumber the vertices of C as
$C = \langle x = x_1, x_2, \ldots, x_V = y \rangle$
STEP 10. Find an index i such that x is adjacent to x_i and y is adjacent to x_{i-1}
STEP 11. Set
$C := \langle x = x_1, x_i, x_{i+1}, \ldots, x_V = y, x_{i-1}, x_{i-2}, \ldots, x_1 \rangle$
STEP 12. Delete the edge (x, y)
STEP 13. Set $k :=$ largest label on an edge of C
End
Until $k = 0$
STEP 14. Output C and stop.

Run HAMCYCLE2 on the two following graphs.

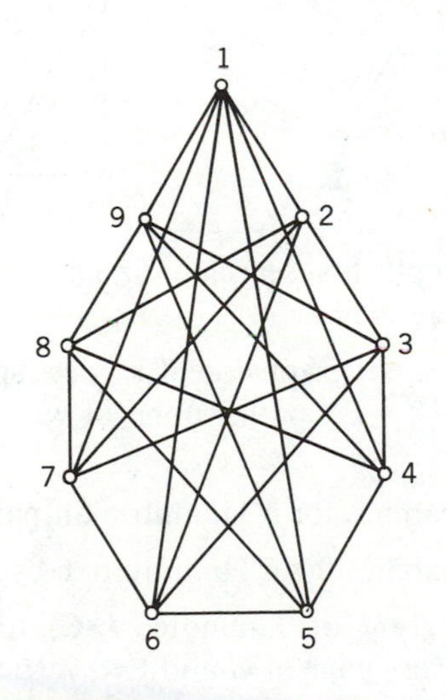

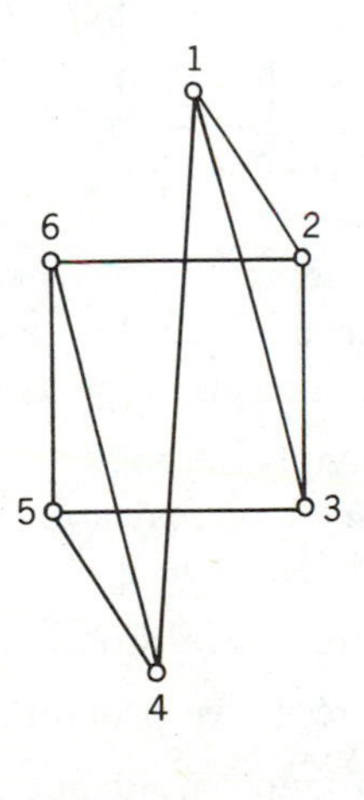

16. Prove that HAMCYCLE2 always produces a Hamiltonian cycle in a graph with all vertices of degree $V/2$ or more.

17. Is HAMCYCLE2 more efficient than HAMCYCLE?

18. Find an example such that the spanning trees created by breadth-first search and depth-first search differ.

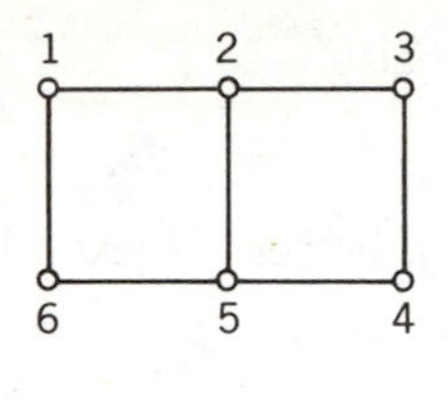

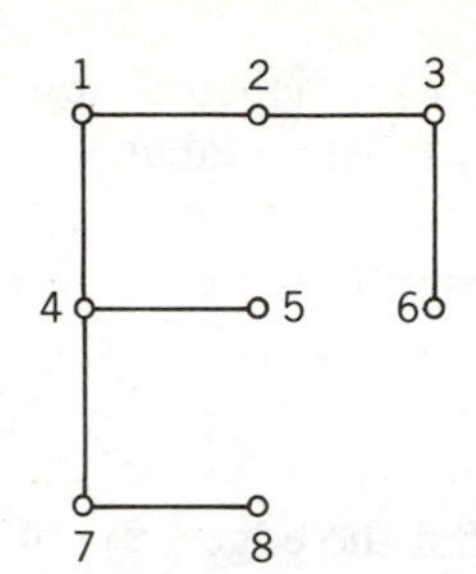

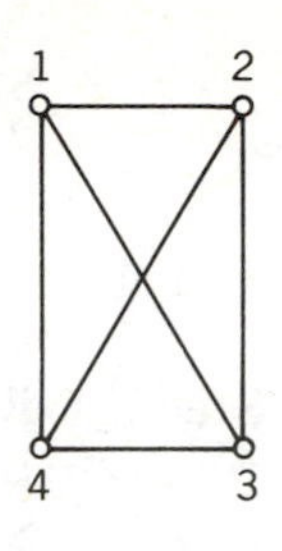

19. Run DEPTHFIRSTSEARCH on the following graphs.

20. (*a*) Modify DEPTHFIRSTSEARCH so that it outputs a spanning tree if the graph is connected and a spanning forest otherwise.
(*b*) Modify DFS further so that it identifies and outputs the vertices in each connected component of the graph.

21. (*a*) Run R-DEPTHFIRSTSEARCH on the graphs of Exercise 19.
(*b*) Prove the analogue of Theorem 3.2 for R-DEPTHFIRSTSEARCH.

22. Run DFS-HAMCYCLE on the following graphs.

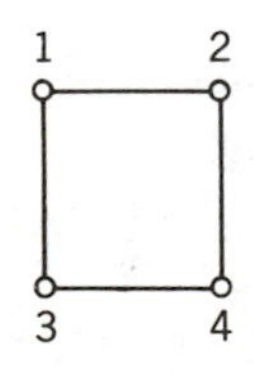

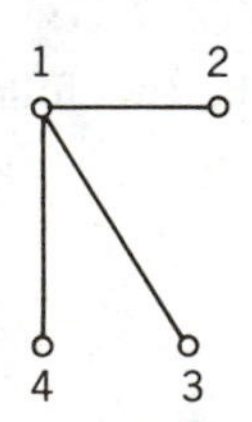

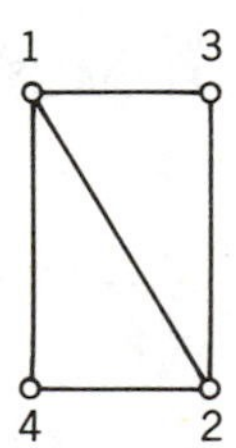

23. Construct an algorithm that uses depth-first search to find a Hamiltonian path in a graph or else reports that there is none.

24. Find a family of (possibly disconnected) graphs on V vertices such that DFS-HAMCYCLE creates more than $(V - 3)!$ paths, none of which extend to a Hamiltonian cycle.

25. Write a recursive program that searches for a Hamiltonian path in a graph.

26. Write a recursive program that searches for a Hamiltonian cycle in a graph.

27. If G is a graph, we define the **line graph** of G, denoted $\boldsymbol{L(G)}$, to be the graph formed with a vertex in $L(G)$ for every edge of G and two vertices adjacent in $L(G)$ if the corresponding edges of G are incident. Prove that if a graph G is the line graph of an Eulerian graph (i.e., there is an Eulerian graph H such that $L(H) = G$), then G is Hamiltonian. Find examples of such graphs G and H.

28. Find an example of a graph G that is not the line graph of any other graph; that is, there is no graph H such that $G = L(H)$.

29. Show that if G is Eulerian, then so is $L(G)$.

30. Prove that G is the line graph of some graph if and only if the edges of G can be divided into a disjoint collection of maximal complete subgraphs.

31. Three mutually adjacent vertices T in a graph G are said to form an **even triangle** if every vertex of G is adjacent to zero or two vertices of T. Otherwise, three mutually adjacent vertices are said to form an odd triangle. Prove that a graph G is the line graph of another graph if and only if (i) G does not contain two odd "overlapping" triangles of the form $\{a, b, c\}$ and $\{a, b, d\}$ and (ii) G does not contain $K_{1,3}$ as an induced subgraph.

8:4 MINIMUM-WEIGHT HAMILTONIAN CYCLES

A common manufacturing task is drilling holes in a sheet of plastic using a laser. Think of the plastic as being in a fixed location in the horizontal plane. The drill is movable and must be located immediately above the target hole while drilling. The time it takes to fabricate one unit equals the time it takes to drill the holes together with the time it takes to move the drill. With a given drill and plastic sheet, the drilling time will be a (small) constant. The time it takes to position the drill will depend on the distance the drill must travel. Thus the manufacturer would like to drill the holes in an order that minimizes the total distance the drill travels.

Question 4.1. Suppose that the drill is initially located above the origin in the x-y plane and that after drilling four holes the drill must return to the origin. Suppose the holes are located at $(0, 0)$, $(1, 0)$, $(0, 1)$, and $(1, 1)$. Find all possible drilling sequences and for each find the total distance that the drill travels.

Efficient fabrication leads to the following graph theory problem. Construct a complete graph whose vertices correspond with the locations of the holes. Since this graph is complete, it contains lots of Hamiltonian cycles. We want a shortest one. Assign to the edge (x, y) a weight that represents the distance from the hole labeled x to the hole labeled y. A minimum-weight Hamiltonian cycle in this graph corresponds with a most efficient ordering of the vertices for drilling.

Problem. Given a weighted complete graph, find a Hamiltonian cycle whose total weight is minimum.

The only known algorithms for this problem are exponential (see Exercise 3). A typical manufacturing application might have hundreds or even thousands of holes to drill. Thus a nonpolynomial algorithm for finding a minimum-weight spanning cycle would be impossibly slow. Still in practice the drilling is done with the holes processed in some order.

This section presents a good approximation algorithm for the laser drilling problem. Specifically, given a weighted complete graph whose edge weights represent distances in the plane, the algorithm will produce a Hamiltonian cycle whose total weight is no more than twice the minimum. Before giving the algorithm formally, we discuss and analyze the major steps in the approximation.

Given a weighted complete graph G, let $H(G)$ denote a minimum-weight Hamiltonian cycle and let $P(G)$ denote a minimum-weight Hamiltonian path. Of course, we don't know a good way to find these subgraphs. However, we can use KRUSKAL (see Section 5.4) to find instead $T(G)$, a minimum-weight spanning tree of G. This will be the first major step in the approximation algorithm.

Example 4.1. Suppose that the vertices of G correspond with the points $(0, 0)$, $(1, 1)$, $(2, 0)$, and $(0, 2)$. Figure 8.23 exhibits $T(G)$, $P(G)$, and $H(G)$. Note that $w(T(G)) = 3\sqrt{2}$, $w(P(G)) = 2 + 2\sqrt{2}$, and $w(H(G)) = 4 + 2\sqrt{2}$.

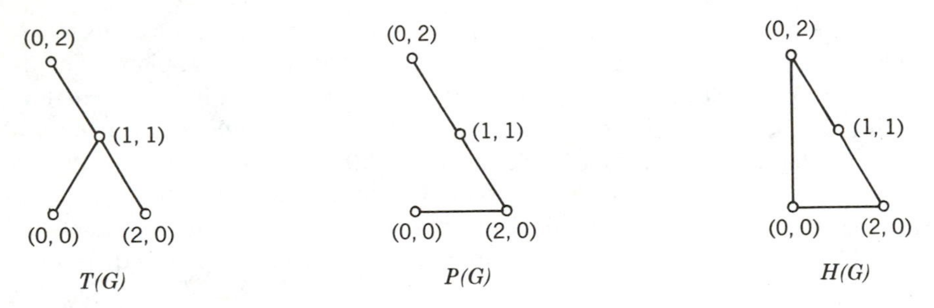

Figure 8.23

Question 4.2. Given the points $(0, 0)$, $(1, 1)$, $(1, -1)$, $(-1, 1)$ and $(-1, -1)$, construct G, $T(G)$, $P(G)$, and $H(G)$ as in the preceding example. Find the weights of each of these graphs.

Proposition 4.1. $w(T(G)) \leq w(P(G)) < w(H(G))$.

Proof. A Hamiltonian path can be obtained from a Hamiltonian cycle by removing an edge (necessarily of positive weight). Thus $w(P(G)) < w(H(G))$. Since $P(G)$ is a path, it is a connected acyclic graph and thus a tree. Since $T(G)$ is a minimum-weight spanning tree of G, its weight can be no more than that of $P(G)$. Thus $w(T(G)) \leq w(P(G))$. □

Note that Proposition 4.1 is true for any weighted graph whose edge weights are positive. In particular, the edge weights don't have to correspond with distances in the plane.

Our second major step will be to find an Eulerian cycle, using material from Section 8.2. (For an alternate explanation, independent of that section, see Exercise 12.) Given any tree T, create a directed graph $D(T)$ by replacing each edge e of T with two directed edges, one in each direction and both of weight $w(e)$. By construction every vertex of $D(T)$ has an equal number of edges directed in and out. By Theorem 2.3 $D(T)$ contains a (directed) Eulerian cycle. We could use either a modified version of EULER or a depth-first search to create such a cycle. Evidently, $w(D(T)) = 2w(T)$. Let D denote an Eulerian cycle in $D(T)$.

Example 4.1 (continued). Figure 8.24 shows the graph $D(T)$. D might consist of $\langle r, c, b, c, d, c, r \rangle$.

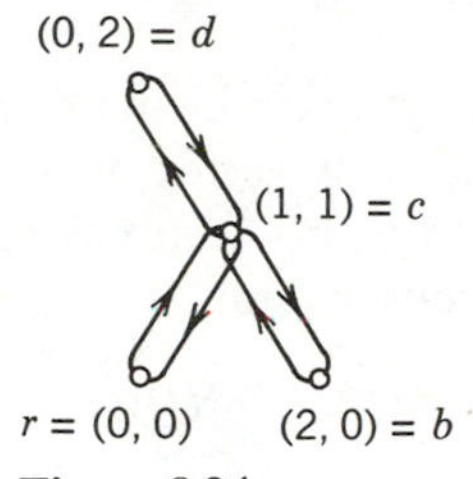

Figure 8.24

Question 4.3. If G is the weighted graph from Question 4.2, find $D(T)$ and some Eulerian cycle D.

The third major step in our approximation algorithm will be to transform D into a Hamiltonian cycle, denoted by C. We need some bookkeeping and a standard look ahead device. Initially, set $C := D$, and let r denote the first vertex of C. Begin traversing C, noting each visited vertex. Continue until you reach a vertex x and an out-directed edge (x, y) such that y is an already visited vertex. Suppose that within D, the edge (x, y) is followed by the edge (y, z). Modify C by replacing (x, y) and (y, z) with (x, z). If z is already visited and (x, z) is followed by (z, w), then replace (x, z) and (z, w) by (x, w), and so on. If the edge (x, y) is not followed by another edge, then $y = r$, and C is the desired Hamiltonian cycle.

Example 4.1 (once again). C initially consists of $\langle r, c, b, c, d, c, r \rangle$. Note that (b, c) visits c for the second time. Thus (b, c) and (c, d) are replaced by (b, d). At this stage $C = \langle r, c, b, d, c, r \rangle$. Now (d, c) visits c for the second time. Thus (d, c) and (c, r) are replaced by (d, r), and $C = \langle r, c, b, d, r \rangle$. Although (d, r) visits r for a second time, it is not followed by another edge. Thus (d, r) is the final edge in the Hamiltonian cycle. Figure 8.25 exhibits the Hamiltonian cycle thus obtained. Note that $w(C) = 2 + 4\sqrt{2}$.

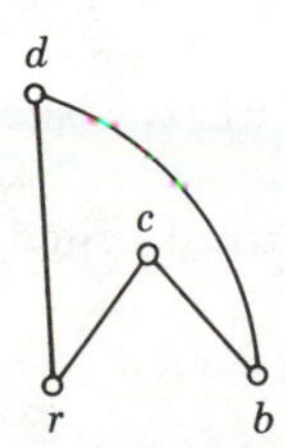

Figure 8.25

Question 4.4. Using the Eulerian cycle from Question 4.3, find a Hamiltonian cycle C. Determine $w(C)$ and using the results of Question 4.2 verify that $w(C) < 2w(H(G))$.

Proposition 4.2. After the edge replacements are complete (as described in the previous paragraph), C is a Hamiltonian cycle and $w(C) < 2w(H(G))$.

Proof. Initially, D is an Eulerian cycle in the directed graph $D(T)$. Since T is a spanning tree of G, D must visit every vertex of G. After all the edge replacements are complete, transforming D to C, no vertex is visited twice other than r. Thus C is a Hamiltonian cycle. By construction

$$w(D) = w(D(T)) = 2w(T(G)) < 2w(H(G)) \qquad \text{by Proposition 4.1.}$$

If $w(C) \leq w(D)$, then the proof is complete. Suppose that (x, y) and (y, z) in D are replaced by (x, z). Since $w(x, z)$ equals the distance from x to z and $w(x, y) + w(y, z)$ equals the distance from x to y plus the distance from y to z, the triangle inequality of plane geometry implies

$$w(x, z) \leq w(x, y) + w(y, z).$$

Thus every time two edges are replaced by one, the weight of the cycle cannot increase. Thus $w(C) \leq w(D)$. □

Note that we have solved the minimum-weight Hamiltonian cycle problem not only in the case where edge weights represent distances in the plane, but also for every weighted complete graph such that $w(x, z) \leq w(x, y) + w(y, z)$ for every triple of vertices x, y, and z. (See also Exercise 7.)

Algorithm APPROXHAM

STEP 1. Input G, a weighted complete graph
STEP 2. Use KRUSKAL to obtain T, a minimum-weight spanning tree of G
STEP 3. Create the digraph $D(T)$ by replacing every edge of T by two oppositely directed edges
STEP 4. Use EULER, modified as in Exercise 2.18, to obtain D an Eulerian cycle in $D(T)$ {Suppose that $D = \langle e_1, \ldots, e_{2V-2} \rangle$ and for all j, $e_j = (x_j, y_j)$.}
STEP 5. Set $C := \langle x_1 \rangle$ and mark x_1 visited
STEP 6. For $j = 1$ to $2V - 3$ do
  STEP 7. If y_j is not marked visited, set $C := C$ followed by y_j; mark y_j visited
STEP 8. Set $C := C$ followed by x_1.
STEP 9. Output C and stop.

Notice that APPROXHAM is a polynomial algorithm: Both KRUSKAL and EULER are polynomial algorithms and the comparisons in step 6 are few. (See also Exercise 10.) Thus APPROXHAM is efficient and guarantees a spanning cycle that is at most twice as long as the shortest one. The ideas of this algorithm have been extended so that a spanning cycle within 50 percent of optimal is always produced. The extended algorithm is more complex and uses the so-called Minimum-Weight Matching algorithm alluded to in Section 8.2.

These results may sound weak, but in practice APPROXHAM is considered useful. Considerable experimental work has been done on APPROXHAM, and on average it appears to produce a cycle fairly close to the shortest. However, there are no theoretical results that establish the quality of its average-case behavior. Another useful approximation algorithm is the greedy algorithm: At each point visit the nearest unvisited neighbor. It also seems to work well on average, but in the worst case it is known only to produce a cycle C for which $w(C) \leq (\lceil 1 + \log(V) \rceil/2)w(H(G))$ if G is a graph with V vertices (see also Exercise 6.)

EXERCISES FOR SECTION 4

1. Suppose that you are required to drill holes at (0, 0), (1, 1), (3, 0), (2, 2), (1, 2), (3, 3), (1, 3), and (0, 2). Assuming that your drill starts and finishes at the origin, find an optimum drilling schedule.
2. For each of the weighted graphs shown, find a minimum-weight Hamiltonian cycle.

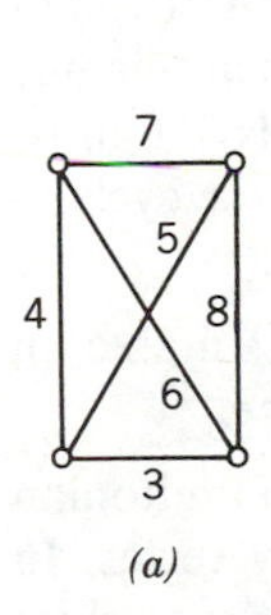

(a)

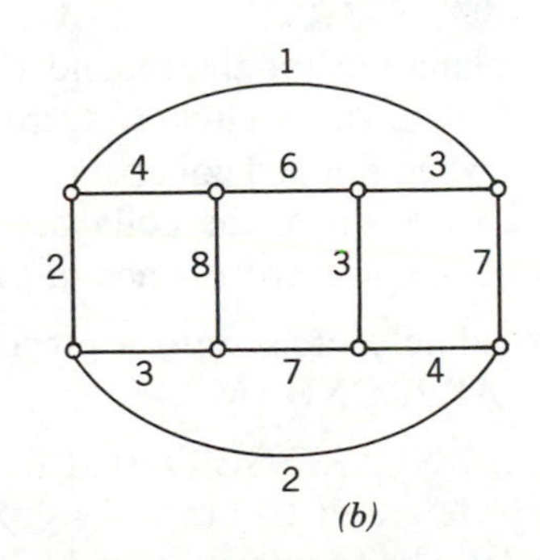

(b)

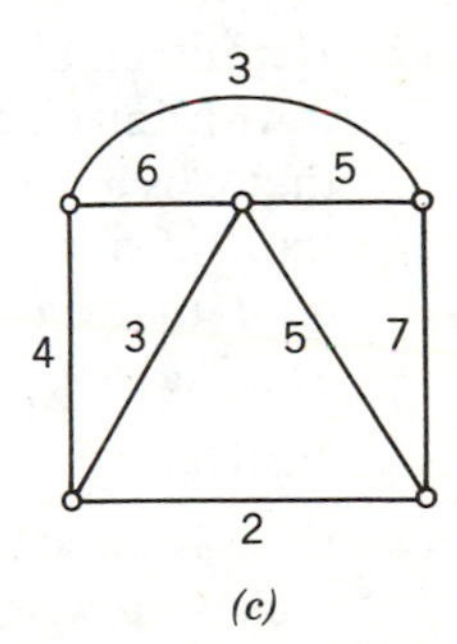

(c)

3. Construct an algorithm that will, given a weighted complete graph, use PERM (see Section 3.4) to generate all possible Hamiltonian cycles and thus find a minimum-weight one. Run your algorithm on the graph shown in part (a) of the preceding problem. Discuss the complexity of your algorithm.
4. Two Hamiltonian cycles on the n-clique K_n are considered the same if one is a cyclic rotation of the other or if one is the reverse of the other. Explain why K_n contains $(n - 1)!/2$ different Hamiltonian cycles.

5. With the same understanding of when two Hamiltonian cycles are different as in the previous problem, determine the number of different Hamiltonian cycles on the complete bipartite graph $K_{n,n}$.

6. Find a set of points in the plane for which the greedy algorithm (see the last paragraph of this section and Section 5.5) does not produce a minimum-weight Hamiltonian cycle. Here the weight of an edge is the distance between the corresponding vertices.

7. Find a weighted graph G such that $w(C) > 2w(H)$, where C is the Hamiltonian cycle produced by APPROXHAM and H is a minimum-weight Hamiltonian cycle.

8. Suppose that G is a weighted graph that represents cities serviced by an airline and where the weight of an edge (x, y) represents the cost of flying from x to y. Explain why the triangle inequality might not hold for this graph.

9. Use APPROXHAM to find Hamiltonian cycles in the graphs whose underlying point sets are
   (*a*) $\{(0,0),(2,0),(4,0),(3,1),(1,3)\}$
   (*b*) $\{(0,0),(3,0),(2,1),(1,2),(0,3)\}$
   (*c*) $\{(0,0),(5,1),(6,3),(4,4),(2,7),(1,8),(3,4)\}$

10. Find an integer d such that APPROXHAM uses $O(V^d)$ comparisons when finding a Hamiltonian cycle on a set of V points in the plane.

11. Find an example of a weighted graph that has two different minimum-weight spanning trees such that starting with these two APPROXHAM produces different-weight Hamiltonian cycles.

12. Here is an alternative explanation for the second major step of APPROXHAM. Imagine a tree T drawn with a circle C surrounding it. Then imagine that C is a balloon, which when popped collapses in and surrounds T tightly. Explain how a clockwise traversal of the collapsed cycle creates a cycle D on T that traverses every edge of T twice, once in each direction.

13. Turn the idea of the preceding exercise into a precise algorithm suitable, in particular, for use within APPROXHAM.

14. Here is an algorithm called NEARINSERT that finds a short Hamiltonian cycle in a weighted complete graph G. Let $C_1 = \langle r \rangle$ an arbitrary vertex. In general, if C_j is a j-cycle, let v be some vertex in $V(G) - V(C_j)$ that is as close as possible to some vertex in C_j, say u. Create C_{j+1} by inserting v into C_j immediately following u. Run NEARINSERT on the graphs of Exercise 9.

15. Create an algorithm FARINSERT that parallels the algorithm from the preceding exercise except that the vertex to be inserted is as far as possible from the already created cycle. Run FARINSERT on the graphs of Exercise 9.

16. Suppose that you are given n points in the plane that are contained in a $\sqrt{n}$ by $\sqrt{n}$ square. It has been conjectured that on average the minimum-weight

spanning cycle has length at most $4\sqrt{n}$. Discuss why this is a plausible conjecture and investigate the examples of this section to see if they support this conjecture.

17. Construct an algorithm that finds a Hamiltonian cycle in a weighted complete graph K_n by first greedily discarding as many heavy edges as possible subject to the condition that the degree of each incident vertex exceeds $n/2$, and then uses HAMCYCLE from the previous section to find a Hamiltonian cycle. Run your algorithm on the graphs from Exercise 9.

8:5 GRAPH COLORING AND AN APPLICATION TO STORAGE ALLOCATION

The algorithms of this book have been presented following the format of the Pascal programming language. Pascal is a "high level" language whose statements can be translated by a compiler into machine language statements. These are the instructions that the central processing unit (cpu) executes. To execute the program the computer must store the machine language instructions plus the values of all variables used in the program. While the program occupies a constant block of memory, there is choice in the storage of variables. Efficient use and reuse of memory locations can save significantly on the total amount of memory needed.

Example 5.1. Consider an algorithm that, among other things, calculates the volume of a rectangular box.

Algorithm VOLUME

STEP 1. Input Length, Width, Height
...
STEP *i*. Area := Length * Width
...
STEP *j*. Volume := Area * Height
...

The values of the variables Length, Width, Height, Area, and Volume could be stored in, say, memory locations 0 through 4, respectively. In a more efficient storage allocation scheme the variable Volume could be assigned to either memory location 0 or 1, provided that there is no use for the variables Length or Width after step *j*.

The compiler assigns variables to memory locations. Two different variables can be assigned to the same memory location, provided that both variables are never needed in the program at the same time. Thus the mapping of the variables to memory locations need not be a one-to-one function, but when appropriate two or more variables can double up in one location.

Definition. Two variables in a program (or algorithm) are said to be **noninterfering** if, regardless of the input values, at no instant during the execution of the program are both variables needed at that instant or needed in memory for the execution of some subsequent step. Otherwise, the two variables are said to be **interfering**.

Example 5.1 (continued). The variables Volume and Length are noninterfering variables, as are Volume and Width. Length and Width are clearly interfering variables. So are Length and Height, since the value of Height must be kept in memory for use in step *j*.

Example 5.2. Suppose that the Post Office offers a choice of rates to magazine publishers. The cost of mailing is a function of either the product of the area of the mailing envelope and its weight, or twice the volume of the envelope.

Algorithm POSTAGE

STEP 1. Input Length, Width, Weight
STEP 2. Area := Length * Width
STEP 3. Cost1 := Area * Weight
... {Some steps that use only Cost1.}
STEP 6. Input Height
STEP 7. Volume := Area * Height
STEP 8. Weight := 2
STEP 9. Cost2 := Volume * Weight
... {Some steps that use only Cost2.}
STEP 12. Print Cost1, Cost2, Volume

Notice that Height and Weight are noninterfering variables, since Height is only used in steps 6 and 7 whereas Weight is needed in steps 1, 2, 3, 8, and 9. However, Cost1 and Height are interfering variables, since the value of Cost1 must be maintained from step 3 to step 12.

Question 5.1. For the variables Length, Width, Height, Weight, Area, Volume, Cost1, and Cost2 from the preceding example, determine which pairs are noninterfering and which interfering. Find a set of four variables, every pair of which is interfering. Are there other sets of four mutually interfering variables? Are there any sets of five variables with every pair interfering?

Ideally, how should the compiler assign variables to memory locations? We assume that all values of a variable should be assigned to just one location during the execution of the program. Next the storage allocation scheme used by the compiler should use as few memory locations as possible. In addition, it should determine the allocation of memory quickly. There is a graph naturally associated with this problem, and we shall see that storage allocation can be accomplished by "coloring" the vertices of this graph. Furthermore, once the corresponding graph is created, allocation can be determined quickly if there is a good algorithm to do the related graph coloring.

The corresponding graph is called the **interference graph**. Specifically, this graph has a vertex for each variable in the program, and two vertices are joined by an edge if they are interfering.

Example 5.2 (continued). Figure 8.26 exhibits the interference graph where the vertices are labeled with abbreviations of the names of the corresponding variables.

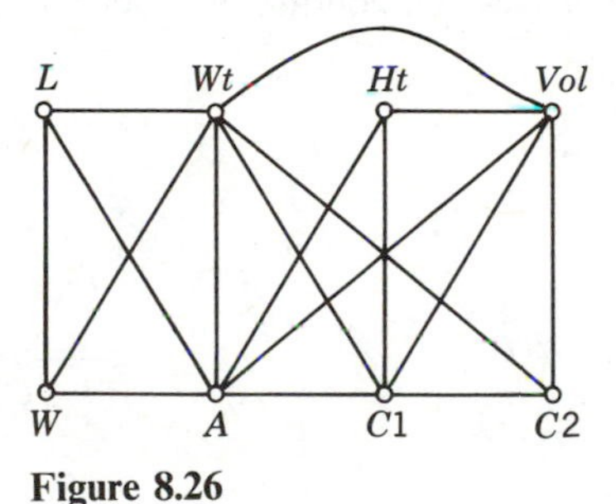

Figure 8.26

Question 5.2. Construct the interference graph for the program from Example 5.1.

Before we proceed with graph coloring, it is appropriate to interject more realism in our storage allocation model. In fact, it is a complex task for a compiler (or for a human) to decide whether or not two variables are interfering. This would involve the potentially infinite task of checking whether or not two variables interfere during the program given any possible input. However, it is possible for a compiler to determine quickly that certain pairs of variables are always noninterfering. These are declared to be noninterfering, and then to be on the safe side all other pairs are declared to be interfering. Thus the "interference graph" derived and used by the compiler may have more edges than the true interference graph. In the simple examples in this section (both algorithms without loops or branching), it is straightforward to determine the real interference graph.

We shift now to graph "colorings." Graph colorings began with the Four-Color Problem, a "puzzle" that asks whether every map can have its countries colored with one of four colors so that no two countries with a border in common receive the same color. This problem intrigued mathematicians for more than a century until it was solved in 1976. (The answer is yes, every map can be 4-colored. The proof due to Kenneth Appel and Wolfgang Haken is long and intricate and includes a computer check of over 1400 cases.)

Definition. For k a positive integer, a graph G is said to be **k-colored** (or **k-colorable**) if each vertex is (or can be) assigned one of k colors so that no two adjacent vertices receive the same color. The **chromatic number of G**, denoted by $\chi(G)$, is the minimum number k such that G can be k-colored. The **clique number of a graph G**, denoted **cl(G)**, is the largest number r such that G contains an r-clique as a subgraph.

We have seen 2-colorability before in Section 5.2: There a 2-colorable graph was called bipartite.

Example 5.3. The graphs G and H in Figure 8.27 are colored with A, B, C, and D. The graph G can be 3-colored (replace the D with a B) as well as 2-colored (in addition replace C by an A); however, H cannot be 2-colored because it contains three mutually adjacent vertices. The clique numbers of these graphs are 2 and 3, respectively.

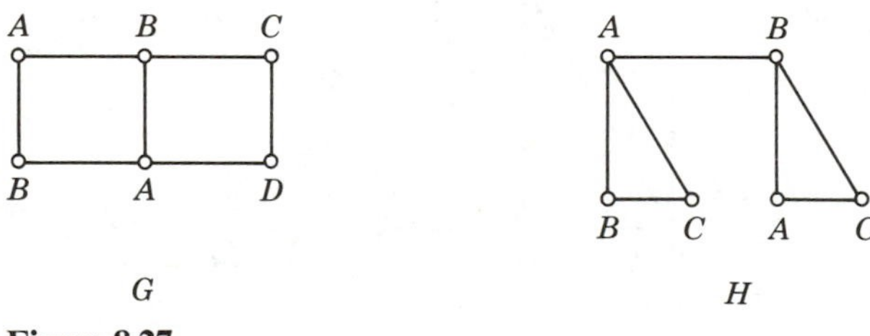

Figure 8.27

Storage allocation problems can be translated into problems about colorings of the interference graph. Associate a different color with each memory location. Then the interference graph receives a coloring by assigning a vertex, labeled by a variable, the color of the memory location to which the variable is assigned. Conversely, a graph coloring (with the same colors as associated with memory locations) prescribes a memory allocation for variables. In both these assignments the graph is correctly k-colored if and only if an assignment of k memory locations is made in which pairs of variables that are interfering are assigned to different memory locations.

Here are some central problems in the theory of graph colorings.

Problem A. Construct an algorithm that, given a graph G and an integer k, determines if G can be k-colored, and if so finds a k-coloring.

Problem B. Construct an algorithm that, given a graph, determines its chromatic number and finds a corresponding coloring.

Problem C. Does the chromatic number of a graph equal its clique number?

These problems translate to important ones for storage allocation. Given a program, suppose that the compiler has declared certain pairs of variables to be noninterfering and all others to be interfering. In this context, Problem A asks for an algorithm to determine if k memory locations are sufficient, and if so, to find such an allocation. Problem B asks for an algorithm to determine the minimum number of storage locations needed, and Problem C asks whether the answer to Problem B is the same as the maximum number of mutually interfering variables. (Look back at Question 5.1.)

Question 5.3. The "interference graph" G created by the compiler may have more edges than the true interferences graph H. Explain why $\chi(G) \geq \chi(H)$. If the compiler assigns $k = \chi(G)$ memory locations to a program based on a coloring of G, explain why two "truly" interfering variables will not be assigned the same memory location.

Question 5.4. Determine the chromatic number and the clique number of each graph shown in Figure 8.28.

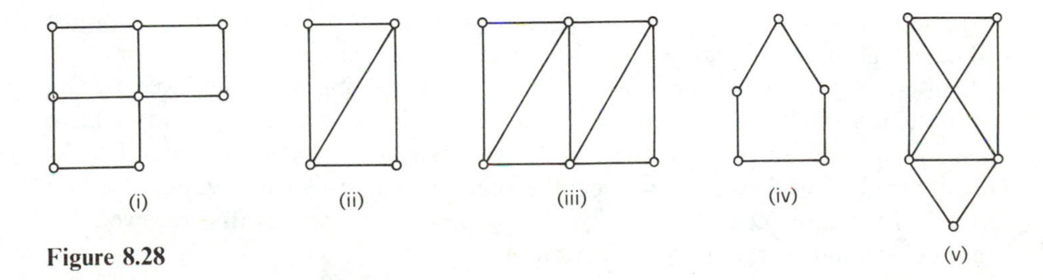

Figure 8.28

Question 5.5. Solve Problem A for $k = 1$.

Problem A for $k = 2$ is not difficult. It is a frequent circumstance that good algorithms follow from insightful theorems. Theorem 5.1, a characterization of 2-colorable graphs, is just such an instance. It should be clear from Example 5.3 and Question 5.4 that a graph containing a cycle with an odd number of vertices cannot be 2-colored. (See Exercise 4.) Thus a 2-colorable graph cannot contain an odd cycle. Surprisingly, this is the only condition needed to ensure that a graph is 2-colorable.

Theorem 5.1. A graph G is 2-colorable if and only if G contains no odd cycle.

Proof. We prove that if G contains no odd cycle, it is 2-colorable. Select any vertex r to be the root and color it red. For each vertex x in G if the distance from r to x is even, color x red. Otherwise, color x blue. If G is not connected, pick a root in each component and repeat. This procedure places 2 colors on the vertices of G, but does each edge join vertices of different colors?

Example 5.4. In the two graphs of Figure 8.29 each vertex, besides the root r, is labeled with its distance from r and the color it receives in the procedure described above.

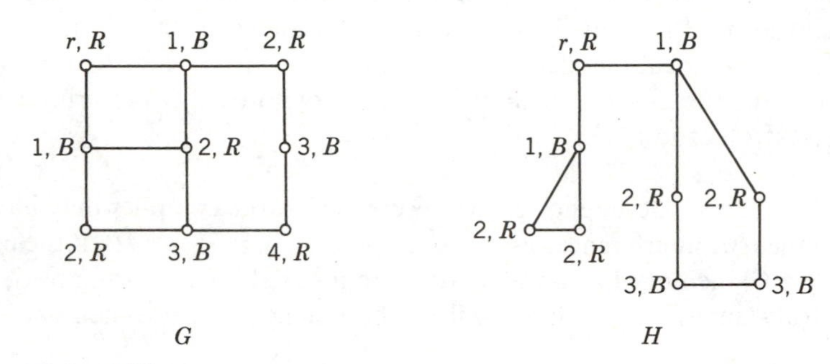

Figure 8.29

To return to the proof, suppose that two vertices, say x and y, with the same color are joined by an edge. If P_x denotes a shortest path from r to x and P_y denotes a shortest path from r to y, then $|P_x|$ and $|P_y|$ are either both even or both odd. In either case their sum is even. The cycle consisting of P_x followed by the edge (x, y) followed by P_y (in reverse order) is an odd cycle unless P_x and P_y have edges in common. In this case these paths are easily seen to contain an odd cycle. This contradiction forces us to conclude that there is no such edge joining two vertices of the same color, that is, every edge joins two vertices that received different colors and so the graph is 2-colored. □

Since the distance from any vertex to the root can be determined by a breadth-first search as described in Section 8.1, this proof suggests an efficient algorithm using BFS to test for 2-colorability (see Exercises 8 and 9).

It would be reasonable to think that there should be an analogue of Theorem 5.1 and the accompanying algorithm for 3-colorable graphs. In fact, there is no known characterization of 3-chromatic graphs or any efficient 3-coloring algorithm. Furthermore, it is unlikely that any will be discovered.

In the remainder of this section we present two algorithms that attempt to k-color a graph, where k is any positive integer. The idea of the first algorithm, known as **Sequential Coloring**, is to dive in and start coloring using as few colors as possible. We consider the vertices from 1 up to V in order and to each vertex v we assign the first available color not already assigned to a neighbor of v.

Algorithm SEQUENTIALCOLOR

STEP 1. Input G with V vertices $x_1, \ldots, x_V$
{Let the potential colors be $1, 2, \ldots, V$.}

STEP 2. For $I = 1$ to V do
    STEP 3. Create $L_I = \langle 1,2,\ldots,I\rangle$ {L_I is the list of colors that might get assigned to x_I.}
STEP 4. For $I = 1$ to V do
    Begin
    STEP 5. Set $c_I :=$ first color in L_I {c_I is the color assigned to x_I.}
    STEP 6. For each J with $I < J$ and (x_I, x_J) in $E(G)$ do
        STEP 7. Set $L_J := L_J - c_I$ {x_J cannot receive the same color as x_I.}
    End
STEP 8. Output each vertex, the color it received, and the total number of colors used; then stop.

Example 5.5. Table 8.6 is a trace of SEQUENTIALCOLOR applied to the graph in Figure 8.30. With the given ordering of the vertices SEQUENTIALCOLOR produces the coloring as shown.

Figure 8.30

Table 8.6

Step No.	I	L_I	c_I	J	L_J
3	1	$\langle 1\rangle$			
3	2	$\langle 1,2\rangle$			
3	3	$\langle 1,2,3\rangle$			
3	4	$\langle 1,2,3,4\rangle$			
5	1		1		
7	1			4	$\langle 2,3,4\rangle$
5	2		1		
7	2			3	$\langle 2,3\rangle$
5	3		2		
7	3			4	$\langle 3,4\rangle$
5	4		3		

Question 5.6. Run SEQUENTIALCOLOR on the labeled graphs shown in Figure 8.31.

SEQUENTIALCOLOR clearly places different colors on adjacent vertices; however, sometimes it uses more than the minimum number of colors. This point is important—we cannot be sure that the results of SEQUENTIALCOLOR are

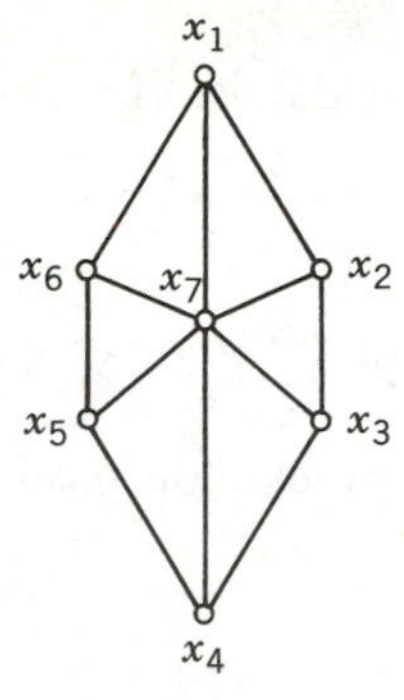

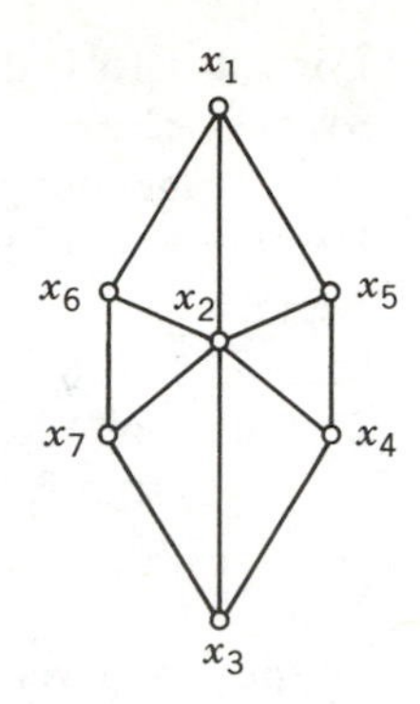

Figure 8.31

the best possible. But does it use just one extra color or just a "few" extra colors (whatever that means)? The answer is no; this algorithm might use far too many colors. Precisely, for every positive integer m there are graphs of chromatic number k on which SequentialColor uses $k + m$ colors. Exercise 29 illustrates this phenomena on 2-chromatic graphs. Thus looking back at the basic coloring problems, this algorithm might or might not solve Problem A (can a graph be k-colored?).

How efficient is SEQUENTIALCOLOR? Here the news is good. The loop in step 2 requires no more than V^2 assignments. The loop at step 4 has two purposes. The coloring (step 5) requires one assignment for each vertex. Updating the lists of possible colors requires one comparison and one assignment for each edge in the graph. Since $E \leq V^2$, SEQUENTIALCOLOR is an $O(V^2)$ algorithm (counting either comparisons or assignments or both.)

However, SEQUENTIALCOLOR can be used to find the chromatic number of a graph and hence to answer Problem B. As you saw in the preceding question, the number of colors used by SEQUENTIALCOLOR depends on the numbering of the vertices not on the graph alone. (The chromatic number depends only on the graph.) If we try every possible numbering of the vertices and with each numbering run SEQUENTIALCOLOR, then the chromatic number of the graph will be the minimum number of colors used in all these runs. This is the content of the next theorem.

Theorem 5.2. Let $\chi(G)$ be the chromatic number of a graph G. Then there is a labeling of the vertices of G with $x_1, x_2, \ldots, x_V$ such that SEQUENTIALCOLOR run on G with this labeling uses $\chi(G)$ colors.

Proof. Suppose that G is colored with colors $1, \ldots, R$. Label the vertices of G so that if $I < J$, then every vertex colored I receives a label before any vertex colored J. One way to accomplish this is to label all the vertices colored 1 (in any order you like), followed by all the vertices colored 2 (in any order you like), continuing until all vertices are labeled. If SEQUENTIALCOLOR is applied to G

with this labeling, then all the vertices originally colored 1 get colored 1. The vertices originally colored 2 might be colored with either 1 or 2. In general, the vertices that were originally colored I will be colored with one of $1, \ldots, I$. Thus no more than R colors are used by SEQUENTIALCOLOR. When $R = \chi(G)$, exactly R colors will be used, by definition of $\chi(G)$. □

The consequence of Theorem 5.2 is that SEQUENTIALCOLOR will find the chromatic number of a graph if all possible orderings of the vertices are tried. There are $V!$ such numberings and so the number of steps of this approach is more than $V!$, making this a slow algorithm. However, we next consider a modification of this approach known as **Backtracking**. This will be more efficient, but how much more we don't divulge yet.

Suppose that we want to know whether a given graph is 3-colorable. The idea of Backtracking is to try systematically all possible 3-colorings of the graph. The algorithm begins by trying to 3-color the graph using SEQUENTIALCOLOR, but if a fourth color is needed, it discontinues this approach. The algorithm backtracks (or backs up) to the last vertex, where there was choice in the coloring, and makes a different choice.

Example 5.6. Figure 8.32 shows a graph that can be 3-colored, but with the given vertex numbering SEQUENTIALCOLOR will 4-color it. A partial 4-coloring is given in Figure 8.32.

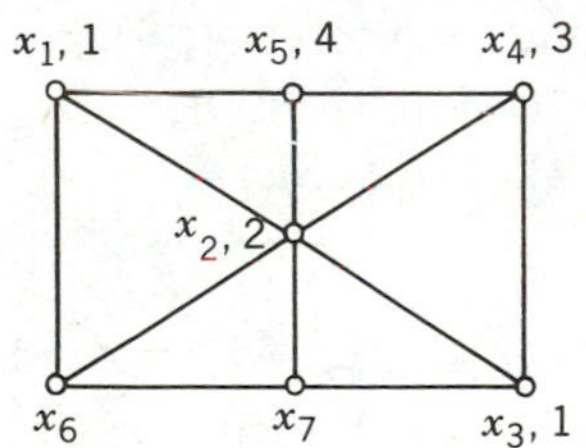

Figure 8.32

Once a fourth color is needed on x_5, we backtrack to x_4. There is no choice for its color if we are to 3-color the graph. However, if we backtrack to x_3, it can be recolored with 3. Then colors 1 and 3 are available for x_4 and x_5, respectively. Furthermore, this partial coloring extends to a 3-coloring of the whole graph by placing colors 3 and 1 on x_6 and x_7, respectively.

In this process of backtracking either a 3-coloring is discovered or all possible 3-colorings fail.

Here are some details of the algorithm BACKTRACKCOLOR. The number of colors that may be used is denoted by N. The possible colors are denoted $1, 2, \ldots, N$. The vertices are denoted by $x_1, x_2, \ldots, x_V$, and c_I contains the color that is used on the vertex x_I. Without loss of generality we may assume that vertex x_1 gets color 1; initially, $c_I := 0$ for $I = 2, \ldots, V$ to denote that these vertices are not colored. The algorithm parallels SEQUENTIALCOLOR: The vertices are colored in increasing numerical order, but if it ever is the case that the coloring cannot be extended to other vertices, the algorithm moves back one vertex and tries to recolor it with the next larger available color. The algorithm repeats this backtracking until either a vertex is found that can receive another color or until it finds that no vertex can be recolored. In the former case the algorithm begins again, trying to extend the partial coloring; in the latter case essentially all colorings fail, that is, the graph cannot be colored with N colors. The variable FORWARD keeps track of when the coloring is being extended (and FORWARD is TRUE) or when backtracking is occurring (and FORWARD is FALSE.)

Algorithm BACKTRACKCOLOR.

STEP 1. Initialize
    Input G with V vertices and a positive integer N {The algorithm will determine if G can be N-colored.}
    Set $c_1 := 1$ {x_1 gets color 1.}
    Set $c_i := 0$ for $i = 2, \ldots, V$
    Set $J := 2$ {J will index the vertex currently being colored.}
    Set FORWARD := TRUE
STEP 2. While $J < V + 1$ do
    STEP 3. If FORWARD = TRUE then
        Call Procedure COLOR
    Else {FORWARD = FALSE}
        Begin
        STEP 4. $c_J := 0$
        STEP 5. $J := J - 1$
        STEP 6. If $J \neq 1$, then
            Call Procedure COLOR
        Else {$J = 1$}
            Output "There is no N-coloring of G" and stop.
        end {Step 3}
STEP 7. Output the coloring and stop.

Procedure COLOR

STEP 1. Find smallest $K > c_J$ such that if $I < J$ and (x_I, x_J) is in $E(G)$, then $c_I \neq K$
STEP 2. If $K \leq N$, then
    Begin
    STEP 3. Set $c_J := K$
    STEP 4. Set $J := J + 1$

Step 5. Set FORWARD := TRUE
End
Else $\{K > N\}$
Step 6. Set FORWARD := FALSE
End.

Note the similarity between BACKTRACKCOLOR and the algorithm DFS-HAMCYCLE of Section 8.3.

Example 5.6 (continued). Table 8.7 is a trace of BACKTRACKCOLOR applied to the graph in Figure 8.32 with $N = 3$.

Table 8.7

Step No.		*J*	*K*	*FORWARD*	$c[1, 2, 3, 4, 5, 6, 7]$
Main	1	2		TRUE	[1, 0, 0, 0, 0, 0, 0]
COLOR	1	2	2		
	3–5	3	2		[1, 2, 0, 0, 0, 0, 0]
COLOR	1	3	1		
	3–5	4		TRUE	[1, 2, 1, 0, 0, 0, 0]
COLOR	1	4	3		
	3–5	5		TRUE	[1, 2, 1, 3, 0, 0, 0]
COLOR	1	5	4		
	6	5		FALSE	
Main	4–5	4			[1, 2, 1, 3, 0, 0, 0]
COLOR	1	4	4		
	6	4		FALSE	
Main	4–5	3			[1, 2, 1, 0, 0, 0, 0]
COLOR	1	3	3		
	3–5	4		TRUE	[1, 2, 3, 0, 0, 0, 0]
COLOR	1	4	1		
	3–5	5		TRUE	[1, 2, 3, 1, 0, 0, 0]
COLOR	1	5	3		
	3–5	6		TRUE	[1, 2, 3, 1, 3, 0, 0]
COLOR	1	6	3		
	3–5	7		TRUE	[1, 2, 3, 1, 3, 3, 0]
COLOR	1	7	1		
	3–5	8		TRUE	[1, 2, 3, 1, 3, 3, 1]
Main	7	Stop.			

Question 5.7. Trace BACKTRACKCOLOR applied to the complete graph K_4 with $N = 3$.

That BACKTRACKCOLOR is not a good algorithm can be seen by imagining an attempt to $(n - 1)$-color K_n for any $n > 1$. To begin with, the colors

$1, 2, \ldots, (n-1)$ are assigned to $x_1, \ldots, x_{n-1}$; however, this coloring cannot be completed. Backtracking to x_{n-2}, we find that it can be recolored with $n-1$ and x_{n-1} with $n-2$, but still there is no color left for vertex n. In general, the algorithm will backtrack to x_I with $I > 1$ and try all colors not on vertices $x_1, \ldots, x_{I-1}$. Since these vertices receive $I-1$ distinct colors, there are

$$(n-1)-(I-1) = n-I$$

colors that are tried on x_I for every fixed coloring of the $x_1, \ldots, x_{I-1}$. This means that precisely

$$(n-2)(n-3)\cdots(n-I)\cdots\cdot 2\cdot 1 = (n-2)!$$

colorings are tried before the algorithm reports that K_n cannot be $(n-1)$-colored. Thus BACKTRACKCOLOR is exponential in the worst case.

Exercises 40 and 41 ask you to verify that BACKTRACKCOLOR is correct. Exercise 43 suggests a recursive version of this coloring scheme.

All known algorithms for deciding whether a graph can be k-colored (for $k > 2$) are exponential. Thus the coloring problem is in the same unresolved state as the Hamiltonian cycle problem, the Traveling Salesrepresentative problem, and the Satisfiability problem. It is also one of the NP-Complete problems (as defined in the end of Section 8.3), which means that there is a polynomial algorithm for one of these four problems if and only if there is a polynomial algorithm for all four of them. However, graph coloring is an important problem with applications to timetable scheduling, routing problems, circuit board testing as well as to storage allocation. Thus coloring algorithms are used in practice. There are ways to seemingly improve the typical running time of these algorithms. For example, vertices of large degree are in some sense the hardest to color, and so numbering the vertices of largest degree with the smallest numbers may lead to a reasonable labeling on which to run sequential coloring. In some applications characteristics of the graphs can be incorporated to increase the efficiency of the algorithm. A variation is considered in Exercise 39.

Curiously, there are no known efficient approximate coloring algorithms, like the approximation algorithm of Section 8.4. Precisely, there is no known algorithm that will color a graph G with $O(\chi(G))$ colors. The best that is known is that there is an algorithm that will color a graph G with $O(\{V/\log(V)\}\chi(G))$ colors, where V is the number of vertices of G. Furthermore, it has been shown that the question of whether a graph is $(c\chi(G))$-colorable for any c less than 2 is also an NP-Complete problem. Thus in the worst case, approximations for graph-coloring algorithms are also very hard.

Look back at Problem C of our fundamental graph-coloring problems; this problem asked whether the chromatic number of a graph equals the size of the largest clique in it. Immediately, we saw that the chromatic number could be greater than the clique number. The clique number of a graph gives a lower bound on the chromatic number and so if an algorithm, like SEQUENTIALCOLOR, ever achieves a coloring with cl(G) colors, then the numbers of colors used is the minimum possible. However, determining the clique number of a graph is also one of the NP-Complete problems. A graph G is called **perfect** if $\chi(H) = \text{cl}(H)$ for every H that is an induced subgraph of G. It has recently been shown that there is a polynomial algorithm to determine whether a perfect graph can be k-colored; this algorithm uses the recent Ellipsoid Method of Linear Programming. On the other hand, there is no known polynomial algorithm to determine whether an arbitrary graph is perfect.

Despite all this bad news from the standpoint of efficient algorithms, the storage allocation problem, with which we began this section, is one that must be confronted by compilers and their designers. In the past, coloring algorithms have been used and modified for special needs, and in these cases they were effective in solving storage allocation problems. Recently, progress has been made by Jeanne Ferrante who has used a process known as "renaming" to get a better algorithm that runs in polynomial time and is applicable in all cases. In renaming, a new variable is created and assumes the value of an old variable at some points in the program. This increases the number of variables (and the number of assignment statements), but the new number of variables remains a polynomial in the number of original variables; the necessary program rewriting can be accomplished in one pass through the program. With this change there is a good algorithm for storage allocation that assigns the minimum possible number of memory locations, which is the maximum number of variables with every pair interfering at some point in the program. The latter number is called **Maxlive**. In graph theory terms, some vertices of the interference graph are split so that the total number of vertices is a polynomial of the original V, and then there is a good algorithm that colors the new graph G' in Maxlive colors. Since Maxlive $= \text{cl}(G) \leq \chi(G)$ for every interference graph G, we see that G' has been colored with the minimum number of colors. (See also Exercise 2.)

If, in addition, a compiler can reorder the statements of a program, then it may be possible to reduce the value of Maxlive and hence the amount of memory needed.

Example 5.7. The following reordering reduces Maxlive from 2 to 1.

Program One: Maxlive = 2	Program Two: Maxlive = 1
1. Define A	1. Define A
2. Define B	2. Use A
3. Use A	3. Define B
4. Use B	4. Use B

However, it has been shown that determining the order of a program for which Maxlive is a minimum is an NP-Complete problem! No matter where you look, there are hard unsolved algorithmic problems.

EXERCISES FOR SECTION 5

1. Here is an algorithm that interchanges the values of two variables (from Chapter 2.)

STEP 1. $x\text{old} := x$
STEP 2. $x := y$
STEP 3. $y := x\text{old}$

Construct the interference graph for this algorithm, find a coloring of the graph with the minimum possible number of colors, and find a storage allocation scheme using this many memory locations.

2. Construct the interference graph G of the following algorithm

STEP 1. Input A, B
STEP 2. $A := A * B$
STEP 3. $C := A^2$
STEP 4. $D := C^2$
STEP 5. $E := D^2$
STEP 6. $B := 3$
STEP 7. $E := E * B$.

Determine $\chi(G)$ and $\text{cl}(G)$. Find a way to "rename" at least one variable so that if G' is the resulting interference graph and $\chi(G')$ memory locations are used, then $\chi(G') = \text{cl}(G')$.

3. Here is one possible algorithm for storage allocation. Assign each variable to a different memory location. Thus, if a program has n variables, say $V_1, V_2, \ldots, V_n$, for $i = 1, 2, \ldots, n$, assign V_i to memory M_{i-1}. Explain why this allocation is quick and avoids all conflicts. What is the disadvantage of this algorithm?

4. Let C_j be a cycle with j vertices. Find $\chi(C_j)$ and $\text{cl}(C_j)$ for all j.

5. Find a 4-coloring of the following graphs if possible. For each graph determine whether its chromatic number is four or not.

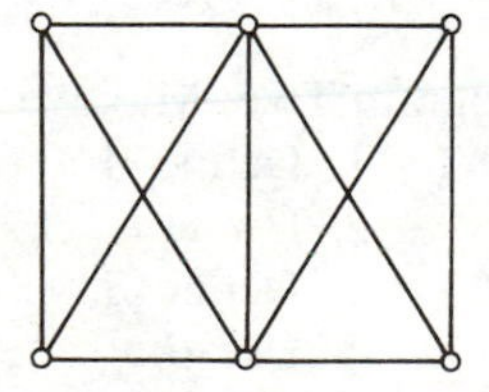

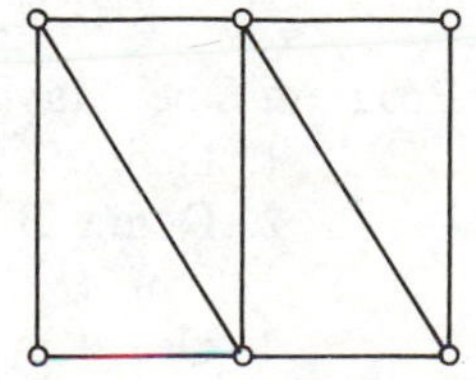

6. Explain why $\chi(G) \geq \text{cl}(G)$ for every graph G. Describe an infinite set of graphs for which $\chi(G) = \text{cl}(G)$ and an infinite set for which $\chi(G) > \text{cl}(G)$.

7. Explain why a graph with V vertices can always be V-colored. Characterize those graphs on V vertices for which $\chi(G) = V$. If G is a graph with V vertices and $\chi(G) = V - 1$, what can you say about cl(G)?

8. Write out the details of a BFS algorithm that tries to 2-color a graph. Run your algorithm on the following graphs.

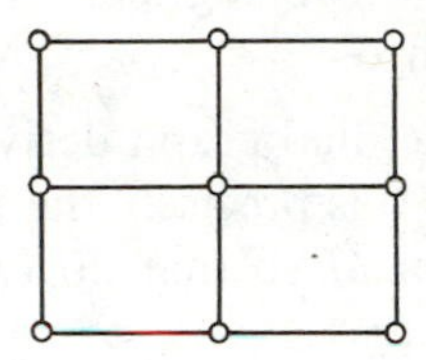

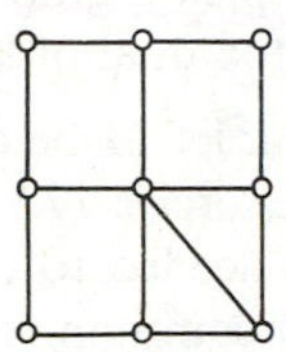

9. Show that the complexity of the algorithm from the preceding problem is $O(V^2)$.

10. Prove that if G is a graph with maximum degree d, then $\chi(G) \leq d + 1$.

11. Prove that if G is a connected graph with maximum degree d ($d > 2$), then $\chi(G) \leq d$ unless $G = K_{d+1}$.

12. For $k = 1, 2, 3$, and 4 find examples of graphs with maximum degree d such that $\chi(G) = d - k$.

13. Suppose that $\alpha(G)$ denotes the largest set of vertices of a graph, no two of which are adjacent. (See also Chapter 5, Supplementary Exercises 20 to 25.) Then prove that if G is any graph with V vertices, then $\chi(G) \geq V/\alpha(G)$.

14. If G is a graph with V vertices and maximum degree d, prove that $\chi(G) \geq V/(V - d)$.

15. If G^c is the complement of the graph G (as defined in Chapter 5, Supplementary Exercise 1), prove that $\chi(G) + \chi(G^c) \leq V + 1$, where V is the number of vertices of G.

16. Color the following map with three colors so that no two regions with a border in common receive the same color. Can fewer colors be used?

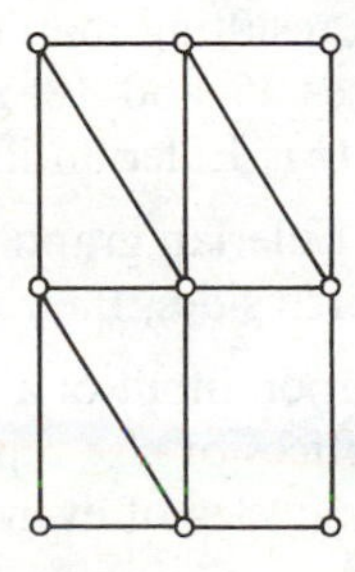

Find a map that can be 2-colored and one that can be 4-colored but not 3-colored.

17. (*a*) Draw a map. From this construct a graph as follows: Create a vertex for every region of the map and join two vertices by an edge if the corresponding regions have a border in common. Show that the graph can be drawn so that no two edges cross. The graph is called the **dual** of the original map.

(*b*) Next pick a graph G drawn so that no two edges cross. Find a map such that the dual of the map is the graph G.

18. In general, let M be a map and $G(M)$ the dual graph derived from it (as defined in Exercise 17.) Explain why $\chi(G(M))$ is precisely the minimum number of colors needed to color every region of M so that no two regions with a border in common receive the same color.

19. A graph is called **planar** if it can be drawn in the plane so that no two edges cross. Prove that if G is a connected, planar graph, drawn with F faces (including the outside face) then $V - E + F = 2$. This result is known as **Euler's formula**. (*Hint:* First prove this for trees. Then use induction on the number of edges of the graph.)

20. Prove that neither K_5 nor $K_{3,3}$ is a planar graph. (*Hint:* Use Euler's formula.)

21. Prove that for a planar graph the average degree, $2E/V$, is less than 6. Use this to conclude that every planar graph contains a vertex of degree 5 or less.

22. Prove that every planar graph can be 6-colored.

23. Prove that every planar graph can be 5-colored.

24. There once was a farmer with a large (square) tract of farm land. The farmer had five children and decided to divide the land into five pieces, one for each child, but to facilitate communication she wanted each piece to have a border in common with all other four pieces. Is such a division possible? If so, give an example of such a division (the pieces don't need to be the same size.) If not, find a division with each piece having at least a corner (or vertex) in common with every other piece.

25. Let G be a connected planar graph, drawn in the plane. Prove that G is Eulerian if and only if the resulting map can have its regions 2-colored. (Or in the notation of Exercises 17 and 18, if G is drawn in the plane and M is the resulting map, then G is Eulerian if and only if $G(M)$ is bipartite.)

26. Prove that if G is a planar Eulerian graph such that every region of the graph in the plane has exactly three sides, then G can be 3-colored.

27. Design an algorithm that upon input of a graph will find a cycle in the graph with an odd number of vertices or else report that there is none. Then design an algorithm to search for cycles of even length. Compare the complexities of your algorithms.

28. Run SEQUENTIALCOLOR on each of the following labeled graphs. Does the algorithm use the minimum possible number of colors?

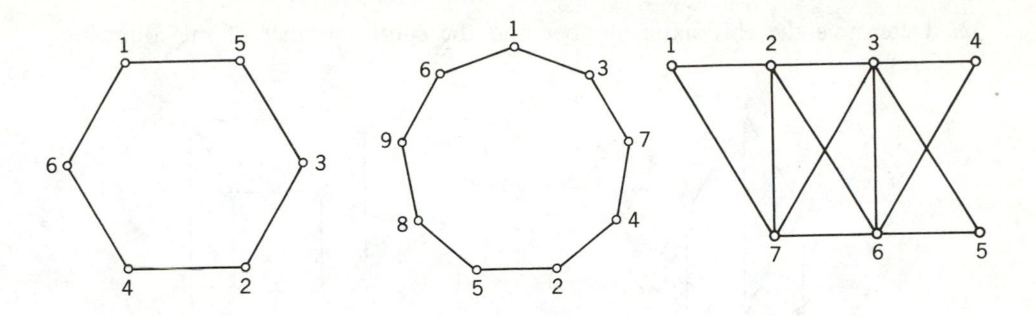

29. Run SEQUENTIALCOLOR on the following graphs.

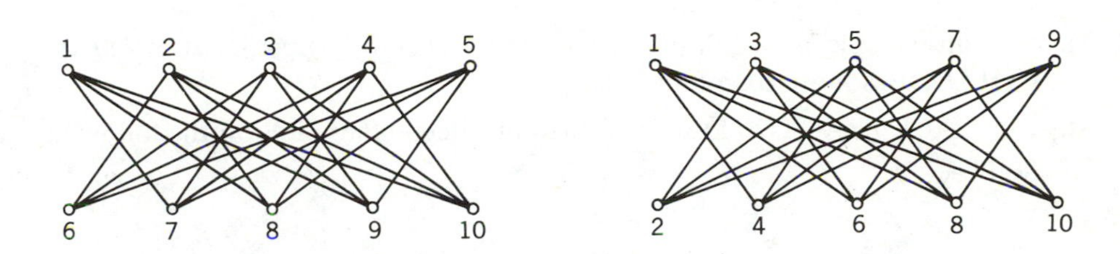

30. Generalize the examples of the preceding problem to show that for every k there are 2-chromatic graphs that when suitably labeled cause SEQUENTIALCOLOR to use $2 + k$ colors on them.

31. Here are some graphs; find the chromatic number of each graph and then find a labeling of them so that SEQUENTIALCOLOR uses that many colors on them.

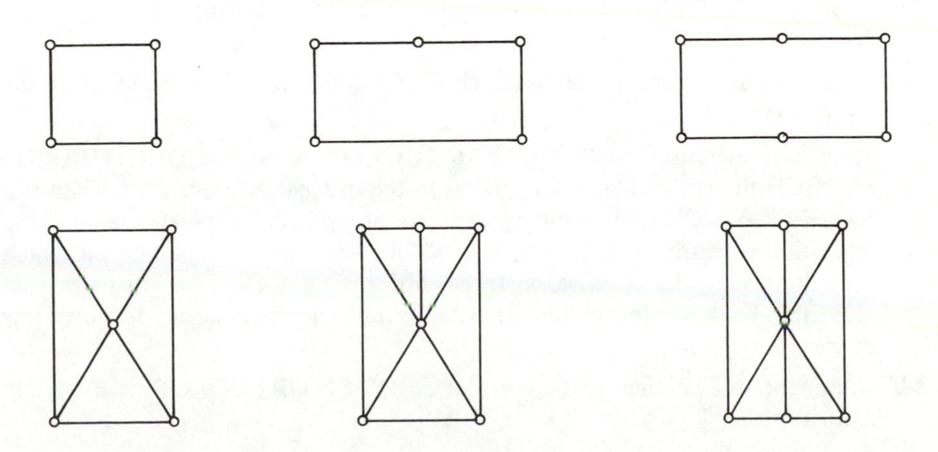

32. Determine the chromatic number and the clique number of the following graphs.

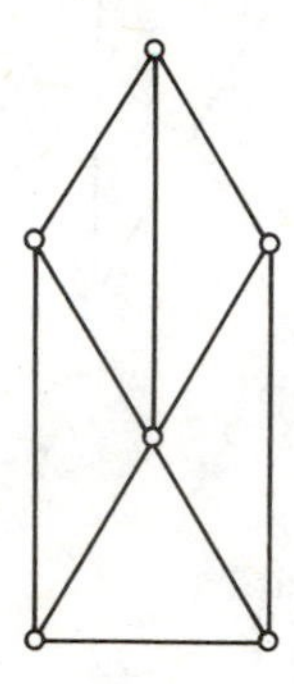

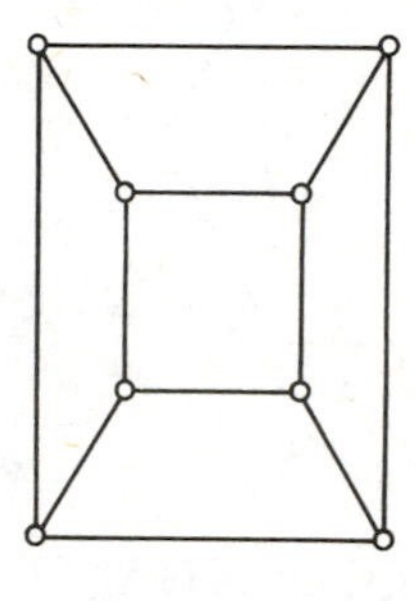

33. Find an example of a 4-colorable graph and a labeling so that SEQUENTIAL-COLOR uses more than four colors on it.

34. Run BACKTRACKCOLOR with the indicated values of N on the following graphs.

$N =$ 3

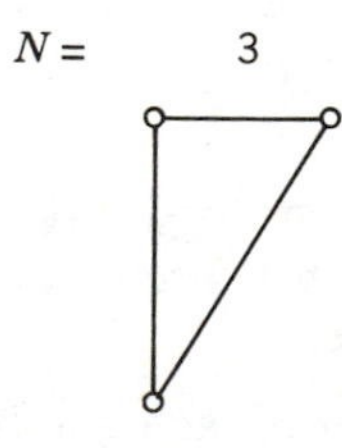

2

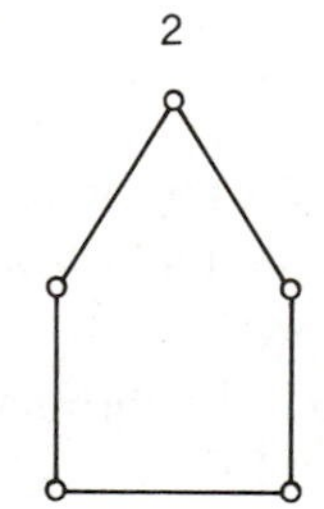

3

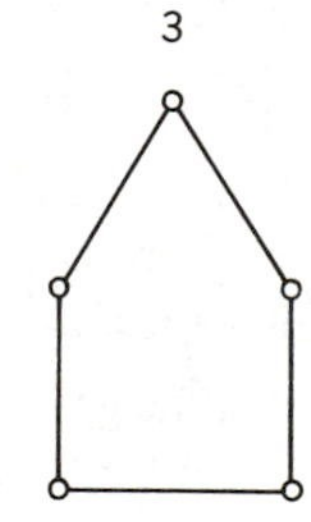

35. Redesign BACKTRACKCOLOR so that it finds all colorings of a graph with N colors.

36. Design an algorithm to determine $\text{cl}(G)$ and determine the complexity of the algorithm.

37. Here is a variation on BACKTRACKCOLOR: First use DEPTHFIRST-SEARCH to visit and label all vertices in the order they are visited. Then run BACKTRACKCOLOR. Find an example of a graph on which the resulting algorithm is more efficient than BACKTRACKCOLOR and one on which they do exactly the same work. In general, on what graphs will this revised algorithm be more efficient and on what graphs equally efficient? Is it ever less efficient?

38. Comment on whether the algorithm SEQUENTIALCOLOR is a greedy algorithm.

39. Design an algorithm that first numbers the vertices of a graph by decreasing degrees (i.e., the vertices of highest degree are numbered first) and then run SEQUENTIALCOLOR. This is known as Largestfirst. Find sets of graphs on which this version is more efficient and sets on which this is no more efficient.

40. Prove that if BACKTRACKCOLOR outputs a coloring, then it has found an N-coloring of G.

41. Prove that if BACKTRACKCOLOR reports "No N-coloring," there is no N-coloring of G. (*Hint:* Proceed by contradiction. Assume that G has an N-coloring and show that BACKTRACKCOLOR must find it.)

42. Find a function $g(V)$ such that BACKTRACKCOLOR requires at most $O(g(V))$ comparisons when run on a graph with V vertices.

43. Write a recursive version of backtrack coloring. (*Hint:* Just as BACKTRACK-COLOR resembles the algorithm DFS-HAMCYCLE, so should the recursive implementations.)

SOLUTIONS TO QUESTIONS IN CHAPTER 1

SECTION 1

1.1 The only number that appears on A, C, and D but not on B is 11. This can be verified by trial and error for each of $0, 1, 2, 3, 4, \ldots, 15$.

1.2 This trick will always work if no two numbers appear on exactly the same set of cards. For example, if both 11 and 5 appeared on cards A, C, and D but not on B, then player 1 could not know which answer, 11 or 5, was correct.

1.3 There is more than one way to design such a pair of cards. Two (of many) examples follow:

(*a*) [2 3] [1 3] **(*b*)** [0 3] [1 3]

Two cards cannot be used to distinguish the numbers 0, 1, 2, 3, and 4. The reasoning is as follows: Write down all possible yes/no responses to a set of two cards:

Card 1	*Card 2*
no	no
no	yes
yes	no
yes	yes

To make the trick work, we must assign to each of these four responses one of the numbers 0, 1, 2, 3, and 4. There are five numbers to assign to four responses, and so two numbers will elicit the same response.

1.4 10,000. There are 4 digits to be assigned with 10 choices for each digit (0, 1, . . . , 9). We apply the Multiplication Principle: There are $10 \cdot 10 = 100$ choices for the first two digits. Regardless of these choices there are 100 choices for the last two digits. Using the Multiplication Principle a second time, there are $100 \cdot 100 = 10{,}000$ choices for all 4 digits.

1.5

Player 2's Number	*Responses*			
	Card A	*Card B*	*Card C*	*Card D*
0	no	no	no	no
1	no	no	no	yes
2	no	no	yes	no
3	no	no	yes	yes
4	no	yes	no	no
5	no	yes	no	yes
6	no	yes	yes	no
7	no	yes	yes	yes
8	yes	no	no	no
9	yes	no	no	yes
10	yes	no	yes	no
11	yes	no	yes	yes
12	yes	yes	no	no
13	yes	yes	no	yes
14	yes	yes	yes	no
15	yes	yes	yes	yes

SECTION 2

2.1 (***a***) $10101 = 1 + 4 + 16 = 21$. (***b***) $100101 = 1 + 4 + 32 = 37$, and (***c***) $11010 = 2 + 8 + 16 = 26$. An even number expressed in binary ends with a 0 and an odd number ends with a 1.

2.2

Decimal Number	*Binary Representation*
0	0
1	1
2	10
3	11
4	100
5	101
6	110
7	111
8	1000
9	1001
10	1010
11	1011
12	1100
13	1101
14	1110
15	1111

If we make each of the binary representations four digits long by adding 0s to the left and replace each 0 with a "no" and each 1 with a "yes," then this table would become the same as that in Question 1.5.

2.3 4 (=100), 5 (=101), 6 (=110), 7 (=111), 12 (=1100), 13 (=1101), 14 (=1110), and 15 (=1111).

2.4

Decimal Number	*Arithmetic*	*Binary Representation*
6	2 + 4	110
19	1 + 2 + 16	10011
52	4 + 16 + 32	110100
84	4 + 16 + 64	1010100
232	8 + 32 + 64 + 128	11101000

SECTION 3

3.1 Step 1. Place one cup of water in the bottom of a double boiler.
Step 2. Place one cup of water in the top of a double boiler.
Step 3. Place one cup of quick oatmeal in the top of the double boiler.
Step 4. Turn on stove burner to medium.
Step 5. Place double boiler on burner and heat for 10 minutes.
Step 6. Remove pot.
Step 7. Turn off burner.

3.2 The successive values assigned to z follow.
(a) 1. **(b)** 20, 10, 5, 16, 8, 4, 2, 1.
(c) 7, 22, 11, 34, 17, 52, 26, 13, 40, 20, 10, 5, 16, 8, 4, 2, 1.

SECTION 4

4.1 **(a)** Algorithm BtoD run on $s = 10101$.

Step No.	j	m	*Is there a jth entry in s?*	*Is the jth entry equal to 1?*
1	0	—		
2	0	0		
3	0	0	yes	
4	0	1		yes
5	1	1		
3	1	1	yes	
4	1	1		no
5	2	1		
3	2	1	yes	
4	2	5		yes
5	3	5		
3	3	5	yes	
4	3	5		no
5	4	5		
3	4	5	yes	
4	4	21		yes
5	5	21		
3	5	21	no STOP	

Result: $m = 21$.

(*b*) Algorithm BtoD run on $s = 11010$.

Step No.	j	m	*Is there a jth entry in s?*	*Is the jth entry Step equal to 1?*
1	0	—		
2	0	0		
3	0	0	yes	
4	0	0		no
5	1	0		
3	1	0	yes	
4	1	2		yes
5	2	2		
3	2	2	yes	
4	2	2		no
5	3	2		
3	3	2	yes	
4	3	10		yes
5	4	10		
3	4	10	yes	
4	4	26		yes
5	5	26		
3	5	26	no STOP	

Result: $m = 26$.

(c) Algorithm BtoD run on $s = 100101$.

Step No.	j	m	*Is there a jth entry in s?*	*Is the jth entry equal to 1?*
1	0	—		
2	0	0		
3	0	0	yes	
4	0	1		yes
5	1	1		
3	1	1	yes	
4	1	1		no
5	2	1		
3	2	1	yes	
4	2	5		yes
5	3	5		
3	3	5	yes	
4	3	5		no
5	4	5		
3	4	5	yes	
4	4	5		no
5	5	5		
3	5	5	yes	
4	5	37		yes
5	6	37		
3	6	37	no STOP	

Result: $m = 37$. These are the same answers as those of Question 2.1.

4.2 Response 1 is not an algorithm because the instruction to stop might not be reached in a finite number of steps, since the binary representation of m might never be written down in step 1.

4.3 Response 2 is an algorithm because **(a)** the instructions are clear; **(b)** after performing an instruction, there is no ambiguity about which instruction is to be performed next; and **(c)** the instruction to stop will be reached after a finite number of instructions. Unlike Response 1, Response 2 finds binary numbers in increasing order (as opposed to at random) so that the mth binary number produced will be the binary representation of the decimal number m. The algorithm is slow because it will consider all n-bit binary numbers, $n \leq 5$, before concluding that 10011 is 19 in binary.

4.4 The algorithm must stop because eventually m must equal zero. Response 3 run on $m = 182$:

Step No.	*m*	*Largest power of 2 that is* $\leq m$	*r*	*Is m equal to 0?*	*Result*
1	182	$2^7 = 128$	7		1_______
2	54			no	
1	54	$2^5 = 32$	5		1_1_____
2	22			no	
1	22	$2^4 = 16$	4		1_11____
2	6			no	
1	6	$2^2 = 4$	2		1_11_1__
2	2			no	
1	2	$2^1 = 2$	1		1_11_11_
2 STOP	0			yes	10110110

4.5 The algorithm must stop because it repeatedly decreases the value of m. Therefore, the value of m must eventually be 0.
Here is algorithm DtoB run on $m = 395$; the values shown are those assigned to the variables after the execution of the given step:

Step No.	*j*	*m*	*q*	*r*	*Answer*
1	0	395	—	—	—
2	0	395	197	1	1
4, 5	1	197	197	1	
2	1	197	98	1	11
4, 5	2	98	98	1	
2	2	98	49	0	011
4, 5	3	49	49	0	
2	3	49	24	1	1011
4, 5	4	24	24	1	
2	4	24	12	0	01011
4, 5	5	12	12	0	
2	5	12	6	0	001011
4, 5	6	6	6	0	
2	6	6	3	0	0001011
4, 5	7	3	3	0	
2	7	3	1	1	10001011
4, 5	8	1	1	1	
2	8	1	0	1	110001011
3 STOP			0		

Result: The binary representation of $m = 395$ is 110001011. Response 4 (DtoB) is easier to use than Response 3 because the user does less and easier arithmetic.

SECTION 5

5.1 (*a*) $A = \{1, 4, 6, 8, 9, 10, 12, 14, 15, 16, 18, 20, 21, 22, 24, 25, 26, 27, 28\}$,
(*b*) $B = \{1, 4, 9, 16, 25\}$, and (*c*) $C = \{4, 8, 9, 12, 16, 18, 20, 24, 25, 27, 28\}$.

5.2 (*a*) $A^c = \{2, 3, 5, 7, 11, 13, 17, 19, 23, 29\}$
(*b*) $B^c = \{2, 3, 5, 6, 7, 8, 10, 11, 12, 13, 14, 15, 17, 18, 19, 20, 21, 22, 23, 24, 26, 27, 28, 29\}$
(*c*) $C^c = \{1, 2, 3, 5, 6, 7, 10, 11, 13, 14, 15, 17, 19, 21, 22, 23, 26, 29\}$

5.3 Every set is a subset of itself. In addition $B \subseteq A$, $C \subseteq A$, $A^c \subseteq B^c$, $A^c \subseteq C^c$.

5.4 Since $B \subseteq A$, $A \cup B = A$ and $A \cap B = B$. Similarly, $A \cup C = A$ and $A \cap C = C$. $B \cup C = \{1, 4, 8, 9, 12, 16, 18, 20, 24, 25, 27, 28\}$ and $B \cap C = \{4, 9, 16, 25\}$.

SECTION 6

6.1 (*a*) (*b*) (*c*) (*d*)

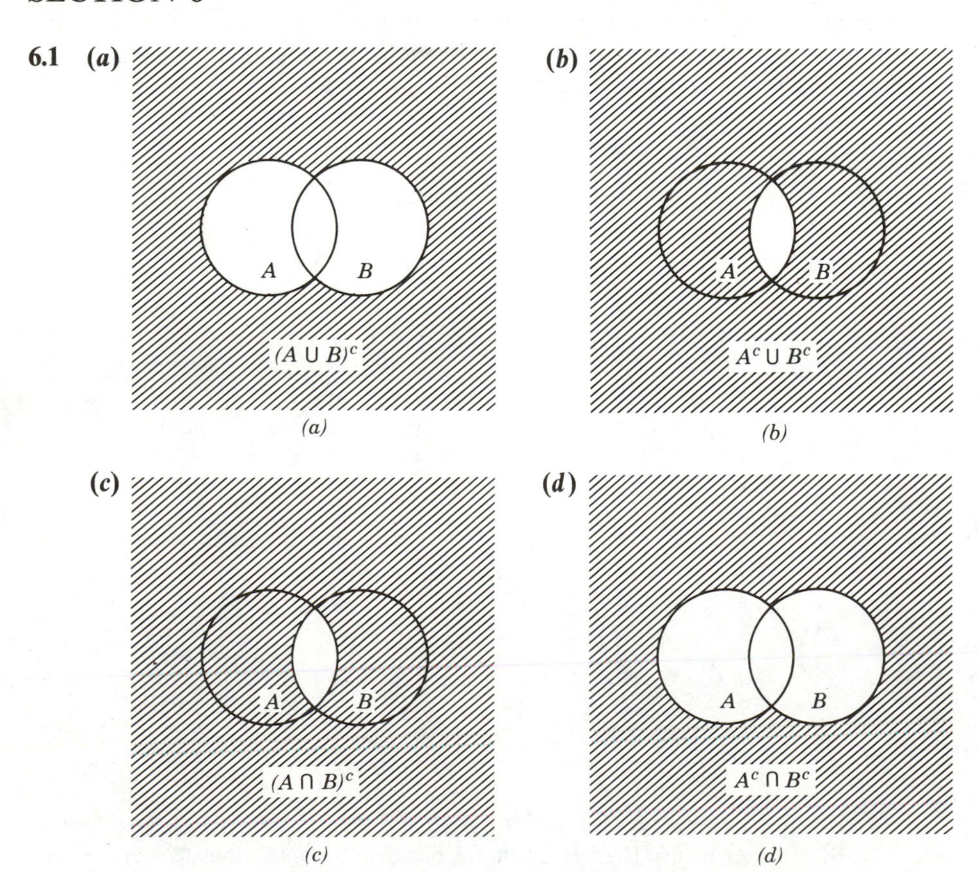

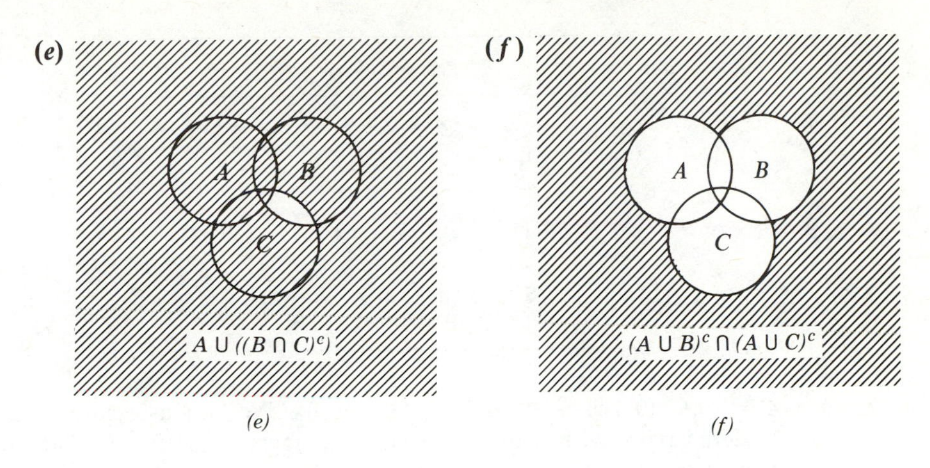

(e) (f)

6.2 **(i)** $(A \cup B)^c = A^c \cap B^c$

Proof. Let x be in $(A \cup B)^c$. Then by the definition of complement, x is not in $A \cup B$. Thus x is not in A and x is not in B. If x is not in A, then x is in A^c. Similarly, if x is not in B, then x is in B^c. Consequently, x is in $A^c \cap B^c$.

Conversely, if x is in $A^c \cap B^c$, then x is in A^c and x is in B^c. In other words, x is not in A and x is not in B. Thus x is not in $A \cup B$. Consequently, x is in $(A \cup B)^c$.

We have shown that $(A \cup B)^c \subseteq A^c \cap B^c$, and that $A^c \cap B^c \subseteq (A \cup B)^c$. Thus $(A \cup B)^c = A^c \cap B^c$, as desired. □

(ii) $(A \cap B)^c = A^c \cup B^c$

Proof. Let x be in $(A \cap B)^c$. Then x is not in $A \cap B$, which means that x is not in A or x is not in B (or both). Then x is in A^c or x is in B^c (or both). Thus x is in $A^c \cup B^c$.

Conversely, let x be in $A^c \cup B^c$. Then x is not in A or x is not in B (or both), which means that x is not in $A \cap B$. Thus x is in $(A \cap B)^c$.

We have shown that $(A \cap B)^c \subseteq A^c \cup B^c$, and that $A^c \cup B^c \subseteq (A \cap B)^c$. Thus $(A \cap B)^c = A^c \cup B^c$, as desired. □

SECTION 7

7.1 $\varnothing$, $\{a\}$, $\{b\}$, $\{c\}$, $\{a,b\}$, $\{a,c\}$, $\{b,c\}$, and $\{a,b,c\}$.

7.2 Algorithm SUBSET run on $A = \{a_1, a_2, a_3\}$ with $n = 3$.

Step No.	j	*List of Subsets*
1		$\varnothing$
2	1	
3	1	$\{a_1\}$
4	2	
3	2	$\{a_2\}, \{a_1, a_2\}$
4	3	
3	3	$\{a_3\}, \{a_1, a_3\}, \{a_2, a_3\}, \{a_1, a_2, a_3\}$
5	3	STOP

7.3 $A \times A = \{(0,0), (0,1), \ldots, (0,9), (1,0), \ldots, (9,9)\}$. We can associate each of these ordered pairs with a unique decimal integer from 0 to 99. Similarly, with each element of $A \times A \times A$ we can associate the decimal numbers from 0 to 999. In general, the set A^n is the set of all ordered n-tuples with entries from $A = \{0, 1, 2, \ldots, 9\}$. Each element corresponds with a unique number from 0 to $10^n - 1$.

7.4 ***(a)*** $A \times A = \{(a,a), (a,b), (a,c), (b,a), (b,b), (b,c), (c,a), (c,b), (c,c)\}$, ***(b)*** r^2, ***(c)*** r^3, and ***(d)*** r^n.

7.5 ***(a)*** The bit vector of T is 110010, since the elements of T are the first, second, and fifth elements in the list of S, ***(b)*** 101001 and 000111, ***(c)*** z is the sixth element in the list of S, and ***(d)*** $\{y\}$ and $\{u, v, w, x\}$.

SECTION 8

8.1 $|A \cup C| = 17$ and $|B \cup C| = 13$.

8.2 The cardinality of each union of two cards is 12.

8.3 ***(a)*** $|A \cup B|$ never equals 4, since $A \subseteq A \cup B$. For ***b)*** to ***e)*** set $A = \{1, 2, 3, 4, 5\}$ and ***(b)*** $B = \{1, 2, 3\}$, ***(c)*** $B = \{1, 2, 6\}$, ***(d)*** $B = \{1, 6, 7\}$, and ***(e)*** $B = \{6, 7, 8\}$. ***(f)*** $|A \cup B|$ never equals 9, since $|A \cup B|$ is the largest when A and B are disjoint and then $|A \cup B| = |A| + |B| = 8$.

8.4 Let $A = \{$students enrolled in Discrete Mathematics$\}$ and $B = \{$students enrolled in Computer Science$\}$. We are given $|A| = 146$, $|B| = 205$, and $|A \cup B| = 232$. From Theorem 8.1

$$
\begin{aligned}
|A \cap B| &= |A| + |B| - |A \cup B| \\
&= 146 + 205 - 232 \\
&= 119 \text{ students in both courses.}
\end{aligned}
$$

SECTION 9

9.1 The range of b is all of B because if q is any binary number and t its decimal equivalent, then $b(t) = q$.

9.2 The map f_1 is not a function from N to B because its range is not contained in B. The map f_2 is a function, since a binary number is either even or odd but never both. The range of f_2 is $\{0, 1\}$. The map f_3 is a function, since for each natural number r there is precisely one string with r 1s. The range of f_3 is all binary numbers that contain no zero. The map f_4 is not a function, since, for example, $f_4(6)$ should equal 0 because 2 divides 6, and yet $f_4(6)$ should equal 1 because 3 divides 6.

9.3 The function b is onto, since the range of b is all of B (see Question 9.1). The function f_2 is not onto, since $\{0, 1\}$ does not include all binary numbers. The function f_3 is not onto, since the range does not include binary numbers containing zeros.

9.4 The function b is one-to-one because if $n \neq n'$ are two different numbers, then their binary representations differ and so $b(n) \neq b(n')$. The function f_2 is not one-to-one because, for example, $f_2(2) = f_2(4) = 0$. The function f_3 is one-to-one because if $r \neq r'$, then $f_3(r)$ and $f_3(r')$ are strings of ones of different length.

9.5 Yes, since $A \neq A'$ implies that $c(A) \neq c(A')$.

9.6 If X is in $P(U)$, then $c \circ c(X) = c(X^c) = (X^c)^c = X$. Thus $c \circ c = i$ and c is its own inverse.

SECTION 10

10.1 (*a*)

x	y	$x \vee y$	$\sim(x \vee y)$	$\sim x$	$\sim y$	$(\sim x) \wedge (\sim y)$
0	0	0	1	1	1	1
0	1	1	0	1	0	0
1	0	1	0	0	1	0
1	1	1	0	0	0	0

(*b*) Note that $(x \wedge y) \wedge z = 1$ if and only if all of x, y, and z equal 1. Similarly, $x \wedge (y \wedge z) = 1$ if and only if all of x, y, and z equal 1.

10.2

x	y	$x \oplus y$	$x \vee y$	$\sim(x \wedge y)$	$(x \vee y) \wedge (\sim(x \wedge y))$	$(x \wedge \sim y) \vee (\sim x \wedge y)$
0	0	0	0	1	0	0
0	1	1	1	1	1	1
1	0	1	1	1	1	1
1	1	0	1	0	0	0

10.3 (*a*) $x \wedge y = 1$ if and only if both x and y equal 1 if and only if $y \wedge x = 1$. **(b)** $(x \vee y) \vee z = 0$ if and only if all of x, y, and z equal 0 if and only if $x \vee (y \vee z) = 0$.

10.4 None.

10.5 (*a*) is a contradiction while both (*b*) and (*c*) are tautologies.

SOLUTIONS TO QUESTIONS IN CHAPTER 2

SECTION 1

1.1

	Value Assigned to x	*Value Assigned to y*	*Value Assigned to* xold	*Value Assigned to* yold
Before step 1	5	2	?	?
After step 1	5	2	5	?
After step 2	5	2	5	2
After step 3	5	5	5	2
After step 4	2	5	5	2

SECTION 2

2.1 Algorithm EXPONENT run with $x = 3$, $n = 4$.

Step No.	i	*ans*
2	0	1
4	0	3
5	1	3
4	1	9
5	2	9
4	2	27
5	3	27
4	3	81
5	4	81
6 STOP	4	81

SECTION 3

3.1 P_n: $2 + 4 + \cdots + 2n = n(n + 1)$ for all positive integers n. Proof by induction on n.

Step 1 (the base case): P_1 is the statement $2 \cdot 1 = 2 = 1 \cdot (1 + 1)$.

Step 2 (the inductive hypothesis): Assume that P_k is true. P_k is the statement $2 + 4 + \cdots + 2k = k(k + 1)$.

Step 3 (the inductive step): Verify that P_{k+1} is true. P_{k+1} is the statement $2 + 4 + \cdots + 2k + 2(k + 1) = (k + 1)(k + 2)$.

$$\begin{aligned} &2 + 4 + \cdots + 2k + 2(k + 1) \\ &\quad = (2 + \cdots + 2k) + 2(k + 1) && \text{by associativity} \\ &\quad = k(k + 1) + 2(k + 1) && \text{by inductive hypothesis} \\ &\quad = (k + 1)(k + 2) && \text{by factoring.} \end{aligned}$$

Therefore, P_{k+1} is true, and P_n is true for all positive n. □

3.2 Suppose that $n = 2$. If $x = -1$, then $1 - x + x^2 = 3$. If $x \neq -1$, then

$$\begin{aligned} 1 - x + x^2 &= \frac{(1 - x + x^2)(1 + x)}{1 + x} \\ &= \frac{1 + x^3}{1 + x}. \end{aligned}$$

Suppose that $n = 3$. If $x = -1$, then $1 - x + x^2 - x^3 = 4$. If $x \neq -1$, then

$$\begin{aligned} 1 - x + x^2 - x^3 &= \frac{(1 - x + x^2 - x^3)(1 + x)}{1 + x} \\ &= \frac{1 - x^4}{1 + x}. \end{aligned}$$

3.3 $$P_n: 1 + x + x^2 + \cdots + x^n = \begin{cases} (1 - x^{n+1})/(1 - x) & \text{if } x \neq 1 \\ n + 1 & \text{if } x = 1 \end{cases}$$

First notice that when $x = 1$, the left-hand side of the equation of P_n is the sum of $(n + 1)$ terms, each equal to 1, and so the equation is valid. Now we focus on the case when $x \neq 1$. Proof by induction on n.

Step 1 (the base case):

$$P_0 \text{ is the statement } 1 = \frac{1 - x^{0+1}}{1 - x} = \frac{1 - x}{1 - x}$$

which is true, since $x \neq 1$.

$$P_1 \text{ is the statement } 1 + x = \frac{1 - x^2}{1 - x}.$$

This statement is also true since

$$\begin{aligned}\frac{1 - x^2}{1 - x} &= \frac{(1 - x)(1 + x)}{1 - x} \\ &= 1 + x, \qquad \text{for } x \neq 1.\end{aligned}$$

Step 2 (the inductive hypothesis): Assume that P_k is true. P_k is the statement

$$1 + x + x^2 + \cdots + x^k = \frac{1 - x^{k+1}}{1 - x} \qquad \text{if } x \neq 1.$$

Step 3 (the inductive step): Show that P_{k+1} is true. P_{k+1} is the statement

$$1 + x + x^2 + \cdots + x^k + x^{k+1} = \frac{1 - x^{k+2}}{1 - x} \qquad \text{if } x \neq 1.$$

$$\begin{aligned} 1 + x + \cdots + x^k + x^{k+1} & \\ &= (1 + \cdots + x^k) + x^{k+1} && \text{by associativity} \\ &= \frac{1 - x^{k+1}}{1 - x} + x^{k+1} && \text{by inductive hypothesis} \\ &= \frac{1 - x^{k+1} + (1 - x)x^{k+1}}{1 - x} && \text{with common denominator} \\ &= \frac{1 - x^{k+1} + x^{k+1} - xx^{k+1}}{1 - x} && \text{by algebra} \\ &= \frac{1 - x^{k+2}}{1 - x} && \text{by more algebra.} \end{aligned}$$

Hence P_{k+1} is true, and P_n is true for all positive integers n. □

SECTION 4

4.1 If A contains one element, then it has one even subset, the empty set. If A contains two elements, then it has two even subsets, the empty set and the whole set A. You should check that a 3-set has four even subsets. Thus it seems as if whenever A is a set with n elements, then the number of even subsets of A is 2^{n-1}. We prove this by induction on n.

Step 1 (the base case): P_1 is the statement that a set A with one element has exactly $2^{1-1} = 2^0 = 1$ even subset. We just checked that this is true.

Step 2 (the inductive hypothesis): We assume that P_k is true. P_k is the statement that a set A with k elements has exactly 2^{k-1} even subsets.

Step 3 (the inductive step): We must verify that P_{k+1} is true. P_{k+1} is the statement that a set A with $k + 1$ elements has exactly $2^{k+1-1} = 2^k$ even subsets.

Consider a set A with $k + 1$ elements. We must show that A has exactly 2^k even subsets. Let x be an element in A and define B to be $A - \{x\}$. By Example 4.1 we know that B has exactly 2^k subsets. We build upon these subsets to obtain all even subsets of A. Namely, let S be an even subset of A. If S does not contain x, then S is an even subset of B. If S does contain x, then $S - \{x\}$ is an odd subset of B, where by an odd subset we mean one containing an odd number of elements. Furthermore, every subset of B is either even or odd. An even subset of B is also an even subset of A, and an odd subset T of B turns into an even subset of A by forming $T \cup \{x\}$. The number of even subsets of B is 2^{k-1} by the inductive hypothesis. The number of odd subsets of B is the total number of subsets (2^k) minus the number of even subsets (2^{k-1}), or $2^k - 2^{k-1}$. Thus the number of odd subsets of B is

$$\begin{aligned} 2^k - 2^{k-1} &= 2^{k-1}(2 - 1) \\ &= 2^{k-1}. \end{aligned}$$

$$\begin{aligned} \text{Thus } \#(\text{even subsets of } A) &= \#(\text{even subsets not containing } x) \\ &\quad + \#(\text{even subsets containing } x) \\ &= \#(\text{even subsets of } B) + \#(\text{odd subsets of } B) \\ &= 2^{k-1} + 2^{k-1} \qquad \text{by inductive hypothesis and the argument given above} \\ &= 2 \cdot 2^{k-1} \\ &= 2^k. \end{aligned}$$

Thus P_{k+1} is established, and P_n is true for every positive integer n. □

4.2 P_n is the proposition that the nth time the comment in algorithm SUM is encountered, it is correct. The last time the comment is encountered j will have the value r, and if the comment is correct, then ans will have the value $r(r + 1)/2$ and the output will be as claimed.

Step 1 (the base case): We check P_1. The variable ans is initially equal to 0, but the first time the comment is encountered, ans has been incremented by $j = 1$ so ans equals 1. The comment asserts that the value of ans is $1 \cdot 2/2 = 1$ and so the comment is correct. You might check also that P_2 is valid.

Step 2 (the inductive hypothesis): We assume P_k, which states that the kth time the comment is encountered, it is true.

Step 3 (the inductive step): We must prove P_{k+1}, which states that the $(k+1)$st time the comment is encountered it is valid. Now the kth time that the comment is reached j has the value of k and by the inductive hypothesis the value in ans is $k(k+1)/2$. The next time j has the value $(k+1)$ and ans has been increased by this value of j:

$$\begin{aligned}\text{ans \{after } k+1 \text{ encounters\}} &= j + \text{ans \{after } k \text{ encounters\}}\\ &= (k+1) + \frac{k(k+1)}{2}\\ &= (k+1)(1+k/2)\\ &= \frac{(k+1)(k+2)}{2},\end{aligned}$$

which is the assertion of P_{k+1}. Thus P_n is true for all positive n. □

4.3 ***(a)*** 4, ***(b)*** 3, ***(c)*** 4, and ***(d)*** 3. The binary representation of 14 can be obtained from that of 7 by adding a 0 at the right. The binary representation of 13 can be obtained from that of 6 by adding a 1 at the right.

SECTION 5

5.1 $n = 7$: four multiplications, since $x^7 = (x^4)(x^2)x$.
$n = 11$: five multiplications, since $x^{11} = (x^8)(x^2)x$
$n = 12$: four multiplications, since $x^{12} = x^8x^4$
$n = 16$: four multiplications, since $x^{16} = (x^8)(x^8)$

5.2 Revised Algorithm DtoB used to find x^{37}:

Variables	*j*	*m*	*q*	*r*	*x*	*ans*	*No. Multiplications and Divisions*
Values	0	37	18	1	x	x	2
After	1	18	9	0	x^2	x	2
Step 2.5	2	9	4	1	x^4	x^5	3
	3	4	2	0	x^8	x^5	2
	4	2	1	0	x^{16}	x^5	2
	5	1	0	1	x^{32}	x^{37}	3
							Total No. 14

Revised Algorithm DtoB used to find x^{52}.

Variables	j	m	q	r	x	*ans*	*No. Multiplications and Divisions*
Values	0	52	26	0	x	1	1
After	1	26	13	0	x^2	1	2
Step 2.5	2	13	6	1	x^4	x^4	3
	3	6	3	0	x^8	x^4	2
	4	3	1	1	x^{16}	x^{20}	3
	5	1	0	1	x^{32}	x^{52}	3
							Total No. 14

SECTION 6

6.1 Steps 1 and 5 were bookkeeping steps in DtoB and are not needed in FASTEXP because of step 2.5.

6.2 $\log(2^2) = \log(4) = 2$ $\quad\log(2^3) = \log(8) = 3$
$\log(2^5) = \log(32) = 5$ $\quad\log(2^{10}) = 10$
$2^{\log(2)} = 2^1 = 2$ $\quad 2^{\log(4)} = 2^2 = 4$
$2^{\log(6)} = 2^{2.584}\ldots = ?$ (Do the next question and then return to finish this.)
$2^{\log(8)} = 2^3 = 8$

6.3 By the definition of logarithm if $\log(2^p) = h$, then $2^h = 2^p$. This implies that $h = p$. Thus $\log(2^p) = p$. If $\log(q) = t$, then by definition $2^t = q$ and by substitution $2^{\log(q)} = 2^t = q$.

6.4 $\lfloor \frac{17}{3} \rfloor = 5$, $\lceil \frac{25}{7} \rceil = 4$, $\lfloor \log(8) \rfloor = 3$, $\lceil \log(13) \rceil = 4$, $\lfloor -\frac{14}{9} \rfloor = -2$, $\lceil \log(25) \rceil = 5$, and $\lfloor \log(13.73) \rfloor = 3$.

SECTION 7

7.1 $N = 17$: $\sqrt{17} = 4.1231\ldots > 4.0874\ldots = \log(17)$. This does not contradict Theorem 7.2, but says that more is true than is stated in the theorem. Namely, it is true that if $n \geq 17$ (see Exercise 7.5), then $\sqrt{n} > \log(n)$. (The bound $n \geq 64$ was used for ease of calculation and proof argument.)

SECTION 8

8.1 Note that $f(n) = 12n^2 - 11 \leq 12n^2$, since subtraction makes things smaller. Thus letting $C = 12$, we have $f(n) = O(n^2)$. Similarly,

$$\begin{aligned} h(n) &= 3n^2 + 4n + 11 \\ &\leq 3n^2 + 4n^2 + 11n^2, \text{ since } n \leq n^2 \text{ and } 1 \leq n^2 \\ &= 18n^2. \end{aligned}$$

Thus letting $C = 18$, we have $h(n) = O(n^2)$.

8.2 An algorithm will be called cubic if there is a function, say $f(n)$, that counts the number of operations given a problem of size n and $f(n) = O(n^3)$. Both L and C ought to take about 16 minutes to solve a problem of size 200. On a problem of size 1000, L should take about 80 minutes and C should take about 2000 minutes.

8.3
$$\begin{aligned} f(n) &= 2n^7 - 6n^5 + 10n^2 - 5 \\ &\leq 2n^7 + 6n^5 + 10n^2 + 5 \\ &\leq 2n^7 + 6n^7 + 10n^7 + 5n^7 = 23n^7. \end{aligned}$$

Thus with $C = 23$, $f(n) = O(n^7)$.

8.4 By Theorem 8.1 $f = O(n^5)$ and $g = O(n^4)$. Thus by Theorem 8.2

$$f + g = O(n^5 + n^4) = O(n^5),$$

and

$$f \cdot g = O(n^5 n^4) = O(n^9).$$

SECTION 9

9.1 (*a*) 353 is a prime number, (*b*) 238 is not an even integer (or 238 is an odd integer).

9.2 (*a*) There exists an integer greater than one that does not have a prime divisor. (*b*) There exists an integer of the form $4n + 1$ that is not a prime. (*c*) There exists a prime greater than 2 that is not odd.

9.3 (*a*) For every integer n, $3n + 1$ is not a prime number. (*b*) For every integer n, $\log(n) \leq n$. (*c*) For every integer n, $n^2 \leq 2^n$.

9.4 (*a*) The hypothesis is that n is even; the conclusion is that $n^2 + n + 1$ is prime. The negation is that there is some even integer n such that $n^2 + n + 1$ is not prime.
(*b*) The hypothesis is that $n^2 + n + 1$ is prime; the conclusion is that n is even. The negation is that there exists an integer n such that $n^2 + n + 1$ is prime and n is not even.
(*c*) The hypothesis is that n is divisible by 6; the conclusion is that n^2 is divisible by 4. The negation is that there is an integer n that is divisible by 6 but n^2 is not divisible by 4.

9.5 (*a*) The converse of 9.4(a) is 9.4(b). The contrapositive of 9.4(a) is that if $n^2 + n + 1$ is not prime, then n is not even.
(*b*) The converse of 9.4(b) is 9.4(a). The contrapositive of 9.4(b) is that if n is not even, then $n^2 + n + 1$ is not prime.
(*c*) The converse of 9.4(c) is that if n^2 is divisible by 4, then n is divisible by 6. The contrapositive of 9.4(c) is that if n^2 is not divisible by 4, then n is not divisible by 6.
(*d*) The converse of Lemma 7.1 is that if $2^r > (r + 1)^2$, then r is greater than 5. The contrapositive of Lemma 7.1 is that if $2^r \leq (r + 1)^2$, then r is no greater than 5.
(*e*) The converse of Theorem 7.2 is that if $\sqrt{n} > \log(n)$, then $n \geq 64$. The contrapositive of Theorem 7.2 is that if $\sqrt{n} \leq \log(n)$, then $n < 64$.

SOLUTIONS TO QUESTIONS IN CHAPTER 3

SECTION 1

1.1 **(1)** There are 10 such paths:

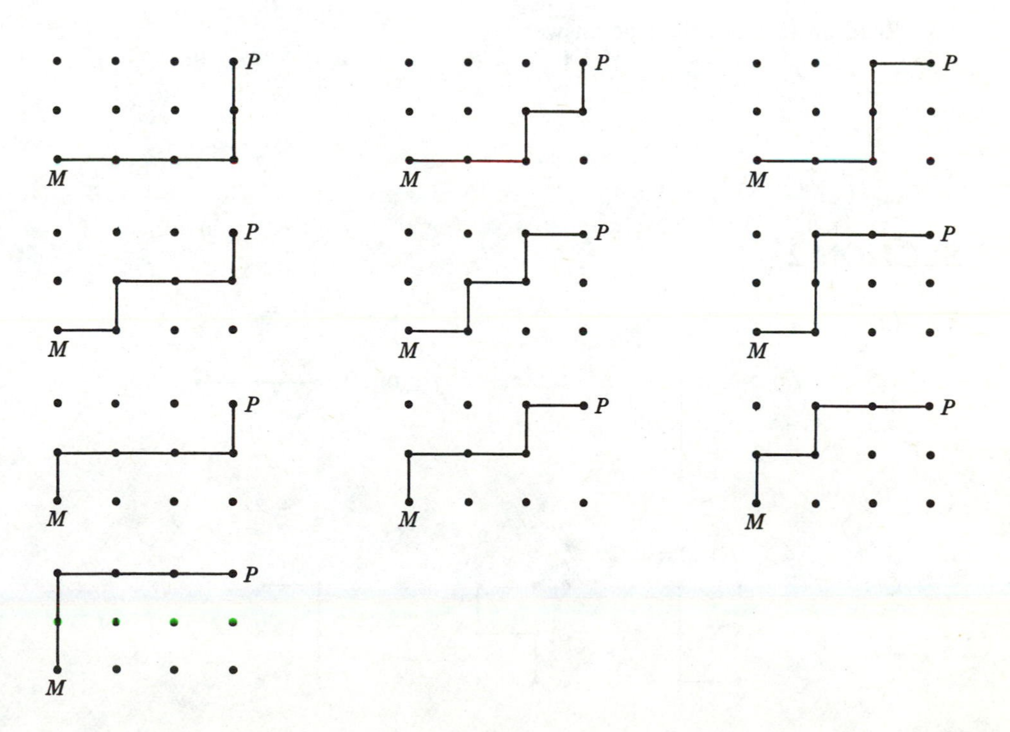

(2) There are 4 such paths:

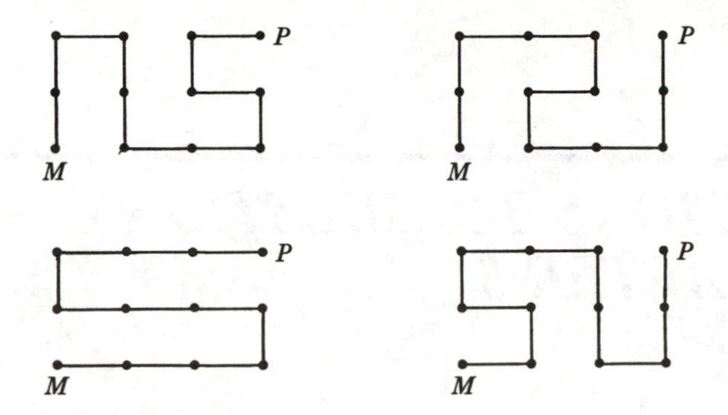

(3) There are no such paths.

1.2 To get outside the rectangle of Figure 3.1 would require either more than 5 Ns or more than 6 Es (or both). Since P is 6 units to the right of M and 5 units above M, any sequence of 6 Es and 5 Ns will correspond with a path from M to P. Any sequence consisting of exactly 3 Es and 2 Ns corresponds with a trip from M to P in Figure 3.2 and any such trip corresponds with such a sequence.

1.3 Read on to learn the true answer.

SECTION 2

2.1 **(*a*)**

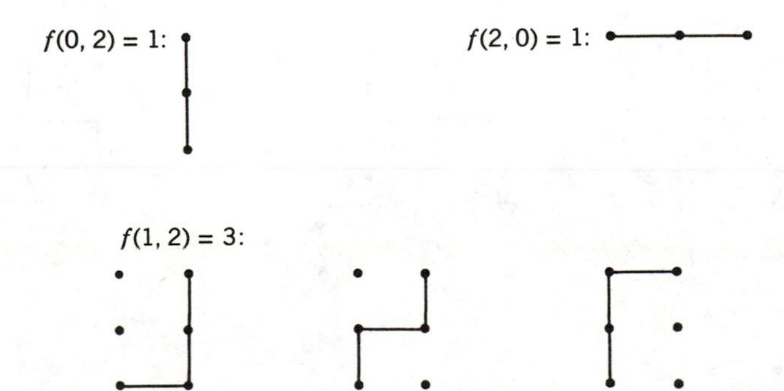

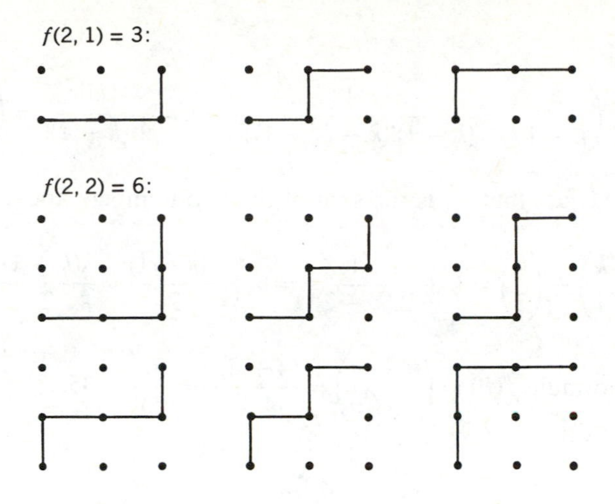

(*b*) $f(0,3) = 1 = f(3,0)$, $f(1,3) = 4 = f(3,1)$, and $f(2,3) = 10 = f(3,2)$.

2.2 There is just one shortest path from $(0,0)$ to $(i,0)$, a straight line consisting of a path of i Es. Thus $f(i,0) = 1$. Similarly, $f(0,j) = 1$.

2.3 Use the fact that $f(i,j) = f(i-1,j) + f(i,j-1)$:

$$f(3,3) = f(2,3) + f(3,2) = 10 + 10 = 20.$$
$$f(4,2) = f(3,2) + f(4,1) = 10 + f(4,1)$$
$$= 10 + f(3,1) + f(4,0) = 10 + 4 + 1 = 15.$$

2.4 The fifth row: 1 5 10 10 5 1. The coordinates of all points that end up on the fifth row of Pascal's triangle is $\{(i,j): i + j = 5\}$. The f values of these points are exactly the corresponding entries of Pascal's triangle.

2.5 $4! = 24$, $5! = 120$, $6! = 720$, $7! = 5040$, and $8! = 40{,}320$. The first n such that $n! > 1{,}000{,}000$ is $n = 10$. This can be determined by continuing to calculate $9! = 362{,}880$ and $10! = 3{,}628{,}800 > 1{,}000{,}000$. Any value greater than 10 would also do.

2.6 $\binom{3}{0} = \binom{3}{3} = 1$, $\binom{3}{1} = \binom{3}{2} = 3$. $\binom{4}{i}$ and $\binom{5}{j}$ turn out to be the fourth and fifth row of Pascal's triangle; see Figure 3.7.

2.7 Use the definition of $\binom{n}{i}$:

$$\binom{k}{0} = \frac{k!}{0!(k-0)!} = \frac{k!}{k!} = 1 = \frac{k!}{k!} = \frac{k!}{k!(k-k)!} = \binom{k}{k}.$$

Similarly,

$$\binom{k}{k-1} = \frac{k!}{(k-1)!(k-(k-1))!} = k = \frac{k!}{1!(k-1)!} = \binom{k}{1}.$$

2.8 We use the fact that x_2 is the sum of the two numbers above:

$$x_2 = \binom{k}{1} + \binom{k}{2} = k + \frac{k!}{2!(k-2)!} = k + \frac{k(k-1)}{2} = \frac{k(k+1)}{2} = \binom{k+1}{2}.$$

2.9 Use the formula $f(P) = \binom{m+n}{m} = \frac{(4+3)!}{4!3!} = \frac{7!}{4!3!} = 35.$

SECTION 3

3.1 The fifteen 4-subsets of A are

$\{a_1, a_2, a_3, a_4\}$	$\{a_1, a_2, a_3, a_5\}$	$\{a_1, a_2, a_3, a_6\}$
$\{a_1, a_2, a_4, a_5\}$	$\{a_1, a_2, a_4, a_6\}$	$\{a_1, a_2, a_5, a_6\}$
$\{a_1, a_3, a_4, a_5\}$	$\{a_1, a_3, a_4, a_6\}$	$\{a_1, a_3, a_5, a_6\}$
$\{a_1, a_4, a_5, a_6\}$	$\{a_2, a_3, a_4, a_5\}$	$\{a_2, a_3, a_4, a_6\}$
$\{a_2, a_3, a_5, a_6\}$	$\{a_2, a_4, a_5, a_6\}$	$\{a_3, a_4, a_5, a_6\}$

Note that

$$\binom{6}{4} = \frac{6!}{4!2!} = 15.$$

The six 5-subsets of A are

$\{a_1, a_2, a_3, a_4, a_5\}$	$\{a_1, a_2, a_3, a_4, a_6\}$
$\{a_1, a_2, a_3, a_5, a_6\}$	$\{a_1, a_2, a_4, a_5, a_6\}$
$\{a_1, a_3, a_4, a_5, a_6\}$	$\{a_2, a_3, a_4, a_5, a_6\}$

Note that

$$\binom{6}{5} = \frac{6!}{5!1!} = 6.$$

3.2

3-subsets of A containing a_6	Remaining 3-subsets of A
$\{a_1, a_2, a_6\}$	$\{a_1, a_2, a_3\}$
$\{a_1, a_3, a_6\}$	$\{a_1, a_2, a_4\}$
$\{a_1, a_4, a_6\}$	$\{a_1, a_2, a_5\}$
$\{a_1, a_5, a_6\}$	$\{a_1, a_3, a_4\}$
$\{a_2, a_3, a_6\}$	$\{a_1, a_3, a_5\}$
$\{a_2, a_4, a_6\}$	$\{a_1, a_4, a_5\}$
$\{a_2, a_5, a_6\}$	$\{a_2, a_3, a_4\}$
$\{a_3, a_4, a_6\}$	$\{a_2, a_3, a_5\}$
$\{a_3, a_5, a_6\}$	$\{a_2, a_4, a_5\}$
$\{a_4, a_5, a_6\}$	$\{a_3, a_4, a_5\}$

3.3 (*a*) $\binom{6}{3} = 6!/(3!3!) = 20$ (*b*) $\binom{11}{5} = 11!/(5!6!) = 462$

(*c*) $\binom{17}{9} = 17!/(9!8!) = 24{,}310.$

3.4 There are $\binom{11}{3} = 165$ 11-letter sequences of Ns and Es with exactly 3 Es. There are $\binom{11}{4} = 330$ 11-letter sequences of Ns and Es with exactly 4 Es. And there are $\binom{11}{7} = \binom{11}{4} = 330$ 11-letter sequences of Ns and Es with exactly 7 Es. There are $2^{11} = 2048$ sequences of Es and Ns containing 11 letters.

3.5 We can transform a j-subset of A to a j-subset of I_n by renaming each element a_i of A as the integer i. Similarly, we can transform a j-subset of I_n to a j-subset of A by renaming the element i of I_n as a_i. The 4-subset of I_n that corresponds with $\{a_1, a_2, a_4, a_{n-1}\}$ is $\{1, 2, 4, n-1\}$. The 5-subset of A that corresponds with $\{1, 2, 3, 5, 8\}$ is $\{a_1, a_2, a_3, a_5, a_8\}$.

3.6 $\{1,3,4,7\}$, $\{1,3,5,6\}$, $\{2,3,4,7\}$, $\{2,3,5,6\}$, $\{3,6,7,8\}$, $\{4,6,7,8\}$

3.7 $\{1,2,3\}$, $\{1,2,4\}$, $\{1,2,5\}$, $\{1,2,6\}$, $\{1,3,4\}$, $\{1,3,5\}$, $\{1,3,6\}$, $\{1,4,5\}$, $\{1,4,6\}$, $\{1,5,6\}$, $\{2,3,4\}$, $\{2,3,5\}$, $\{2,3,6\}$, $\{2,4,5\}$, $\{2,4,6\}$, $\{2,5,6\}$, $\{3,4,5\}$, $\{3,4,6\}$, $\{3,5,6\}$, $\{4,5,6\}$.

3.8 $\{1,2,3\}$, $\{1,2,4\}$, $\{1,2,5\}$, $\{1,2,6\}$, $\{1,2,7\}$, $\{1,3,4\}$, $\{1,3,5\}$, $\{1,3,6\}$, $\{1,3,7\}$, $\{1,4,5\}$, $\{1,4,6\}$, $\{1,4,7\}$, $\{1,5,6\}$, $\{1,5,7\}$, $\{1,6,7\}$, $\{2,3,4\}$, $\{2,3,5\}$, $\{2,3,6\}$, $\{2,3,7\}$, $\{2,4,5\}$, $\{2,4,6\}$, $\{2,4,7\}$, $\{2,5,6\}$, $\{2,5,7\}$, $\{2,6,7\}$, $\{3,4,5\}$, $\{3,4,6\}$, $\{3,4,7\}$, $\{3,5,6\}$, $\{3,5,7\}$, $\{3,6,7\}$, $\{4,5,6\}$, $\{4,5,7\}$, $\{4,6,7\}$, $\{5,6,7\}$.

3.9 Algorithm JSET run on $j = 4$ and $n = 6$:

Step No.	h	b_h	$n+h-j$	*FOUND*	k	b_k	*SUBSET*
2							{1, 2, 3, 4}
3	5			false			
4, 5, 6	4	4	6	true			
8	4	5					
9–12					5		{1, 2, 3, 5}
3	5			false			
4, 5, 6	4	5	6	true			
8	4	6					
9–12					5		{1, 2, 3, 6}
3	5			false			
4, 5, 6	4	6	6	false			
	3	3	5	true			
8	3	4					
9, 10					4	5	
11, 12							{1, 2, 4, 5}
3	5			false			
4, 5, 6	4	5	6	true			
8	4	6					
9–12					5		{1, 2, 4, 6}
3	5			false			
4, 5, 6	4	6	6	false			
	3	5	5	true			
8	3	5					
9, 10					4	6	
11, 12							{1, 2, 5, 6}
3	5			false			
4, 5, 6	4	6	6	false			
	3	5	5	false			
	2	2	4	true			
8	2	3					
9, 10					3	4	
					4	5	
11, 12							{1, 3, 4, 5}
3	5			false			
4, 5, 6	4	5	6	true			
8	4	6					
9–12					5		{1, 3, 4, 6}
3	5			false			
4, 5, 6	4	6	6	false			
	3	4	5	true			
8	3	5					
9–12					4	6	{1, 3, 5, 6}

3.9 (continued)

Step No.	h	b_h	$n+h-j$	*FOUND*	k	b_k	*SUBSET*
3	5			false			
4, 5, 6	4	6	6	false			
	3	5	5	false			
	2	3	4	true			
8	2	4					
9–12					3	5	$\{1,4,5,6\}$
3	5			false			
4, 5, 6	4	6	6	false			
	3	5	5	false			
	2	4	4	false			
	1	1	3	true			
8	1	2					
9–12					2	3	$\{2,3,4,5\}$
3	5			false			
4, 5, 6	4	5	6	true			
8	4	6					
9–12					5		$\{2,3,4,6\}$
3	5			false			
4, 5, 6	4	6	6	false			
	3	4	5	true			
8	3	5					
9–12					4	6	$\{2,3,5,6\}$
3	5			false			
4, 5, 6	4	6	6	false			
	3	5	5	false			
	2	3	4	true			
8	2	4					
9–12					3	5	$\{2,4,5,6\}$
3	5			false			
4, 5, 6	4	6	6	false			
	3	5	5	false			
	2	4	4	false			
	1	2	3	true			
8	1	3					
9–12					2	4	$\{3,4,5,6\}$
3	5			false			
4, 5, 6	4	6	6	false			
	3	5	5	false			
	2	4	4	false			
	1	3	3	false			
7				false STOP			

3.10 The algorithm JSET produces $\binom{n}{3} = n(n-1)(n-2)/6 = O(n^3)$ subsets when $j = 3$.

3.11 Note $\binom{n}{j} = \dfrac{n(n-1)(n-2)\cdots(n-j+1)}{j!} \leq \dfrac{n^j}{j!} = O(n^j)$.

SECTION 4

4.1 The 4! permutations of the set $\{1, 2, 3, 4\}$ are
$\langle 1\ 2\ 3\ 4\rangle$, $\langle 1\ 2\ 4\ 3\rangle$, $\langle 1\ 4\ 2\ 3\rangle$, $\langle 4\ 1\ 2\ 3\rangle$, $\langle 1\ 3\ 2\ 4\rangle$,
$\langle 1\ 3\ 4\ 2\rangle$, $\langle 1\ 4\ 3\ 2\rangle$, $\langle 4\ 1\ 3\ 2\rangle$, $\langle 3\ 1\ 2\ 4\rangle$, $\langle 3\ 1\ 4\ 2\rangle$,
$\langle 3\ 4\ 1\ 2\rangle$, $\langle 4\ 3\ 1\ 2\rangle$, $\langle 2\ 1\ 3\ 4\rangle$, $\langle 2\ 1\ 4\ 3\rangle$, $\langle 2\ 4\ 1\ 3\rangle$,
$\langle 4\ 2\ 1\ 3\rangle$, $\langle 2\ 3\ 1\ 4\rangle$, $\langle 2\ 3\ 4\ 1\rangle$, $\langle 2\ 4\ 3\ 1\rangle$, $\langle 4\ 2\ 3\ 1\rangle$,
$\langle 3\ 2\ 1\ 4\rangle$, $\langle 3\ 2\ 4\ 1\rangle$, $\langle 3\ 4\ 2\ 1\rangle$, $\langle 4\ 3\ 2\ 1\rangle$.

4.2 A 3-subset can be formed by filling in the three blanks $\{\_, \_, \_\}$ with distinct elements of the n-set. There are n choices for the first blank, $n - 1$ choices for the second blank, and $n - 2$ choices for the third blank. By the Multiplication Principle, there are a total of $n(n-1)(n-2)$ ways to fill in the blanks. However, once a subset is "filled in," then every permutation of the elements in that set will produce the same set. There are 3! permutations of each set and so in the total of $n(n-1)(n-2)$, each set is listed 3! times. Thus there are $n(n-1)(n-2)/6$ different 3-subsets of an n-set.

4.3 Trace of algorithm PERM run on $S = \{1, 2, 3, 4\}$:

Values of j	*Permutations*
1	$\langle 1\rangle$
2	$\langle 1\ 2\rangle$ $\langle 2\ 1\rangle$
3	$\langle 1\ 2\ 3\rangle$ $\langle 1\ 3\ 2\rangle$ $\langle 3\ 1\ 2\rangle$
	$\langle 2\ 1\ 3\rangle$ $\langle 2\ 3\ 1\rangle$ $\langle 3\ 2\ 1\rangle$
4	$\langle 1\ 2\ 3\ 4\rangle$ $\langle 1\ 2\ 4\ 3\rangle$ $\langle 1\ 4\ 2\ 3\rangle$ $\langle 4\ 1\ 2\ 3\rangle$
	$\langle 1\ 3\ 2\ 4\rangle$ $\langle 1\ 3\ 4\ 2\rangle$ $\langle 1\ 4\ 3\ 2\rangle$ $\langle 4\ 1\ 3\ 2\rangle$
	$\langle 3\ 1\ 2\ 4\rangle$ $\langle 3\ 1\ 4\ 2\rangle$ $\langle 3\ 4\ 1\ 2\rangle$ $\langle 4\ 3\ 1\ 2\rangle$
	$\langle 2\ 1\ 3\ 4\rangle$ $\langle 2\ 1\ 4\ 3\rangle$ $\langle 2\ 4\ 1\ 3\rangle$ $\langle 4\ 2\ 1\ 3\rangle$
	$\langle 2\ 3\ 1\ 4\rangle$ $\langle 2\ 3\ 4\ 1\rangle$ $\langle 2\ 4\ 3\ 1\rangle$ $\langle 4\ 2\ 3\ 1\rangle$
	$\langle 3\ 2\ 1\ 4\rangle$ $\langle 3\ 2\ 4\ 1\rangle$ $\langle 3\ 4\ 2\ 1\rangle$ $\langle 4\ 3\ 2\ 1\rangle$

4.4 To find an integer N so that for $n \geq N$, $n! > 10 \cdot 2^n$, proceed as in the proof of Theorem 4.2: Suppose that $N = Cr^r = 10 \cdot 2^2 = 40$. Then for $n > 40$, we know that $n! > 10 \cdot 2^n$.

SECTION 5

5.1 The four colors of the code are r, y, b, and w. It is not possible to determine their order.

5.2 We may assume, by Question 5.1, that the colors in the code are r, y, b, and w. The secret code can be determined from the guesses given in Example 5.1 by the following reasoning:

Deduction	*Guesses and Deductions Used*
1. r and y are not in the first and second positions, respectively	Guess 2
2. Either b is in the third position or w is in the fourth position (but not both)	Guess 1 and Guess 2
3. Either w is in the second position or y is in the third position (but not both)	Guess 2 and Guess 3
4. w is in the second position	Deduction 3 and Guess 4
5. b is in the third position	Deductions 2 and 4
6. r is in the fourth position	Deductions 1, 4, and 5
7. y is in the first position Result: Code = $y\ w\ b\ r$	Deductions 1, 4, 5, and 6

Note that this reasoning is only one of many ways to arrive at the above result.

5.3 Left to the reader.

5.4 The number of codes is $4! = 24$. Use PERM or the results of Question 4.3.

5.5 $g\,g\,b\,p \quad g\,g\,p\,b \quad g\,b\,g\,p \quad g\,p\,g\,b \quad g\,p\,b\,g \quad g\,b\,p\,g$
$b\,g\,g\,p \quad p\,g\,g\,b \quad p\,g\,b\,g \quad b\,g\,p\,g \quad p\,b\,g\,g \quad b\,p\,g\,g$

5.6 We have seen that if the code has 4 colors, then there are 24 possible codes, and if the code has 3 colors, there are 12 codes (see Question 5.5). If there are just 2 colors, there are 4 ways to have 3 of one color and 1 of the other color, and there are $\binom{4}{2} = 6$ ways to have 2 colors occurring twice each. If there is one color, there is only one possible code. Thus the maximum number of possible codes and hence guesses in step 2 of Algorithm 1 is 24, when there are 4 different colors.

SECTION 6

6.1 $n = 3$: $\binom{3}{0} - \binom{3}{1} + \binom{3}{2} - \binom{3}{3} = 1 - 3 + 3 - 1 = 0.$

$n = 4$: $\binom{4}{0} - \binom{4}{1} + \binom{4}{2} - \binom{4}{3} + \binom{4}{4} = 1 - 4 + 6 - 4 + 1 = 0.$

$n = 5$:

$$\binom{5}{0} - \binom{5}{1} + \binom{5}{2} - \binom{5}{3} + \binom{5}{4} - \binom{5}{5} = 1 - 5 + 10 - 10 + 5 - 1 = 0.$$

6.2 $\binom{2}{2} + \binom{3}{2} + \binom{4}{2} = 1 + 3 + 6 = 10 = \binom{5}{3}.$

$\binom{2}{2} + \binom{3}{2} + \binom{4}{2} + \binom{5}{2} = 1 + 3 + 6 + 10 = 20 = \binom{6}{3}.$

$\binom{2}{2} + \binom{3}{2} + \binom{4}{2} + \binom{5}{2} + \binom{6}{2} = 1 + 3 + 6 + 10 + 15 = 35 = \binom{7}{3}.$

From this evidence it seems reasonable to conjecture that

$$\binom{2}{2} + \binom{3}{2} + \cdots + \binom{i}{2} + \cdots + \binom{n}{2} = \binom{n+1}{3}.$$

For a proof of this fact, read the rest of this section.

SOLUTIONS TO QUESTIONS IN CHAPTER 4

SECTION 1

1.1 $\frac{3}{12} = \frac{1}{4}$, $\frac{13}{121}$ is simplified, $\frac{65}{130} = \frac{1}{2}$, and $\frac{34,567}{891,011}$ is simplified. $\frac{1}{3} + \frac{1}{2} = \frac{5}{6}$, $\frac{1}{4} + \frac{1}{3} = \frac{7}{12}$, and $\frac{1}{15} + \frac{1}{65} = \frac{16}{195}$.

1.2 (***a***) 65, (***b***) 8, and (***c***) 1.

1.3 (***a***) $\gcd(5,7) = 1$, (***b***) $\gcd(4,6) = 2$, and (***c***) $\gcd(5,10) = 5$. Since $\gcd(b,c)$ must divide both b and c, it cannot be larger than either.

1.4 Algorithm GCD1 run on the pair (3, 4):

Step No.	b	c	*Are both b/g and c/g integers?*	g	gcd
1	3	4		3	
3, 4	3	4	no	2	
3, 4	3	4	no	1	
5 STOP	3	4			1

Algorithm GCD1 run on the pair (3, 12):

Step No.	b	c	*Are both b/g and c/g integers?*	g	gcd
1	3	12		3	
3 STOP	3	12	yes	3	3

Algorithm GCD1 run on the pair (6, 20):

Step No.	b	c	*Are both b/g and c/g integers?*	g	gcd
1	6	20		6	
3, 4	6	20	no	5	
3, 4	6	20	no	4	
3, 4	6	20	no	3	
3, 4	6	20	no	2	
3 STOP	6	20	yes		2

1.5 ***(a)*** (5, 7) or any pair for which gcd = 1, ***(b)*** (4, 6) or any pair with 1 < gcd, and ***(c)*** (4, 12) or any pair where gcd = b.

1.6 *Algorithm ADDFRACT1*

STEP 1. Input a, b, c, d {The sum $a/b + c/d$ is to be calculated and output as a simplified fraction.}

STEP 2. Set numer := $a * d + b * c$; set denom := $b * d$

STEP 3. Use SIMPLIFY on the pair (numer, denom)

STEP 4. Output the result of step 3 and stop.

1.7 Let m = minimum $\{a, b\}$. Then upon input of a and b, the algorithm SIMPLIFY performs (at most) $2(m - 1)$ divisions (in GCD1) plus two divisions (in step 3). Let m' = minimum $\{(ad - bc), bd\}$. ADDFRACT1 performs three multiplications in step 2 for a total of $2m' + 3$ multiplications and divisions.

SECTION 2

2.1 Using $\log(n) < B$, (*a*) $(\log(n))^2 < B^2$, (*b*) $\log(n^2) = 2\log(n) < 2B$, and (*c*) $\log(\log(n)) < \log(B)$.

2.2 Since by (1), $\log(n) \geq B/2$, $n = 2^{\log(n)} \geq 2^{B/2} = \sqrt{2}^B$.

SECTION 3

3.1 $\gcd(18, 30) = \gcd(12, 18) = \gcd(6, 12) = \gcd(0, 6) = 6$; $\gcd(18, 48) = \gcd(18, 30) = 6$; and $\gcd(18, 66) = \gcd(18, 48) = 6$.

3.2 (*a*) 1, (*b*) 3, and (*c*) 5.

3.3 (*a*) $q_1 = 4$, $r_1 = 0$, and $\gcd(3, 12) = \gcd(0, 3) = 3$. (*b*) $q_1 = 9$, $r_1 = 4$, and $\gcd(13, 121) = \gcd(4, 13) = 1$. (*c*) $q_1 = 1$, $r_1 = 144$, and $\gcd(233, 377) = \gcd(144, 233) = 1$. (*d*) $q_1 = 25$, $r_1 = 26{,}836$, and $\gcd(34{,}567; 891{,}011) = \gcd(26{,}836; 34{,}567) = 1$. (See solution to Question 3.4(d).)

3.4 (*a*) (12, 20):

$$\begin{aligned} 20 &= 1 \cdot 12 + 8 \\ 12 &= 1 \cdot 8 + 4 \\ 8 &= 2 \cdot 4 + 0 \qquad \text{so } \gcd(12, 20) = 4. \end{aligned}$$

It took three divisions to find that $\gcd(12, 20) = 4$.

(*b*) (5, 15): $15 = 3 \cdot 5 + 0$ so $\gcd(5, 15) = 5$.

It took one division to find that $\gcd(5, 15) = 5$.

(*c*) (377, 610):

$$\begin{aligned} 610 &= 1 \cdot 377 + 233 \\ 377 &= 1 \cdot 233 + 144 \end{aligned}$$

The remaining equations are identical to those in the second part of Example 3.1, so $\gcd(377, 610) = 1$. It took 13 divisions to find that $\gcd(377, 610) = 1$.

(*d*) (34,567; 891,011):

$$\begin{aligned}
891{,}011 &= 25 \cdot 34{,}567 + 26{,}836\\
34{,}567 &= 1 \cdot 26{,}836 + 7731\\
26{,}836 &= 3 \cdot 7731 + 3643\\
7731 &= 2 \cdot 3643 + 445\\
3643 &= 8 \cdot 445 + 83\\
445 &= 5 \cdot 83 + 30\\
83 &= 2 \cdot 30 + 23\\
30 &= 1 \cdot 23 + 7\\
23 &= 3 \cdot 7 + 2\\
7 &= 3 \cdot 2 + 1\\
2 &= 2 \cdot 1 + 0 \qquad \text{so } \gcd(34{,}567; 891{,}011) = 1.
\end{aligned}$$

It took 11 divisions to find that gcd (34,567; 891,011) = 1.

3.5 **(*a*)** Algorithm EUCLID run on the pair (6, 20):

Step No.	b	c	q	r	gcd
1	6	20		6	
3, 4			3	2	
5	2	6			
3, 4			3	0	
5					2
6 STOP					

The Euclidean equations are $20 = 3 \cdot 6 + 2$ and $6 = 3 \cdot 2 + 0$. Using the first equation, we obtain $2 = -3 \cdot 6 + 1 \cdot 20$.

(*b*) Algorithm EUCLID run on the pair (3, 4):

Step No.	b	c	q	r	gcd
1	3	4		3	
3, 4			1	1	
5	1	3			
3, 4			3	0	
5					1
6 STOP					

The Euclidean equations are $4 = 1 \cdot 3 + 1$ and $3 = 3 \cdot 1 + 0$. Using the first equation, we obtain $1 = -1 \cdot 3 + 1 \cdot 4$.

(*c*) Algorithm EUCLID run on the pair (55, 89):

Step No.	*b*	*c*	*q*	*r*	gcd
1	55	89		55	
3, 4			1	34	
5	34	55			
3, 4			1	21	
5	21	34			
3, 4			1	13	
5	13	21			
3, 4			1	8	
5	8	13			
3, 4			1	5	
5	5	8			
3, 4			1	3	
5	3	5			
3, 4			1	2	
5	2	3			
3, 4			1	1	
5	1	2			
3, 4			2	0	
5					1
6 STOP					

The Euclidean equations are

1. $89 = 1 \cdot 55 + 34$
2. $55 = 1 \cdot 34 + 21$
3. $34 = 1 \cdot 21 + 13$
4. $21 = 1 \cdot 13 + 8$
5. $13 = 1 \cdot 8 + 5$
6. $8 = 1 \cdot 5 + 3$
7. $5 = 1 \cdot 3 + 2$
8. $3 = 1 \cdot 2 + 1$
9. $2 = 2 \cdot 1 + 0$.

We must start with equation 8 and work our way backward to express the gcd as a linear combination of 55 and 89:

$1 = -1 \cdot 2 + 1 \cdot 3 = -1 \cdot (-1 \cdot 3 + 1 \cdot 5) + 1 \cdot 3$	using equation 7
$= 2 \cdot 3 - 1 \cdot 5 = 2 \cdot (-1 \cdot 5 + 1 \cdot 8) - 1 \cdot 5$	using equation 6
$= -3 \cdot 5 + 2 \cdot 8 = -3 \cdot (-1 \cdot 8 + 1 \cdot 13) + 2 \cdot 8$	using equation 5
$= 5 \cdot 8 - 3 \cdot 13 = 5 \cdot (-1 \cdot 13 + 1 \cdot 21) - 3 \cdot 13$	using equation 4
$= -8 \cdot 13 + 5 \cdot 21 = -8 \cdot (-1 \cdot 21 + 1 \cdot 34) + 5 \cdot 21$	using equation 3
$= 13 \cdot 21 - 8 \cdot 34 = 13 \cdot (-1 \cdot 34 + 1 \cdot 55) - 8 \cdot 34$	using equation 2
$= -21 \cdot 34 + 13 \cdot 55 = -21 \cdot (-1 \cdot 55 + 1 \cdot 89) + 13 \cdot 55$	using equation 1
$= 34.55 - 21.89.$	

SECTION 4

4.1

n	2	3	4	5	6	11	13
F_{n-2}	0	1	1	2	3	34	89
F_{n-1}	1	1	2	3	5	55	144
sum	1	2	3	5	8	89	233

4.2

n	16	17	18	19	20
F_{n-2}	377	610	987	1,597	2,584
F_{n-1}	610	987	1,597	2,584	4,181
sum $= F_n$	987	1,597	2,584	4,181	6,765
2^n	65,536	131,072	262,144	524,288	1,048,576

In each case listed above, $F_n < 2^n$.

4.3 We claim that the Principle of Complete Induction is valid, by which we mean that if assertions (i) and (ii) are both verified, then the proposition P_n is proved for all $n \geq N$: Suppose that we verify (by hand) that $P_N, P_{N+1}, \ldots,$ and P_{N+i} are all true. Then setting $k = N + i$ in (ii) shows that $P_{k+1} = P_{N+i+1}$ is true. Then we can repeat (ii) with $k = N + i + 1$. Since we've just demonstrated that $P_N, \ldots, P_{N+i+1}$ are all true, we get that P_{N+i+2} is true, and so on. In general, we can work our way up to the truth of P_n for any integer $n \geq N$.

4.4 Using a calculator, one can check that

$$F_{10} = 55 < (\tfrac{3}{2})^{10} < (\tfrac{3}{2})^{11} < F_{11} = 89.$$

We must prove for $n \geq 11$ that $F_n \geq (\frac{3}{2})^n$: For the base cases we notice that

$$F_{11} = 89 > (\tfrac{3}{2})^{11} = 86.4\ldots$$

and

$$F_{12} = 144 > (\tfrac{3}{2})^{12} = 129.7\ldots.$$

As in Example 4.2 we require base cases with two consecutive integers (or $j = 1$) because the proof uses the fact that $F_{k+1} = F_k + F_{k-1}$. We use complete induction and so assume that $P_{11}, P_{12}, \ldots, P_k$ are all true for some arbitrary value of k. That is, $F_i > (\tfrac{3}{2})^i$ for all $11 \le i \le k$. Notice that $i \ge 11$, since otherwise the claim that $F_i > (\tfrac{3}{2})^i$ is not true. We must prove that $F_{k+1} > (\tfrac{3}{2})^{k+1}$.

$$\begin{aligned} F_{k+1} &= F_k + F_{k-1} > (\tfrac{3}{2})^k + (\tfrac{3}{2})^{k-1} \qquad \text{by inductive hypothesis} \\ &= (\tfrac{3}{2})^{k-1}(\tfrac{3}{2} + 1) > (\tfrac{3}{2})^{k-1} \cdot (\tfrac{9}{4}) = (\tfrac{3}{2})^{k+1}. \end{aligned}$$

Thus $F_n > (\tfrac{3}{2})^n$ for all $n \ge 11$.

4.5 We begin with the equation $x - 1 = 1/x$ and multiply both sides by x to obtain $x^2 - x = 1$ or $x^2 - x - 1 = 0$. Using the quadratic formula, we find the roots to be $(1 + \sqrt{5})/2 = \phi$ and $(1 - \sqrt{5})/2 = \phi'$. Since these are not zero, they are also solutions to the original equation. Alternatively,

$$\begin{aligned} \frac{1}{\phi} &= \frac{2}{1+\sqrt{5}} = \frac{2(1-\sqrt{5})}{(1+\sqrt{5})(1-\sqrt{5})} \\ &= \frac{2(1-\sqrt{5})}{1-5} = \frac{1-\sqrt{5}}{-2} = \frac{-1+\sqrt{5}}{2} \\ &= \frac{1+\sqrt{5}}{2} - 1 = \phi - 1. \end{aligned}$$

4.6

$$\begin{aligned} \frac{\phi^2 - \phi'^2}{\sqrt{5}} &= \frac{((1+\sqrt{5})/2)^2 - ((1-\sqrt{5})/2)^2}{\sqrt{5}} \\ &= \frac{(1 + 2\sqrt{5} + 5)/4 - (1 - 2\sqrt{5} + 5)/4}{\sqrt{5}} \\ &= \frac{4\sqrt{5}/4}{\sqrt{5}} = 1 = F_2. \end{aligned}$$

4.7 The algorithm segment

Set $C := A + B$

Set $B := A$

Set $A := C$

placed inside the appropriate loop will calculate Fibonacci numbers if A and B are initially assigned the values F_1 and F_0, respectively. This uses three memory locations. It is possible to use only two memory locations with the segment

Set $B := A + B$

Set $A := A + B$

placed inside the appropriate loop.

SECTION 5

5.1 **(i)** $c = F_8 = 21$ and $b = F_7 = 13$

$$\begin{aligned} 21 &= 1 \cdot 13 + 8, &\quad q_1 &= 1, r_1 = 8 \\ 13 &= 1 \cdot 8 + 5, &\quad q_2 &= 1, r_2 = 5 \\ 8 &= 1 \cdot 5 + 3, &\quad q_3 &= 1, r_3 = 3 \\ 5 &= 1 \cdot 3 + 2, &\quad q_4 &= 1, r_4 = 2 \\ 3 &= 1 \cdot 2 + 1, &\quad q_5 &= 1, r_5 = 1 \\ 2 &= 2 \cdot 1 + 0, &\quad q_6 &= 2, r_6 = 0 \end{aligned}$$

(ii) $c = F_{10} = 55$ and $b = F_9 = 34$

$$\begin{aligned} 55 &= 1 \cdot 34 + 21, &\quad q_1 &= 1, r_1 = 21 \\ 34 &= 1 \cdot 21 + 13, &\quad q_2 &= 1, r_2 = 13 \\ 21 &= 1 \cdot 13 + 8, &\quad q_3 &= 1, r_3 = 8 \end{aligned}$$

... as in the preceding part.

5.2 The maximum number of Euclidean equations occurs when $b = 3$ and $c = 5$, and this number is three.

5.3 If $b = 77$ and $c = 185$, the first two Euclidean equations are
1. $185 = 2 \cdot 77 + 31, \quad q_1 = 2, r_1 = 31$
2. $77 = 2 \cdot 31 + 15, \quad q_2 = 2, r_2 = 15$

5.4 (When $b = 26$ and $c = 32$, there is no value of t such that r_{2t+2} and r_{2t} are defined.) When $b = 233$ and $c = 377$, the largest integer t for which r_{2t+2} is defined is $t = 5$. Thus we compute the quantity r_{2t+2}/r_{2t} for $t = 1, 2, 3, 4, 5$: $r_4/r_2 = \frac{34}{89}$, $r_6/r_4 = \frac{13}{34}$, $r_8/r_6 = \frac{5}{13}$, $r_{10}/r_8 = \frac{2}{5}$, and $r_{12}/r_{10} = 0$. These fractions are all less than $\frac{1}{2}$.

SECTION 6

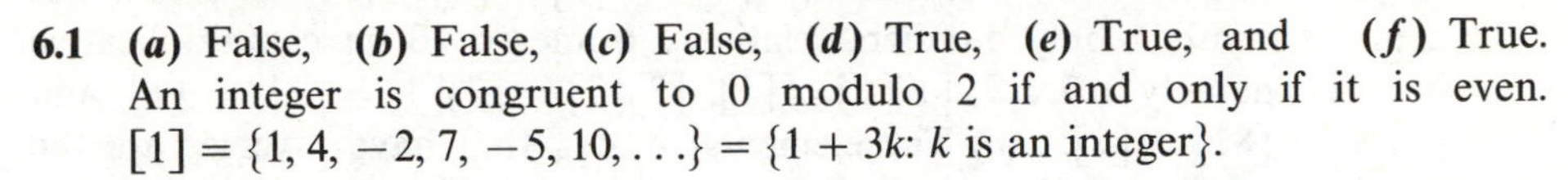

6.1 (*a*) False, (*b*) False, (*c*) False, (*d*) True, (*e*) True, and (*f*) True. An integer is congruent to 0 modulo 2 if and only if it is even. $[1] = \{1, 4, -2, 7, -5, 10, \ldots\} = \{1 + 3k\colon k \text{ is an integer}\}$.

6.2 (*a*) $n + i$, (*b*) $-n + i$.

6.3 **(i)** $a \equiv a \pmod{n}$:

Proof. $a - a = 0 = 0 \cdot n$. Since $a - a$ is divisible by n, we have $a \equiv a \pmod{n}$. □

(ii) If $a \equiv b \pmod{n}$, then $b \equiv a \pmod{n}$:

Proof. If $a \equiv b \pmod{n}$, then there is an integer i such that $(a - b) = in$. But then $(b - a) = -in$, which implies that $b - a$ is divisible by n. Thus $b \equiv a \pmod{n}$. □

6.4 Both $\leq$ and $\subseteq$ are relations on Z and on all subsets of Z, respectively. If two numbers are called related when their difference is even, then this gives a relation on Z, but not on R, since we do not know what it means for an arbitrary real number to be even.

6.5 **(i)** Given that $a \equiv c \pmod{n}$ and $b \equiv d \pmod{n}$, we know that there exist integers i and j such that $a = c + in$ and $b = d + jn$. Thus

$$a + b = c + d + in + jn = c + d + (i + j)n.$$

Thus $a + b - (c + d)$ is divisible by n and so $a + b \equiv c + d \pmod{n}$. □
(ii) Proceeding as in (i) we have $a - b = c - d + in - jn$. Thus $a - c - (b - d)$ is divisible by n and so $a - c \equiv b - d \pmod{n}$. □

6.6 If x is in $[a]$ and y is in $[b]$, then

$$a \equiv x \pmod{n} \quad \text{and} \quad b \equiv y \pmod{n}.$$

By Lemma 6.3, part (iii)

$$ab \equiv xy \pmod{n} \quad \text{and} \quad [ab] = [xy].$$

Thus multiplication is well defined.

6.7 **(i)** $a = 3$, $b = 4$, $c = 8$, $d = 9$, and $n = 5$. Note that $a \cdot b = 3 \cdot 4 = 12 \equiv 2 \pmod{5}$ and that $c \cdot d = 8 \cdot 9 = 72 \equiv 2 \pmod{5}$. Further, $3 \equiv 8 \pmod{5}$ and $\gcd(3, 5) = 1$. Finally, $4 \equiv 9 \pmod{5}$.
(ii) $a = 3$, $b = 4$, $c = 15$, $d = 8$, and $n = 12$. Note that $a \cdot b = 3 \cdot 4 \equiv 0 \pmod{12}$ and that $c \cdot d = 15 \cdot 8 = 120 \equiv 0 \pmod{12}$. Further, $\gcd(3, 12) = 3$, and $4 \not\equiv 8 \pmod{12}$.

6.8 (*a*) All nonzero elements of Z_5 have multiplicative inverses: $[1][1] = [1]$, $[2][3] = [3][2] = [6] = [1]$, and $[4][4] = [16] = [1]$. (*b*) Since 10 is not a prime number, only numbers relatively prime to 10 have multiplicative inverses, namely 1, 3, 7, 9: $[1][1] = [1]$, $[3][7] = [7][3] = [21] = [1]$, and $[9][9] = [81] = [1]$. (*c*) The elements of Z_{18} that have inverses are the numbers relatively prime to 18: $[1], [5], [7], [11], [13], [17]$.

6.9 If $p = 11$ and $b = 4$, the equivalence classes (mod 11) are $[4]$, $[2 \cdot 4] = [8]$, $[3 \cdot 4] = [12] = [1]$, $[4 \cdot 4] = [16] = [5]$, $[5 \cdot 4] = [20] = [9]$, $[6 \cdot 4] = [24] = [2]$, $[7 \cdot 4] = [28] = [6]$, $[8 \cdot 4] = [32] = [10]$, $[9 \cdot 4] = [36] = [3]$, $[10 \cdot 4] = [40] = [7]$. We also note that

$$4^{11-1} = 4^{10} = 1{,}048{,}576 = 1 + 95{,}325 \cdot 11 \equiv 1 \pmod{11}.$$

Finally, if $c = 11$, $11^{10} \equiv 0 \pmod{11}$.

SECTION 7

7.1 In ASCII "HOWDY" = 7279876889. The message 83858270327383328580 represents "SURF IS UP."

7.2 ZZ produces 9090. With $B = 4$ the smallest number is 3232.

7.3 (*a*) $323 = 17 \cdot 19$, (*b*) $4087 = 61 \cdot 67$, and (*c*) $8633 = 89 \cdot 97$.

7.4 Here are all numbers between 2 and 76 that are relatively prime to 60: 7, 11, 13, 17, 19, 23, 29, 31, 37, 41, 43, 47, 49, 53, 59, 61, 67, 71, 73.

7.5 We know that $9991 = 97 \cdot 103$. To show that $\gcd(7676, 9991) = 1$ and that $\gcd(7932, 9991) = 1$ we check that neither 97 nor 103 divides 7676 or 7932. Next we calculate R_2 and R_3:

$$R_2 \equiv M_2^{11} \equiv M_2 M_2^2 M_2^8 \pmod{9991}.$$
$$M_2^2 \equiv 7676^2 \equiv 58920976 \equiv 4049 \pmod{9991}.$$
$$M_2^4 \equiv 4049^2 \equiv 16394401 \equiv 9161 \pmod{9991}.$$
$$M_2^8 \equiv 9161^2 \equiv 83923921 \equiv 9512 \pmod{9991}.$$

Thus $R_2 \equiv 7676 \cdot 4049 \cdot 9512 \equiv 9884 \pmod{9991}$.

$$R_3 \equiv M_3^{11} \equiv M_3 M_3^2 M_3^8 \pmod{9991}.$$
$$M_3^2 \equiv 7932^2 \equiv 62916624 \equiv 3297 \pmod{9991}.$$
$$M_3^4 \equiv 3297^2 \equiv 10870209 \equiv 1 \pmod{9991}.$$
$$M_3^8 \equiv 1^2 \equiv 1 \pmod{9991}.$$

Thus $R_3 \equiv 7932 \cdot 3297 \cdot 1 \equiv 5357 \pmod{9991}$.

7.6 Assume that there are 30 days $= 30 \cdot 24 \cdot 60 = 43{,}200$ minutes in a month. Then to check all N B digit numbers from 0 to N with $e = 11$ requires roughly

$N15B^2 = 10^{B+1}15B^2$ single-digit operations. The problem can be restated as follows: For what value of B is $10^{B+1}15B^2/17{,}800 \geq 43{,}200$? The answer is $B \geq 6$. Thus in order to keep Eve calculating for a month, the value of N must be at least 10^7.

7.7 The multiplicative inverse (mod 8) of $e = 7$ is $d = 7$. The encryption of 2: $2^7 = 128 \equiv 8 \pmod{15}$. The decryption of 8: $8^7 = 2097152 \equiv 2 \pmod{15}$. The encryption of 7: $7^7 = 823543 \equiv 13 \pmod{15}$. The decryption of 13: $13^7 = 62748517 \equiv 7 \pmod{15}$.

7.8 First we show that $R_2^{4451} \equiv M_2$:

$$R_2^{4451} \equiv (9884^{4096})(9884^{256})(9884^{64})(9884^{32})(9884^{2})(9884)$$

With a total of 12 multiplications we find

$$\begin{aligned}
9884^2 &\equiv 97693456 \equiv 1458 \pmod{9991}\\
9884^4 &\equiv 1458^2 \pmod{9991}\\
&\equiv 2125764 \pmod{9991} \equiv 7672 \pmod{9991}\\
9884^8 &\equiv 7672^2 \pmod{9991}\\
&\equiv 58859584 \pmod{9991} \equiv 2603 \pmod{9991}\\
9884^{16} &\equiv 2603^2 \pmod{9991}\\
&\equiv 6775609 \pmod{9991} \equiv 1711 \pmod{9991}\\
9884^{32} &\equiv 1711^2 \pmod{9991}\\
&\equiv 2927521 \pmod{9991} \equiv 158 \pmod{9991}\\
9884^{64} &\equiv 158^2 \pmod{9991}\\
&\equiv 24964 \pmod{9991} \equiv 4982 \pmod{9991}\\
9884^{128} &\equiv 4982^2 \pmod{9991}\\
&\equiv 24820324 \pmod{9991} \equiv 2680 \pmod{9991}\\
9884^{256} &\equiv 2680^2 \pmod{9991}\\
&\equiv 7182400 \pmod{9991} \equiv 8862 \pmod{9991}\\
9884^{512} &\equiv 8862^2 \pmod{9991}\\
&\equiv 78535044 \pmod{9991} \equiv 5784 \pmod{9991}\\
9884^{1024} &\equiv 5784^2 \pmod{9991}\\
&\equiv 33454656 \pmod{9991} \equiv 4788 \pmod{9991}\\
9884^{2048} &\equiv 4788^2 \pmod{9991}\\
&\equiv 22924944 \pmod{9991} \equiv 5590 \pmod{9991}\\
9884^{4096} &\equiv 5590^2 \pmod{9991}\\
&\equiv 31248100 \pmod{9991} \equiv 6243 \pmod{9991}
\end{aligned}$$

With five more multiplications we find

$$\begin{aligned}
9884^{4451} &\equiv 6243 \cdot 8862 \cdot 4982 \cdot 158 \cdot 1458 \cdot 9884 \pmod{9991}\\
&\equiv (6243 \cdot 8862) \cdot (4982 \cdot 158) \cdot (1458 \cdot 9884) \pmod{9991}\\
&\equiv 5299 \cdot 7858 \cdot 3850 \pmod{9991}\\
&\equiv (5299 \cdot 7858) \cdot 3850 \pmod{9991}\\
&\equiv 7045 \cdot 3850 \pmod{9991}\\
&\equiv 27123250 \pmod{9991}\\
&\equiv 7676 \pmod{9991}\\
&= M_2.
\end{aligned}$$

Next we show that $R_3^{4451} \equiv M_3$:

$$R_3^{4451} \equiv (5357^{4096})(5357^{256})(5357^{64})(5357^{32})(5357^{2})(5357)$$

With a total of two multiplications we find that

$$\begin{aligned}
5357^2 &\equiv 28697449 \equiv 3297 \pmod{9991}\\
5357^4 &\equiv 3297^2 \pmod{9991}\\
&\equiv 10870209 \pmod{9991} \equiv 1 \pmod{9991}
\end{aligned}$$

All of the remaining powers of 5357 will equal 1 modulo 9991. Then with one more multiplication we find that

$$\begin{aligned}
R_3^{4451} &\equiv 5357^{4451} \pmod{9991}\\
&\equiv 1 \cdot 1 \cdot 1 \cdot 1 \cdot 3297 \cdot 5357 \pmod{9991}\\
&\equiv 17662029 \pmod{9991}\\
&\equiv 7932 \pmod{9991}\\
&= M_3.
\end{aligned}$$

SOLUTIONS TO QUESTIONS IN CHAPTER 5

SECTION 1

1.1 There are $\binom{4}{2} = 6$ pairs of possible direct connections among the four buildings A, C, M, and S. At least three direct connections are needed so that communication is possible between every pair of buildings. Not every set of three direct connections will ensure that each pair of buildings can communicate. See the following illustrations.

There are $\binom{5}{2} = 10$ pairs of possible direct connections among five buildings. At least four direct connections are required to ensure communications among every pair of buildings. Not every set of four direct connections will guarantee communications between each pair of buildings. See the following illustrations.

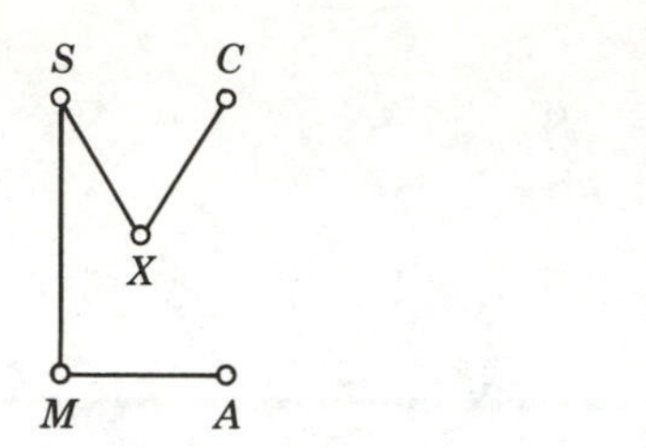

1.2 There are 16 possible LAN configurations. Four of these have one building directly linked to each of the other three buildings. In the remaining 12 the buildings are linked in a path of three cables. By checking all possibilities, one can determine that joining Stoddard with each of the other three buildings has a minimum total cost of \$148,000. Note that the total cost with Stoddard is less than the cost without it.

SECTION 2

2.1 Here are three graphs with $V = 4$ and $E = 3$.

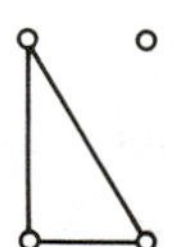

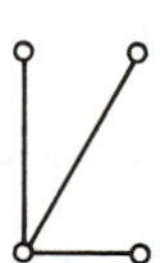

2.2 We display the graph in Figure 5.4 with each vertex labeled with its degree: $V = 8$, $E = 10$, and the sum of the degrees of all the vertices is 20.

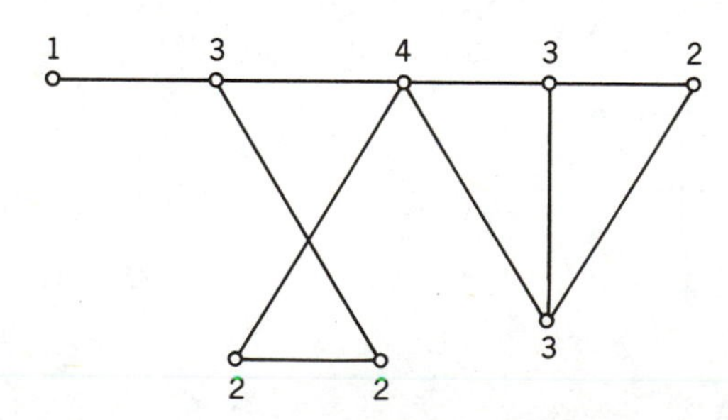

2.3 From Theorem 2.1 the sum of all of the degrees is an even number. The contribution to this sum made by the vertices of even degree is even. Therefore the contribution to this sum made by the vertices of odd degree also must be even. The only way this can occur is if the number of odd vertices is even.

2.4 The 11 different graphs on 4 vertices are as follows. (Note that in the solution to Question 2.1 we listed all different graphs with $V = 4$ and $E = 3$.)

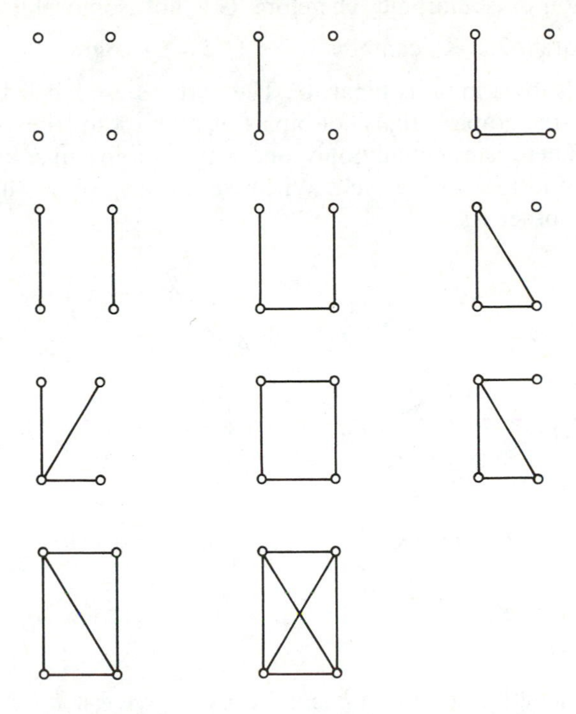

2.5 We show the graphs, G and H, from Figure 5.7 with each vertex labeled, and then proceed with an adjacency argument.

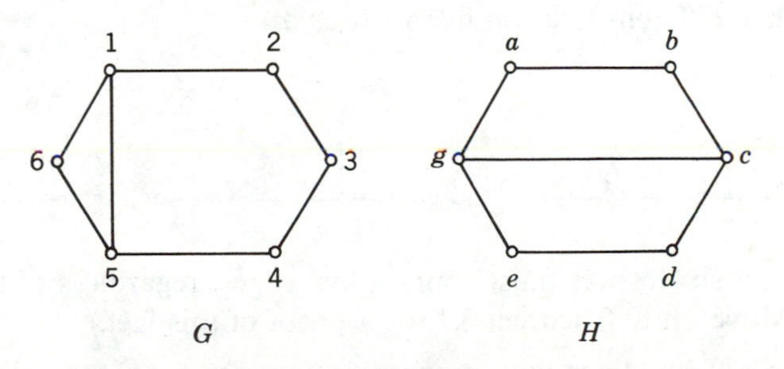

If f were an isomorphism from G to H, it would satisfy the relation $\deg(x, G) = \deg(f(x), H)$ for each vertex x of G. Thus we must have that $f(1) = g$ or c and $f(5) = c$ or g. Without loss of generality, we choose $f(1) = g$ and $f(5) = c$. The question is, what is $f(6)$? (This is where we get stuck and can conclude that G and H are not isomorphic.) In order to preserve adjacency, f must map the vertex 6 to a vertex in H that is adjacent to both

$f(1) = g$ and $f(5) = c$. But there is no vertex in H that is adjacent to both g and c. Thus we cannot find $f: V(G) \to V(H)$ that satisfies property (ii) of the definition of isomorphic. Therefore, G is not isomorphic to H.

2.6 From Theorem 2.2, K_7 contains $7(7-1)/2 = 21$ edges.

2.7 The graph shown in (a) is bipartite. The vertices are labeled with R and B. To see why the graph in (b) is not bipartite, attempt to label the vertices with R and B. There is essentially only one way of doing this, by alternating R with B around the outside cycle. When we do this, we see that some Rs are adjacent to other Rs.

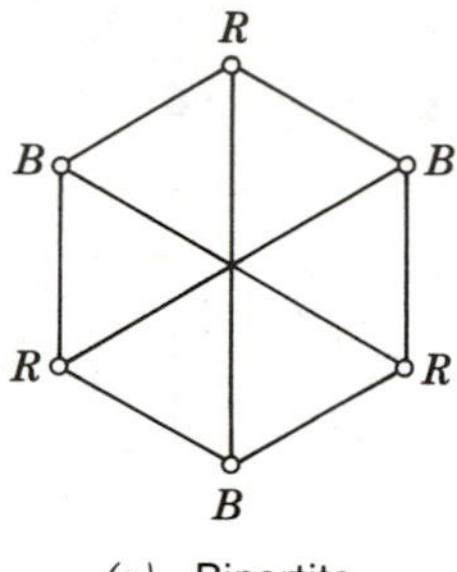

(a) Bipartite

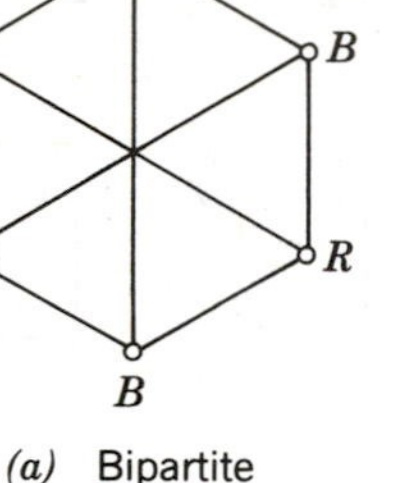

(b) Not bipartite

SECTION 3

3.1 A path of length 5 from a to b is given by $\langle a, x, r, e, w, b \rangle$. A path of length 3 from z to r is given by $\langle z, a, x, r \rangle$. A 4-cycle through b is given by $\langle b, w, c, z, b \rangle$.

3.2 The three different trees on five vertices are

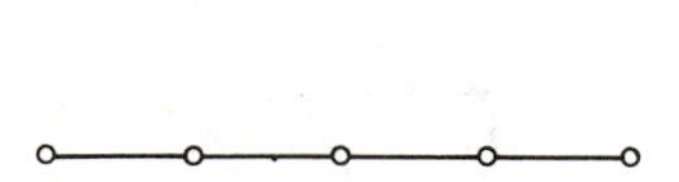
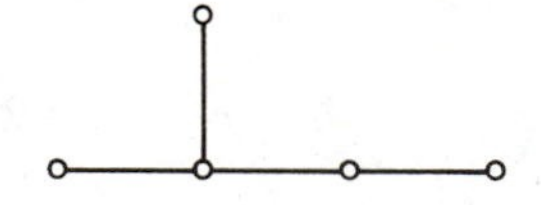
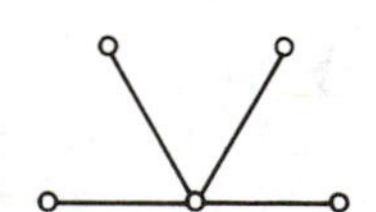

3.3 A tree on six vertices must contain five edges, regardless of the particular tree. (Move on to Theorem 3.1 for a proof of this fact.)

3.4 A set of 40 buildings with every pair connected by coaxial cable can be thought of as a 40-clique. From Theorem 2.2, a 40-clique contains $40(40-1)/2 = 780$ edges, or in this case, cables. A good guess as to the minimum number of cables needed to connect 40 buildings is 39. The reasoning behind this guess is as follows: A graph that models the LAN (i) should be connected (since every pair of buildings must be able to communicate), and (ii) should not contain any cycles (since these introduce unnecessary connections). Thus the model for the LAN that uses the fewest number of edges is a tree with 40 vertices, which necessarily has 39 edges.

3.5 (*a*) The forest in Figure 5.15 contains 5 components, 14 vertices and 9 edges. Thus $E = V - C$. (*b*) Let the component trees be labeled $T_1, T_2, \ldots, T_C$ with the number of vertices and edges of each component given by $V_1, V_2, \ldots, V_C$ and $E_1, E_2, \ldots, E_C$, respectively. Then, in total, $E = E_1 + E_2 + \cdots + E_C$ and $V = V_1 + V_2 + \cdots + V_C$. Since each component is a tree, we know that $E_i = V_i - 1$ for $i = 1, 2, \ldots, C$. Our goal is to compute E, the total number of edges in the forest F, and we hope the result will be that $E = V - C$:

$$E = E_1 + E_2 + \cdots + E_C = (V_1 - 1) + (V_2 - 1) + \cdots + (V_C - 1)$$
$$= V_1 + V_2 + \cdots + V_C - (1 + 1 + \cdots + 1) = V - C.$$

3.6 We shall show that after the removal of edge e, every pair of vertices in the graph $G - e$ is still connected by some path, and thus the graph $G - e$ is connected.

Proof. Suppose that e is an edge of the cycle C. Pick two vertices, say z and w in $V(G)$. Since G is connected, there is a path $P = P(z, w)$ from z to w. If P does not include e, then there is a path from z to w in $G - e$. Otherwise, P uses e and thus intersects with C. Suppose that u is the first vertex of P that is a vertex of C and v is the last vertex of P that is a vertex of C. Thus P consists of three segments, $P(z, u)$ from z to u, $P(u, v)$ from u to v, and $P(v, w)$ from v to w. If $u = v$, then we can construct a new path P' consisting of $P(z, u)$ followed by $P(v, w)$. P' is a path from z to w in $G - e$. If $u \neq v$, then within C there is a path $P^{\#}$ that joins u with v but does not contain the edge e. Let P' consist of $P(z, u)$ followed by $P^{\#}$ followed by $P(v, w)$. P' is a path from z to w in $G - e$. Thus $G - e$ is connected. □

3.7 A connected graph with V vertices and $V - 1$ edges is a tree.

Proof. If G is acyclic, then by definition, G is a tree. If G contains a cycle, by the preceding question it is possible to remove an edge from G, leaving a connected graph. Continue removing edges from cycles until you are left with a connected, acyclic graph. Such a graph has $V - 1$ edges, the original number of edges. Thus no edges were removed, there can be no cycle in G, and G is a tree. □

3.8 One possible spanning tree of the graph in Figure 5.19 is as follows.

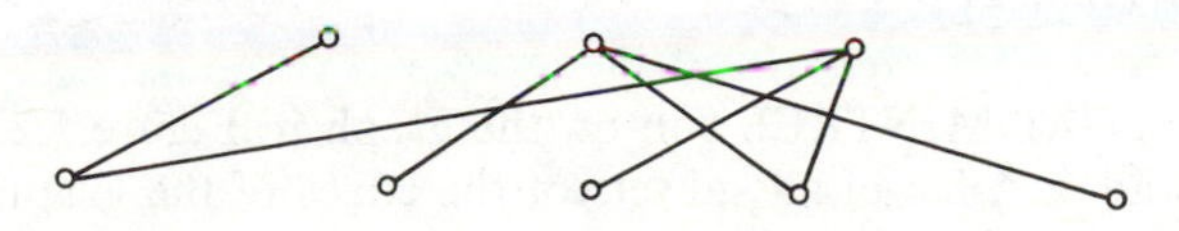

3.9 All spanning trees of the graph shown in Figure 5.21, along with the weight of each, are as follows.

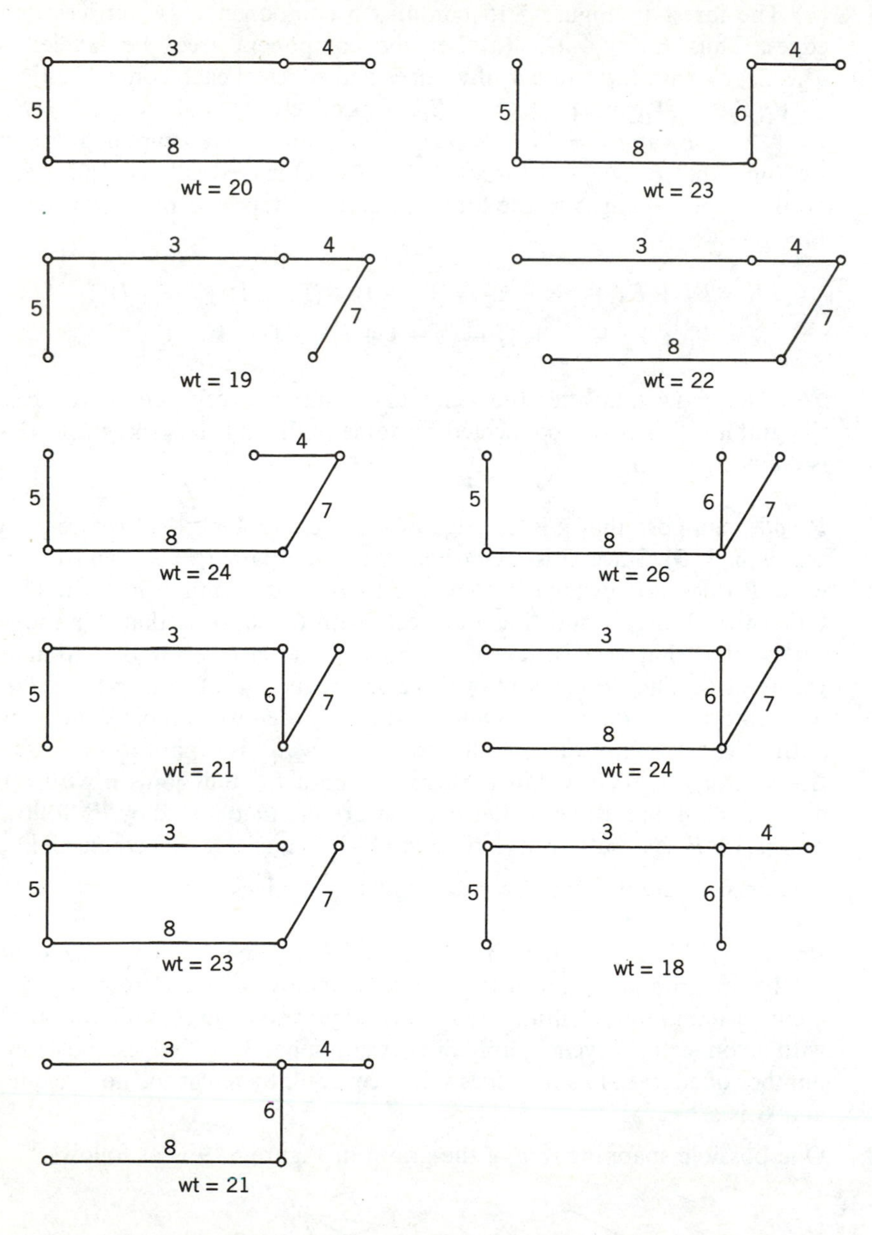

3.10 Algorithm BADMINTREE run on the graph in Figure 5.22:
Steps 2 and 3. A list of all subsets of the edges of the graph in Figure 5.22

with exactly three edges follows. For those graphs that are trees, we give the total weight:

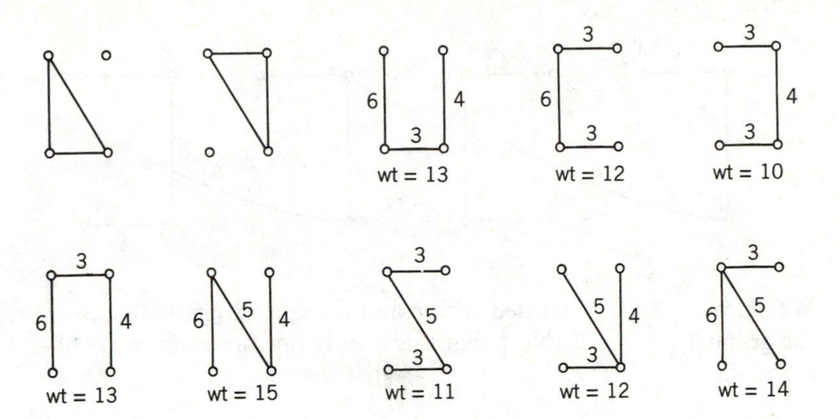

Step 4. The spanning tree of minimum weight is the tree shown above whose weight is 10.

3.11 $V = 3$: $\binom{3(3-1)/2}{3-1} = \binom{3}{2} = 3$

$V = 4$: $\binom{4(4-1)/2}{4-1} = \binom{6}{3} = 20$

$V = 5$: $\binom{5(5-1)/2}{5-1} = \binom{10}{4} = 210$

$V = 6$: $\binom{6(6-1)/2}{6-1} = \binom{15}{5} = 3003$

$V = 7$: $\binom{7(7-1)/2}{7-1} = \binom{21}{6} = 54264$

SECTION 4

4.1 (*a*) The result of running KRUSKAL on the first graph in Figure 5.24.

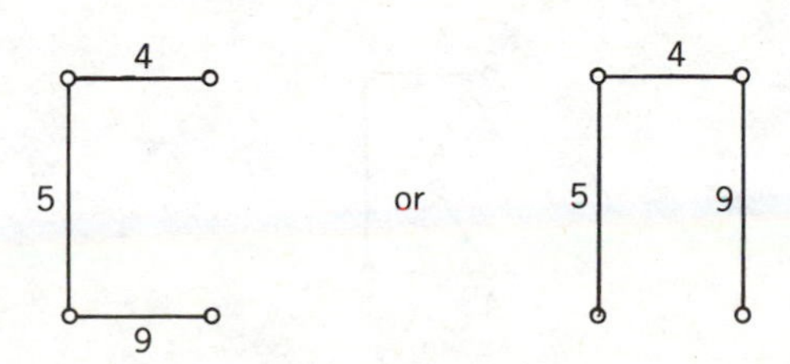

(*b*) The result of running KRUSKAL on the second graph in Figure 5.24 is a spanning forest. KRUSKAL reports failure, since the graph is not connected.

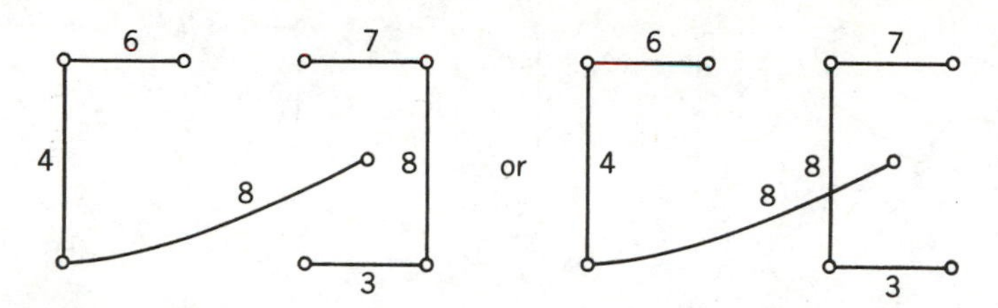

4.2 We show T_1 and T_2 labeled with c and d, respectively, and then we exhibit the graph $T_2 + c - d$. (Note that this is only one of several ways of choosing c and d.)

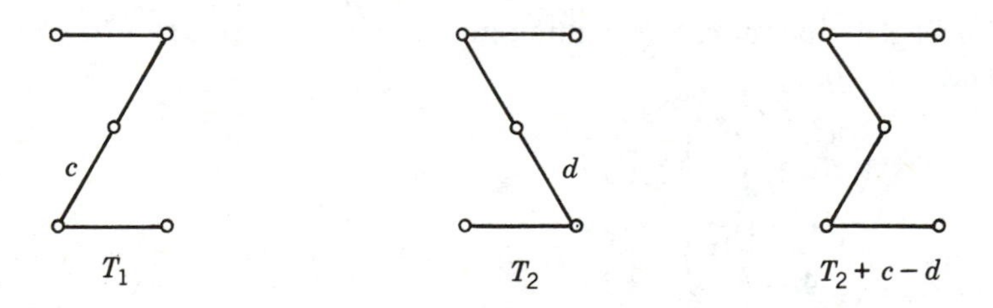

4.3 Since $E = O(V^2)$, $O(E^2) + O(E \cdot V) = O((V^2)^2) + O(V^2 V) = O(V^4) + O(V^3) = O(V^4)$. Similarly, $O(E \log(E)) = O(V^2 \log(V^2)) = O(V^2 \log(V))$.

4.4 (*a*) $\begin{bmatrix} 0 & 1 & 1 & 0 \\ 1 & 0 & 0 & 1 \\ 1 & 0 & 0 & 0 \\ 0 & 1 & 0 & 0 \end{bmatrix}$ (*b*) $\begin{bmatrix} 0 & 1 & 0 & 1 & 1 & 0 & 0 \\ 1 & 0 & 1 & 1 & 0 & 1 & 0 \\ 0 & 1 & 0 & 0 & 0 & 1 & 1 \\ 1 & 1 & 0 & 0 & 1 & 1 & 0 \\ 1 & 0 & 0 & 1 & 0 & 0 & 0 \\ 0 & 1 & 1 & 1 & 0 & 0 & 1 \\ 0 & 0 & 1 & 0 & 0 & 1 & 0 \end{bmatrix}$

4.5

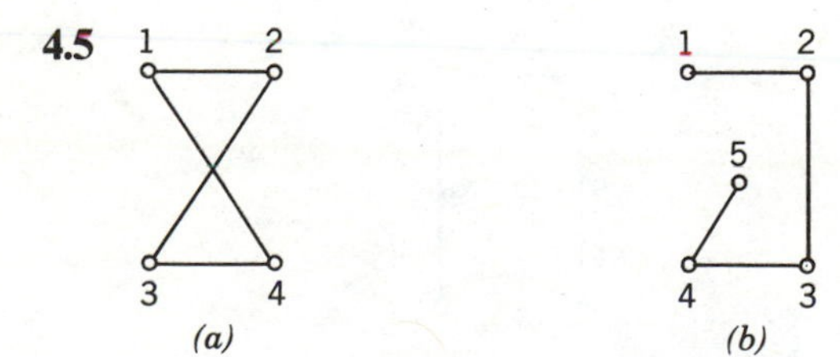

4.6 If $f(n) = O(n^k)$, there are constants C and N such that

$$\begin{aligned} f(n) &\leq Cn^k && \text{for all } n \geq N \\ &= C(B^{1/2})^k && \text{since } B = n^2 \\ &= C(B^{k/2}) \\ &= O(B^{k/2}). \end{aligned}$$

SECTION 5

5.1 *Algorithm GREEDYMAX*

STEP 1. Order the objects of E in order of decreasing weight; assume E contains m objects $e_1, \ldots, e_m$
STEP 2. Set $j := 1$ {j will index the objects.}
STEP 3. Set T to be empty {T will contain the desirable subset being created.}
STEP 4. Repeat
Begin
STEP 5. If $T + e_j$ is desirable, set $T := T + e_j$
STEP 6. $j := j + 1$
End
Until $j > m$
STEP 7. Output T and stop.

If E is the set of weighted edges in a graph and desirability is defined as being acyclic, then the algorithm GREEDYMAX adds the heaviest weight edges to T unless a cycle is formed. Thus at the end T contains a maximum weight spanning forest.

5.2 *Algorithm GREEDYCYCLE*

STEP 1. Set C to be empty; set $j := 0$
STEP 2. Repeat
Begin
STEP 3. Find the lightest edge e such that $C + e$ is a path; set $C := C + e$
STEP 4. Set $j := j + 1$
End
Until $j = V - 1$
STEP 5. Set $C := C + (x, y)$, where x and y are the end vertices of the path of C
STEP 6. Output C and stop.

An example of a weighted K_4 followed by the result of running GREEDYCYCLE on this K_4 with $V = 4$ is shown in the following figure. The final graph shown is the true minimum weight 4-cycle.

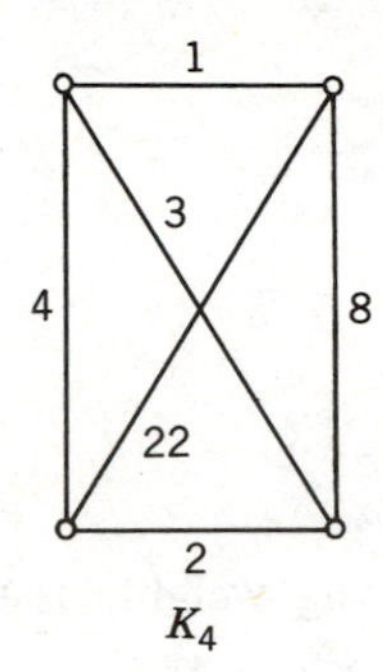

K_4

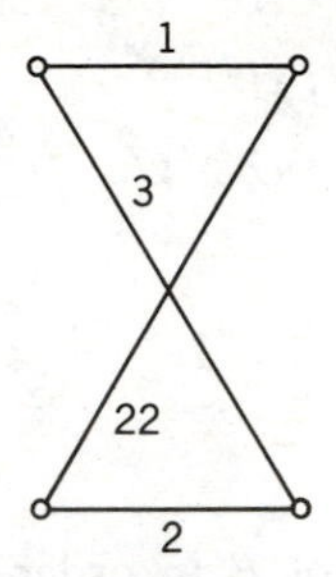

Result of GREEDYCYCLE

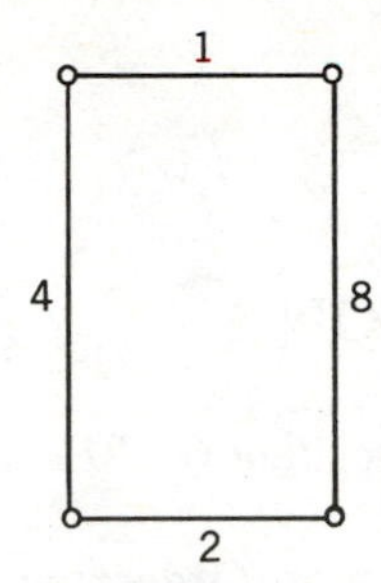

True minimum-weight 4-cycle

SOLUTIONS TO QUESTIONS IN CHAPTER 6

SECTION 1

1.1 **(*a*)** 1, the social security number might be first in the card file. **(*b*)** 20, the social security number might be last in the card file. **(*c*)** On average you might expect the director to check about one-half or $(1 + 20)/2 = 10.5$ cards.

1.2 **(*a*)** To find the card with the social security number that is alphabetically first in the card file, the director must make one comparison. To find the card with the social security number that is alphabetically second in the card file, the director must make two comparisons. In general, to find the card that is alphabetically in the ith position in the card file, the director must make i comparisons. In total, the director must make

$$1 + 2 + 3 + \cdots + 19 + 20 = \frac{20(20 + 1)}{2} = 210 \text{ comparisons.}$$

(*b*) It takes 420 seconds = 7 minutes to make all the comparisons. It takes 20 minutes to record all the information. Thus the director spends more time recording than comparing.

1.3 The director will have to examine all cards to be certain of finding the one with the smallest social security number. Thus it will require 19 comparisons if she compares the first card with the second, the smaller with the third, the smallest with the fourth, and so on.

1.4 Trace of Algorithm SELECTSORT run on $\langle 6,4,2,3\rangle$:

Step No.	i	j	a_1	a_2	a_3	a_4	*TN*
4	1	2	6	4	2	3	6
5	1	2	6	6	2	3	4
4	1	3	6	6	2	3	4
5	1	3	6	6	4	3	2
4	1	4	6	6	4	3	2
5	1	4	6	6	4	3	2
6	1	4	2	6	4	3	2
2, 3, 4	2	3	2	6	4	3	6
5	2	3	2	6	6	3	4
4	2	4	2	6	6	3	4
5	2	4	2	6	6	4	3
6	2	4	2	3	6	4	3
2, 3, 4	3	4	2	3	6	4	6
5	3	4	2	3	6	6	4
6	3	4	2	3	4	6	4
7 STOP							

SECTION 2

2.1

$n =$	136	68	34	17	9	5	3
$\lfloor (n+1)/2 \rfloor =$	68	34	17	9	5	3	2

2.2 If n is odd, then there are exactly $(n-1)/2$ records before and after the mth record. Otherwise, if n is even, there are $n/2 - 1$ records before the mth record and $n/2$ after it. The largest number of records that still must be searched is no more than $n/2$.

2.3

pair =	(6, 8)	(10, 17)	(18, 33)	(35, 67)	(69, 136)
mid =	7	13	25	51	102

2.4 $A = \langle 2, 3, 5, 7, 11, 13, 17, 19 \rangle$

(*a*) Trace of algorithm BINARYSEARCH run with $S = 5$:

Step No.	*first*	*last*	*mid*	a_{mid}
4	1	8	4	7
5	1	8	4	7
6	1	3	4	7
4, 5	1	3	2	3
6	3	3	2	3
4	3	3	3	5
5	Found S at location 3 and STOP.			

We examined and compared S with three entries of A.

(*b*) With $S = 10$:

Step No.	*first*	*last*	*mid*	a_{mid}
4, 5	1	8	4	7
6	5	8	4	7
4, 5	5	8	6	13
6	5	5	6	13
4, 5	5	5	5	11
6	5	4	5	11
7	S is not in A and STOP.			

We examined three elements in A.

(*c*) With $S = 17$:

Step No.	*first*	*last*	*mid*	a_{mid}
4, 5	1	8	4	7
6	5	8	4	7
4, 5	5	8	6	13
6	7	8	6	13
4	7	8	7	17
5	Found S at location 7 and STOP.			

We compared S with three entries of A.

2.5

n	S	$3\lfloor \log(n) \rfloor + 4$	*Array*	*No. of Comparisons Required*
2	3	7	$\langle 1,2 \rangle$	7
2	1	7	$\langle 1,2 \rangle$	2
3	4	7	$\langle 1,2,3 \rangle$	7
3	2	7	$\langle 1,2,3 \rangle$	2
4	5	10	$\langle 1,2,3,4 \rangle$	10
4	2	10	$\langle 1,2,3,4 \rangle$	2

2.6 (***a***) Because SEQSEARCH sequentially searches the card file, 1000 comparisons are required in the worst case, the case where the card being searched for is last in the file. (***b***) From Theorem 2.1, BINARYSEARCH requires at most $3\lfloor \log(1000) \rfloor + 4 = 31$ comparisons.

SECTION 3

3.1 Trace of algorithm BININSERT run on $A = \langle 2, 5, 7, 9, 13, 15, 19 \rangle$

(***a***) With $D = 1$

Step No.	*first*	*last*	*mid*	a_{mid}	A
1	1	7	?	?	$\langle 2,5,7,9,13,15,19,1 \rangle$
3	1	7	4	9	
4	1	3	4	9	
3	1	3	2	5	
4	1	1	2	5	
3	1	1	1	2	
4	1	0	1	2	
8	1				$\langle 2,2,5,7,9,13,15,19 \rangle$
9	1				$\langle 1,2,5,7,9,13,15,19 \rangle$

(***b***) With $D = 4$

Step No.	*first*	*last*	*mid*	a_{mid}	A
1	1	7	?	?	$\langle 2,5,7,9,13,15,19,4 \rangle$
3	1	7	4	9	
4	1	3	4	9	
3	1	3	2	5	
4	1	1	2	5	
3	1	1	1	2	
4	2	1	1	2	
8	2				$\langle 2,5,5,7,9,13,15,19 \rangle$
9	2				$\langle 2,4,5,7,9,13,15,19 \rangle$

(*c*) With $D = 14$

Step No.	*first*	*last*	*mid*	a_{mid}	*A*
1	1	7	?	?	⟨2, 5, 7, 9, 13, 15, 19, 14⟩
3	1	7	4	9	
4	5	7	4	9	
3	5	7	6	15	
4	5	5	6	15	
3	5	5	5	13	
4	6	5	5	13	
8	6				⟨2, 5, 7, 9, 13, 15, 15, 19⟩
9	6				⟨2, 5, 7, 9, 13, 14, 15, 19⟩

(*d*) With $D = 23$

Step No.	*first*	*last*	*mid*	a_{mid}	*A*
1	1	7	?	?	⟨2, 5, 7, 9, 13, 15, 19, 23⟩
3	1	7	4	9	
4	5	7	4	9	
3	5	7	6	15	
4	7	7	6	15	
3	7	7	7	19	
4	8	7	7	19	
5	8				⟨2, 5, 7, 9, 13, 15, 19, 23⟩

3.2 Trace of BININSERT with $A = \langle 2, 5, 9, 13, 15, 19, 16 \rangle$:

Step No.	*first*	*last*	*mid*	a_{mid}	*A*
1	1	7	?	?	⟨2, 5, 7, 9, 13, 15, 19, 16⟩
3	1	7	4	9	
4	5	7	4	9	
3	5	7	6	15	
4	7	7	6	15	
3	7	7	7	19	
4	7	6	7	19	
8	7				⟨2, 5, 7, 9, 13, 15, 19, 19⟩
9	7				⟨2, 5, 7, 9, 13, 15, 16, 19⟩

Trace of BINARYSEARCH with $A = \langle 2, 5, 7, 9, 13, 15, 19 \rangle$ and $S = 16$:

Step No.	*first*	*last*	*mid*	a_{mid}
4, 5	1	7	4	9
6	5	7	4	9
4, 5	5	7	6	15
6	7	7	6	13
4, 5	7	7	7	19
6	7	6	7	19
7 *S* is not in *A*				

The algorithm BININSERT is an extended version of BINARYSEARCH. The variables in the algorithm take on exactly the same values and similar comparisons of elements are made. However, instead of just announcing that "*S* is not in *A*," BININSERT continues by shifting part of *A* and inserting *S* into the array *A*.

3.3 Trace of BINARYSORT with $A = \langle 13, 23, 17, 19, 18, 28 \rangle$:

Step No.	*m*	*n*	*A*
1	?	6	$\langle 13, 23, 17, 19, 18, 28 \rangle$
2	2		
3			$\langle 13, 23, 17, 19, 18, 28 \rangle$
2	3		
3			$\langle 13, 17, 23, 19, 18, 28 \rangle$
2	4		
3			$\langle 13, 17, 19, 23, 18, 28 \rangle$
2	5		
3			$\langle 13, 17, 18, 19, 23, 28 \rangle$
2	6		
3			$\langle 13, 17, 18, 19, 23, 28 \rangle$

3.4 In BININSERT every execution of step 2, except for the final one, forces an execution of step 4. Steps 2 and 4 each require one comparison. The final execution of step 2 requires one additional comparison and step 5 requires one additional comparison. In total then, BININSERT requires 2(the number of executions of step 4) + 2. In each part of Question 3.1, step 4 is executed three times. Thus the total number of comparisons is $2 \cdot 3 + 2 = 8$. Further,

$$8 \le 2\lfloor \log(7) \rfloor + 4 = 2 \cdot 2 + 4 = 8.$$

3.5 In Question 3.3, BINARYSORT is performed on an array of size 6. Within BINARYSORT, the procedure BININSERT is called five times on arrays of sizes 2, 3, 4, 5 and 6, respectively. From Theorem 3.1 we have that BININSERT requires at most $2\lfloor \log(r) \rfloor + 4$ comparisons to insert the $(r + 1)$st item into a sorted array of r items. Thus the total number of comparisons in Question 3.3 is given by

$$(2\lfloor \log(1) \rfloor + 4) + (2\lfloor \log(2) \rfloor + 4) + (2\lfloor \log(3) \rfloor + 4) + (2\lfloor \log(4) \rfloor + 4) + (2\lfloor \log(5) \rfloor + 4) = 4 + 6 + 6 + 8 + 8 = 32.$$

Note that when $n = 6$, $(n - 1)(2\lfloor \log(n - 1) \rfloor + 4) = 5(2\lfloor \log(5) \rfloor + 4) = 40$.

SECTION 4

4.1 A search tree illustrating a binary search of an array of 15 elements follows.

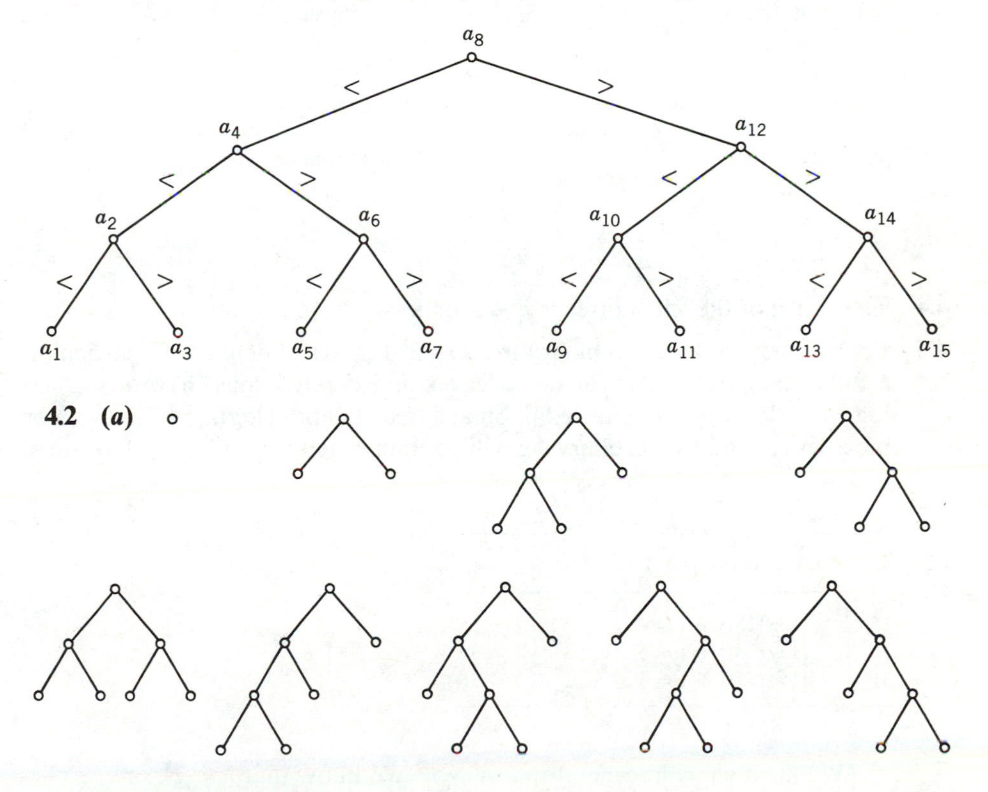

4.2 (*a*)

(*b*) There is one binary tree with two leaves: the one above with three vertices. There are two binary trees with three leaves: These are the binary trees with five vertices. There are five binary trees with four leaves: These are the binary trees with seven vertices.

4.3 (*a*) Left subtree

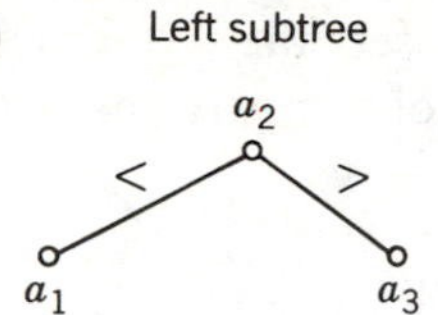

Right subtree

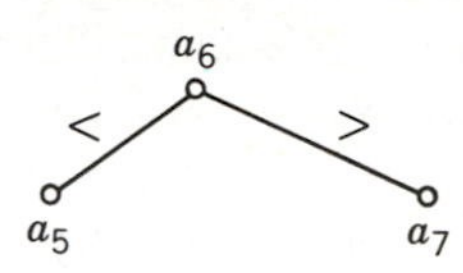

(*b*) Left subtree

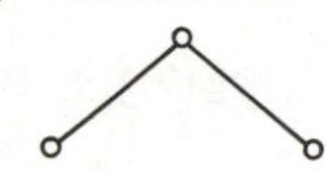

Right subtree

(*c*) Left subtree

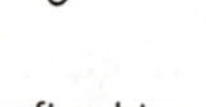

Right subtree

(*d*) Left subtree

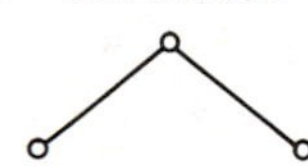

Right subtree

(*e*) Left subtree

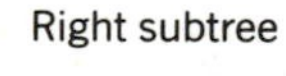

Right subtree

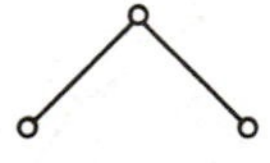

4.4 The depth of the left subtree is $d - 1$ or less.

4.5 From Theorem 4.1, a full binary tree has 2^k vertices at depth k. In particular, a full binary tree with n or more leaves and depth k must have $n \leq 2^k$ or $\log(n) \leq \lceil \log(n) \rceil = k$. Thus a full binary tree of depth $\lceil \log(n) \rceil$ will have n or more leaves and by Corollary 4.2 will contain exactly $2^{\lceil \log(n) \rceil + 1} - 1$ vertices.

4.6

n	$k = \lfloor \log(n) \rfloor + 1$	$n' = 2^k - 1$
15	$\lfloor \log(15) \rfloor + 1 = 4$	$2^4 - 1 = 15$
26	$\lfloor \log(26) \rfloor + 1 = 5$	$2^5 - 1 = 31$
31	$\lfloor \log(31) \rfloor + 1 = 5$	$2^5 - 1 = 31$

We must show in general that $n' \geq n$. We know that $n < 2^k$, or

$$n \leq 2^k - 1 = 2^{(\lfloor \log(n) \rfloor + 1)} - 1 = n'.$$

4.7 For convenience, we replace the label a_i with i. A binary search tree for a 23-element array follows. Note that the tree must have 31 vertices (see Question 4.6).

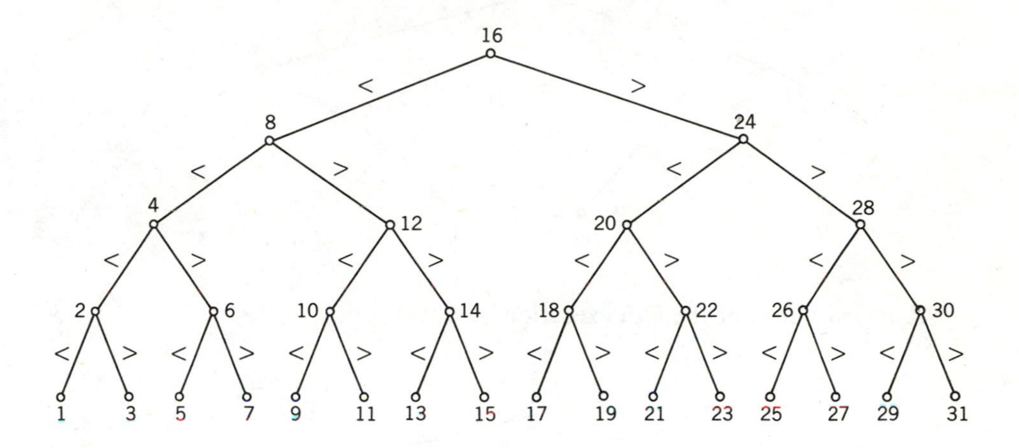

4.8 Labeled tree after the second execution of step 7:

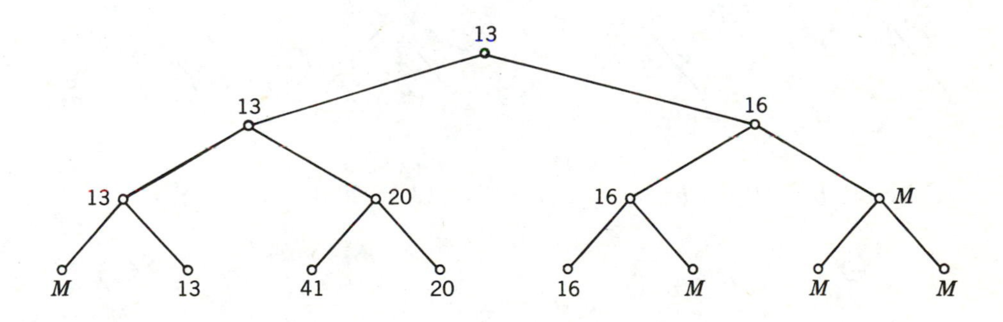

Labeled tree after the third execution of step 7:

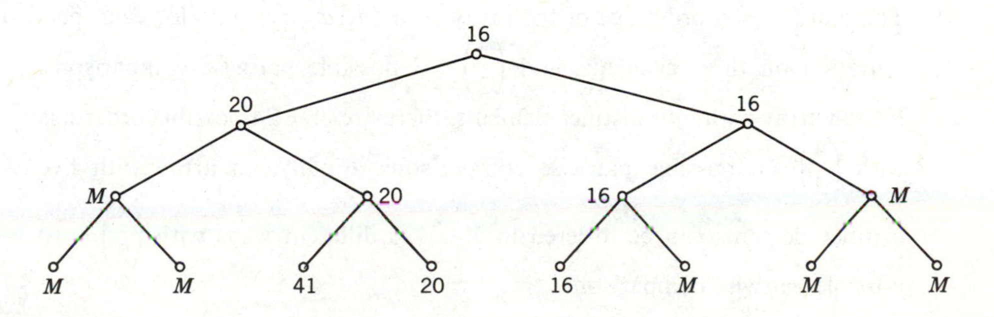

Labeled tree after the fourth execution of step 7:

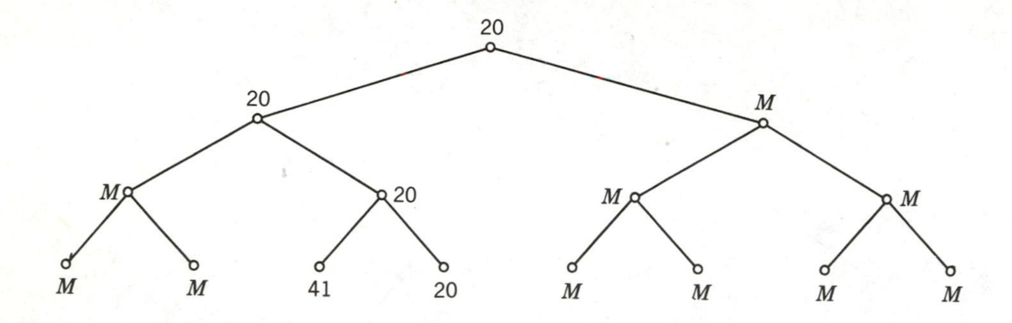

Labeled tree after the fifth execution of step 7:

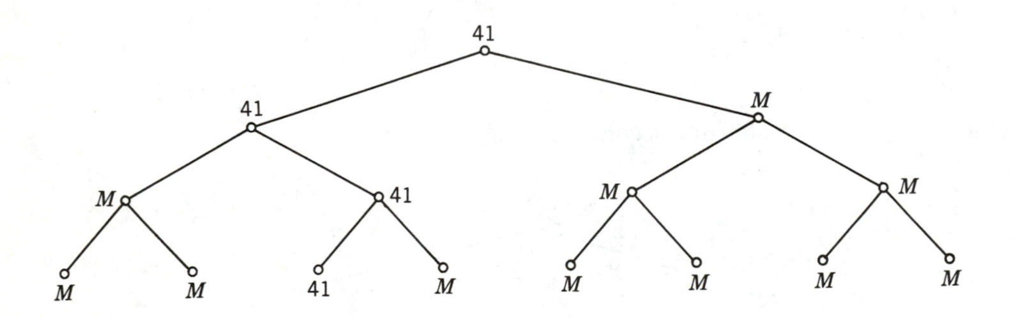

Result: $B = \langle b_1, b_2, b_3, b_4, b_5, b_6 \rangle = \langle 7, 9, 13, 16, 20, 41 \rangle$.

SECTION 5

5.1 There are $3! = 6$ orderings of the array $A = \langle a_1, a_2, a_3 \rangle$, one for each permutation on three elements, and $\binom{3}{2} = 3$ possible pairwise comparisons. For an array with four distinct elements, there are $4! = 24$ possible orderings with $\binom{4}{2} = 6$ possible pairwise comparisons. Finally, an array with five distinct elements can be ordered in $5! = 120$ different ways with $\binom{5}{2} = 10$ possible pairwise comparisons.

5.2 (*a*)

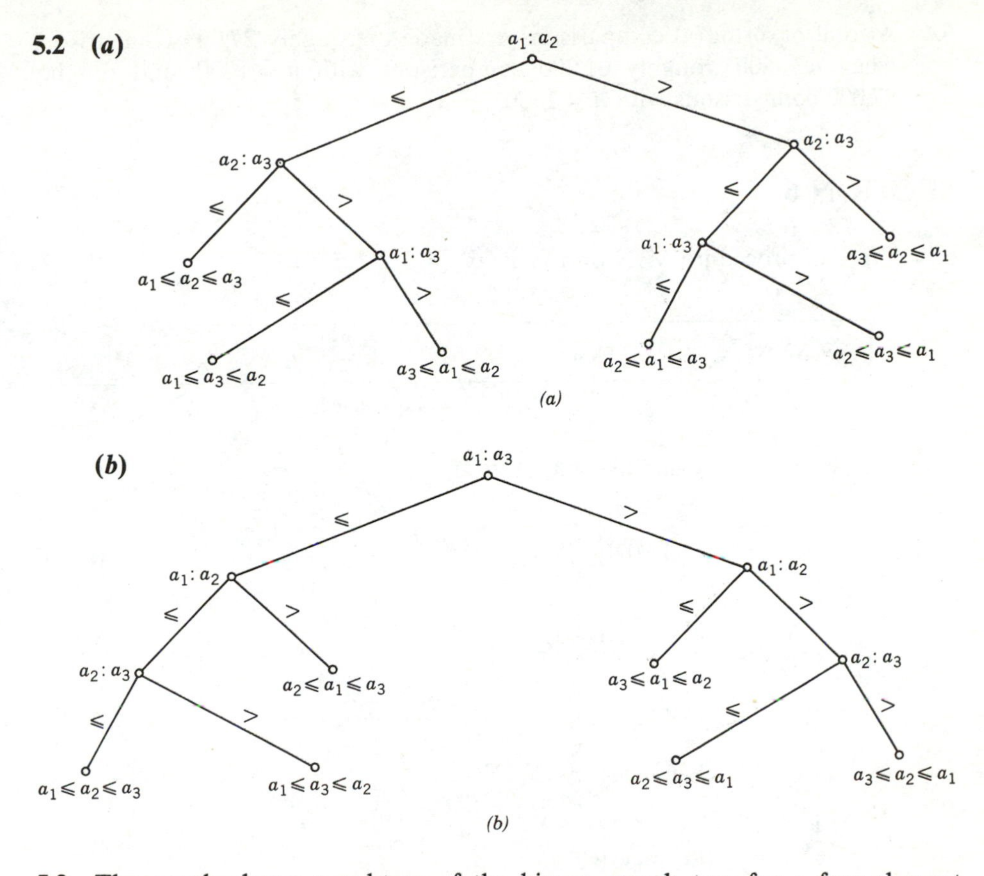

(b)

5.3 The graph shows a subtree of the binary search tree for a four-element array as specified by Example 5.3. In particular, the root of the subtree is labeled with $a_4{:}a_2$ and we have assumed that $a_1 \leq a_2 \leq a_3$. If the whole binary search tree were drawn, it would look like the tree in Figure 6.6, only this subtree would be hanging in place of the leaf labeled "$a_1 \leq a_2 \leq a_3$." And a similar subtree would be hanging under every other leaf. In the worst case five comparisons will be made.

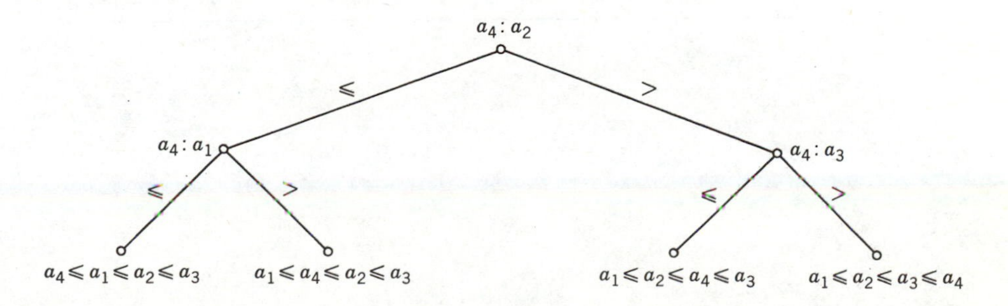

5.4 A total of $5n\log(n)$ comparisons are made or roughly 27,000 comparisons when $n = 600$, roughly 61,000 comparisons with $n = 1200$, and roughly 97,000 comparisons with $n = 1800$.

SECTION 6

6.1 Trace of procedure MIN on the array $\langle 4,3,2,1,5\rangle$:

Step No.	n	k	a_k	a_{n+1}	
1	5				
2	4				
3	{Call MIN $(\langle 4,3,2,1\rangle, 4, k)$}				(A)
1	4				
2	3				
3	{Call MIN $(\langle 4,3,2\rangle, 3, k)$}				(B)
1	3				
2	2				
3	{Call MIN $(\langle 4,3\rangle, 2, k)$}				(C)
1	2				
2	1				
3	{Call MIN $(\langle 4\rangle, 1, k)$}				(D)
1	1	1			
5	{Return to (D)}				
	1	1	4	3	
4	1	2			
5	{Return to (C)}				
	2	2	3	2	
4	2	3			
5	{Return to (B)}				
	3	3	2	1	
4	3	4			
5	{Return to (A)}				
	4	4	1	5	
4	4	4			
5	{Return with $k = 4$}				

6.2 Trace of procedure GCD run on $b = 13$ and $c = 21$:

Step No.	b	c	r	g	
1	13	21	?	?	
2	13	21	8		
3	{Call GCD$(8, 13, g)$}				(A)
2	8	13	5		
3	{Call GCD$(5, 8, g)$}				(B)
2	5	8	3		
3	{Call GCD$(3, 5, g)$}				(C)
2	3	5	2		
3	{Call GCD$(2, 3, g)$}				(D)
2	2	3	1		
3	{Call GCD$(1, 2, g)$}				(E)
2	1	2	0		
3	{Call GCD$(0, 1, g)$}				(F)
1				1	
	{Return to (F)}				
				1	
	{Return to (E)}				
				1	
	{Return to (D)}				
				1	
	{Return to (C)}				
				1	
	{Return to (B)}				
				1	
	{Return to (A)}				
				1	

Result: $\gcd(13, 21) = 1$. Note that the procedure GCD is called six times.

6.3 Trace of MIN on the array $A = \langle -1, 0.333, 5.2, -10, 6.001, 17\rangle$
(*a*) With start = 2, finish = 3

Step No.	*start*	*finish*	a_{finish}	*k*	a_k	
1	2	3				
2	{Call MIN$(A, 2, 2, k)$}					(A)
1	2	2		2		
4	{Return to (A)}					
	2	3	5.2	2	0.333	
3	2	3	5.2	2	0.333	
4	{Return with $k = 2$}					

Result: The index of the smallest entry in $\langle a_2, a_3\rangle$ is 2.

(*b*) With start = 3, finish = 6, $A = \langle -1, 0.333, 5.2, -10, 6.001, 17\rangle$

Step No.	*start*	*finish*	a_{finish}	*k*	a_k	
1	3	6				
2	{Call MIN$(A, 3, 5, k)$}					(A)
1	3	5				
2	{Call MIN$(A, 3, 4, k)$}					(B)
1	3	4				
2	{Call MIN$(A, 3, 3, k)$}					(C)
1	3	3		3		
4	{Return to (C)}					
	3	4	-10	3	5.2	
3	3	4	-10	4	-10	
4	{Return to (B)}					
	3	5	6.001	4	-10	
3	3	5	6.001	4	-10	
4	{Return to (A)}					
	3	6	17	4	-10	
3	3	6	17	4	-10	
4	{Return with $k = 4$}					

Result: The index of the smallest entry in $\langle a_3, a_4, a_5, a_6\rangle$ is 4.

(c) With start = 1, finish = 6

Step No.	*start*	*finish*	a_{finish}	*k*	a_k	
1	1	6				
2	{Call MIN $(A, 1, 5, k)$}					(A)
1	1	5				
2	{Call MIN $(A, 1, 4, k)$}					(B)
1	1	4				
2	{Call MIN $(A, 1, 3, k)$}					(C)
1	1	3				
2	{Call MIN $(A, 1, 2, k)$}					(D)
1	1	2				
2	{Call MIN $(A, 1, 1, k)$}					(E)
1	1	1		1		
4	{Return to (E)}					
	1	2	0.333	1	−1	
3	1	2	0.333	1	−1	
4	{Return to (D)}					
	1	3	5.2	1	−1	
3	1	3	5.2	1	−1	
4	{Return to (C)}					
	1	4	−10	1	−1	
3	1	4	−10	4	−10	
4	{Return to (B)}					
	1	5	6.001	4	−10	
3	1	5	6.001	4	−10	
4	{Return to (A)}					
	1	6	17	4	−10	
3	1	6	17	4	−10	
4	{Return with $k = 4$}					

Result: The index of the smallest entry in the array A is 4.

SECTION 7

7.1 Each vertex of the following tree is labeled with the subarray of A to be sorted at that vertex.

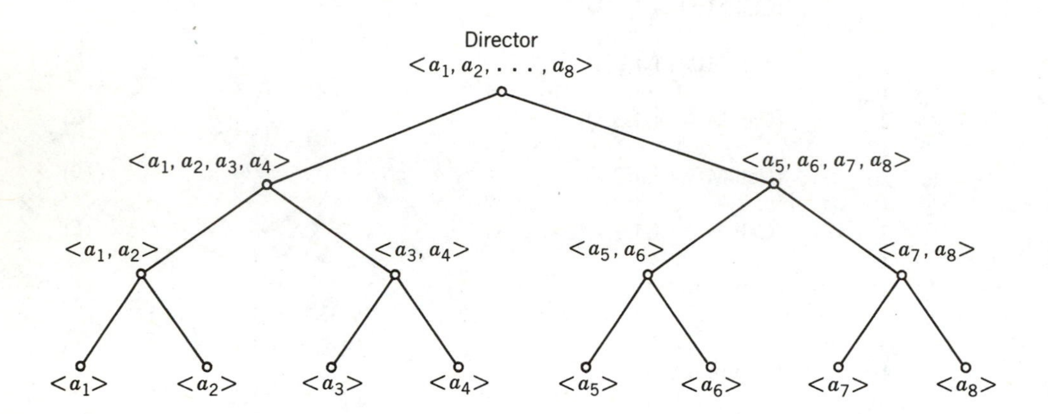

The tree above has depth 3 and 15 vertices. Every vertex except the root corresponds with one assistant.

7.2 Trace of MERGE run on $C = \langle 0.1, 0.2, 0.3, 0, 0.09, 0.19, 0.29, 0.39, 0.49 \rangle$ with start = 1, mid = 3, and finish = 9.

Step No.	i	j	k	D
1	1	4	1	
3	1	5	2	$\langle 0, \ldots$
	1	6	3	$\langle 0, 0.09, \ldots$
	2	6	4	$\langle 0, 0.09, 0.1, \ldots$
	2	7	5	$\langle 0, 0.09, 0.1, 0.19, \ldots$
	3	7	6	$\langle 0, 0.09, 0.1, 0.19, 0.2, \ldots$
	3	8	7	$\langle 0, 0.09, 0.1, 0.19, 0.2, 0.29, \ldots$
	4	8	8	$\langle 0, 0.09, 0.1, 0.19, 0.2, 0.29, 0.3, \ldots$
10				$\langle 0, 0.09, 0.1, 0.19, 0.2, 0.29, 0.3, 0.39, 0.49 \rangle$
15				$C = \langle 0, 0.09, 0.1, 0.19, 0.2, 0.29, 0.3, 0.39, 0.49 \rangle$.

7.3 **(*a*)** Trace of procedure MERGESORT run on the array $C = \langle 1, 0 \rangle$ with start = 1 and finish = 2:

Step No.	*C*	*start*	*mid*	*finish*	
1, 2	⟨1, 0⟩	1	1	2	
3	{Call MERGESORT(*C*, 1, 1)}				(*A*)
1, 2	⟨1⟩	1		1	
	{Return to (*A*)}	1	1	2	
4	{Call MERGESORT(*C*, 2, 2)}				(*B*)
1, 2	⟨0⟩	2		2	
	{Return to (*B*)}	1	1	2	
5	{Call MERGE(*C*, 1, 1, 2)}	1	1	2	
	⟨0, 1⟩				
6	Return.				

(*b*) Trace of procedure MERGESORT run on the array $C = \langle 22, 24, 23 \rangle$ with start = 1 and finish = 3:

Step No.	*C*	*start*	*mid*	*finish*	
1, 2	⟨22, 24, 23⟩	1	2	3	
3	{Call MERGESORT(*C*, 1, 2)				(*A*)
1, 2	⟨22, 24⟩	1	1	2	
3	{Call MERGESORT(*C*, 1, 1)}				(*B*)
1, 2	⟨22⟩	1		1	
	{Return to (*B*)}	1	1	2	
4	{Call MERGESORT(*C*, 2, 2)}				(*C*)
1, 2	⟨24⟩	2		2	
	{Return to (*C*)}	1	1	2	
5	{Call MERGE(*C*, 1, 1, 2)}	1	1	2	
	⟨22, 24⟩				
	{Return to (*A*)}	1	2	3	
4	{Call MERGESORT(*C*, 3, 3)}	1	2	3	(*D*)
1, 2	⟨23⟩	3		3	
	{Return to (*D*)}	1	2	3	
5	{Call MERGE(*C*, 1, 2, 3)}	1	2	3	
	⟨22, 23, 24⟩				
6	Return.				

(*c*) Trace of procedure MERGESORT run on the array $C = \langle 1.1, 3.3, 2.2, 4.4 \rangle$ with start = 1 and finish = 4:

Step No.	*C*	*start*	*mid*	*finish*	
1, 2	$\langle 1.1, 3.3, 2.2, 4.4 \rangle$	1	2	4	
3	{Call MERGESORT $(C, 1, 2)$}				(*A*)
1, 2	$\langle 1.1, 3.3 \rangle$	1	1	2	
3	{Call MERGESORT (C, 1, 1)}				(*B*)
1, 2	$\langle 1.1 \rangle$	1		1	
	{Return to (*B*)}	1	1	2	
4	{Call MERGESORT $(C, 2, 2)$}	1	1	2	(*C*)
1, 2	$\langle 3.3 \rangle$	2		2	
	{Return to (*C*)}	1	1	2	
5	{Call MERGE $(C, 1, 1, 2)$}	1	1	2	
	$\langle 1.1, 3.3 \rangle$				
	{Return to (*A*)}	1	2	4	
4	{Call MERGESORT $(C, 3, 4)$}				(*D*)
1, 2	$\langle 2.2, 4.4 \rangle$	3	3	4	
3	{Call MERGESORT $(C, 3, 3)$}				(*E*)
1, 2	$\langle 2.2 \rangle$	3		3	
	{Return to (*E*)}	3	3	4	
4	{Call MERGESORT $(C, 4, 4)$}				(*F*)
1,2	$\langle 4.4 \rangle$	4		4	
	{Return to (*F*)}	3	3	4	
5	{Call MERGE $(C, 3, 3, 4)$}	3	3	4	
	$\langle 2.2, 4.4 \rangle$				
	{Return to (*D*)}	1	2	4	
5	{Call MERGE $(C, 1, 2, 4)$}	1	2	4	
	$\langle 1.1, 2.2, 3.3, 4.4 \rangle$				
6	Return.				

7.4 (***a***) Since Mergesort was called three times and Merge once on an array of size 2, $3 + 3 \cdot 2 = 9$ comparisons were performed. $9 = 3 \cdot 2 \cdot \log(2) + 2 \cdot 2 - 1$.
(***c***) Since Mergesort was called seven times and Merge three times on arrays of sizes 2, 2, and 4, $7 + 3(2 + 2 + 4) = 31$ comparisons were performed.
$31 = 3 \cdot 4 \cdot \log(4) + 2 \cdot 4 - 1$.

7.5 As in Question 7.3 (b) Mergesort is called five times and Merge is called twice on arrays of sizes 2 and 3. Thus $5 + 3(2 + 3) = 20$ comparisons are performed.
$3 \cdot 3 \cdot \log(3) + 2 \cdot 3 - 1 < 20 < 6 \cdot 3 \cdot \log(3) + 10 \cdot 3 - 1$.

SOLUTIONS TO QUESTIONS IN CHAPTER 7

SECTION 1

1.1 (*a*) 64, (*b*) 21, (*c*) 13, (*d*) 217, (*e*) 17.

1.2 (*a*) $S_4 : f_n = n!$, (*b*) $S_5 : g_n = \binom{n}{2}$ (*c*) $S_6 : h_n = 2^n - 1$,

$$(d)\ S_7 : j_n = \begin{cases} n + 1 & \text{if } 1 \le n \le 2 \\ 2n - 1 & \text{if } 3 \le n \le 4 \\ 2n + 1 & \text{if } 5 \le n \le 6 \\ 2n + 3 & \text{if } n = 7 \end{cases}$$

is one of many "creative" solutions to the problem of finding a function that generates the first seven prime numbers. (*e*) $S_8 : k_n = n^2$, (*f*) $S_9 : m_n = (-1)^{n-1} 3^{n-1}$

1.3 The functions $f_n = 2^{n+1} + 1$ and $g_n = 2n^2 - 2n + 5$ are two of many that produce the values $f_1 = g_1 = 5$, $f_2 = g_2 = 9$ and $f_3 = g_3 = 17$.

1.4 $S_1 : a_n = a_{n-1} + 1$ with $a_1 = 1$

$S_2 : a_n = 2a_{n-1}$ with $a_1 = 2$

1.5

$n =$	1	2	3	4	5	6	7	8
$a_n =$	0	1	3	6	10	15		
$M_n =$	1	2	2	3	3	3	3	4

1.6

$n =$	1	2	3	4	5
$H'_n =$	1	$\frac{3}{2}$	$\frac{11}{6}$	$\frac{25}{12}$	$\frac{137}{60}$
$H''_n =$	1	$\frac{3}{2}$	$\frac{11}{6}$	$\frac{25}{12}$	$\frac{137}{60}$

We see that for $n = 1, 2, \ldots, 5$, $H'_n = H''_n$.

1.7

$n =$	1	2	3	4	5
$C_n =$	1	1	2	5	14

SECTION 2

2.1 S_6 satisfies $a_n = 2a_{n-1} + 1$ as does 9, 19, 39, 79, 159, ... S_9 satisfies $a_n = (-3)a_{n-1}$ as does -6, 18, -54, 162, -486, ...

2.2 If one of a_1 and a_3 is unspecified, then a_5 and all subsequent "odd" entries of the sequence will be undefined. Similarly, if one of a_2 and a_4 is unspecified, then a_6 and all subsequent "even" entries will be undefined.

$n =$	1	2	3	4	5	6	7	8	9	10
$a_n =$	1	1	1	1	2	2	3	3	5	5

The sequence listed above can be obtained from the Fibonacci sequence by listing each term twice. Since, by Theorem 4.3.1,

$$F_n = \frac{\phi^n - \phi'^n}{\sqrt{5}}$$

where

$$\phi = \frac{1 + \sqrt{5}}{2} \text{ and } \phi' = \frac{1 - \sqrt{5}}{2},$$

$$a_n = F_{\lceil n/2 \rceil} = \frac{\phi^{\lceil n/2 \rceil} - \phi'^{\lceil n/2 \rceil}}{\sqrt{5}}$$

2.3 **(i)** 2, **(ii)** 3, and **(iii)** 1.

2.4 **(i)** $a_n = na_{n-1} = n(n-1)a_{n-2} = n(n-1)(n-2)a_{n-3} = \cdots$
$= n(n-1)(n-2) \cdots (n-(n-2))a_1 = n(n-1)(n-2) \cdot \cdots \cdot 2 \cdot 1 = n!$.

(ii) $b_n = b_{n-1} + 2 = b_{n-2} + 2 + 2 = b_{n-3} + 2 + 2 + 2 = \cdots$
$= b_{n-(n-1)} + 2 + 2 + \cdots + 2 \qquad \{n - 1\ 2s\}$
$= b_1 + (n-1)2 = 1 + 2n - 2 = 2n - 1.$

2.5 **(i)** For the base case we have $a_1 = 1! = 1$. Then

$$\begin{aligned} a_{k+1} &= (k+1)a_k \\ &= (k+1)k! && \text{by the inductive hypothesis} \\ &= (k+1)!, && \text{as desired.} \end{aligned}$$

(ii) For the base case we have $b_1 = 2 \cdot 1 - 1 = 1$. Then

$$\begin{aligned} b_{k+1} &= b_k + 2 \\ &= (2k-1) + 2 && \text{by the inductive hypothesis} \\ &= 2k + 1 = 2(k+1) - 1, && \text{as desired.} \end{aligned}$$

2.6 The relations $a_n = na_{n-2}$, $a_n = a_{n-1} + a_{n-3}$, $a_n = 2a_{\lfloor n/2 \rfloor}$, and $a_n = na_{n-1}$ are all homogeneous, since each is satisfied by the sequence that is identically 0. The relation $b_n = b_{n-1} + 2$ is inhomogeneous, since when we replace each b_j with 0, the result, $0 = 0 + 2$, is not valid.

SECTION 3

3.1 **(i)** Not: inhomogeneous; **(ii)** not: not linear and **(iii)** not: inhomogeneous.

3.2 **(i)** The characteristic equation is $x - 2 = 0$ and the characteristic root is $q_1 = 2$. **(ii)** The characteristic equation is $x^2 - x - 6 = 0$ and the characteristic roots are $q_1 = 3$ and $q_2 = -2$. **(iii)** The characteristic equation is $x^2 - 2x + 1 = 0$ and the characteristic root is $q_1 = 1$.

3.3 The base cases are $a_1 = 2^1 - 1 = 1$ and $a_2 = 2^2 - 1 = 3$. The inductive hypothesis is that $a_k = 2^k - 1$. We substitute this in the given recurrence relation

$$\begin{aligned} a_{k+1} &= 3a_k - 2a_{k-1} \\ &= 3(2^k - 1) - 2(2^{k-1} - 1) \\ &= 3 \cdot 2^k - 3 - 2^k + 2 \\ &= 2^k(3-1) - 1 \\ &= 2^{k+1} - 1, && \text{as desired.} \end{aligned}$$

3.4 **(i)** $a_0 = 0$, **(ii)** $a_0 = 1$, and **(iii)** $a_0 = 0$.

3.5 **(i)** From Question 3.2 the characteristic equation for $a_n = a_{n-1} + 6a_{n-2}$ is $x^2 - x - 6 = 0$ with characteristic roots $q_1 = 3$ and $q_2 = -2$. Thus the general formula that solves this recurrence relation is given by $a_n = c3^n + d(-2)^n$ for some constants c and d. We determine the constants c and d from the initial conditions:

$$2 = a_0 = c3^0 + d(-2)^0 = c + d$$
$$1 = a_1 = c3^1 + d(-2)^1 = 3c - 2d.$$

Adding twice the first equation to the second, we obtain $5c = 5$. Thus $c = 1$ and then $d = 1$ by substitution. Thus $a_n = 3^n + (-2)^n$.

(ii) We have accomplished most of the work for this part of the problem above; the only difference is in the initial conditions. Hence we have

$$1 = a_0 = c3^0 + d(-2)^0 = c + d$$
$$3 = a_1 = c3^1 + d(-2)^1 = 3c - 2d.$$

Again adding twice the first equation to the second we obtain $5c = 5$. Thus $c = 1$ and $d = 0$. Thus $a_n = 3^n$.

(iii) From Question 3.2, the characteristic equation for $a_n = 2a_{n-1} - a_{n-2}$ is $x^2 - 2x + 1 = 0$, which has a root of multiplicity 2.

SECTION 4

4.1 For the base cases we have $b_1 = 1$ and $b_2 = 2$. Then

$$\begin{aligned} b_{k+1} &= 2b_k - b_{k-1} && \text{the given recurrence} \\ &= 2k - (k-1) && \text{by the inductive hypothesis} \\ &= k + 1, && \text{as desired.} \end{aligned}$$

4.2 For $p(x) = x^2 - 2x + 1$, we construct $D(x)$ as follows:

$$\begin{aligned} D(x) &= \frac{x^2 - 2x + 1 - (q^2 - 2q + 1)}{x - q} \\ &= \frac{(x^2 - q^2) - 2(x - q) + (1 - 1)}{x - q} && \text{by regrouping} \\ &= \frac{(x - q)(x + q) - 2(x - q)}{x - q} && \text{by algebra} \\ &= x + q - 2 && \text{by division.} \end{aligned}$$

4.3 The characteristic equation of $b_n = 4b_{n-1} - 4b_{n-2}$ is $x^2 - 4x + 4 = 0$ and its characteristic root is $q_1 = 2$. If $b_k = 2^k$, then

$$\begin{aligned} 4b_{n-1} - 4b_{n-2} &= 4 \cdot 2^{n-1} - 4 \cdot 2^{n-2} \\ &= 2^{n+1} - 2^n = 2^n(2-1) = 2^n = b_n. \end{aligned}$$

If $b_k = k2^k$, then

$$\begin{aligned} 4b_{n-1} - 4b_{n-2} &= 4(n-1)2^{n-1} - 4(n-2)2^{n-2} \\ &= 2^n[2(n-1) - (n-2)] = n2^n = b_n. \end{aligned}$$

4.4 The characteristic equation of $c_n = -3c_{n-1} - 3c_{n-2} - c_{n-3}$ is

$$x^3 + 3x^2 + 3x + 1 = (x+1)^3 = 0$$

and the characteristic root is $q_1 = -1$ which has multiplicity 3.

If $c_k = (-1)^k$, then

$$\begin{aligned} -3c_{n-1} - 3c_{n-2} - c_{n-3} &= -3(-1)^{n-1} - 3(-1)^{n-2} - (-1)^{n-3} \\ &= (-1)^{n-3}[-3(-1)^2 - 3(-1) - 1] \\ &= (-1)^{n-3}[-3 + 3 - 1] \\ &= (-1)^{n-3}(-1) = (-1)^{n-2} = (-1)^n = c_n. \end{aligned}$$

If $c_k = k(-1)^k$, then

$$\begin{aligned} &-3c_{n-1} - 3c_{n-2} - c_{n-3} \\ &= -3(n-1)(-1)^{n-1} - 3(n-2)(-1)^{n-2} - (n-3)(-1)^{n-3} \\ &= (-1)^{n-3}[-3(n-1)(-1)^2 - 3(n-2)(-1) - (n-3)] \\ &= (-1)^{n-3}[-3n + 3 + 3n - 6 - n + 3] \\ &= (-1)^{n-3}[-n] = n(-1)^{n-2} = n(-1)^n = c_n. \end{aligned}$$

Finally, if $c_k = k^2(-1)^k$, then

$$\begin{aligned} &-3c_{n-1} - 3c_{n-2} - c_{n-3} \\ &= -3(n-1)^2(-1)^{n-1} - 3(n-2)^2(-1)^{n-2} - (n-3)^2(-1)^{n-3} \\ &= (-1)^{n-3}[-3(n^2 - 2n + 1) + 3(n^2 - 4n + 4) - (n^2 - 6n + 9)] \\ &= (-1)^{n-3}[-3n^2 + 6n - 3 + 3n^2 - 12n + 12 - n^2 + 6n - 9] \\ &= (-1)^{n-3}[-n^2] = (-1)^{n-2}n^2 = n^2(-1)^n = c_n. \end{aligned}$$

4.5 From Theorem 4.2, a solution of the recurrence relation given in Question 4.3 is of the form

$$a_n = c_1(-1)^n + c_2n(-1)^n + c_3n^2(-1)^n.$$

With the initial conditions $a_0 = 1$, $a_1 = -2$ and $a_2 = 1$, we can solve for the constants c_1, c_2, and c_3:

$$\begin{aligned} 1 &= a_0 = c_1 + 0c_2 + 0c_3 \\ -2 &= a_1 = -c_1 - c_2 - c_3 \\ 1 &= a_2 = c_1 + 2c_2 + 4c_3 \end{aligned}$$

The solution to this system of equations is $c_1 = 1$, $c_2 = 2$, and $c_3 = -1$. Thus a solution to the recurrence relation with the given initial conditions is $a_n = (-1)^n + 2n(-1)^n - n^2(-1)^n$.

SECTION 5

5.1 Reread Section 6.2.

$n =$	2	3	4	5
$B_n =$	7	7	10	10

We note for $n = 2, 3, 4$, and 5 that $B_n = 3\lfloor \log(n) \rfloor + 4$.

5.2 Reread Exercise 7 in Chapter 6, Section 7.

$n =$	2	4	8
$M_n =$	9	31	87
$3n \log(n) + 2n - 1 =$	9	31	87

5.3 **(a)** $k = 1$, $d = 2$, $c = 0$ and $e = 3$. **(b)** $k = 1$, $d = 2$, $c = 0$ and $e = 2$. **(c)** $k = 2$, $d = 2$, $c = 3$ and $e = 1$.

5.4 1 initial condition. If $a_0 = 1$, then

$n =$	1	2	3	4	5	6	7
$a_n =$	2	2	3	3	3	3	3

5.5 The proof is by induction on i, where $n = 2^i$. The base case is $i = 0$: $a_{2^0} = a_1 + \log(1)c = a_1 + 0 = a_1$. Assuming the result for $i = k$, let $i = k + 1$.

$$
\begin{aligned}
a_n &= a_{2^{k+1}} && \text{since } n = 2^i = 2^{k+1}\\
&= a_{\lfloor n/2 \rfloor} + c && \text{by the recurrence relation}\\
&= a_{2^k} + c && \text{since } n/2 = 2^k\\
&= a_1 + \log(2^k)c + c && \text{by the inductive hypothesis}\\
&= a_1 + kc + c = a_1 + (k+1)c\\
&= a_1 + \log(2^{k+1})c && \text{by properties of log}\\
&= a_1 + \log(n)c.
\end{aligned}
$$

5.6 With $c = 2$, Theorem 5.1 implies that $C_n \leq 2\lfloor \log(n) \rfloor + 4$. This is the same result as Theorem 3.1 from Chapter 6.

5.7
$$
\begin{aligned}
M_{2^k} &\leq 2M_{(2^k)/2} + 4 \cdot 2^k \qquad \text{by } (C')\\
&= 2M_{2^{k-1}} + 2^{k+2}.\\
&\leq 2(2M_{2^{k-2}} + 2^{k+1}) + 2^{k+2}\\
&= 2^2 M_{2^{k-2}} + 2^{k+2} + 2^{k+2}\\
&= 2^2 M_{2^{k-2}} + 2 \cdot 2^{k+2}\\
&\quad \ldots\\
&\leq 2^i M_{2^{k-i}} + i2^{k+2}\\
&\quad \ldots\\
&\leq 2^k M_{2^{k-k}} + k2^{k+2}\\
&= 2^k M_1 + 4(2^k)k
\end{aligned}
$$

Since $\log(2^k) = k$ and $n = 2^k$, the previous expression can be rewritten as $nM_1 + 4n\log(n)$. Next we verify this formula by induction on k, where $n = 2^k$. That is, if $n = 2^k$ and M_n satisfies (C'), then we must show that $M_n \leq 4n\log(n) + M_1 n$. For the base case $k = 0$ and $n = 2^0 = 1$. Then $M_n = M_1 \leq 4 \cdot 1 \cdot 0 + M_1 \cdot 1 = M_1$. We assume the result for $n = 2^k$ and check $n = 2^{k+1}$:

$$
\begin{aligned}
M_n &= M_{2^{k+1}}\\
&\leq 2M_{2^k} + 4 \cdot 2^{k+1} && \text{by } (C')\\
&\leq 2(4 \cdot 2^k k + M_1 2^k) + 4 \cdot 2^{k+1} && \text{by the inductive hypothesis}\\
&= 4 \cdot 2^{k+1}k + M_1 2^{k+1} + 4 \cdot 2^{k+1}\\
&= 4 \cdot 2^{k+1}(k+1) + M_1 2^{k+1}\\
&= 4n\log(n) + M_1 n.
\end{aligned}
$$

SOLUTIONS TO QUESTIONS IN CHAPTER 8

SECTION 1

1.1 The union of shortest paths forms a minimum-distance spanning tree, shown in (*a*). With the root specified to be 5, we obtain the different tree shown in (*b*).

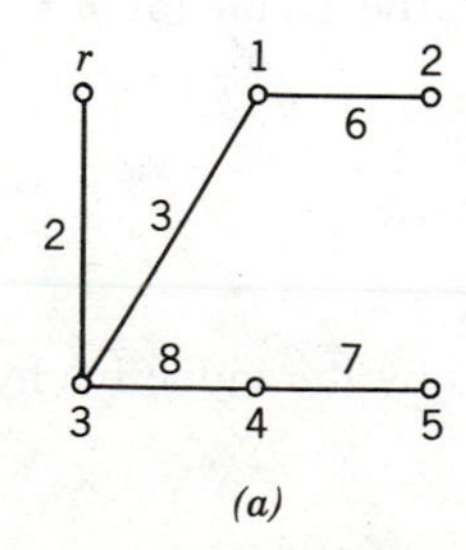

(*a*)

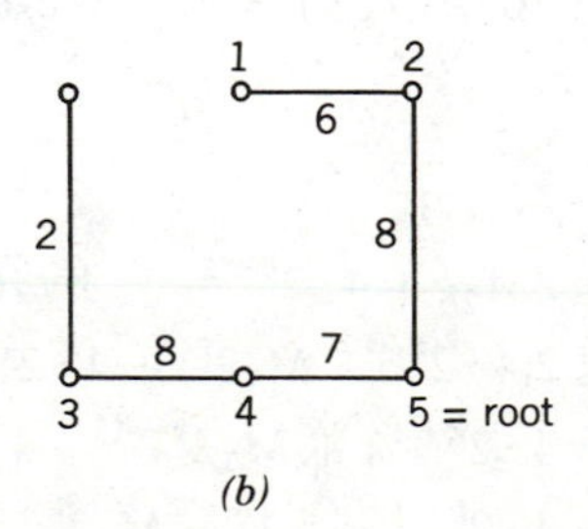

(*b*)

1.2 Here is a trace of DIJKSTRA on the weighted graph from Figure 8.4.

Step No.	*j*	*z*	*V(T)*	*E(T)*
2	?	?	$\{r\}$	$\varnothing$
4	1	t		
5	1	t	$\{r,t\}$	$\{(r,t)\}$
4	2	b		
5	2	b	$\{r,t,b\}$	$\{(r,t),(r,b)\}$
4	3	s		
5	3	s	$\{r,t,b,s\}$	$\{(r,t),(r,b),(t,s)\}$
4	4	c		
5	4	c	$\{r,t,b,s,c\}$	$\{(r,t),(r,b),(t,s),(r,c)\}$
4	5	a		
5	5	a	$\{r,t,b,s,c,a\}$	$\{(r,t),(r,b),(t,s),(r,c),(b,a)\}$
4	6	q		
5	6	q	$\{r,t,b,s,c,a,q\}$	$\{(r,t),(r,b),(t,s),(r,c),(b,a),(a,q)\}$
4	7	p		
5	7	p	$\{r,t,b,s,c,a,q,p\}$	$\{(r,t),(r,b),(t,s),(r,c),(b,a),(a,q),(q,p)\}$

1.3 If G is not connected, then step 4 cannot be executed $V-1$ times as required. If some edge weights were negative, then the minimum distance from the root to the first attached vertex x might be less than the weight of the first edge $e=(r,x)$. Thus the tree T in the base case might not have a shortest path in it. Later in the proof, when we add the edge (v,x), we claim that x is closer to the root than v. This would not be true if the weight of (v,x) were negative. If DIJKSTRA is run on the graph in Figure 8.5, then the distance from the root to any of the other vertices is not well defined, since every time you traverse a cycle around the triangle you add a total of -1 to your path length.

SECTION 2

2.1 The graph shown in (*a*) is the only Eulerian graph with four vertices. The graph in (*b*) has an Eulerian path from x to y but is not Eulerian. The simplest such graph would just be a path with eight vertices.

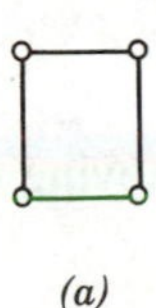

(*a*)

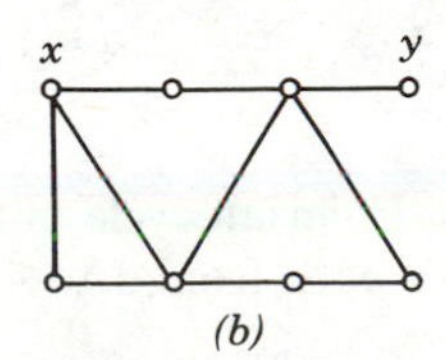

(*b*)

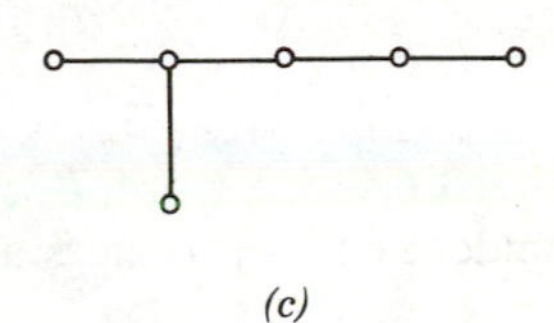

(*c*)

2.2 The first and last graphs are Eulerian (See Theorem 2.1). The second and fourth contain Eulerian paths but not Eulerian cycles.

2.3 A graph with four vertices all of whose degrees are even must have every degree either 0 or 2. To be connected there cannot be any vertices of degree 0. There is only one graph, the 4-cycle, which is Eulerian. Similarly, a graph with five vertices must have every degree either 2 or 4. We list them together with one Eulerian cycle.

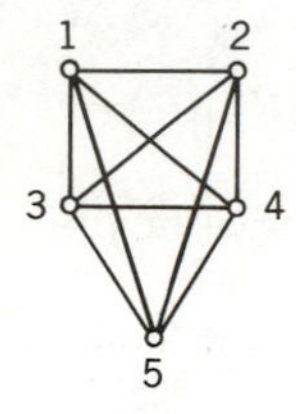

<1, 2, 4, 5, 3, 1, 4, 3, 2, 5, 1>

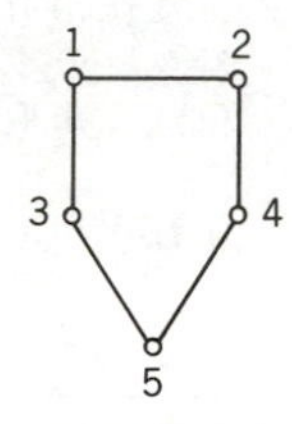

<1, 2, 4, 3, 5, 1>

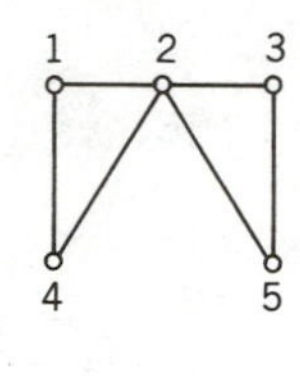

<1, 2, 3, 5, 2, 4, 1>

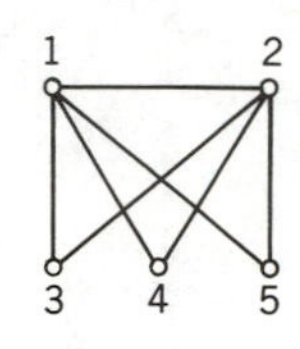

<1, 3, 2, 4, 1, 5, 2, 1>

2.4

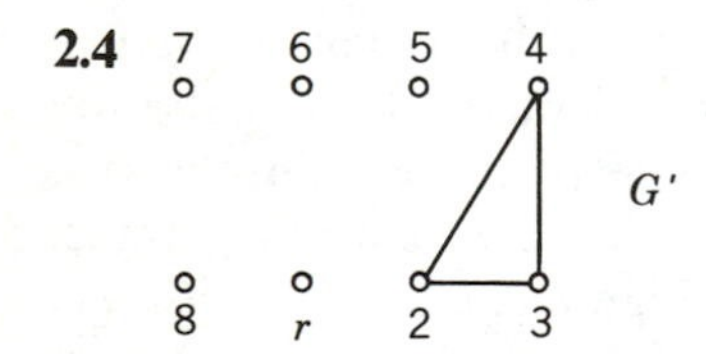

2.5 The vertices 3 and 4 have odd degree. To construct G' we create vertex r adjacent to 3 and 4 as shown in the following figure.

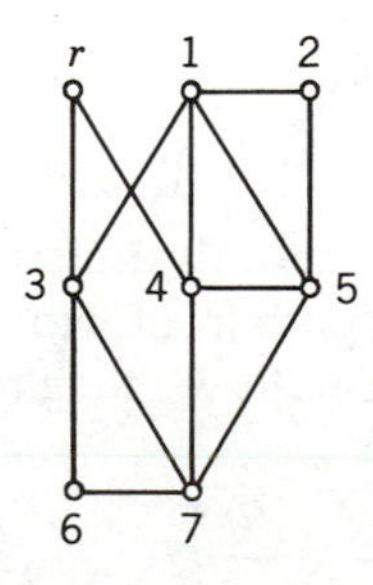

$\langle 4, r, 3, 6, 7, 5, 2, 1, 5, 4, 7, 3, 1, 4 \rangle$ is an Eulerian cycle in G'. Removing r and its incident edges produces an Eulerian path from 3 to 4.

2.6 For convenience we label the graph.

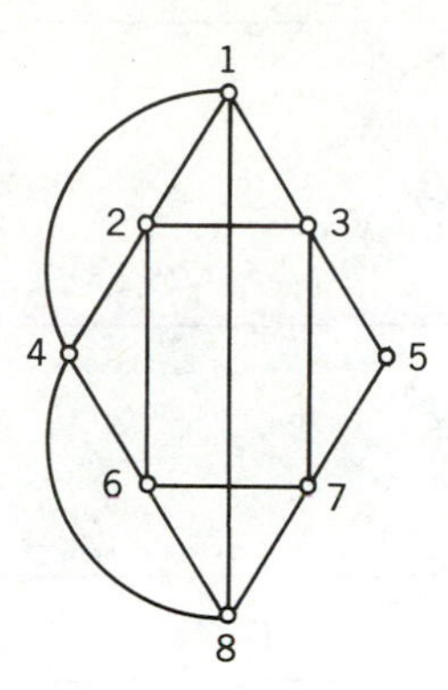

Step No.	*x*	*C*	*D*
1	1	$\langle 1 \rangle$	
3	1		
4	1		$\langle 1,2,3,1,4,6,7,8,1 \rangle$
6		$\langle 1,2,3,1,4,6,7,8,1 \rangle$	
3	2		
4	2		$\langle 2,4,8,6,2 \rangle$
6		$\langle 1,2,4,8,6,2,3,1,4,6,7,8,1 \rangle$	
3	3		
4			$\langle 3,5,7,3 \rangle$
6		$\langle 1,2,4,8,6,2,3,5,7,3,1,4,6,7,8,1 \rangle$	

2.7 $\langle 1,2,4,3,1 \rangle$ is an Eulerian cycle in the first graph. The second graph does not contain an Eulerian path or cycle. $\langle 2,3,1,2,4,3 \rangle$ is an Eulerian path in the third graph as is $\langle 2,4,3,1,2,3 \rangle$. The fourth graph contains lots of Eulerian cycles, for example, $\langle 1,2,1,3,4,2,4,3,1 \rangle$.

SECTION 3

3.1 The first and third do; $\langle 1,4,2,3,1 \rangle$ and $\langle 1,2,3,4,5,1 \rangle$, respectively.

3.2

3.3

Step No.	z	P	C
2		$\varnothing$	
4	x	$\langle x, v, t, w, y\rangle$	
5			$\langle x, v, t, w, y, x\rangle$
4	u	$\langle u, v, t, w, y, x\rangle$	
5			$\langle u, t, w, y, x, v, u\rangle$

3.4 First graph

Step No.	J	K	T	$E(T)$
1	1		{1}	$\varnothing$
3	1	2		
4–6	2	2	{1, 2}	{(1, 2)}
3	2	3		
4–6	3	3	{1, 2, 3}	{(1, 2), (2, 3)}
3	3	4		
4–6	4	4	{1, 2, 3, 4}	{(1, 2), (2, 3), (3, 4)}
3	4	{no K}		
7	3			
3	3	{no K}		
7	2			
3	2	5		
4–6	5	5	{1, 2, 3, 4, 5}	{(1, 2), (2, 3), (3, 4), (2, 5)}
9 STOP				

Second graph

Step No.	J	K	T	$E(T)$
1	1		{1}	$\varnothing$
3	1	2		
4–6	2	2	{1, 2}	{(1, 2)}
3	2	6		
4–6	6	6	{1, 2, 6}	{(1, 2), (2, 6)}
3	6	{no K}		
7	2			
3	2	{no K}		
7	1			
3	1	{no K}		
8 STOP			{1, 2, 6}	{(1, 2), (2, 6)}

3.5 First graph

Step No.	*J*	*K*	*FORWARD*	*PATH*
Main 1	2		TRUE	⟨1,0,0,0,0⟩
Build 1	2	2		
3–5	3	2	TRUE	⟨1,2,0,0,0⟩
Build 1	3	3		
3–5	4	3	TRUE	⟨1,2,3,0,0⟩
Build 1, 2	4	{no *K*}	FALSE	
Main 4, 5	3			⟨1,2,3,0,0⟩
Build 1	3	4		
3–5	4	4	TRUE	⟨1,2,4,0,0⟩
Build 1, 2	4	{no *K*}	FALSE	
Main 4, 5	3			⟨1,2,4,0,0⟩
Build 1, 2	3	{no *K*}	FALSE	
Main 4, 5	2			⟨1,2,0,0,0⟩
Build 1	2	3		
3–5	3	3	TRUE	⟨1,3,0,0,0⟩
Build 1	3	2		
3–5	4	2	TRUE	⟨1,3,2,0,0⟩
Build 1	4	4		
3–5	5	4	TRUE	⟨1,3,2,4,0⟩
Main 7, 8				⟨1,3,2,4,1⟩
Main 9 STOP				

Second graph

Step No.	*J*	*K*	*FORWARD*	*PATH*
Main 1	2		TRUE	⟨1,0,0,0,0⟩
Build 1	2	2		
3–5	3	2	TRUE	⟨1,2,0,0,0⟩
Build 1	3	3		
3–5	4	3	TRUE	⟨1,2,3,0,0⟩
Build 1, 2	4	{no *K*}	FALSE	
Main 4, 5	3			⟨1,2,3,0,0⟩
Build 1	3	4		
3–5	4	4	TRUE	⟨1,2,4,0,0⟩
Build 1, 2	4	{no *K*}	FALSE	
Main 4, 5	3			⟨1,2,4,0,0⟩
Build 1, 2	3	{no *K*}	FALSE	
Main 4, 5	2			⟨1,2,0,0,0⟩

continued

Second graph (continued)

Step No.	*J*	*K*	*FORWARD*	*PATH*
Build 1	2	4		
3–5	3	4	TRUE	$\langle 1,4,0,0,0\rangle$
Build 1	3	2		
3–5	4	2	TRUE	$\langle 1,4,2,0,0\rangle$
Build 1	4	3		
3–4	5	3	TRUE	$\langle 1,4,2,3,0\rangle$
Main 7, 8			FALSE	
Main 4, 5	4			$\langle 1,4,2,3,0\rangle$
Build 1, 2	4	{no K}	FALSE	
Main 4, 5	3			$\langle 1,4,2,0,0\rangle$
Build 1, 2	3	{no K}	FALSE	
Main 4, 5	2			$\langle 1,4,0,0,0\rangle$
Build 1, 2	2	{no K}	FALSE	
Main 4, 5	1			$\langle 1,0,0,0,0\rangle$
Main 6	NO HAM CYCLE, STOP			

SECTION 4

4.1 Denote the locations by $O = (0,0)$, $P = (1,0)$, $Q = (0,1)$, and $R = (1,1)$. Since the drill must start and end at O, there are $3! = 6$ possible drilling sequences: *OPRQ* and *OQRP* have total distance 4, *OPQR* and *ORQP* have total distance $2 + 2\sqrt{2}$ as do *ORPQ* and *OQPR*. Note that the second sequence of each of the preceding pairs is the reverse of the first.

4.2

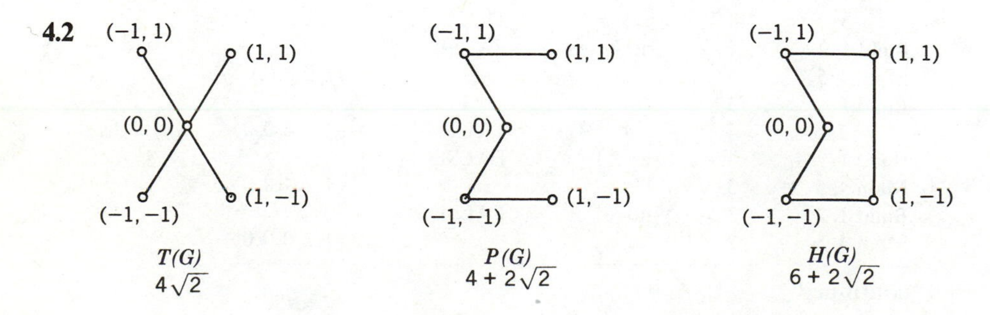

4.3 D might consist of $\langle O, P, O, Q, O, S, O, R, O\rangle$.

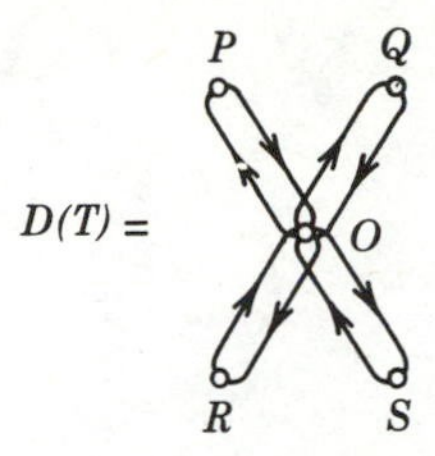

4.4 The Hamiltonian cycle C produced from D is $\langle O, P, Q, S, R, O\rangle$. $W(C) = 6 + 2\sqrt{2} = w(H(G))$.

SECTION 5

5.1 See Figure 8.26: Each interfering pair of variables is represented by an edge of this graph. Sets of four mutually interfering variables: $\{L, W, Wt, A\}$, $\{Wt, Vol, A, C1\}$, $\{Ht, A, C1, Vol\}$, $\{Wt, Vol, C1, C2\}$. There is no set of five mutually interfering variables.

5.2

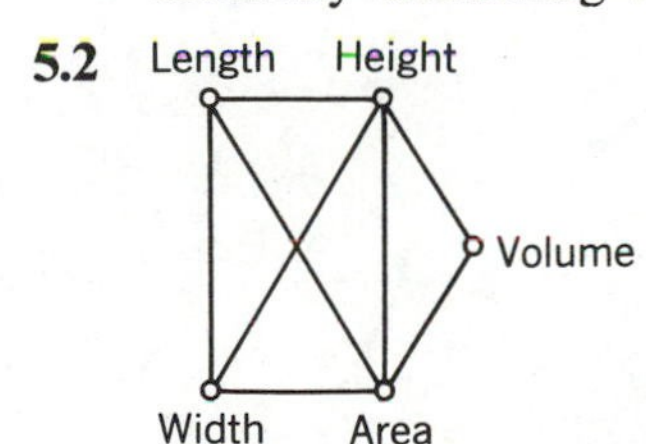

5.3 Adding edges may increase the chromatic number: If x and y are not adjacent in H, but are adjacent in G, then a coloring of H is a coloring of G unless x and y are assigned the same color. In that case, an additional color may be needed to color G. If the compiler assigns $k = \chi(G)$ memory locations to a program based on a coloring of G, then no two variables, joined by an edge in G, receive the same color and so are not assigned to the same memory location. Edges join every pair of "truly" interfering variables and maybe more.

5.4 **(i)** $\chi(G) = \text{cl}(G) = 2$; **(ii)** $\chi(G) = \text{cl}(G) = 3$; **(iii)** $\chi(G) = \text{cl}(G) = 3$; **(iv)** $\chi(G) = 3$, $\text{cl}(G) = 2$; and **(v)** $\chi(G) = \text{cl}(G) = 4$.

5.5 A graph is 1-colorable if and only if it does not contain an edge. An algorithm to 1-color could check that the graph contains no edges and then assign the color 1 to every vertex.

5.6 **(*a*)**

Step No.	I	c_I	J	L_J
5	1	1		
7	1		2	$\langle 2\rangle$
7	1		6	$\langle 2,3,4,5,6\rangle$
7	1		7	$\langle 2,3,4,5,6,7\rangle$
5	2	2		
7	2		3	$\langle 1,3\rangle$
7	2		7	$\langle 3,4,5,6,7\rangle$
5	3	1		
7	3		4	$\langle 2,3,4\rangle$
7	3		7	$\langle 3,4,5,6,7\rangle$
5	4	2		
7	4		5	$\langle 1,3,4,5\rangle$
7	4		7	$\langle 3,4,5,6,7\rangle$
5	5	1		
7	5		6	$\langle 2,3,4,5,6\rangle$
7	5		7	$\langle 3,4,5,6,7\rangle$
5	6	2		
7	6		7	$\langle 3,4,5,6,7\rangle$
5	7	3		

(*b*)

Step No.	I	c_I	J	L_J
5	1	1		
7	1		2	$\langle 2\rangle$
7	1		5	$\langle 2,3,4,5\rangle$
7	1		6	$\langle 2,3,4,5,6\rangle$
5	2	2		
7	2		3	$\langle 1,3\rangle$
7	2		4	$\langle 1,3,4\rangle$
7	2		5	$\langle 3,4,5\rangle$
7	2		6	$\langle 3,4,5,6\rangle$
7	2		7	$\langle 1,3,4,5,6,7\rangle$
5	3	1		
7	3		4	$\langle 3,4\rangle$
7	3		7	$\langle 3,4,5,6,7\rangle$
5	4	3		
7	4		5	$\langle 4,5\rangle$
5	5	4		
5	6	3		
7	6		7	$\langle 4,5,6,7\rangle$
5	7	4		

5.7 Here is a trace of BACKTRACKCOLOR applied to K_4 with $N = 3$. Let the vertices be x_1, x_2, x_3, and x_4.

Step No.	*J*	K	*FORWARD*	$c[1,2,3,4]$
Main 1	2		TRUE	[1,0,0,0]
Color 1	2	2		
3–5	3	2	TRUE	[1,2,0,0]
Color 1	3	3		
3–5	4	3	TRUE	[1,2,3,0]
Color 1	4	4		
6	4	4	FALSE	
Main 4, 5	3			[1,2,3,0]
Color 1	3	4		
6	3	4	FALSE	
Main 4, 5	2			[1,2,0,0]
Color 1	2	3		
3–5	3	3	TRUE	[1,3,0,0]
Color 1	3	2		
3–5	4	2	TRUE	[1,3,2,0]
Color 1	4	4		
6	4	4	FALSE	
Main 4, 5	3			[1,3,2,0]
Color 1	3	4		
6	3	4	FALSE	
Main 4, 5	2			[1,3,0,0]
Color 1	2	4		
6	2	4	FALSE	
Main 4, 5	1			[1,0,0,0]
6	THERE IS NO 3-COLORING OF *G*, STOP			

INDEX

Acyclic graph, 253
Adjacency matrix, 268
Adjacent transportation, 160
Adjacent vertices, 242
Algorithm, 8. *See also Algorithms and Procedures following Index.*
  approximation, 389
  bad, 120
  complexity, 119, 186–187
  correct, 12
  cubic, 104, 469
  divide-and-conquer, 325
  efficient, 118
  exponential, 120
  good, 11–12, 120
  greedy, 273
  input to, 11
  linear, 95, 104
  logarithmic, 95, 108
  output, 11
  polynomial, 120
  quadratic, 104
  recursive, 318
  relative efficiency, 119
All Pairs Problem, 399
Arithmetic:
  congruence, 211
  modular, 211
  modulo *n*, 211
Arithmetic progression, 124
Array, 33
  2-dimensional, 289
ASCII code, 222

Backtracking, 439
Bad algorithm, 120
Ballot problem, 387
Base case of proof by induction, 72
Basic solution of recurrence relation, 356
Base 3 representation, 20
Bernoulli numbers, 345, 386–387
"Big oh" notation:
  definition 1, 103
  definition 2, 108
Binary:
  coded decimal, 61
  fraction, 8, 20
  notation, 6
  number, 6
  tree, 303
    full, 305
Binomial coefficient, 134
Binomial theorem, 169
Bipartite graph, 247
  complete, 247
Bit (binary digit), 17
  vector, 33
Boolean function, 51
  AND, 51
  associative law, 54

commutative law, 54
distributive law, 54
NOT, 51
OR, 51
satisfiable, 56
variable, 148
XOR (exclusive or), 52
Breadth-first-search, 395, 436
spanning tree, 395
Busy Beaver N-game, 63

Cartesian product, 31
Catalan numbers, 342, 387–388
Ceiling function, 94
Center of a tree, 272
Characteristic equation, 355
Characteristic function, 33, 49
Characteristic polynomial, 355
Characteristic root, 355
Chromatic number of graph, 434
Clause, 55
Clique:
in graph, 246
number of graph, 434
Collatz problem, 11
Complement of graph, 277
Complete bipartite graph, 247
Complete graph, 246
Complete residue system, 220
Complexity of algorithm, 119, 186–187
Component of graph, 251, 253
Composite number, 140
Composition of functions, 45
Conclusion, 112
Congruent numbers, 211
Conjunctive normal form (CNF), 55
Connected directed graph, 410
strongly, 410
weakly, 410
Connected graph, 251, 253
Contradiction, 56
Contrapositive, 115
Converse, 116
Cube, generalized, 251
Cycle, 252
in directed graph, 406

Decimal number, 6
Decrypting key, 228
Decryption, 223
Degree of vertex, 243
De Morgan's laws, 52
Depth:
of tree, 304
of vertex, 304
Depth-first search, 414
spanning tree, 415
Diameter of graph, 277
Difference quotient, 366
Directed graph, 406
Disjoint union, 28
Disjunctive normal form (DNF), 58
Distance:
in graph, 253
in weighted graph, 272
Divisor, 182
Do loop, 83
nested, 145
Domain of function, 40
Dual graph, 446

Eccentricity of vertex, 272
Edge of graph, 241
Encryption, 223
Equivalence class:
of equivalence relation, 214
modulo n, 211–212
Equivalence relation, 213
Euclidean algorithm, 190–193, 223
Euclidean equations, 192
Eulerian cycle, 400
in directed graph, 406, 426
Eulerian graph, 400
arbitrarily traceable, 407
Eulerian path:
in directed graph, 406
Euler phi function, 238
Euler's formula, 446
Euler's theorem, 238, 401
Exclusive or (XOR), 52
Exponent of encryption scheme, 224
Exponential algorithm, 120
Exponentiation, 68

Factorial function, 134
Factorial representation, 140
Fermat's last theorem, 71–72
Fermat's little theorem, 218–219, 238
Fibonacci numbers, 198, 360
and complexity of Euclidean algorithm, 206–210
File, 286
Floor function, 94
For . . . do loop, 83

For . . . down to . . . do, 156
Forest, 253
Four Color Problem, 433
Function(s), 40
  characteristic, 33, 49
  domain, 40
  equal, 42
  image, 40
  inverse, 46
  one-to-one (or 1–1), 43
  onto, 42
  range, 40
  target, 40

Generating function, 174
Geometric series, 74
  alternating, 74
Good algorithm, 11–12, 120
Graph, 241
  connected, 251, 253
  directed, 406
  Eulerian, 406
  grid, 242
  Hamiltonian, 411
  isomorphism, 244
  k-colorable, 434
  k-colored, 434
  perfect, 443
  planar, 446
  regular, 249
  2-colorable, 247, 435
Greatest common divisor (gcd), 182, 185
Grid:
  graph, 242
  rectangular, 129

Hamiltonian cycle, 411
  minimum-weight, 425
Hamiltonian graph, 411
Hamiltonian path, 411
Harmonic numbers, 342
Hereditary property, 273
Hierarchy of functions, 108
Hilbert's tenth problem, 205
Hypothesis, 112

i ($\sqrt{-1}$), 358
Identity map (or permutation), 45
If and only if, 59, 116
Incident vertex and edge, 242
Independence number, 280
Independent set of vertices, 280
Induced subgraph, 262
Induction, 72
  complete, 198
Inductive hypothesis, 73
Inductive step, 73
Information theoretic bound (on sorting), 314
Integers modulo n, 215
Interference graph, 433
Intractable problem, 120
Inverse function, 46
Isomorphic graphs, 244
Isomorphism:
  of graphs, 244
  of labeled graphs, 278
Iteration, 348

j-subset, 30

Key, 286
Kruskal's algorithm, 264, 426

Lamé's theorem, 208
Leaf (in a tree), 272
Least common multiple (lcm), 185
Level of vertex, 304
Lexicographic ordering, 145–148
LHRRWCC (linear homogenous recurrence relation with constant coefficients), 353
Linear combination, 194, 357
Linear ordering, 221
Line graph, 424
Literal, 55
Local area network (LAN), 239
Logarithm (to the base 2), 92
Logically equivalent, 115
Loglog(n), 102
Loop, 9
  do, 83
    nested, 145
  for . . . do, 83
    down to, 156
  repeat . . . until, 164
  while . . . do, 69
Lucas numbers, 364

Magic trick, 1
Mastermind, 161
Mathematical induction, 72
  complete, 198
Matrix, 268
Maximal subset, 273
Minimum-distance spanning tree, 390
Minimum-weight matching problem, 405
Minimum-weight spanning tree, 258, 264

Modular arithmetic, 211
Modulo n, 211
Multigraph, 408
Multiple:
  of an integer, 182
  root, 358
Multiplication principle, 2
  generalization, 87
Multiplicative inverse modulo n, 217
Multiplicity of root, 358
Multiset, 35, 164

Natural numbers, 21
Negation, 111
NP-Complete problem, 420
n-set, 29
n-tuple, 32
Null set, 21

One-to-one function, 43
Onto function, 42
Ordered pair, triple, 32

Pancake problem, 338
Partition, 28, 214
Pascal's triangle, 133–134
Path, 252
  in directed graph, 406
  length, 252, 390
Permutation(s), 44, 153
  distance between, 160
  and English change ringing, 159–160
  even, 160
  identity, 45
  inversions, 180
  odd, 160
Pigeonhole principle, 44
Polynomial algorithm, 120
Polynomial identity, 74
Prime number, 21
Principle of inclusion and exclusion, 38
Procedure, 296
Proof by contradiction, 110
Public key encryption scheme, 223

Quadratic formula, 362
Quotient, 191

Radius of graph, 398
Range of function, 40
Record, 286
Recurrence relation, 346
  constant coefficient, 353
  divide-and-conquer, 373
  homogeneous, 349
  inhomogeneous, 349
  initial conditions, 347
  linear, 353
  order, 353
Recursive algorithm, procedure, 318
Reducing modulo n, 225
Reflexive property of relation, 213
Relation, 212–213
  graph of symmetric, 245
Relatively prime integers, 218
Remainder, 191
Repeat . . . until, 164
Residue, least nonnegative, 214
Root of tree, 303
RSA scheme, 223

Satisfiability problem, 57, 122
Search tree, 303
  binary, 311
Sequence, 339
  integer, 339
  nth term, 340
  symmetric, 140
  unimodal, 140
Set(s), 21
  cardinality, 35
  complement, 22
  difference, 25
  disjoint, 25
  element, 21
  empty, 21
  equality, 21
  finite, 35
  intersection, 22
  null, 21
  relative complement, 25
  subset(s), 21
    number, 80
    number of j-subsets of n-set, 142
  union, 22
Sieve of Eratosthenes, 236
Simplified fraction, 181
Spanning forest, 257
Spanning-in-tree, 410
Spanning subgraph, 256
Spanning tree, 256
Stable sorting algorithm, 334
Stirling's formula, 158
Storage allocation scheme, 432–433
Subgraph, 256
  induced, 262

Subtrees (left and right), 303
Symmetric property of relation, 213

Target of function, 40
Tautology, 56
Ternary representation, 20
Total ordering, 221
Towers of Hanoi, 382
Trace of an algorithm, 17
Transitive closure of a graph, 282
Transitive property of a relation, 213
Trapdoor function, 223
Traveling Salesrepresentative Problem, 274
Tree, 253
  binary, 303
  planted planar, 388
  search, 303
Trivial programming language, 62–63

Universe, 20

Variables:
  Boolean, 148
  interfering, 432
  noninterfering, 432
Vector, 33
Venn diagram, 26
Vertex of a graph, 241

Weighted graph, 257
Well defined operation, 215
While . . . do, 69
Wilson's theorem, 238
Worst-case analysis, 92

ALGORITHMS AND PROCEDURES

ADDRACT1, 482
APPROXHAM, 428

BACKTRACKCOLOR, 440–441
BADMINTREE, 258
BINARYSEARCH, 291
BINARYSORT, 297
BININSERT, 296
Borůvka's algorithm, 280
BREADTHFIRSTSEARCH (BFS), 395
BtoD, 16
BUBBLES, 87
BUBBLESORT, 288
BUCKETSORT, 332, 338

COLLATZ, 11

DEPTHFIRSTSEARCH (DFS), 415
DFS-HAMCYCLE, 418–419
DIJKSTRA, 392
DIJKSTRA2, 398–399
DIVISORSEARCH, 234
DtoB, 18, 84, 89

EUCLID, 195
EULER, 408–409
EXPONENT, 68, 82, 188–189

FASTEXP, 91, 187
FIB, 319
FIB2, 323
FOURSUM, 85
FUN, 15

GCD, 320
GCD1, 183, 187–188
GCD3, 335
GREEDYCYCLE, 501
GREEDYMAX, 501
GREEDYMIN, 273

HAMCYCLE, 413
HAMCYCLE2, 422–423

IND, 281
INDUCTION, 73
INDUCTIONC, 199
INSERTIONSORT, 334

JSET, 148

KRUSKAL, 264, 426

LABGPHISO, 278

MAX, 86
MERGE, 326
MERGESORT, 327

ALGORITHMS AND PROCEDURES

MIN, 318, 320
MYSTERY, 337

ODDSUM, 85

PAIR, 145
PERM, 156
POSTAGE, 432
Prim's, 280

R-DEPTHFIRSTSEARCH, 417–418
R-DtoB, 335
R-JSET, 324
RSA, 232
R-SELECTSORT, 321
R-SUBSET, 324

SELECTSORT, 285
SEQSEARCH, 284
SEQUENTIALCOLOR, 436–437
SIMPLIFY, 184
SPEEDY, 118
SPTREE, 262–263
SQUARESUM, 86
SUBSET, 30–31
SUM, 83

TREESORT, 307

VOLUME, 431

NOTATIONS

□ end of proof, 27–28

Algorithmic:
  := assignment statement, 15
  * multiplication symbol, 15
  / division symbol, 15

Set theory:
  $A \times A$, A^n $A \times B$ Cartesian product, 31–33
  $\subseteq$ containment, 21
  A^c the complement of A, 22
  $\{\ldots\}$ curly brace notation for sets, 21
  $A–B$ difference, 25
  $\in$ element of, 21
  $\emptyset$ empty set, null set, 21
  $\cap$ intersection, 22
  $(a_1,a_2,\ldots,a_n)$ n-tuple, 31
  (a,b) ordered pair, 31
  (a,b,c) ordered triple, 31
  $\langle\ldots\rangle$ permutation, 153
    array, 284
  $\cup$ union, 22

Functions:
  $\binom{k}{i}$ binomial coefficient, 135

Boolean functions:
  $\wedge$ AND, 51
  $\vee$ OR, 51
  $\sim$ NOT, 51
  $\oplus$ XOR, 53
  $\lceil\ \rceil$ ceiling function, 94
  χ_s characteristic function, 49
  $(g \circ f)$ composition of functions, 45
    f^2,f^3,f^n, 49–50
  $n!$ factorial function, 134
  $\lfloor\ \rfloor$ floor function, 94
  $\log(n)$ logarithm to the base, 2, 92
  $\log_d(n)$ logarithms to the base d, 124, 376
  $O(g)$ big oh of g, 103, 108

Number theory:
  $a \equiv b \pmod{n}$ equivalence modulo n, 211
  $[x]$ equivalence class modulo n, 211
  F_k kth Fibonacci number, 198
  $\gcd(b,c)$ greatest common divisor, 182
    $\gcd(a,b,c)$, 185
  $\operatorname{lcm}(b,c)$ least common multiple, 185
  ϕ phi, 202
  ϕ', 203
  $\phi(m)$ Euler phi function, 238
  $\sim$ relation, 212–213
  Z_n integers modulo n, 215

Graph theory:
  $\alpha(G)$ independence number of G, 280
  $A(G)$ adjacency matrix, 268
  C_k k-cycle, 253
  $\chi(G)$ chromatic number of G, 434

Graph theory (*Continued*)

cl(G) clique number of G, 434
deg(x), deg(x,G) degree of vertex x, 243
d(G) diameter of the graph G, 398
d(x,y) distance from x to y, 253, 272
E(G) the set of edges of the graph G, 242
G^c the complement of G, 277
K_r the complete graph on r vertices, 246
$K_{p,q}$ the complete bipartite graph, 247
L(G) line graph of G, 424
Nbor(v) neighbor of the vertex v, 408
P_k k-path, 253
Q_n generalized cube or n-cube, 251
r(G) radius of G, 398
V(G) the set of vertices of the graph G, 242
w(e) weight of an edge e, 257
w(T) weight of a tree T, 257